AF621248

# Brain Dynamics and the Striatal Complex

**Conceptual Advances in Brain Research**

A series of books focusing on brain dynamics and information processing systems of the brain.

Edited by *Robert Miller, University of Otago, New Zealand (Editor-in-chief). Günter Palm, Universität Ulm, Germany* and *Gordon Shaw, University of California at Irvine, USA.*

---

**Volume 1**

Brain Dynamics and the Striatal Complex
edited by *R. Miller* and *J.R. Wickens*

**Forthcoming Volumes**

Complex Brain Functions—Conceptual Advances in Russian Neuroscience
edited by *A.M. Ivanitsky and P. Balaban*

Time and the Brain
edited by *R. Miller*

Sex Differences in Brain Lateralization
edited by *V.L. Bianki* and *E.B. Filipova*

Cortical Areas: Unity and Diversity
edited by *A. Schüz* and *R. Miller*

**Volumes in Preparation**

The Female Brain
Neural Determinism
Dynamics of Brain Function
Brain Oscillations and Functional Memory

---

# Brain Dynamics and the Striatal Complex

Edited by

**R. Miller**

and

**J.R. Wickens**

*School of Medical Sciences*
*University of Otago*
*New Zealand*

CRC Press
Taylor & Francis Group
Boca Raton London New York

CRC Press is an imprint of the
Taylor & Francis Group, an **informa** business

First published 2000 by Harwood Academic Publishers

Published 2021 by CRC Press
Taylor & Francis Group
6000 Broken Sound Parkway NW, Suite 300
Boca Raton, FL 33487-2742

© 2000 by Taylor & Francis Group, LLC
CRC Press is an imprint of Taylor & Francis Group, an Informa business

No claim to original U.S. Government works

ISBN-13: 978-90-5702-478-8 (hbk)

ISSN: 1029-2136

This book contains information obtained from authentic and highly regarded sources. Reasonable efforts have been made to publish reliable data and information, but the author and publisher cannot assume responsibility for the validity of all materials or the consequences of their use. The authors and publishers have attempted to trace the copyright holders of all material reproduced in this publication and apologize to copyright holders if permission to publish in this form has not been obtained. If any copyright material has not been acknowledged please write and let us know so we may rectify in any future reprint.

Except as permitted under U.S. Copyright Law, no part of this book may be reprinted, reproduced, transmitted, or utilized in any form by any electronic, mechanical, or other means, now known or hereafter invented, including photocopying, microfilming, and recording, or in any information storage or retrieval system, without written permission from the publishers.

For permission to photocopy or use material electronically from this work, please access www. copyright.com (http://www.copyright.com/) or contact the Copyright Clearance Center, Inc. (CCC), 222 Rosewood Drive, Danvers, MA 01923, 978-750-8400. CCC is a not-for-profit organiza-tion that provides licenses and registration for a variety of users. For organizations that have been granted a photocopy license by the CCC, a separate system of payment has been arranged.

**Trademark Notice:** Product or corporate names may be trademarks or registered trademarks, and are used only for identification and explanation without intent to infringe.

**Visit the Taylor & Francis Web site at**
**http://www.taylorandfrancis.com**

**and the CRC Press Web site at**
**http://www.crcpress.com**

---

**British Library Cataloguing in Publication Data**

A catalogue record for this book is available from the British Library.

# CONTENTS

# SERIES PREFACE

The workings of the brain, including the human brain are a source of endless fascination. In the last generation, experimental approaches to brain research have expanded massively, partly as a result of the development of powerful new techniques. However, the development of concepts which integrate and make sense of the wealth of available empirical data has lagged far behind the experimental investigation of the brain. This series of books entitled Conceptual Advances in Brain Research (CABR) is intended to provide a forum in which new and interesting conceptual advances can be presented to a wide readership in a coherent and lucid way.

The series will encompass all aspects of the sciences of brain and behaviour, including anatomy, physiology, biochemistry and pharmacology, together with psychological approaches to defining the function of the intact brain. In particular, the series will emphasise modern attempts to forge links between the biological and the psychological levels of describing brain function. It will explore new cybernetic interpretations of the structure of nervous tissue; and it will consider the dynamics of brain activity, integrated across wide areas of the brain and involving vast numbers of nerve cells. These are all subjects which are expanding rapidly at present. Subjects relating to the human nervous system as well as clinical topics related to neurological or psychiatric illnesses will also make important contributions to the series.

These volumes will be aimed at a wide readership within the neurosciences. However, brain research impinges on many other areas of knowledge. Therefore, some volumes may appeal to a readership, extending beyond the neurosciences. Books suitable for the series are monographs, edited multiauthor collections or books deriving from conferences, provided they have a clear underlying conceptual theme. In order to make these books widely accessible within the neurosciences and beyond, the style will emphasise broad scholarship comprehensible by readers in many fields, rather than descriptions in which technical detail of a particular speciality is dominant.

The next decades promise to provide major new revelations about brain function, with far-reaching impact on the way we view ourselves. These great breakthroughs will require a broad interchange of ideas across many fields. We hope that the CABR series plays a significant part in the exploration of this important frontier of knowledge.

# PREFACE

This book is the first volume in a new series of books, entitled Conceptual Advances in Brain Research. Relatively few books have been written addressing the way in which dynamic functions are related to cellular structure and synaptic organization in the basal ganglia. At present this is a somewhat speculative relation. Nevertheless a number of relatively solid themes have emerged in recent years. Indeed, there has probably been greater progress in relating structure to function within the basal ganglia than in some other areas of the mammalian forebrain. The striatum is the largest and best characterised nucleus within the basal ganglia, and understanding of the basal ganglia as a whole cannot get very far without a detailed model of the striatum, relating structure to function. In compiling this volume we have used a rather broad definition of the striatum, including the ventral (so-called limbic) parts of it. We have also included material concerning the structures and processes on the afferent side of the striatum (the cerebral cortex, the dopamine rich divisions of the midbrain, and the amygdala, and their respective functions). In later chapters, the two-way interplay between striatum and cerebral cortex is discussed from a wider perspective.

Major issues addressed by the various contributing authors are as follows: Dominating the early chapters of the book are several chapters dealing with the role of the dopaminergic input to the striatum. These include consideration of the behavioural role of dopamine, the functions of dopamine as seen in electrophysiological experiments in behaving animals, and the actions of dopamine at the cellular and synaptic level. The link between psychological concepts of reward-related learning (including some other aspects of instrumental conditioning paradigms) and the underlying biology is explicitly discussed in some of these early chapters. The middle section of the book deals with cellular, synaptic and network organization of the striatum, considered by itself. Of some importance are attempts to derive descriptions of network dynamics in the striatum from cytological and single-neurone electrophysiological evidence. Another issue dealt with in some chapters of the middle section is the relative roles of the so-called 'direct' and 'indirect' outflow pathways from the striatum, and the possibility that they are modulated by different dopamine receptors. Both these issues have theoretical significance well beyond the striatum, indeed for the dynamics of the forebrain as a whole. However, empirical evidence relating to both these issues is at present insufficient for making definitive conclusions. The last part of the book contains two chapters exploring the mutual interaction between striatum and cerebral cortex. A distinctive feature of the book is a discussion section which was organized by electronic mail. This consists of an edited dialogue between some of the chapter authors (and other invited discussants). This section seeks consensus on some issues, and explores some of the remaining controversies.

# LIST OF CONTRIBUTORS

**Beninger, R.J.**
Department of Psychiatry
Queen's University
Kingston
Ontario K7L 3N6
Canada

**Bennett, B.D.**
Department of Anatomy and Neurobiology
College of Medicine
University of Tennessee, Memphis
855 Monroe Avenue
Memphis, TN 38163
USA

**Gerfen, C.R.**
Section of Neuroanatomy
Laboratory of Neurophysiology
National Institute of Mental Health
Bethesda, MD 20892
USA

**Hernández, A.**
Instituto de Fisiología Celular
Universidad Nacional Autónoma de México
04510 México DF
Mexico

**Hyland, B.I.**
Department of Physiology
School of Medical Sciences
University of Otago
PO Box 913, Dunedin
New Zealand

**Joel, D.**
Department of Psychology
Tel Aviv University
Ramat-Aviv
Tel Aviv 69978
Israel

**Kitai, S.T.**
Department of Anatomy and Neurobiology
College of Medicine
University of Tennessee, Memphis
855 Monroe Avenue
Memphis, TN 38163
USA

**Miller, R.**
Department of Anatomy and Structural Biology
School of Medical Sciences
University of Otago
PO Box 913, Dunedin
New Zealand

**Olmstead, M.C.**
Department of Psychology
Queen's University
Kingston
Ontario K7L 3N6
Canada

**Oorschot, D.E.**
Department of Anatomy and Structural Biology
School of Medical Sciences
University of Otago
PO Box 913, Dunedin
New Zealand

**Opris, I.**
Instituto de Fisiología Celular
Universidad Nacional Autónoma de México
04510 México DF
Mexico

**Plenz, D.**
Laboratory of Systems Neuroscience
National Institute of Mental Health
Bethesda, MD 20892
USA

**Rebec, G.V.**
Program in Neural Science
Department of Psychology
Indiana University
Bloomington, IN 47405
USA

**Romo, R.**
Instituto de Fisiología Celular
Universidad Nacional Autónoma de México
04510 México DF
Mexico

**Salinas, E.**
Instituto de Fisiología Celular
Universidad Nacional Autónoma de México
04510 México DF
Mexico

**Surmeier, D.J.**
Department of Physiology
North Western Medical School
Searle 5-447
320 E-Superior Street
Chicago, IL 60611
USA

**Weiner, I.**
Department of Psychology
Tel Aviv University
Ramat-Aviv
Tel Aviv 69978
Israel

**Wickens, J.R.**
Department of Anatomy and Structural Biology
School of Medical Sciences
University of Otago
PO Box 913, Dunedin
New Zealand

**Wilson, C.J.**
Department of Anatomy and Neurobiology
College of Medicine
University of Tennessee, Memphis
855 Monroe Avenue
Memphis, TN 38163
USA

**Zainos, A.**
Instituto de Fisiología Celular
Universidad Nacional Autónoma de México
04510 México DF
Mexico

# 1 Relationships of Substantia Nigra Dopamine Neurone Activity to Behaviour

**Brian I. Hyland**

*Department of Physiology, School of Medical Sciences, University of Otago, P.O. Box 913 Dunedin, New Zealand.*
*Phone (0064) 3-479-7342; FAX: (0064) 3 479 7323;*
*e mail: brian.hyland@stonebow.otago.ac.nz*

Available data on the activity of single dopamine cells during behaviour in the monkey, cat and rat is reviewed. A paucity of directly comparable studies in terms of behavioural paradigms and analysis methods makes across-species conclusions difficult. However, analysis of results for similar behavioural situations within the monkey literature do suggest some recurring themes, at least for the primates studied. These centre on relationships of cell activity to events related to reward, but not simply as markers of reward; rather the neural-behavioural relationships change with learning of associations between behavioural acts, sensory-signals and reward. These findings implicate dopamine cells in the monkey in reward-driven learning and have recently been combined with theoretical models of such learning to provide a framework for considering the pattern of cell activity seen in the behaviourally conditioned monkey. Major challenges for future research are identified, including the need to determine the extent to which the monkey results can be generalized across species, and to understand the functional implications of single nigro-striatal dopamine neurone recording data *vis a vis* neurochemically derived measures of dopamine release in the striatum.

KEYWORDS: dopamine neurone; substantia nigra; single cell recording; behaving animal; reward; learning; monkey; cat; rat.

## 1. INTRODUCTION

The dopaminergic projection to the striatum plays a vital, if poorly understood role in normal functioning of the basal ganglia. Knowledge of the situations in which pulses of dopamine are released into the striatum is important information for models attempting to integrate cellular actions of dopamine with the broader functioning of the striatum. There are two ways of acquiring this information. Dialysis and electrochemical detection methods provide information concerning the level of synaptic overflow of dopamine over relatively large regions of the striatum. On the other hand, recording from the dopamine cells themselves potentially offers a desirable combination of high specificity (provided dopamine cells can be identified in behaving animals) and the best possible temporal resolution, without damage to the striatum. Recent work has thrown some light on the kind of events which are able to excite dopamine cells into increased activity during normal behaviour, and it is this information that is reviewed here. Dopamine neurones of the substantia nigra are the main focus, as it is this group about which most is known.

## 2. IDENTIFICATION OF DOPAMINE CELLS IN BEHAVING ANIMALS

Dopamine cells occupy a well defined and fairly restricted area approximated by the *pars compacta* of the substantia nigra, the retrorubral area, and the ventral tegmental area (Dahlström and Fuxe, 1964; German and Manaye, 1993). However, criteria other than simple histological location are required to identify dopamine neurones, since they are scattered in neighbouring regions such as the *pars reticulata* (German and Manaye, 1993), and there are other, non-dopaminergic cells among the dopaminergic cell groups (Guyenet and Aghajanian, 1978; Yung, Hausser and Jack, 1991; Silva and Bunney, 1988; Matsuda *et al.*, 1987; Lacey, Mercuri and North, 1989). With intracellular recording methods, it is possible to label the recorded cell and identify it as dopaminergic by histochemical means. However, in behaving animals, intracellular recordings are not practicable, so direct histochemical identification of recorded neurones has not been possible. What has made the exploration of dopaminegic cell activity possible in the behaving animal is the finding that dopaminergic neurones have a specific electrophysiological and pharmacological profile that, to a reasonable extent, sets them apart from neighbouring cells, and that can be recognized in extracellular recordings.

These criteria were initially developed from studies in anaesthetized or paralyzed animals, in which it was noted that cells recorded in regions known to contain dopamine cells had slow (< 10 Hz) rates of activity, a tendency to fire occasionally in bursts, and an inhibitory response to dopamine agonists (Bunney *et al.*, 1973). Furthermore, such cells were not found after treatment with 6-hydroxydopamine, which selectively destroys dopaminergic cells, consistent with the proposal that these features marked dopaminergic neurones. Cells in the region with these characteristics were noted to also have long duration action potentials (Aghajanian and Bunney, 1973, 1977). Subsequent studies in reduced preparations employing intracellular recording methods, labeling, and histochemical identification have confirmed that cells with this combination of firing rate and pattern, waveform duration, and pharmacological characteristics are indeed dopaminergic (Grace and Bunney, 1980, 1983; Grace and Onn, 1989; Yung, Hausser and Jack, 1991).

In contrast, non-dopaminergic cells in the vicinity generally have higher rates (although some do fall within the dopamine cell range), and narrower action potentials (although the low end of the distribution of dopamine cell action potential durations may well overlap [Chiodo *et al.*, 1980]). They are also generally not inhibited by dopamine agonists, although partial inhibition of non-dopamine cells has been reported (Grace and Bunney, 1985). Most studies of dopamine cell activity in behaving animals have therefore employed the electrophysiological criteria of rate less than 10 Hz, an action potential waveform generally in excess of around 1.5 ms duration, together with histological localization of recording sites within the substantia nigra. Although an inhibitory response to apomorphine offers perhaps the most clear cut identification of dopaminergic cells (see Figure 1.1), use of this criterion is not widespread in the behaving animal literature, owing to the technical difficulties of holding cells during drug administration, and the pharmacological implications inherent in giving repeated doses of dopaminergic drugs.

Of course, these electrophysiological features relate not only to dopaminergic neurones. Cells in several regions of the brain have low spontaneous firing rates and broad action potentials. Such cells include the 5-HT containing neurones of the dorsal raphe (Heym, Steinfels and Jacobs, 1982) and presumed interneurones in the striatum (Aosaki *et al.*, 1994).

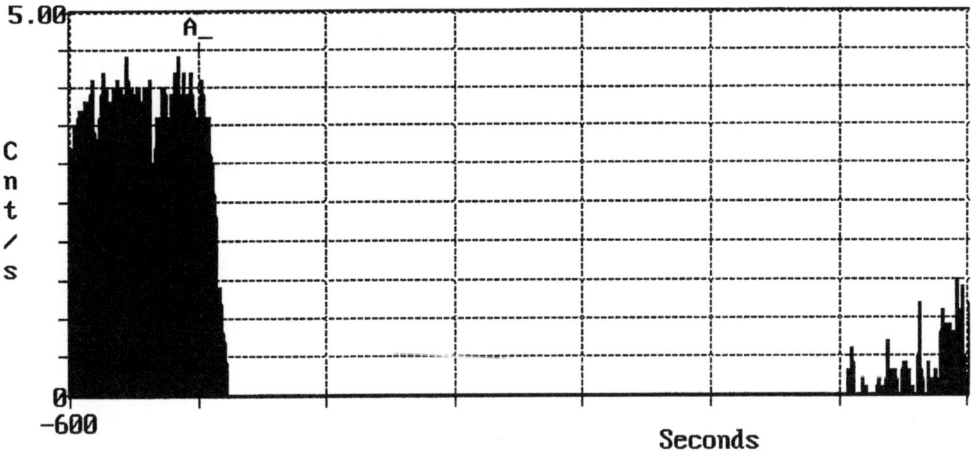


**Figure 1.1.** Response of a putative dopamine cell in the substantia nigra of an awake rat to 750 $\mu$g/kg i.p. apomorphine (A). Peri-event time histogram shows firing rate in Hz. Complete suppression of activity was accompanied by stereotypic behaviour (licking at a wall). Recovery of cell activity began after 50 minutes. (Hyland, Perk, Hay and Miller, unpublished data).

Given that the electrophysiological criteria impose arbitrary limits on what are in fact continuously graded variables, there remains the chance for both positive and negative errors in such experiments, i.e. the inclusion of non-dopaminergic cells, and the exclusion of dopamine cells. It is not possible at this stage to quantify the likely extent of false positive and negative errors in classification, but these are likely to be few. In studies where overlapping electrophysiological properties were reported in cells identified by response to dopamine agonists, the numbers of neurones involved were only a small proportion of the total (Guyenet and Aghajanian, 1978, Chiodo *et al.*, 1980; Steinfels *et al.*, 1983a; Aebischer and Schultz, 1984). Furthermore, at least in the case of behaving animals, cells having an apparent mismatch between electrophysiological and pharmacological criteria may have arisen due to failure of drug delivery. On balance, therefore, given the constraints of working with conscious, moving animals, the available criteria are believed to offer reasonable confidence that the majority of reported neurones are indeed dopaminergic. Nevertheless, the possibility that a specific subgroup of dopaminergic cells with narrow action potentials (Chiodo *et al.*, 1980) are being systematically excluded by the use of purely electrophysiological criteria might deserve further investigation, to rule out the slight possibility that such cells form a special group in terms of response properties.

## 3. BEHAVIOURAL CORRELATES OF CHANGES IN DOPAMINE CELL ACTIVITY

### 3.1. Studies in Primates

The largest body of data on the activity of dopaminergic cells in behaving animals comes from work in monkeys, by W. Schultz and collaborators in Fribourg, Switzerland. Over several years, these workers have amassed a large amount of detailed information on the responses of hundreds of dopamine neurones in a large number of behavioural contingencies, using several different tasks and behavioural paradigms. Early studies by this group involved monkeys reaching out to grasp and retrieve food rewards, while more recently tasks were used in which liquid primary rewards were delivered to the animal's mouth after correct performance of conditioned movements. Figure 1.2 shows a summary diagram of the main types of behavioural paradigm employed in these studies, but does not include every variant.

The main thrust of the interpretations these authors make of their data has been summarized in several reviews (Schultz, 1994,1997; Schultz, Dayan and Montague, 1997). For the purposes of this chapter an attempt has been made to present the main findings as they relate to proposed theories of dopamine cell function which have arisen. Of necessity, many details of the original works have been omitted, and readers are recommended to consult the original papers for more in-depth information (Schultz, Ruffieux and Aerbischer, 1983; Schultz, 1986; Schultz and Romo, 1990; Romo and Schultz, 1990; Ljungberg, Apicella and Schultz, 1991, 1992; Schultz, Apicella and Ljungberg, 1993a; Mirenowicz and Schultz, 1994, 1996).

The basic electrophysiological properties of dopamine cells reported in these studies have shown remarkable consistency. In all studies, no cell fired at greater than 9 Hz during spontaneous activity, and the minimum level of spontaneous activity was 0.5 Hz, except in the first study where the rate of resting activity ranged to 0 Hz. Action potential durations ranged from 1.4 to 5.5 ms in all studies except the first (Schultz *et al.*, 1983: 1–2 ms), in which a higher high-pass filter setting was employed (300 *vs.* 100 Hz). Although apomorphine testing was not performed in most of these studies, overall the electrophysiological parameters described match well with those displayed by samples of apomorphine-inhibited cells recorded by the same group in anaesthetized or awake animals (Aebischer and Schultz, 1984; Schultz, 1986).

### *3.1.1. Analysis of the proportion of cell responding in various situations*

On the assumption that reasonably similar methodologies will have been used within the group across the studies, a form of meta-analysis was performed on the original papers, and is presented in Figure 1.3. This analysis is confined to percentages of neurones responding, since this can be summarized as a single figure. Changes in the measured amplitude of responses are not shown, but do tend to follow the same trends. Some of these are noted in the text below. Results from different tasks that are interpreted (by the present author) as reflecting similar processes have been grouped together, in order to obtain an overview of dopamine cell responsiveness to the various task-related events. Only excitations are included, because in most cases excitatory responses dominated. Nevertheless, in several conditions dominated by excitation, inhibited neurones were also found, albeit in small number. Where inhibitory responses were considered important they are described below in the text. It must be noted that it is inherently more difficult to detect inhibition in neurones with such low firing-rate. In some studies, different sensory modalities were tested within the same paradigm; where the results of these variants were similar, the mean of the values across modalities is given to simplify the figure.

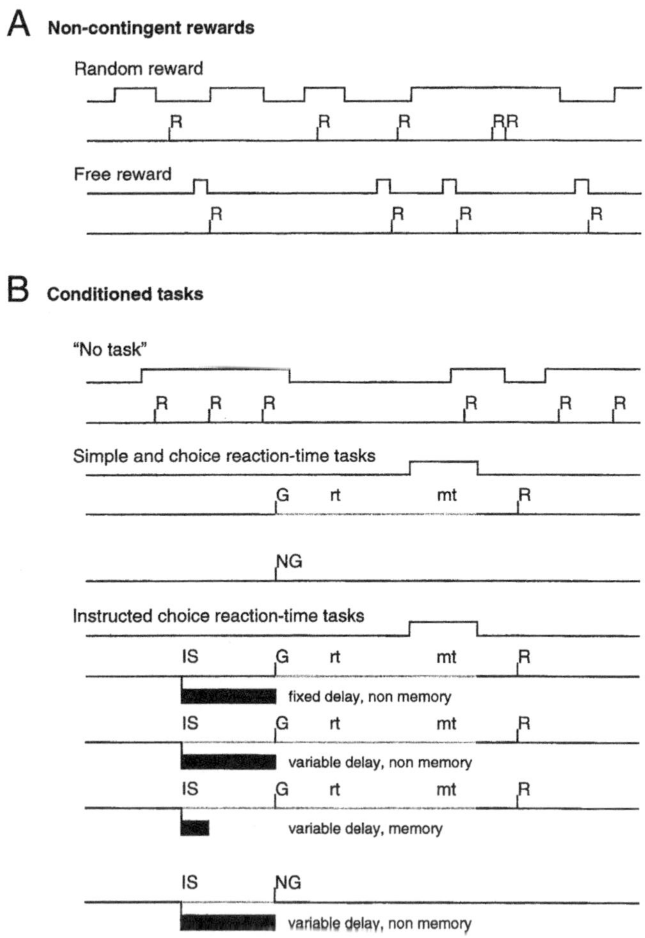


**Figure 1.2.** A schematic diagram of some behavioural situations in which dopamine cells have been recorded by the Schultz group. In each, the top line indicates behaviour emitted by the animal, and lines underneath show timing of signals (G = GO, NG = NOGO, IS = instruction signal, rt = reaction time, mt = movement time) and reward delivery (R). Dotted lines indicate variable time intervals, and solid bars show stimulus duration. **(A)** Situations in which rewards were obtained without prior cues or any particular *conditioned* behaviour. Random rewards were delivered by injecting fluid into the mouth at unpredictable intervals, independent of what the monkey was doing (Mirenowicz and Schultz, 1994, 1996). Free rewards were obtained by the monkey reaching into a covered enclosure at its own pace to grasp food rewards without prior indication that reward would be present (Romo and Schultz, 1990). Common to both, obtaining of reward was unpredictable. **(B)** Tasks in which certain behaviour was required before rewards would become available. Reward delivery could thus be considered predictable. The "no task" condition was labeled as such by the authors (Ljungberg, Apicella and Schultz, 1992) but in fact required that the monkey keep the hand on a key; while this position was maintained, fluid rewards were delivered at regular intervals. Simple and choice reaction time tasks involve a signal (usually after some holding period during which the monkey must maintain a certain position) indicating to the animal what it should do. In the "simple" variant, there is only one thing to be done; in the "choice" situation the signal indicates which of two (in these experiments) possible responses are required. In some situations, the alternative was to not respond (NOGO). In one study correct NOGO performance was rewarded (Schultz, Ruffieux and Aebischer, 1983), while in others it was not (Schultz, 1986; Schultz and Romo, 1990). In instructed choice reaction-time tasks the correct choice is indicated (Instruction signal) to the animal prior to the GO signal. Variants of this include fixed *versus* variable instructed delay intervals, and situations in which the instruction cue is available continuously (non-memory) or only at the beginning of the delay period (memory).

Although no attempt has been made to test whether the differences in percentages shown are statistically significant, the conclusions suggested by the median values given in Figure 1.3 can be summarized as follows:

*(1) High proportions of dopamine cells respond to the unpredictable obtaining of primary reward, but less respond when reward delivery is predictable.*
Figure 1.3 A shows that a high proportion of dopamine neurones are excited by delivery or attainment of primary rewards, when this is not predicted by a discrete external cue or simple behaviour. This includes situations where an animal reaches (at its own pace) into a hidden space, which may or may not contain a food reward (Romo and Schultz, 1990), and when liquid rewards are delivered purely randomly (Mirenowicz and Schultz, 1994, 1996). In most cases, once a predictable relationship between a particular behaviour or stimulus and the obtaining of reward has been established through training, fewer neurones respond to the reward itself. The condition of predictable rewards usually occurred in situations where a conditioned trigger stimulus induced a movement that was rewarded. However, a low rate of responsiveness also occurred when reward was delivered at regular intervals, so long as the animal remained with the limb in a fixed position, without any external stimulus (Ljungberg, Apicella and Schultz, 1992). This can also be considered a form of conditioned response with a predictable reward outcome. It should be noted that the case in which 100% of neurones responded during training represents only 9 cells from one monkey. The data points which do not follow this trend, remaining around 50% of neurones responding during and after training and overtraining, all come from the only study in which a delayed-alternation task was used. It has been argued that specific aspects of performance of this kind of task may explain the apparent discrepancy (Ljungberg, Apicella and Schultz, 1991).

In contrast, few neurones respond to aversive outcomes (Figure 1.3A, middle panel), and even these may have done so through a process of "stimulus generalization" (Mirenowicz and Schultz, 1996), suggesting that the valence and not just the alerting value of the outcome is important. The somatic-control data point refers to an experiment involving the monkey obtaining hidden food reward, by reaching out and grasping it; only few neurones were excited at the touch of a non-reward object. However, in this experiment 31% of the tested neurones were *inhibited* where an expected reward was not present (Romo and Schultz, 1990). Similarly, in studies involving liquid reward delivered passively to the animal, neurones that had been previously found to be excited by reward delivery were inhibited if rewards were not delivered (either due to experimenter intervention or erroneous performance by the animal) (Ljungberg, Apicella and Schultz, 1991; Schultz, Apicella and Ljungberg, 1993a). Thus, dopamine neurones are capable of signaling not only presence of reward, but also the lack of expected reward. Furthermore, the inhibitions occurred at the time that the reward would normally have been delivered, implying that the dopamine cells have access to information about the time at which rewarding events are expected, relative to external or internal cues. This, together with the fact that responses to rewards decline when they are predictable has led to the hypothesis that dopamine cells might signal the difference between expectation and outcome with regard to reward, rather than simply the presence of reward *per se* (Schultz, 1997).

*(2) Few dopamine neurones are activated by movement alone.*
Although 44% of neurones were found to show some modulation throughout arm movement in the first study (Schultz, Ruffieux and Aebischer, 1983) the responses were small,

and tonic rather than phasic in nature. In subsequent studies the proportion showing such activity progressively declined (Figure 1.3A), and none have been reported by this group since 1990. Similarly, another study, which probably included a majority of dopamine cells in a sample of substantia nigra *pars compacta* neurones with low firing rate (DeLong, Crutcher and Georgopoulos, 1983), failed to find significant numbers modified during movement performance. One other study did note slight activation of nigral dopamine neurones during movements in a simple unsignalled repetitive back and forth arm movement task (Magarinos-Ascone, Buno and Garcia-Austt, 1992). Overall, however, the generally unremarkable response of dopamine neurones during movement has been considered paradoxical, given that lack of dopamine activity is responsible for disorders of movement in Parkinson's disease. It has been suggested that only tonic rather than phasic release of dopamine from the nigro-striatal terminals is needed to support normal motor function of the striatum (DeLong, Crutcher and Georgopoulos, 1983).

*(3) A high proportion of dopamine neurones respond to arbitrary, conditioned GO and NOGO signals.*

The left-most panels of Figure 1.3 B and C summarize responses, during performance of operantly conditioned tasks, of dopamine cells to auditory and visual GO signals that occur without any warning. Note however that the monkey must have achieved some prior steady state behavioural criterion in order for the GO signal to be delivered. It is clear that a high proportion of dopamine neurones respond to these imperative signals. Only few dopamine neurones respond to similar sensory cues prior to learning of conditioned association (Ljungberg, Apicella and Schultz, 1992) or after training, but when tested out of task context (Schultz and Romo, 1990) (Figure 1.3 B, right-most panel). This seems to be a fundamental difference compared to data from dopamine cells in cats (see below).

The left-most panel in Figure 1.3 C shows that few neurones respond to a stimulus informing the animal it must perform an act to avoid an unpleasant stimulus. This suggests that only signals imbued with "positive" meaning for the animal activate the dopamine cells, and shows that the response to sensory signals is not indicative of a general preparation to make a movement. On the other hand, the next panel (which incorporates 2 data points from 2 different studies which happen to be equal) shows that dopamine neurones do respond in high proportion to NOGO signals. This is of interest, since in the behavioural paradigms employed, correct NOGO performance was not rewarded in any way. Thus, these neurones could be said to be responding to stimuli that are task-relevant but completely neutral in terms of reward value. In both of these studies the NOGO stimuli were nearly identical to GO stimuli used at the same time, and so it is possible that these responses reflect processes of "stimulus generalization" (Schultz and Romo, 1990; Mirenowicz and Schultz, 1996). However, if dopamine neurones do indeed generalize so efficiently in their responses, their value as indicators of impending reward would have to be called into question. One possibility is that the *intensity* of the response is important; in both studies employing NOGO stimuli, responses of the majority of neurones to these signals were smaller in amplitude than to the GO stimuli for the same neurone (Schultz, 1986; Schultz and Romo, 1990). Another feature of responses to NOGO signals is that many of the responding neurones showed a late inhibition, at the time reward would normally be obtained in GO trials. In this case, the initial excitation might be only part of a complex multiphasic signal involved in conveying information relating to mismatch of

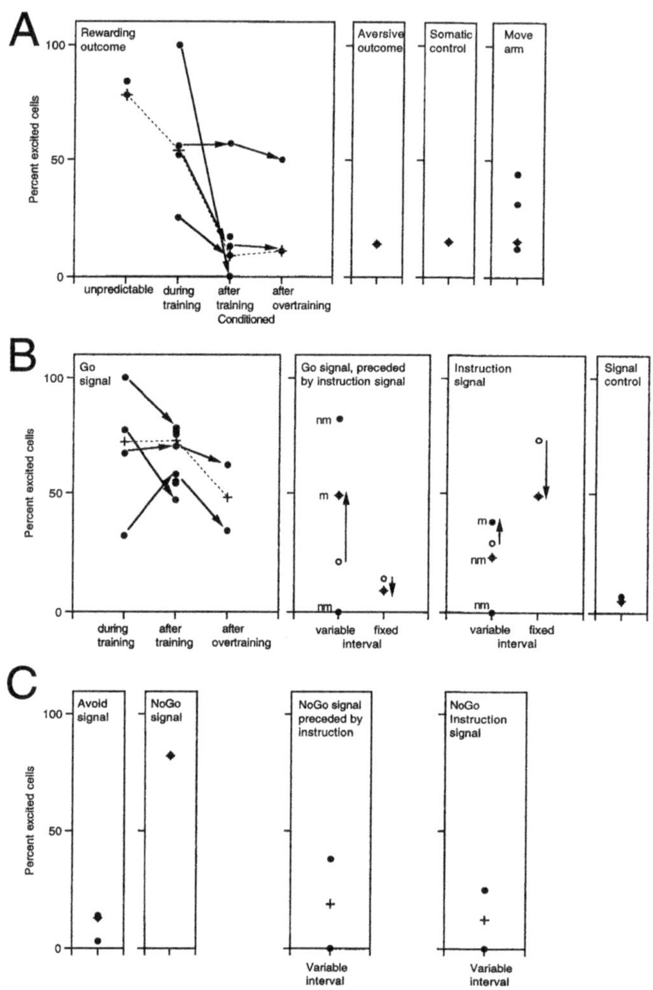


**Figure 1.3.** Percentages of monkey dopamine cells excited by various behavioural contingencies. Data from (Schultz, Ruffieux and Aebischer, 1983; Schultz, 1986; Schultz and Romo, 1990; Romo and Schultz, 1990; Ljungberg, Apicella and Schultz,1991 1992; Schultz, Apicella and Ljungberg, 1993a; Mirenowicz and Schultz, 1994, 1996). Dots show values from individual studies, while crosses show median values across studies (joined together by dashed lines, within a panel). Points derived from the same study over stages of training are joined by arrows. Where several data points are available across different stages of training, these are separated into different columns, while when only single study values are available the early stage training data is shown as open circles in the same column. **Part (A)** shows, from left to right, responses to delivery of primary rewards (either touch of food in reaching tasks, or delivery of fluid to the mouth), to aversive stimuli, to a non-food tactile stimulus, and to whole arm movements. **Part (B)** shows responses to signals related to GO responses in conditioned movement tasks, from left to right, to GO signals when these occurred alone, to GO signals when these were preceded by an instruction stimulus, to the instruction stimulus itself, and to similar signals outside of task context. Instructed tasks either had a fixed or variable delay between the instruction and GO signals, these data being shown in separate columns. In addition, in some studies the instruction cue was removed during the delay period, implicating memory mechanisms (m), while in others the cue was continuously available (non-memory, nm). Studies using the fixed interval used a non-memory condition. **(C)** The left most panel shows responses to a signal indicating that the monkey should act to avoid an aversive stimulus. The remaining panels show responses to NOGO related signals. All NOGO studies used variable interval and non-memory condition. Note that in the study that showed responses to the NOGO signal no reward was offered for correct NOGO performance.

expectation and outcome referred to in (1) above. It would be of interest to know what the responses to NOGO signals would be if they were very different in nature from GO signals, to avoid the possibility of stimulus generalization. Note that in the studies where this apparent stimulus generalization occurred, the monkeys were still successfully discriminating the signals, since they performed the NOGO task successfully, i.e. they did not make a movement. In contrast, dopamine neurones that respond to touch of food do not generalize their response to touch of non-food items (Romo and Schultz, 1990). Perhaps in this case the stimuli were sufficiently different ; alternatively, generalization may only be a property of responses to arbitrary, conditioned stimuli.

The effect of stage of training is also illustrated in Figure 1.3B. Across all studies, there is not a particularly clear effect of the level of training on the numbers responding to the GO signal, although within studies a trend for responses to decline with over-training has been emphasized, indicated on the graph by arrows connecting data points from the same studies. It should be noted that the case in which 100% of neurones responded during training represents only 9 cells from one monkey (Mirenowicz and Schultz, 1994).

A point emphasized by the authors of these papers is that after training, at the same time that responses to the primary reward have fallen to low levels, responses to the preceding GO signal are at their highest level. This has led to the suggestion that there is a transfer of responsiveness during training from the reward to the preceding signal (Ljungberg, Apicella and Schultz, 1992; Schultz, Apicella and Ljundberg, 1993a). This transfer is inferred by study of different populations of neurones at various stages of training. However, neurones studied after training which respond to the GO signal but not the reward itself in the conditioned task have been shown to respond to rewards when these are delivered randomly, without the prior signal (Mirenowicz and Schultz, 1994) (Figure 1.4).

One aspect of the transfer hypothesis that awaits resolution is whether such transfer will occur to any signal, which reliably predicts reward, or only to signals requiring a response by the animal. For instance, in experiments in which fluid is delivered to the mouth as reward, an audible click preceded the fluid arrival by about 55 ms (Schultz, Apicella and Ljungberg, 1993a). Simple Pavlovian conditioning principles might predict that the animal's behaviour, and possibly cellular responses, would become aligned to this indicator during the course of training. Only four neurones appear to have been tested with the control experiment in which the click occurs but fluid is not delivered (Ljungberg, Apicella and Schultz, 1991). None of these tested neurones were activated by the sound of the solenoid. (All were in fact inhibited at the time reward would normally have arrived.) However, neurones were tested after training, and that study was unusual in that the delayed alternation paradigm was used, and the responses to primary reward stayed high, despite overtraining (see above). It would be of interest to know whether responses to the solenoid occur at some intermediate stage during learning of the more typical conditional signaled tasks. However, it remains possible that in addition to the salience of a signal, dopamine neurones are sensitive to the simple amplitude of sensory signals. There is evidence for an additive effect of stimuli, with mixed stimuli producing larger responses than any one of the component stimuli alone (Schultz and Romo, 1990). Thus, perhaps the solenoid click, while audible and motivationally significant, was nevertheless insufficient to activate them.

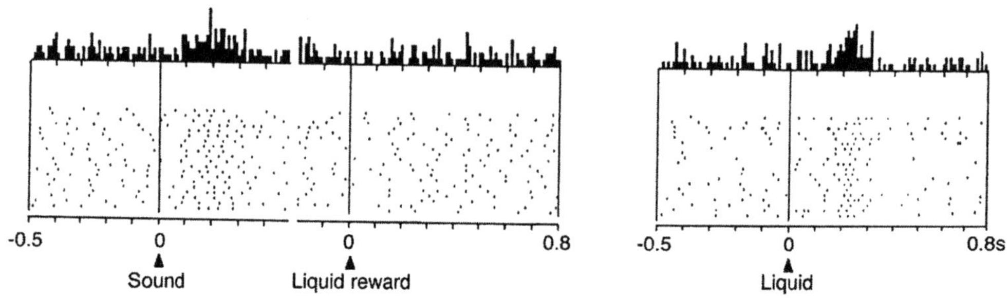


**Figure 1.4.** Responses of a monkey substantia nigra dopamine neurone to behavioural events, following training in a rewarded reaction-time movement task. Lines of dots show times of action potential occurrence on each trial, while histograms show summed activity over all trials. The right panel shows responses to randomly delivered liquid rewards, given out of task context. A period of increased activity in the cell is seen after liquid delivery. The left panel shows the response of the same neurone during task performance, in which an imperative sound signal triggered a movement response by the monkey, with reward delivered after successful task performance. Here, the neurone responds to the movement-triggering sound, but not the reward itself. Adapted and reproduced with permission from Mirenowicz and Schultz 1994, *Journal of Neurophysiology*, **72**, 1024–1027).

*(4) Level of responsiveness to both GO and NOGO stimuli can be influenced by the presence of a preceding "instruction" signal.*

Figure 1.3B and C also show data for responses of dopamine neurones to GO and NOGO signals in "instructed choice tasks", where the imperative signal is preceded by an "instruction stimulus", which indicates to the animal what response will be required. Filled circles show post-training data, which has been used to calculate medians for the figure, while open circles show values obtained during training, and arrows connect points from the same study. Two delay paradigms have been used, one in which the interval from instruction signal to GO signal is fixed, and another where this interval is varied randomly from trial to trial. The results from these two types of study are shown separately, as they appear to differ.

In the fixed-delay experiment, very few dopamine cells continued to respond to the GO signal, whilst responses to the instruction signal were seen (see below). This has been taken as another example of "transfer" of dopamine cell sensitivity to the earliest available reward-predicting sensory cue. In this study behavioural responses to the instructed GO signal sometimes occurred at "impossibly" short latencies, implying anticipation by the animal. It is therefore possible that in fixed delay situations the monkey uses the instruction cue as a form of behavioural trigger to release a delayed movement (Schultz, Apicella and Ljungberg, 1993a). This reliance on an internal timing mechanism might be expected to reduce the importance of the GO signal.

In contrast, when the instruction-GO interval was variable, responses to GO signals appear in most cases to remain within the range seen when they are not associated with previous instruction signals. It could be argued that in this case the GO signal remains the earliest reliable predictor of the expected time of reward, whereas the instruction signal time is not a reliable temporal marker.

In variable delay experiments, there were two additional parameters that differed: whether a warning signal occurred, and whether memory processes were required during the delay period. The study that found no responding cells at all was the first in the series

(Schultz, Ruffieux and Aebischer, 1983), and differed from all subsequent experiments by having an additional warning signal at a fixed interval before the instruction. This study also found no responding cells to the conditioned reward, instructed NOGO signals, or to the instruction signals, and in each of these cases inclusion of this data point has reduced the median values shown in Figure 1.3 to some extent. There was also no response to the warning signal, ruling out the possibility that the dopamine cells had completely transferred their responding to this earliest event in the sequence. On the other hand, this study had the highest level of movement related activity of any in the series. Another early study of nigral cells from a different laboratory, which probably included dopamine cells in the sample, also found no responses to task related stimuli (DeLong, Crutcher and Georgopoulos, 1983). The reasons for the lack of responses to task stimuli in these early studies remain uncertain, but it would be interesting if they could be identified, if only to ensure that future studies by other workers do not miss responses for the same reasons.

The other difference between the studies that used a variable duration delay period, was whether memory processes were necessarily invoked. The study that produced the intermediate proportion of responding cells (Schultz, Apicella and Ljungberg, 1993a) used a "memorized" delay task, in which the instruction light was turned off for much of the delay period, while the other two both had the instruction signal illuminated for the duration of the delay. Since the data point from the memorized delay study falls between the values from the other studies, it is difficult to assess whether this parameter had any effect on the result. If the zero value from (Schultz, Ruffieux and Aebischer, 1983) is put to one side as a possible anomaly, then the higher value for the remaining non-memorized study (Schultz and Romo, 1990) might suggest that having to memorize the required response reduces the impact of the GO signal upon dopamine cells. Conversely, the importance of the *instruction* signal at a discrete moment in time might be expected to be greater when the information it contains is only available for a limited period, compared to when it is continually available (see next section).

Another aspect of the data relating to GO signal responses in the presence of instruction signals is the effect of training. Because this has been examined in only one study, these results are overlaid on the same graph, using open circles to represent data obtained during training. In the experiment where a variable interval was used, the level of responsiveness to the GO signal increased with training, whereas with a fixed interval there was already a low level during training and if anything this fell further. Perhaps in the variable case the additional light was initially a distraction, but, with training, the monkey again learnt to attend specifically to the GO signal. With the fixed interval the perceived importance of the GO signal might have declined as confidence grew that the instruction provided all information required.

Responses to NOGO signals that are preceded by instruction signals are shown in Figure 1.3 C, below those for GO signals. Only variable delay studies have been performed, and one of these was the study that produced no responsive cells in any condition. The other utilized a non-memorized type of task, and in comparison to the response to the instructed GO signals in this paradigm, the response to (unrewarded) instructed NOGO signals was less. Thus prior information that the upcoming NOGO signal was of no reward value does seem to have reduced the response of the dopamine cells to it. Among many mechanisms, this could simply reflect a reduction in attentiveness by the animal.

*(5) Dopamine cells respond to instruction signals in instructed choice tasks*
The third panel in Figure 1.3B shows the proportion of tested dopamine neurones that were excited by instruction signals for GO responses in instructed delayed response tasks. Conventions are as for the previous panel.

When data from variable and fixed interval paradigms are compared, it appears that the instruction signal in the variable paradigm may activate less neurones than in the fixed one, opposite to that seen for the GO signal itself. Following the reasoning set out above, in the variable interval paradigm the instruction would be less useful as a marker. This apparent difference holds even if the zero value (from the 1983, possibly anomalous paper) is ignored.

Within the variable interval group, the data suggest that an instruction signal in a memory condition may activate more dopamine cells than in a non-memory condition. Again, since in the former case there is no opportunity for information to be gleaned through delay, the animal must attend more closely to the instruction signal at its onset, compared to the condition where the instruction remains illuminated through the delay.

Finally, the effect of learning is again in opposite directions for fixed and variable conditions, as it was for the GO signal. In the variable-delay memory case, responses to the instruction signal increase with learning (Schultz, Apicella and Ljungberg, 1993a), perhaps reflecting acquisition of insight into the importance of the instruction signal in this more difficult task. Learning effects in the non-memory variant of the task has not been formally assessed, but might be in the opposite direction. When instructed and non-instructed trials were intermingled, it was noted that more responses (77%) were obtained when instructed delay trials were rare, but lessened in frequency as instructed delay trials became more common (Schultz and Romo, 1990). This may have reflected a learning effect. This drop in responsiveness is consistent with the monkey learning that it did not need to attend to the instruction at its onset, since it would always be available. It is however unclear why the level of responsiveness should be so high initially, compared to when instruction signals were introduced at the beginning of learning in the memorized-delay study.

For fixed interval delays (which also did not require memory), the drop in responsiveness to the instruction stimulus with learning presumably reflects the fact that the time of onset of the signal is of little importance, since the information is available for the duration of a predictable delay period.

The fourth panel in Figure 1.3 C shows the available data for responses to instruction signals indicating that the upcoming imperative signal would indicate a NOGO response requirement. One data point was from the early study in which no responses were noted. The other reveals a proportion of excited cells (25%) rather similar to that for GO instruction signals in the same type of task (non-memory, variable interval), although correct performance of the NOGO trials was not rewarded. Perhaps this is another example of stimulus generalization in dopamine neurones. However, NOGO instruction stimuli do have a differential effect on dopamine neurones, in that almost as many (21%) of the tested dopamine cells were found to be *inhibited* (as the sole response) by the NOGO instruction signals (Schultz and Romo, 1990). In contrast, none of the tested cells was inhibited, as the only response, following instruction signals for GO in that study.

### *3.1.2. Latency of dopamine cell responses in monkey*

In studies involving reaching for food rewards (Romo and Schultz, 1990; Schultz and Romo, 1990; Schultz, 1986), latencies of dopamine cell responses to the various stimuli were very short, with medians of 50–65 ms and ranges of 25–130 ms reported. This is particularly impressive given that the cells are responding selectively; for instance, cells activated by touch of unseen food held on a wire with median latency of 65 ms were not activated by touch of the wire alone (Romo and Schultz, 1990). The selective nature of this response must have been based on processing of shape and or texture of the touched item, presumably by the somatosensory cortex, and activation of excitatory inputs to dopamine cells only if this processing resulted in recognition of the item as food. The latter step may have required additional processing, e.g. by the amygdala (Mirenowicz and Schultz, 1996). In comparison, responses to simple tactile stimuli in the secondary somatosensory cortex of monkeys fall into two groups, those with latencies of around 30 ms, and those with latencies greater than 70 ms, the latter being considered to reflect intracortical processing (Burton and Sinclair, 1990). Subsequent studies of dopamine neurones employed more arbitrary signals, and movements not directly relating to the position or active acquisition of rewards. Here the response latencies occupy higher ranges (of the order of 60–215 ms) and so probably higher median values (only mean values were reported in the later studies) (Ljungberg, Apicella and Schultz, 1991, 1992; Schultz *et al.*, 1993b; Mirenowicz and Schultz, 1994). This might reflect longer processing times for arbitrary stimuli. Curiously however, in these studies, responses to fluid delivery to the mouth, which might be expected to be the equivalent of touching food rewards, appear to have a longer latency than the somatosensory input from the digits (even after subtracting the delay from reward delivery command to the actual arrival of fluid at the mouth). In fact, in three of the four studies, the response to fluid in the mouth had a higher mean value than the responses to visual and auditory cue signals.

### *3.1.3. Theories of dopamine cell function in monkeys*

Recently, attention has been drawn to similarities between some of the above properties of dopamine cells in monkeys, and the behaviour of particular elements in a mathematical model of artificial systems capable of learning on the basis of match/mismatch of predictions and outcomes (Montague, Dayan and Sejnowski, 1996; Schultz, Dayan and Montague, 1997). In the model, a single sensory cue has a "temporal representation", such that at each moment following the cue there is an output from this representation, which depends on previous experience of the likelihood of encountering a reward at that moment relative to the signal. This predictive output is then compared with a signal bringing information about the presence or absence of reward at that moment, and an "error signal" is generated on the basis of this comparison. With each "trial", this error output is fed back to the "temporal representation", adjusting the prediction at each moment in time that will be generated following the next occurrence of the sensory cue. Three main features (illustrated in Figure 1.5) of the error signal are of interest in the present context: First, on initial "trials" by the algorithm, there is a large error output following the actual reward, since this was not predicted, but in the presence of a predictive stimulus, this

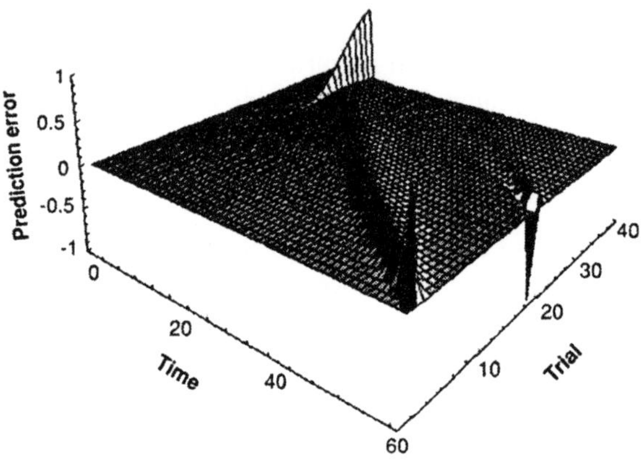


**Figure 1.5.** "Prediction error" output of a recently presented simulated network capable of learning stimulus-reward associations (Montague, Dayan and Sejnowski, 1996; Schultz, Dayan and Montague, 1997). In this model, prediction error is equated with dopamine cell activity. The surface shows changes in the activity within single trials (Time axis) and across trials, as training proceeds (Trial axis). "Sensory cues" are presented at time step 10 and 20, followed by "reward" at time step 60. Initially, there is a large response to the reward, but this declines with training to be replaced by a small positive response which progressively shifts in time until it follows the earliest sensory cue. Failure of reward delivery on a trial after some training causes a negative inflection at the "expected" time of reward delivery. This causes a negative response to also progressivley move backwards in time as it fades on successive trials. Note that no "action" is required by the model to obtain the reward, there is no response to the second sensory cue at any stage, and the response to the first cue starts small and becomes progressively stronger with more training. Reprinted, with permission from Schultz, Dayan and Montague, 1997, *Science, New York*, **275**, 1593–1599. Copyright (1997) American Association for the Advancement of Science.

response fades with repeated trials. In the monkey, responses to rewards are reduced if there is a predictive GO signal triggering the monkey's action to obtain a reward. Second, during subsequent trials the response progressively shifts backwards in time, until it follows the earliest available sensory cue. In the monkey, responses to GO signals occur when responses to rewards have declined; and when there is a preceding instruction stimulus, neurones do respond to this. Notably, in the model there appears to be no response to intervening predictive stimuli at any stage of the learning; only the earliest stimulus evokes a response. This matches well with the low level of responsiveness to GO signals in instructed tasks in which a fixed interval between the stimuli was used. Third, the absence of a predicted reward produces a large negative prediction signal, again as reported in monkey dopamine cells. The similarity of these features with properties of dopamine cells summarized above has led to the proposition that the dopamine system may provide the error signal in a biological instantiation of this kind of learning model (Montague, Dayan and Sejnowski, 1996; Schultz, Dayan and Montague, 1997).

Schultz, Dayan and Montague (1997) go on to discuss how such a signal could be used to guide behaviour, and point out that the dopamine signal could be used as an aid to solve the "temporal credit assignment problem". This refers to the fact that actions at one point in time can affect the acquisition of rewards in the future; the problem for the animal is then to identify which actions were crucial in the obtaining of reward. In this sense the dopamine signal can be looked upon as a "retrospective spotlight", enabling those actions which were favourable to be picked out and repeated. This is related to the idea that activity in the dopamine system marks relevant external stimuli (Wickelgren,

1997), but places emphasis on the marking of internal states associated with the animal's own actions. A similar idea has been formulated before, in a non-mathematical way by Miller (1981), who suggests that a "state of readiness" can be formed in striatal synapses and endures for a short time, during which period synapses are eligible for modification under the influence of dopamine. Here, dopamine cells are considered to be signaling that "what was just done was a good idea". In this conception, responses to random reward in the monkey studies would be seen as part of a search procedure, attempting to identify the act which led to the delivery of the reward. It is a requirement (for this to be useful to the animal) that such responses be able to retroactively mark for future reference those pathways which were active in the generation of the movement, and select those which repeat; the mechanism by which this could be achieved is at present unknown. Once the association has been learnt, the dopamine signal would be no longer required. On the other hand, when rewards are randomly delivered independent of behavioural state, the prior conditions are always different, and so the response should never habituate. Such a model can also account for responses to the predictive stimuli. During training in conditioned tasks, the monkey learns that certain responses to a GO signal will bring reward; however, in addition *presentation of the GO signal is contingent* upon the monkey achieving some steady state condition for a certain period of time. Thus during task acquisition it is necessary for the animal to identify which preceding act is associated with delivery of the GO signal. Consistent with the original purpose of the reward responses, dopamine activation now signals that a preceding act led to successful acquisition of the GO signal. A decline in responsiveness with learning might also be expected to occur, and has been seen to some extent. When there is an instruction stimulus, the whole sequence is moved one step back, and the monkey must learn what it has to do to generate the instruction stimulus, hence dopamine cells begin to respond to that stimulus as a retrospective marker that whatever was just done was a good idea, in the sense of identifying necessary conditions for the trial to proceed.

It is certainly of interest to find biological and modeling data intersecting in this way. Of course, there are areas where the data do not match precisely with what might be expected on the basis of the model. The power of an integrated physiological-modeling approach is that these mismatches can guide biological research and improvements in the model. One general difficulty of such a model, pointed out by the authors, is how a "temporal representation" of a discrete signal could be spread out over time in the nervous system. However this is not a problem of this model *per se*. Whether or not the particular type of representation modeled here is in fact correct, some kind of temporal representation must exist, to explain behaviour which seems to be based on estimation of elapsed time (such as anticipation errors in reaction time tasks, for instance). A similar general problem exists for the proposed retrospective identification of neural states associated with particular behavioural acts, as noted above.

Areas of apparent difference between the current model and physiological data which may be deserving of further research include the following:

(1) The model is of a relatively simple conditioned-unconditioned stimulus pairing, in which there was a constant interval between stimulus and reward delivery, whereas in the reaction-time tasks used in the monkey studies the timing of reward delivery would depend on the monkey's own reaction and movement times. This might pose difficulties for a model in which predictions at specific future times are set at the

moment of stimulus delivery: How would a biological system know in advance what the reaction and movement times will be? To precisely match the model parameters, it would also be of general interest to confirm that dopamine cells respond to arbitrary, predictive conditioned stimuli in tasks in which rewards delivery does not depend on any action by the animal.

(2) In the monkey, responses to unrewarded NOGO stimuli occur, even in well trained animals, albeit of lesser intensity. In addition, responses to stimuli predicting aversive outcomes occur, if the stimuli are similar. Can such "stimulus generalization", in the face of continued accurate task performance, be successfully incorporated in this kind of model?

(3) The model predicts that, after a few trials, the initial large response following the reward shrinks away, to be replaced by an earlier smaller peak which at first occurs immediately before the reward (i.e. at long latency following the signal), and then as trials progress, at progressively shorter and shorter latencies, until it immediately follows the earliest predictive stimulus, at which point it grows in size (See Figure 1.5). There is no suggestion in the monkey data that latency of response to GO signals changed during training, but such an effect might have not been noticed due to the fact that any one cell was recorded only on a particular day during training, and the data pooled across cells. Since the model data reflect changes in activity of a single cell over training, such an effect would be best looked for in a simple conditioned-unconditioned stimulus paradigm (as used in the model). Ideally this paradigm would be one in which individual cells could be tracked during the period over which the association was learnt. Unfortunately, only one cell would be able to be studied per animal.

(4) In the simulation generated by the model, the response to the earliest predictive stimulus appears to grow with repeated trials. In the closest matching monkey experiment, in which a fixed delay between the Instruction and GO signals was used, the number of dopamine cells responding to the instruction signal was high in initial stages of training, and then declined. One possibility is that, in the model, the time of occurrence of the earliest predictive stimulus was in itself unpredictable, whereas in the monkey work presentation of this stimulus was contingent upon the monkey first meeting some specific behavioural criterion (e.g. holding a key). Again, dopamine cell recording studies using the simpler associative paradigm are needed to clarify this discrepancy.

Such models, which implicate the nigro-striatal dopamine system in learning, derived as they are from experiments in highly constrained conditioning experiments, still face the challenge of being related to real life. It remains to be seen whether the signals and responses occurring in a trained task, involving flashing lights and so on, can be mapped onto an ethologically plausible scenario, such as learning to reach out for a certain fruit and not another. Furthermore, it remains to be established whether they can be generalized to dopamine cells in all species (see below).

### 3.2. Studies in Cats

Studies in cats dominated the early reports of dopamine activity in conscious behaving animals. Most work in this species has concerned the possibility of state-related changes, such as relations to particular phases of the sleep-waking cycle. These reports confirmed that cat substantia nigra cells were similar in electrophysiological and pharmacological characteristics to those of identified dopamine cells in anaesthetized rats (Steinfels, Heym and Jacobs, 1981; Trulson, Preussler and Howell, 1981). However, in these and similar later studies only slight increases were found in mean rate during awake periods *vs* sleep, or period of movement *vs* inactivity (Steinfels *et al.*, 1983a, b; Trulson, 1985a). No clear relationship of tonic rate to phases of sleep or wakefulness was seen, and no phasic activations related to particular components of walking, grooming or eating were noted.

Steinfels and co-workers have however identified a situation in which dopamine cells in the behaving cat are activated (Steinfels *et al.*, 1983a, b). They found that dopaminergic cells responded robustly and reliably with short latency (around 60 ms) to delivery of non-contingent auditory (clicks) or visual (flashes) stimuli. Remarkably, when compared to more recent data from monkeys described above, these responses were never associated with reward, did not depend on the animal orienting to the stimulus (although they were depressed during sleep), and showed no diminution after 2000 trials. However, some habituation to auditory (but not visual) meaningless stimuli was noted in a later study (Rasmussen, Strecker and Jacobs, 1986). Similar phasic responses to simple, unconditioned, auditory, visual, olfactory and tactile stimuli have also been noted in conscious cats in other studies (Trulson, 1985a; Strecker and Jacobs, 1985). Orientation to the stimulus cannot account for these results since the behavioural response habituated completely, whereas the neural response did not. However, it was noted that if the cat is attending to something else at the same time, then the response to the simple sensory stimulus is attenuated, suggesting some attentional process is involved or required for this response (Strecker and Jacobs, 1985).

Other dopamine neurones responded during orientation of the cat to stimuli with more meaning, such as cage door opening, or the appearance of the experimenter (Steinfels *et al.*, 1983a). These responses were often inhibitory, and disappeared once the cat stopped orienting to the stimulus, which occurred rapidly, within four repetitions. Sustained high levels of attention toward arousing stimuli (highly palatable food, inaccessible rats) failed to activate these cells (Strecker and Jacobs, 1985).

Two studies (Strecker, Steinfels and Jacobs, 1983; Trulson, Crisp and Trulson, 1983) looked at the responses of nigral dopamine cells in food deprived cats to eating, which, for purposes of comparison with primate data reviewed above, might be considered a "rewarding" event. Comparisons were made of mean firing rate while the cat was eating, *versus* periods when it was not. No change in mean activity across the periods was noted in either study. Similarly, handling that might be considered "rewarding", since it induced purring in the cats, was found to produce no sustained change in dopamine cell activity (Strecker and Jacobs, 1985). However, a caveat applying to many of these studies is that transient changes associated with the delivery of rewards may have been missed by the analysis used, i.e. calculation of mean firing rate over extended periods. One study specifically compared the results of integrating activity over 30 seconds following auditory and visual sensory stimulus delivery, *versus* peri-event time histograms analysis at a resolution in the ms range (Trulson, 1985a). This showed that in the former case only small percentages of neurones showed significant change compared to baseline, whereas in the latter analysis, almost all neurones were shown to have responses. Nevertheless, it would seem likely that any strong brief bursts occurring consistently at the time of reward delivery might have been noticed during the experiments.

The response of dopamine cells to direct elevation of blood glucose above that expected following a meal has also been investigated in conscious cats, but no responses were found (Strecker, Steinfels and Jacobs, 1983; Trulson *et al.*, 1983).

Substantia nigra dopamine cells in cats also failed to respond to aversive or stressful stimuli (tail pinch, cold, white noise), at least as detected by analysis of average rate over an extended time period (Strecker and Jacobs, 1985).

In summary, studies in cats have mainly looked for sustained "state-related" changes in activity, and were generally not well designed for detecting phasic responses. However, nigral neurones in cat have been found to respond robustly to "meaningless" sensory stimuli, without habituation but with some dependence on attention processes, and during orienting to complex "meaningful" stimuli but only if orientation occurs. These sensory responses appear to stand in stark contrast to the profile of dopamine cells in behaving monkeys, suggesting the possibility of a major species difference.

### 3.3. Studies in Rats

In rats, there are few studies of behavioural correlates of dopamine cell activity. This may relate to a relative difficulty in finding active dopamine cells in awake rats, and to the fact that that they may be more difficult to hold for long periods of time than other types of neurone (Miller, Sanghera and German, 1981 and personal observations). This stands in contrast to cat and monkey studies, where there seems to have been less problem in holding dopamine cells over extended periods. The reasons for this are unclear, but a lower relative stability of the headpiece on the smaller and thinner rat skull may contribute, and methods and materials used in electrode manufacture may be important (Diana, Garcia-Munoz and Freed, 1987).

The first published study in awake behaving rats was amongst the earliest for any species (Miller, Sanghera and German,1981), and the results in fact to a large degree foreshadowed later work in primates. In this study, almost all of a small sample of nigral dopamine cells responded (the majority with inhibition) to signals indicating that an appetitive reward will be forthcoming, and/or to the delivery of the reward itself. Responses of dopamine cells were examined in two tasks, a signaled lever press to obtain reward (operant task), and a simple classical conditioning paradigm, in which a signal was followed by reward at a fixed delay, independent of what the animal did. However "most" of the cells were recorded in the classical conditioning paradigm. Interestingly, this paradigm actually matches that used in the modeling studies (referred to above) more closely than the monkey work with which the model has been compared. The fact that, in the nigra, inhibitory responses seem to have dominated is not consistent with the model proposed. However only relatively few cells were recorded, the number of trials for each cell was very small (3–5), and the majority of cells were obtained in animals pretreated with haloperidol (to increase the number of active cells found with each electrode penetration). The increase in firing rates that would be expected in response to haloperidol might have increased the chance of detecting inhibitory responses. Whether this pharmacological manipulation would have any additional qualitative effect upon responses to task-related stimuli is unknown.

Surprisingly, no subsequent published studies have followed up these observations in the rat. However, in an ongoing study we have obtained confirmatory evidence that some nigral dopamine cells in the rat can be excited both by delivery of primary food rewards, and by signals indicating their imminent arrival (Hyland, Hay, Perk and Miller, unpublished observations; see Figure 1.6).

As in the cat, dopamine cells in the rat appear to show no relationship as a population to any phase of the sleep waking cycle, although individual neurones were noted to have slight changes (Miller *et al.*, 1983). A mild increase in firing rate and a tendency to burst was noted in a small population of dopamine cells during locomotor activity compared to rest (Diana *et al.*, 1989), but the extent to which individual cells were involved and the temporal relationship of activity changes to movement was not quantified.

Somewhat akin to results in cats, an increased tendency for nigral dopamine cells in rats to fire in bursts when animals oriented to noise or visual stimuli was noted in passing, in two studies (Freeman, Meltzer and Bunney, 1985; Freeman and Bunney, 1987). The possible relationship of these observations to the reward signal and delivery responses noted earlier (Miller *et al.*, 1981) has not been explored.

In summary, based on limited data, dopamine cells in rats appear to respond to rewarding events and signals preceding them. These responses might be similar to those described above in monkeys, although in the one published study, and our own experience, responses to predictable primary rewards seem to have persisted in well-trained animals. Further work is required to fully establish the similarity of rat reward

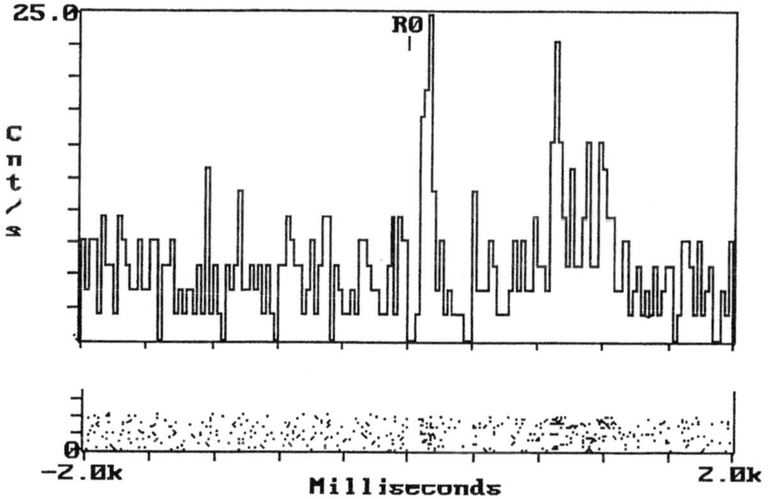


**Figure 1.6.** Activity of a rat substantia nigra dopamine neurone during performance of a trained task. The rat was trained to press one of two levers, depending on the position of a signal light, in order to receive a food pellet reward. Correct performance was signaled by a light flash at the feeder and the noise of the feeder motor, with food arriving via an elongated tube approximately 1 second later. The dot raster display of action potential occurrences on each trial and the peri-event time histogram are centered on the time of the light flash/motor sound ("R0"), and show two seconds on each side of this time. Following this signal there was a brief increase in activity, and then further activity once the food arrived at the animal and was taken into the mouth. In this cell there was no detectable change in activity at the time of the trigger stimulus or during performance of the lever press movement. (Hyland, Perk, Hay and Miller, unpublished data).

responses to those seen in monkeys, in terms of the requirement for unpredictability and the proposed “transfer” of responsiveness from primary rewards to reward-related stimuli. Rather anecdotal evidence suggests the possibility of “orienting” responses in rat dopamine neurones. These responses sound similar to the rapidly habituating orienting response noted in cats and monkeys, but further characterization of these is required. Responses to “meaningless” auditory or visual sensory stimuli have not been tested in the substantia nigra of the conscious rat (although they appear to be present in the ventral tegmental area—see below).

## 4. PROBLEMS AND FUTURE DIRECTIONS

### 4.1. the Possibility of Species Differences

General theories of the role of dopamine cell activity in behaviour have largely assumed that there is considerable homogeneity of response of dopamine cells, and so refer to functions of “the dopamine system” using selected data from different animal models. There is a huge body of behavioural data available on the rat dopamine system, and in the future it is likely that genetically altered mouse lines might offer new approaches to the study of dopamine systems. On the other hand, the most complete exploration of dopamine cell responses to specific behavioural contingencies has been performed in monkeys. It would therefore seem to be a priority to establish whether rodent dopamine neurones perform in similar fashion to those of primates, especially given the apparent strong species difference between primates and cats, at least in terms of responses to sensory stimuli. It remains to be seen whether a single theory of dopaminergic cell function can be formulated at a general enough level to account for cell properties across all experimental animals.

### 4.2. Substantia Nigra Compared to Ventral Tegmental Area

Within species, a question of some interest is whether the different anatomical groupings of dopamine neurones have different functions. For instance, the substantia nigra dopamine neurones are held to project largely to the striatum, and to be more involved in motor control, while ventral tegmental area neurones project more prominently ventrally within the basal ganglia and to frontal cortex, and are widely believed to be more involved in cognitive or affective function. However, the available data on dopamine cell activity in relation to behaviour do not provide much support for such a division. Firstly, nigral dopamine neurones do not seem to be particularly “motoric” in function, at least with regards to bursts of high frequency activity. Secondly, in the majority of studies by Schultz’s group in which dopamine neurones from the ventral tegmental area were recorded, these were considered sufficiently similar in their responses to be pooled with the nigral cells (Romo and Schultz, 1990; Schultz and Romo, 1990; Ljungberg , Apicella and Schultz, 1991, 1992; Mirenowicz and Schultz, 1994, 1996). Only two possible differences have been noted by these workers. In one study, medially placed neurones were significantly more related to trigger stimuli than to movement execution, while the opposite relation held for lateral neurones, although this difference was not significant when the comparison was made using anatomical boundaries (Schultz, 1986). In another study, a significantly higher proportion of ventral tegmental area cells responded during learning of a task than did nigral cells, but this difference disappeared once the task had been learnt (Schultz, Apicella and Ljungberg, 1993a). The significance of these isolated differences remains to be established.

Although other groups have recorded from monkey VTA during behaviour, either no attempt was made to identify dopamine cells (Fabre *et al.*, 1983), or the data were pooled in such a way that it is not possible to tell which responses were found particularly in dopamine cells (Nishino *et al.*, 1987). This makes it hard to know if the neurones related to the signal (Fabre *et al.*, 1983), bar-pressing and arm movement (Nishino *et al.*, 1987) described in these studies reflect dopaminergic activity. Suppression of activity in cells was noted once food rewards were obtained and being consumed (Nishino *et al.*, 1987).

In the cat, ventral tegmental area dopamine neurones were recently found to respond to simple, unconditioned sensory stimuli not associated with reward in a manner identical to previous observations on nigral cells (Horvitz, Stewart and Jacobs, 1997). On the other hand, ventral tegmental dopamine neurones have been noted to increase firing in response to an aversive stimulus in the cat (Trulson and Preussler, 1984), in contrast to nigral neurones which were not activated by such stimuli (Strecker and Jacobs, 1985).

In the rat, the original study of dopamine cell activity during behaviour by Miller, Sanghera and German (1981) included cells from both nigral and tegmental dopamine populations. Tegmental cells seemed overall less responsive (in that most of the unaffected cells were from this region) but when they did respond, tegmental area cells were more likely to be excited than nigral cells (the majority of which were inhibited). Another study in the rat found several ventral tegmental area cells that displayed a robust and non-habituating increase of activity in relation to whisker movement, either spontaneously during sniffing behaviour or during manipulation by the experimenter (Freeman and Bunney, 1987). It is not clear in the report whether substantia nigra cells showed similar whisker movement relations, although one nigral cell was said to respond when the rat oriented to the arrival of the experimenter.

Several studies have only examined tegmental dopamine neurones in the rat, so it is not known how nigral cells in this species would have responded to the various stimuli employed. However, some of the stimuli have been tested on nigral cells in other species. In the head-fixed but conscious rat, all tested tegmental dopamine cells responded (mostly with excitation) to aversive stimuli (Kiyatkin, 1988), as found in cat tegmentum (see above), but not, to date, in monkey. In addition, over a third of neurones were excited by meaningless, unrewarded light or sound stimuli, again in contrast to monkey but similar to cat tegmental (and in this case nigral) cells. In a study of the ventral tegmental area in which dopamine neurones were not specifically identified (Kosobud, Harris and Chapin,1994) some cells with low firing rate were found to be inhibited during consumption of appetitive rewards, a finding replicating an observation made in monkey ventral tegmental area in one study (Nishino *et al.*, 1987) but not reported for identified dopaminergic cells by Schultz's group. Depression of dopamine cell activity following heroin self-administration has also been noted in rat ventral tegmental area neurones identified as dopaminergic on the basis of action potential duration (Kiyatkin and Rebec, 1997).

In summary, when dopamine cells from both the substantia nigra and ventral tegmental area have been recorded in the same studies, few consistent differences have been noted in their relation to behaviour. Comparisons across studies are difficult, and across species even more so, but the inconsistent mixture of apparent similarities and contradictions in the data suggest that more work is needed to clarify the issue.

### 4.3. Are Substantia Nigra Dopamine Cells a Homogenous Population?

Even within the substantia nigra of one species, it is not totally correct to assume that all dopamine cells are the same. Certainly, some stimuli (e.g. unpredicted reward in monkeys, meaningless sensory stimuli in cats) do activate many nigral dopamine neurones. However, the possibility of subgroups forming other functions cannot be discounted. For example in many situations, a few cells are inhibited by stimuli that excite the majority. Are these aberrant, or is their "message" of importance for the system as a whole? Another feature of the monkey data is the "missing 20%". This refers to the fact that in all cases where a large number of randomly selected neurones has been studied, no more than around 80% of the cells respond to any of the task stimuli employed. One possibility is that they are different neurones in each study—i.e., the unresponsive ones in study A would have been among cells responding to a different stimulus used in study B; nevertheless, it is an obvious heterogeneity. Finally, there is the question of whether the neurones showing tonic and rather minor activation in relation to the movement, rather than in response to specific signals, seen in up to 40% of neurones in early studies, form a particular functional grouping.

Specific evidence for heterogeneity of dopamine neurones at the cellular level has been obtained in many studies of anaesthetized or reduced preparations, largely in rats. For example, different groups of substantia nigra dopamine neurones respond to different degrees to apomorphine (Freeman and Bunney, 1987; Shepard and German, 1988), and in opposite directions to oestrogen (Chiodo and Caggiula, 1983), morphine (Ostrowski and Caggiula, 1991), and sensory stimuli (Chiodo *et al.*, 1980). In primates, differential sensitivity of specific nigral dopamine populations to the toxin MPTP has been noted (Schneider, Yuwiler and Markham, 1987; Varastet *et al.*, 1994).

The meaning in terms of functions of release of dopamine within the striatum during behaviour, if any, of variation among the responses and properties of dopamine cells remains to be established.

### 4.4. Are Data from Single Cell Recordings Consistent with Neurochemical Measurements of Dopamine Release During Behaviour?

An underlying assumption of work involving recording from dopamine cells during behaviour is that the observed pattern of action potentials gives an indication of the relative degree of dopamine release from axon terminals. With the availability of methods such as dialysis and electrochemical detection to directly measure substances in the brain of behaving animals, an obvious question is whether the data are congruent. If they are, then there are important implications for the design of future experiments. The neurochemical methods seem to generate data more rapidly, since one is detecting population activity by definition; there is no need to sample many cells to achieve this measure of population response. However, dialysis, while specific, suffers from problems with temporal precision, and electrochemistry, while offering greater temporal resolution still operates at broader time scales than cell recording methods. Despite improvements in methodology, there also remain some difficulties with identifying the source of the electrochemical signal. Furthermore, since these methods detect overflow, the relationship of the measured values to local release at the synaptic level can only be inferred, and any regional variation can only be assessed on a relatively large spatial scale. It is clear that the neurochemical and electrophysiological approaches measure quite different aspects of dopamine dynamics in the striatum, with the cell recording method resolving dopamine activation at a much finer-grain, both temporally and spatially. A major challenge for the future is to determine what relation the measures taken have to synaptic functioning in the striatum, and whether both or only one are needed to provide the necessary information to understand the role of dopamine in this structure. Given these theoretical issues, and the vast literature that has accumulated on dopamine neurochemistry, an exhaustive review has not been attempted. However, a selection of papers of particular relevance to the dorsal striatum (presumed target of nigral dopamine neurones) is briefly discussed.

An early study of the monkey striatum found a sharp rise in voltammetric signal in association with exposure to a stressor (associated with a prolonged behavioural response) and during consumption of food (Lindsay *et al.*, 1981). Dopamine cells in the monkey are said not to respond to stressful or aversive stimuli (see above), and the duration of the increase in the electrochemical signal during eating seems longer than the bursts seen in response to primary rewards. If these electrochemical findings can be confirmed, it might suggest that dopamine is being released in such situations from terminals in the striatum, independent of pulsatile activity in the dopamine neurone cell bodies and axons, perhaps by transmitters acting directly on dopamine terminals (Grace, 1991; Salamone, 1996).

One brief report has appeared in which voltammetry in the caudate nucleus was combined with single unit recording of dopamine neurones in the substantia nigra, in the cat (Trulson, 1985b). Data were averaged over 5 minutes epochs and the results suggest that electrochemical signal in the striatum can indeed be dissociated from nigral dopamine cell activity. While both measures increased during periods of movement compared to no-movement, during sleep, cell activity stayed constant while the electrochemical signal fell. Unfortunately, this preliminary but potentially powerful work does not appear to have been replicated since the development of more selective electrochemistry methods.

More data is available for rats, but is still surprisingly sparse for the dorsal striatum. A recent study in rats found that in this species, like the monkey, stressful stimuli increase electrochemical (presumed dopamine) signal in the striatum (Doherty and Gratton, 1992). However, in the rat an intra-striatal mechanism may not need to be invoked to explain this release, since rat dopamine cells may be activated by such stimuli (Kiyatkin, 1988).

Electrochemical detection of dopamine in target structures following signals and rewards in conditioned tasks has also been attempted in the rat. In the dorsal striatum, presumed dopamine levels were found to rise after a GO signal, and at or before the time of the operant lever-press response (Shinba *et al.*, 1998). The level remained high during the lever press and the subsequent food consumption, with a less prominent rise during passive feeding. This result seems consistent with the monkey data on dopamine cell responses to cue signals, except that the latency (mean 400 ms) and duration (mean 7 sec) of the rise in dopamine were much longer than for the bursts in monkey dopamine cells seen in these situations. Perhaps the duration of eating was longer, or another possibility is that detection of dopamine metabolites extended the time course of the signal. Another apparent difference with the monkey single-cell recording work is that after "error" trials in which no reward was delivered, the electrochemical signal rose to the same extent as during rewarded trials, whereas an inhibition might have been expected from the monkey data.

There is a much larger literature on the detection of dopamine release in the ventral striatum (nucleus accumbens), almost exclusively in the rat. This work has been reviewed elsewhere (Salamone, 1996). Overall, these studies suggest that presumed dopamine activity in the ventral striatum can be increased by aversive conditions, as well as during appetitive reward, again unlike the recorded monkey neurones. However there is considerable controversy even within the rat neurochemistry literature over such simple issues as whether consumption of food raises, does not change, or lowers dopamine levels in the accumbens. It is thus easy to pick results that support or contradict the single unit studies in monkeys. For instance, a recent study found that, as noted above for the rat dorsal striatum, presumed dopamine levels increased in the accumbens when reward was withheld (Richardson and Gratton, 1996) seemingly contradicting the monkey single neurone responses in such situations. On the other hand the same study did show reduced responsiveness to reward, and increased responses to a warning signal with progression of experience in the task, which could be argued to be similar to reported effects of training on responses of monkey dopamine cells.

Generally, the usefulness of comparing electrochemistry data with the single cell results is severely compromised by the fact that the most systematic neural data is from monkey, while the neurochemistry comes largely from rat accumbens. Further studies of monkey electrochemical responses are required to investigate the extent to which the apparent conflicts between electrochemical and neuronal recording data are species specific, relate to dorsal *vs* ventral striatum, or reflect differences between the two methodologies discussed above. Even within species, comparing across studies is difficult, because of possibly significant effects of small differences in task design. Therefore, combined studies in which cell activity and neurochemistry are simultaneously performed in the same animal would seem to be ideal.

## 5. CONCLUSIONS

Across species, there are some similarities in activity of single dopaminergic neurones in the substantia nigra during behaviour, which hint at a common functional relationship of these cells in all animals. However there are also differences, which point to the possibility of species-specific signalling roles for these neurones. In the monkey, a reasonably coherent picture emerges from the concerted efforts of one laboratory. Here, dopamine neurones of the substantia nigra are rather selectively activated by rewarding events (not generally alerting) or by the earliest cue which allows prediction of such events. This activity might provide a signal marking certain emitted behaviours as being important for the obtaining of rewards. If so, this would suggest a role for dopamine cell activity in the learning of motor acts or components of behavioural sequences.

## REFERENCES

Aebischer, P. and Schultz, W. (1984) The activity of pars compacta neurons of the monkey substantia nigra is depressed by apomorphine, *Neuroscience Letters*, **50**, 25–29.

Aghajanian, G.K. and Bunney, B.S. (1973) In *Frontiers in catecholamine research*, edited by E. Usdin, and S.H. Snyder, pp. 643–648, New York: Pergamon Press.

Aghajanian, G.K. and Bunney, B.S. (1977) Dopamine "autoreceptors": pharmacological characterization by microiontophoretic single cell recording studies, *Naunyn-Schmiedebergs Archives of Pharmacology*, **297**, 1–7.

Aosaki, T., Tsubokawa, H., Ishida, A., Watanabe, K., Graybiel, A.M. and Kimura, M. (1994) Responses of tonically active neurons in the primate's striatum undergo systematic changes during behavioral sensorimotor conditioning, *Journal of Neuroscience*, **14**, 3969–3984.

Bunney, B.S., Walters, J.R., Roth, R.H. and Aghajanian, G.K. (1973) Dopaminergic neurons: effect of antipsychotic drugs and amphetamine on single cell activity, *Journal of Pharmacology & Experimental Therapeutics*, **185**, 560–571.

Burton, H. and Sinclair, R.J. (1990) Second somatosensory cortical area in macaque monkeys. I. Neuronal responses to controlled, punctate indentations of glabrous skin on the hand, *Brain Research*, **520**, 262–271.

Chiodo, L.A., Antelman, S.M., Caggiula, A.R. and Lineberry, C.G. (1980) Sensory stimuli alter the discharge rate of dopamine (DA) neurons: evidence for two functional types of DA cells in the substantia nigra, *Brain Research*, **189**, 544–549.

Chiodo, L.A. and Caggiula, A.R. (1983) Substantia nigra dopamine neurons: alterations in basal discharge rates and autoreceptor sensitivity induced by estrogen, *Neuropharmacology*, **22**, 593–599.

Dahlström, A. and Fuxe, K. (1964) Evidence for the existence of monoamine-containing neurones in the central nervous system.I. Demonstration of monoamines in the cell bodies of brain stem neurons. *Acta Physiologica Scandinavica*, Suppl. **232**, 1–55.

DeLong, M.R., Crutcher, M.D. and Georgopoulos, A.P. (1983) Relations between movement and single cell discharge in the substantia nigra of the behaving monkey, *Journal of Neuroscience*, **3**, 1599–1606.

Diana, M., Garcia-Munoz, M. and Freed, C.R. (1987) Wire electrodes for chronic single unit recording of dopamine cells in substantia nigra pars compacta of awake rats, *Journal of Neuroscience Methods*, **21**, 71–79.

Diana, M., Garcia-Munoz, M., Richards, J. and Freed, C.R. (1989) Electrophysiological analysis of dopamine cells from the substantia nigra pars compacta of circling rats, *Experimental Brain Research*, **74**, 625–630.

Doherty, M.D. and Gratton, A. (1992) High-speed chronoamperometric measurements of mesolimbic and nigrostriatal dopamine release associated with repeated daily stress, *Brain Research*, **586**, 295–302.

Fabre, M., Rolls, E.T., Ashton, J.P. and Williams, G. (1983) Activity of neurons in the ventral tegmental region of the behaving monkey, *Behavioural Brain Research*, **9**, 213–235.

Freeman, A.S. and Bunney, B.S. (1987) Activity of A9 and A10 dopaminergic neurons in unrestrained rats: further characterization and effects of apomorphine and cholecystokinin, *Brain Research*, **405**, 46–55.

Freeman, A.S., Meltzer, L.T. and Bunney, B.S. (1985) Firing properties of substantia nigra dopaminergic neurons in freely moving rats, *Life Sciences*, **36**, 1983–1994.

German, D.C. and Manaye, K.F. (1993) Midbrain Dopaminergic Neurons (Nuclei A8, A9, and A10): 3-Dimensional Reconstruction in the Rat, *Journal of Comparative Neurology*, **331**, 297–309.

Grace, A.A. (1991) Phasic versus tonic dopamine release and the modulation of dopamine system responsivity: a hypothesis for the etiology of schizophrenia, *Neuroscience*, **41**, 1–24.

Grace, A.A. and Bunney, B.S. (1980) Nigral dopamine neurons: intracellular recording and identification with L-dopa injection and histofluorescence, *Science*, **210**, 654–656.

Grace, A.A. and Bunney, B.S. (1983) Intracellular and extracellular electrophysiology of nigral dopaminergic neurons—1. Identification and characterization, *Neuroscience*, **10**, 301–315.

Grace, A.A. and Bunney, B.S. (1985) Low doses of apomorphine elicit two opposing influences on dopamine cell electrophysiology, *Brain Research*, **333**, 285–298.

Grace, A.A. and Onn, S.P. (1989) Morphology and electrophysiological properties of immunocytochemically identified rat dopamine neurons recorded *in vitro*, *Journal of Neuroscience*, **9**, 3463–3481.

Guyenet, P.G. and Aghajanian, G.K. (1978) Antidromic identification of dopaminergic and other output neurons of the rat substantia nigra, *Brain Research*, **150**, 69–84.

Heym, J., Steinfels, G.F. and Jacobs, B.L. (1982) Activity of serotonin-containing neurons in the nucleus raphe pallidus of freely moving cats, *Brain Research*, **251**, 259–276.

Horvitz, J.C., Stewart, T. and Jacobs, B.L. (1997) Burst activity of ventral tegmental dopamine neurons is elicited by sensory stimuli in the awake cat, *Brain Research*, **759**, 251–258.

Kiyatkin, E.A. (1988) Functional properties of presumed dopamine-containing and other ventral tegmental area neurons in conscious rats, *International Journal of Neuroscience*, **42**, 21–43.

Kiyatkin, E.A. and Rebec, G.V. (1997) Activity of presumed dopamine neurons in the ventral tegmental area during heroin self-administration, *Neuroreport*, **8**, 2581–2585.

Kosobud, A.E., Harris, G.C. and Chapin, J.K. (1994) Behavioral associations of neuronal activity in the ventral tegmental area of the rat, *Journal of Neuroscience*, **14**, 7117–7129.

Lacey, M.G., Mercuri, N.B. and North, R.A. (1989) Two cell types in rat substantia nigra zona compacta distinguished by membrane properties and the actions of dopamine and opioids, *Journal of Neuroscience*, **9**, 1233–1241.

Lindsay, W.S., Herndon, J.G., Jr., Blakely, R.D., Justice, J.B., Jr. and Neill, D.B. (1981) Voltammetric recording from neostriatum of behaving rhesus monkey, *Brain Research*, **220**, 391–396.

Ljungberg, T., Apicella, P. and Schultz, W. (1991) Responses of monkey midbrain dopamine neurons during delayed alternation performance, *Brain Research*, **567**, 337–341.

Ljungberg, T., Apicella, P. and Schultz, W. (1992) Responses of monkey dopamine neurons during learning of behavioral reactions, *Journal of Neurophysiology*, **67**, 145–163.

Magarinos-Ascone, C., Buno, W. and Garcia-Austt, E. (1992) Activity in monkey substantia nigra neurons related to a simple learned movement, *Experimental Brain Research*, **88**, 283–291.

Matsuda, Y., Fujimura, K., Yoshida, S., Hausser, M.A., Yung, W.H. and Lacey, M.G. (1987) Two types of neurons in the substantia nigra pars compacta studied in a slice preparation, *Neuroscience Research*, **5**, 172–179.

Miller, J.D., Farber, J., Gatz, P., Roffwarg, H. and German, D.C. (1983) Activity of mesencephalic dopamine and non-dopamine neurons across stages of sleep and walking in the rat, *Brain Research*, **273**, 133–141.

Miller, J.D., Sanghera, M.K. and German, D.C. (1981) Mesencephalic dopaminergic unit activity in the behaviorally conditioned rat, *Life Sciences*, **29**, 1255–1263.

Miller, R. (1981) *Meaning and Purpose in the Intact Brain*, Oxford University Press, Oxford.

Mirenowicz, J. and Schultz, W. (1994) Importance of unpredictability for reward responses in primate dopamine neurons, *Journal of Neurophysiology*, **72**, 1024–1027.

Mirenowicz, J. and Schultz, W. (1996) Preferential activation of midbrain dopamine neurons by appetitive rather than aversive stimuli, *Nature*, **379**, 449–451.

Montague, P.R., Dayan, P. and Sejnowski, T.J. (1996) A framework for mesencephalic dopamine systems based on predictive Hebbian learning, *Journal of Neuroscience*, **16**, 1936–1947.

Nishino, H., Ono, T., Muramoto, K., Fukuda, M. and Sasaki, K. (1987) Neuronal activity in the ventral tegmental area (VTA) during motivated bar press feeding in the monkey, *Brain Research*, **413**, 302–313.

Ostrowski, N.L. and Caggiula, A.R. (1991) Correlation between locomotor stimulation and the electrophysiological effects of low doses of morphine on substantia nigra dopamine neurons. I. Acute drug administration, *Journal of Pharmacology & Experimental Therapeutics*, **257**, 72–81.

Rasmussen, K., Strecker, R.E. and Jacobs, B.L. (1986) Single unit response of noradrenergic, serotonergic and dopaminergic neurons in freely moving cats to simple sensory stimuli, *Brain Research*, **369**, 336–340.

Richardson, N.R. and Gratton, A. (1996) Behavior-relevant changes in nucleus accumbens dopamine transmission elicited by food reinforcement: an electrochemical study in rat, *Journal of Neuroscience*, **16**, 8160–8169.

Romo, R. and Schultz, W. (1990) Dopamine neurons of the monkey midbrain: contingencies of responses to active touch during self-initiated arm movements, *Journal of Neurophysiology*, **63**, 592–606.

Salamone, J.D. (1996) The behavioral neurochemistry of motivation: methodological and conceptual issues in studies of the dynamic activity of nucleus accumbens dopamine, *Journal of Neuroscience Methods*, **64**, 137–149.

Schneider, J.S., Yuwiler, A. and Markham, C.H. (1987) Selective loss of subpopulations of ventral mesencephalic dopaminergic neurons in the monkey following exposure to MPTP, *Brain Research*, **411**, 144–150.

Schultz, W. (1986) Responses of midbrain dopamine neurons to behavioral trigger stimuli in the monkey, *Journal of Neurophysiology*, **56**, 1439–1461.

Schultz, W. (1994) Behavior-related activity of primate dopamine neurons, *Revue Neurologique*, **150**, 634–639.

Schultz, W. (1997) Dopamine neurons and their role in reward mechanisms, *Current Opinion in Neurobiology*, **7**, 191–197.

Schultz, W., Apicella, P. and Ljungberg, T. (1993a) Responses of monkey dopamine neurons to reward and conditioned stimuli during successive steps of learning a delayed response task, *Journal of Neuroscience*, **13**, 900–913.

Schultz, W., Apicella, P., Ljungberg, T., Romo, R. and Scarnati, E. (1993b) Reward-related activity in the monkey striatum and substantia nigra. In: *Chemical Signalling in the Basal Ganglia, (Progress in Brain Research, vol. 99)*, edited by G.W. Arbuthnott and P.C. Emson, pp. 227–235, North Holland: Elsevier.

Schultz, W., Dayan, P. and Montague, P.R. (1997) A neural substrate of prediction and reward, *Science, New York*, **275**, 1593–1599.

Schultz, W. and Romo, R. (1990) Dopamine neurons of the monkey midbrain: contingencies of responses to stimuli eliciting immediate behavioral reactions, *Journal of Neurophysiology*, **63**, 607–624.

Schultz, W., Ruffieux, A. and Aebischer, P. (1983) The activity of pars compacta neurons of the monkey substantia nigra in relation to motor activation, *Experimental Brain Research*, **51**, 377–387.

Shepard, P.D. and German, D.C. (1988) Electrophysiological and pharmacological evidence for the existence of distinct subpopulations of nigrostriatal dopaminergic neuron in the rat, *Neuroscience*, **27**, 537–546.

Shinba, T., Andow, Y., Shinozaki, T., Ozawa, N. and Yamamoto, K. (1998) Phasic increase of monoamine-related electrochemical signal in the rat caudate nucleus following conditioned auditory stimulation during the reaction-time task, *Brain Research*, **781**, 284–290.

Silva, N.L. and Bunney, B.S. (1988) Intracellular studies of dopamine neurons in vitro: pacemakers modulated by dopamine, *European Journal of Pharmacology*, **149**, 307–315.

Steinfels, G.F., Heym, J. and Jacobs, B.L. (1981) Single unit activity of dopaminergic neurons in freely moving cats, *Life Sciences*, **29**, 1435–1442.

Steinfels, G.F., Heym, J., Strecker, R.E. and Jacobs, B.L. (1983a) Behavioral correlates of dopaminergic unit activity in freely moving cats, *Brain Research*, **258**, 217–228.

Steinfels, G.F., Heym, J., Strecker, R.E. and Jacobs, B.L. (1983b) Response of dopaminergic neurons in cat to auditory stimuli presented across the sleep-waking cycle, *Brain Research*, **277**, 150–154.

Strecker, R.E. and Jacobs, B.L. (1985) Substantia nigra dopaminergic unit activity in behaving cats: effect of arousal on spontaneous discharge and sensory evoked activity, *Brain Research*, **361**, 339–350.

Strecker, R.E., Steinfels, G.F. and Jacobs, B.L. (1983) Dopaminergic unit activity in freely moving cats: lack of relationship to feeding, satiety, and glucose injections, *Brain Research*, **260**, 317–321.

Trulson, M.E. (1985a) Activity of dopamine-containing substantia nigra neurons in freely moving cats, *Neuroscience and Biobehavioral Reviews*, **9**, 283–297.

Trulson, M.E. (1985b) Simultaneous recording of substantia nigra neurons and voltammetric release of dopamine in the caudate of behaving cats, *Brain Research Bulletin*, **15**, 221–223.

Trulson, M.E., Crisp, T. and Trulson, V.M. (1983) Dopamine-containing substantia nigra units are unresponsive to changes in plasma glucose levels induced by dietary factors, glucose infusions or insulin administration in freely moving cats, *Life Sciences*, **32**, 2555–2564.

Trulson, M.E. and Preussler, D.W. (1984) Dopamine-containing ventral tegmental area neurons in freely moving cats: activity during the sleep-waking cycle and effects of stress, *Experimental Neurology*, **83**, 367–377.

Trulson, M.E., Preussler, D.W. and Howell, G.A. (1981) Activity of substantia nigra units across the sleep-waking cycle in freely moving cats, *Neuroscience Letters*, **26**, 183–188.

Varastet, M., Riche, D., Maziere, M. and Hantraye, P. (1994) Chronic MPTP treatment reproduces in baboons the differential vulnerability of mesencephalic dopaminergic neurons observed in Parkinson's disease, *Neuroscience*, **63**, 47–56.

Wickelgren, I. (1997) Getting the brain's attention.*Science*, **278**, 35–37.

Yung, W.H., Hausser, M.A. and Jack, J.J.B. (1991) Electrophysiology of Dopaminergic and Non-Dopaminergic Neurones of the Guinea Pig Substantia Nigra Pars Compacta Invitro, *Journal of Physiology, London*, **436**, 643–667.

# 2 The Role of Dopamine in the Control of Locomotor Activity and Reward-Related Incentive Learning

**Richard J. Beninger†* and Mary C. Olmstead***

*(Departments of *Psychology and †Psychiatry, Queen's University, Kingston ON K7L 3N6 Canada)*
*Correspondence to: Dr. Richard J. Beninger, Department of Psychology, Queen's University Kingston ON K7L 3N6, Canada*
*Tel: 613 533 2486; FAX: 613 533 2499; email: beninger@psyc.queensu.ca*

The neurotransmitter dopamine (DA) acts at a number of terminal regions in the brain including the nucleus accumbens, caudate nucleus and frontal cortex and at a number of different receptor subtypes grouped into the D1- and D2-like families. In general, increases in DA neurotransmission lead to increases in locomotor activity, while decreases in DA neurotransmission lead to decreases in activity. The contribution of DA terminal regions is not uniform, however, and neither is the contribution of receptor subtypes. DA also plays an important and possibly critical role in reward-related incentive learning: In tasks where conditioning occurs to a specific cue, D1-like agonists impair performance whereas D2-like agonists augment performance in many cases. In tasks where conditioning is to contextual cues, D1- and D2-like agonists similarly lead to conditioning. Some recent studies using antagonists show that a block of D1-like receptors may produce a greater effect on reward-related learning than a block of D2-like receptors. Taken together, results suggest a critical role for D1-like receptors in reward-related learning. Animal learning studies have revealed a number of details of the effects of motivational state on the incentive value of primary rewarding stimuli. These effects influence incentive learning. It may be possible to integrate results from psychopharmacological and behavioural studies within the framework of a neuronal model of dopamine-mediated incentive learning involving the modification of corticostriatal synapses activated in close temporal contiguity with the presentation of reward.

KEY WORDS: Dopamine, D1-Like Receptors, Incentive learning, Locomotor activity, Reward

## 1. INTRODUCTION

A relatively small number of cell bodies located in the region of the ventral midbrain use dopamine (DA) as their neurotransmitter substance. These cells send axons to several forebrain sites including the dorsal and ventral regions of the striatum and the frontal cortex. Their axons branch extensively, each dopaminergic neurone forming hundreds of thousands of synaptic specializations in its terminal region. Although small in number, these elaborately branched cells are in a unique position to modulate connections in the brain between incoming sensory/perceptual messages and outgoing motor signals. In this chapter, we will discuss the role played by these dopaminergic cells in the control of locomotor activity and reward-related incentive learning.

Midbrain dopaminergic neurones often are grouped into two systems. One consists of cell bodies located in the substantia nigra of the midbrain that project primarily to dorsal striatal structures, the caudate and putamen; this system is termed the nigrostriatal system. The other consists of cell bodies located in the ventral tegmental area of the midbrain that project mainly to two regions, the ventral striatal area termed the nucleus accumbens and the medial regions of the frontal cortex; this system is termed the mesolimbic-mesocortical system. For detailed anatomy of these systems see Lindvall (1979). We will review data showing that these two dopaminergic systems play different roles in the control of behaviour (see also chapter 12 by Joel and Weiner).

In recent years it has been found that receptors for DA are of at least five different subtypes. They have been classified into two main families based on the differential effects of DA receptors on the receptor-associated enzyme adenylyl cyclase. DA receptors classified as D1-like are those that stimulate this enzyme, and those classified as D2-like inhibit the enzyme (Kebabian and Calne, 1979; Stoof and Kebabian, 1981). D1-like receptors include the D1 and D5 subtypes and D2-like receptors include the D2, D3 and D4 subtypes (Niznik and Van Tol, 1992; Civelli, Bunzow and Grandy, 1993; Sibley, Monsma and Shen, 1993). We will review findings showing that pharmacological manipulations of these receptor subtypes produces different behavioural effects that suggest a differential contribution of DA receptor subtypes to the control of locomotor activity and learning.

## 2. DOPAMINE AND LOCOMOTOR ACTIVITY

In general, increases in dopaminergic neurotransmission lead to increases in locomotor activity and decreases in dopaminergic neurotransmission lead to decreases in locomotor activity. However, different terminal regions and different receptor subtypes do not contribute in the same way (for reviews see: Clark and White, 1987; Beninger, Mazurski and Hoffman, 1991).

Locomotion is measured in a variety of ways (Robbins, 1977). Observational techniques include counting crossings of lines painted on the floor of a chamber, or rating scales that rate the level of activation of animals from lying down with eyes closed to vigorous stereotyped activity (Ellinwood and Balster, 1974). There are a number of automated approaches. These include tilt cages fixed on a fulcrum with micro switches under either end that detect movement back and forth over the fulcrum, and jiggle cages set on foam blocks with micro switches under each corner that quantify activity. Many activity monitoring systems use one or more infrared emitters and detectors fixed at various points around the cage to detect movement indicated by beam breaks. Some modern approaches involve creating a digital video record of the animals' movement in the cage and then using sophisticated computer software to analyse a variety of movement-related variables such as distance moved, percentage of time spent moving, rate of movement, movement along walls versus movement in the central region of the cage, etc. Some studies have compared the profile of activational effects of some stimulant drugs as measured by different techniques (for example using a rating scale *versus* infrared emitters and detectors) and have shown that the different approaches yield similar results (Beninger, Cooper and Mazurski, 1985).

One of the agents most commonly used to stimulate locomotor activity is amphetamine, an indirectly-acting DA agonist that stimulates release and blocks uptake (Scheel-Krüger, 1971; Westerink, 1979; Grace, 1991). Although amphetamine affects other neurotransmitters (e.g., norepinephrine) pharmacological studies have shown that its locomotor effects depend on DA (Scheel-Krüger, 1971). Amphetamine produces a dose-dependent increase in locomotion at the low end of the dose-effect curve with locomotor activity decreasing again as doses rise. This latter effect is a result of stereotypy, that is activities in a single place, such as repetitive movements of the head or licking or gnawing behaviour (Ungerstedt, 1979). Systemically injected D2-like agonists such as bromocriptine similarly stimulate locomotor activity at lower doses, and stereotypy at higher doses (Jackson, Jenkins and Ross, 1988). D1-like agonists such as SKF 38393, on the other hand, although having been found to stimulate locomotion in animals, produce less intense effects and do not produce stereotypy like that seen with D2-like agonists (Waddington and O'Boyle, 1989). D1-like agonists but not D2-like agonists have been found to lead to grooming behaviour at higher doses (Molloy and Waddington, 1987). Thus, both D1- and D2-like receptors appear to be involved in the stimulation of locomotor behaviour but stimulation of D2-like receptors leads to larger motor effects.

Since the recent identification of the several members of the D1- and D2-like receptor families, some old and new pharmacological agents have been identified that are relatively specific for one of these receptor subtypes (Sibley, Monsma and Shen, 1993). Some recent studies have evaluated the locomotor effects of agents such as 7-OH-DPAT or PD 128,907 that are somewhat selective for D3 receptors. These agents produced a decrease in locomotor activity at low doses, and an increase at high doses (Daly and Waddington, 1993; Ahlenius and Salmi, 1994). The low dose effects are thought to be mediated by D3 receptors located presynaptically, stimulation of these receptors with a D3 agonist would decrease the release of DA and therefore decrease activity. However, some evidence suggests that there also may be post-synaptic D3 receptors that normally inhibit locomotion, and that direct stimulation of these receptors may contribute to the decreases in locomotion produced by D3 agonists (Svensson, Carlsson and Waters, 1994; Svensson *et al.*, 1994). The discovery of new agents specific for individual receptors in each DA receptor family, and investigations of their behavioural effects are eagerly awaited.

DA receptor antagonists produce decreases in locomotor activity. This effect has been seen with a wide range of DA receptor blockers (Niemegeers and Janssen, 1979). Agents with relative specificity for D1-like receptors such as SCH 23390, and those such as substituted benzamides that are quite specific for D2-like receptors similarly produce decreases in locomotor activity (Beninger, Mazurski, and Hoffman, 1991). As was concluded from studies of agonists with relative selectively for D1- or D2-like receptors, the results of studies with antagonists suggest that both receptor families are involved in the control of locomotor behaviour.

Some studies have evaluated the effects of D1- or D2-like DA receptor antagonists on agonist-induced locomotor activity. In general, antagonists at either D1- or D2-like receptors reduce the stimulant effects of amphetamine or D1- or D2-like receptor-specific agonists (Clark and White, 1987; Beninger, Mazurski and Hoffman, 1991). Thus, for example, the stimulant effects of the D2-like agonist RU24213 were blocked by the D2-like antagonist Ro22-2586 or by the D1-like antagonist SCH 23390 (Pugh *et al.*, 1985). These and many related observations have led researchers to conclude that D1- and D2-like receptors act in a synergistic manner in the control of locomotor activity (Clark and White, 1987).

The contribution of DA terminal areas has also been studied. Locomotor stimulation, sometimes indicated by contralateral rotation subsequent to unilateral injections (Miller and Beninger, 1991), has been reported following injections of amphetamine into the caudate nucleus (Pycock, 1980), nucleus accumbens (Colle and Wise, 1991; Messier, Mrabet and Destrade, 1991; Messier *et al.*, 1991) or medial frontal cortex (Stewart, Morency and Beninger, 1985). Others, however, have reported no locomotor activation following bilateral intra-frontal cortical injections of amphetamine (Vezina *et al.*, 1991). D1-like agonists produce locomotor stimulation following injection into the nucleus accumbens (Dreher and Jackson, 1989) but are without effect in the caudate nucleus (Smith, Sutton and Beninger, 1997) or frontal cortex (Beninger, Musgrave and Dickson, 1990). D2-like agonists similarly produce locomotor stimulation following injection into the nucleus accumbens (Dreher and Jackson, 1989) and are without effect when injected into the caudate (Smith, Sutton and Beninger, 1997). However, we have found that D2-like agonists injected into the frontal cortex, like amphetamine, produce mild contralateral rotation, an effect consistent with locomotor stimulation (Stewart, Morency and Beninger, 1985; Beninger, Musgrave and Dickson, 1990). Taken together, these results point to both D1- and D2-like receptors in the nucleus accumbens as important for the control of locomotor activity with some contribution from striatal and frontal cortical regions.

It is interesting to note the recent finding that the relatively selective D3 receptor agonist 7-OH-DPAT, when injected into the nucleus accumbens, produced decreased locomotor activity (Gilbert and Cooper, 1995). We have replicated this finding in unpublished studies from our labs. However, there also has been one recent report of an increase in activity (movement time) following microinjections within the accumbens of 7-OH-DPAT (Meyer, 1996). Perhaps the discrepant results relate to the specific dependent measure used. Results suggest that D3 receptors in the nucleus accumbens may inhibit locomotor activity.

The apparent co-operative interaction of D1- and D2-like receptors in the control of locomotor activity that has been described above is not seen in rodents chronically depleted of DA. In rats such depletion is produced with bilateral 6-OHDA lesions of the medial forebrain bundle, after which there is a chronic loss of DA neurons in the ventral midbrain. In these animals, unlike normosensitive animals, D1-like antagonists fail to block the stimulant effects of D2-like agonists and D2-like antagonists fail to block the stimulant effects of D1-like agonists. However, both types of antagonists continue to block the effects of agonists at their own receptor subtype, as expected (Beninger, Mazurski and Hoffman, 1991). Thus, there is an uncoupling of the usual synergistic interaction of D1- and D2-like receptors following chronic DA depletion.

In summary, DA plays an important role in the control of locomotor activity. Both D1- and D2-like receptors are involved, and the nucleus accumbens appears to be a region of considerable importance with some role for the caudate-putamen and frontal cortex. There is a co-operative interaction between D1- and D2-like receptors in normosensitive animals that is lost in animals with supersensitive DA receptors following chronic DA depletion. Preliminary studies of D3 receptors suggest an inhibitory influence on locomotor activity possibly in the nucleus accumbens and possibly at pre-synaptic

DA receptors. Further studies with agents that are relatively selective for DA receptor subtypes are awaited eagerly.

## 3. DOPAMINE AND CONDITIONED BEHAVIOURS

There is general agreement that DA plays an important and perhaps critical role in reward-related learning (Beninger, 1983, 1993; Wise and Rompré, 1989; LeMoal and Simon, 1991). Behaviours are conditioned by making the presentation of reward contingent upon those behaviours. Thus, for example, the process of training a rat to press a lever involves the shaping of behaviour by presentations of reward for successive approximations of the target response (Skinner, 1938). Eventually, reward is presented only when the desired response is made and the behaviour then has been shaped. One way of describing the learning of lever pressing responses for reward is to emphasize the stimuli that come to control the behaviour. Thus, in the above example, the stimulus lever and other stimuli associated with it (e.g. its place on the wall of the chamber) come to elicit approach and other responses including the lever press.

The role of DA in conditioned behaviours has been studied using a number of paradigms. It is useful and informative to divide those paradigms into two broad categories for the purposes of understanding the effects of manipulations of dopaminergic neurotransmission on reward-related learning. These categories are determined by identifying the stimuli that come to control the behaviour being tested. Tasks like lever pressing form one category which will be termed the *specific cue category*. The defining characteristic of this category is that a specific subset of stimuli from within the experimental environment comes to control behaviour. In the example given above, it was the lever stimulus and related stimuli that came to control approach and other responses, and not other stimuli or places in the test apparatus.

The other category of paradigm will be termed the *contextual category*. The defining characteristic of this category is that the stimuli that come to control behaviour can be the environment itself, not a particular subset of stimuli from within that environment. This type of paradigm would include place conditioning and conditioned activity. In each case, the experimental environment is paired with a rewarding stimulus, usually a drug (but conditioning also occurs with food or water as reward). The test involves an assessment of the time spent in the reward-related place in place conditioning, and simply the level of activity in conditioned activity studies. Unlike lever pressing tasks, there is no need for the animal to make a specific response to a specific subset of stimuli from within the test environment.

It should be noted that the distinction between the specific cue *versus* the contextual categories is being made to provide a basis for understanding the differential behavioural effects of amphetamine *versus* apomorphine and D1- *versus* D2-like agonists in different paradigms. In tasks of the specific cue category it is possible (and likely) that contextual cues acquire some control over behaviour in addition to the control exerted by specific cues. Similarly, in tasks of the contextual category it is possible that a specific feature of the test chamber, for example floor texture, comes to control behaviour. However, the distinction being made between the two task categories has face validity, provides a basis for understanding a large number of pharmacological effects, and makes testable predictions about drug effects in additional paradigms.

Another important consideration is the point in training at which the dopaminergic manipulation takes place. In tasks of the specific cue category, DA receptor agonists or antagonists can be given during the shaping or acquisition period or during responding in trained animals. There are many examples of both approaches in the literature. In tasks of the contextual category, dopaminergic agents can be given during conditioning when the test environment is being paired with reward or during the test phase, when the effects of conditioning are being assessed. In almost all published studies in this category, dopaminergic agents are given during the conditioning phase, not the test.

Two more points of information are needed to provide the background necessary to evaluate the results of studies of dopaminergic agonists and antagonists given to animals trained in tasks of the specific cue or contextual categories. The first is that there is a dopaminergic signal associated with the presentation of a rewarding stimulus to an animal. Thus, dopaminergic neurones have a low tonic level of activity but show bursts of activity when reward occurs (Schultz, Dayan and Montague, 1997). This aspect of dopaminergic neuronal function is discussed in detail in Chapter 2 of this volume (Hyland, 1999). The second point is that dopaminergic neurotransmission can be augmented by agents like amphetamine that enhance release and block uptake (Scheel-Krüger, 1971; Westerink, 1979; Grace, 1991) or by agents like apomorphine that directly stimulate DA receptors (Colpaert, vanBevan and Leysen, 1976); apomorphine stimulates DA receptors of both the D1- and D2-like families (Neve and Neve, 1997). On the one hand, amphetamine at moderate doses would augment the reward signal by increasing the release of DA and prolonging its action in the synapse. On the other hand, apomorphine at doses that lead to a high level of occupation of post-synaptic DA receptors, would mask the DA signal associated with reward. These different mechanisms of action of DA agonists provide valuable insights into the relative contribution of D1- and D2-like receptors to reward-related learning, as discussed in the following sections.

### 3.1. Dopamine Agonists and Studies in the Specific Cue Category

When animals are trained to lever press for food or water, and are then treated with amphetamine or apomorphine, differential effects are seen. The effects of amphetamine and related drugs have been described as rate-dependent; these compounds enhance low rates of responding and suppress high rates (Harris, Snell and Loh, 1978; Lucki, 1983). In contrast, treatment with apomorphine results in a uniform reduction regardless of baseline rate (Harris, Snell and Loh, 1978). Perhaps when response rates are high, the additional locomotor activation produced by amphetamine leads to a disruption of response organization, resulting in the observed decrease in lever pressing (Robbins, 1981). With respect to the actions of these drugs on low rates of well-trained responding, these findings are consistent with the argument that amphetamine enhances the DA signal produced by reward, whereas apomorphine masks the signal.

Another task in this category is lever pressing for brain stimulation reward (BSR). In this paradigm, both amphetamine and apomorphine have been found to enhance reward (Gallistel and Karras, 1984; Fouriezos and Francis, 1992). This has been seen as leftward shifts in rate-intensity functions, demonstrating reduced thresholds for reward following either of these drugs. These results would appear to be at odds with the differential effects of amphetamine *versus* apomorphine in lever pressing tasks maintained by natural rewards.

Unlike lever pressing for food or water reward, where a natural stimulus is presented contingent upon responding and requires consumatory acts like eating or drinking, BSR reward directly stimulates brain circuits thought to be activated normally by the consumption of natural rewards. *In vivo* voltammetry and microdialysis experiments have shown increased DA release in the nucleus accumbens following either natural rewards or BSR (Phillips, Blaha and Fibiger, 1989; Young and Michael, 1993; Kiyatkin, 1995; Mas, Fumero and Gonzalez-Mora, 1995; Westerink, 1995). However, the speed of onset of the effect is dramatically different for natural rewards versus BSR. Thus, natural rewards produce a gradual increase (Kiyatkin and Gratton, 1994) whereas BSR produces a large and immediate increase in accumbens DA (Phillips, Blaha and Fibiger, 1989). Perhaps the intensity of the DA signal produced by BSR is so great that it cannot be masked by doses of apomorphine that are argued to be effective in masking the weaker signal produced by natural rewards. This provides a basis for understanding the apparently contradictory effects of apomorphine on responding for food or water versus BSR.

Another paradigm in the specific cue category is lever pressing to self-administer drugs. However, almost all of the work with agonists in this paradigm involves using discrete injections of small self-administered volumes as rewarding stimuli, rather than evaluating the effects of experimenter-administered systemic injections of larger volumes and doses. Thus, with either amphetamine or apomorphine the stimulation of DA receptors would be timed to correspond to the period immediately following the response. Usually, there is a stimulus light or other cue that remains on for a period following the response. Presumably, the onset of action of amphetamine or apomorphine would correspond with the presence of this stimulus which would acquire conditioned rewarding properties that then would serve to maintain the response-controlling properties of the lever and related stimuli. It has been found that both amphetamine and apomorphine are self-administered (Pickens and Harris, 1968; Baxter *et al.*, 1974).

A number of studies have evaluated the effects of amphetamine and apomorphine on the acquisition of responding for conditioned reward. Results have shown that amphetamine produces a dose-dependent and selective enhancement of acquisition of pressing a lever that produces a stimulus previously associated with food or water, in a test situation with two levers available (Robbins *et al.*, 1983; Mazurski and Beninger, 1986; Beninger and Ranaldi, 1992). In the same test situation, apomorphine led to a non-specific enhancement of responding on both levers, the control of responding by the conditioned reward being lost (Robbins *et al.*, 1983; Beninger and Ranaldi, 1992). These findings agree with the differential effects of amphetamine and apomorphine on low rates of well-trained lever pressing for food or water that were discussed above. Results support the idea that amphetamine enhances the reward signal while apomorphine masks it.

In recent years, studies similar to those reviewed above have been done to evaluate the effects of D1- versus D2-like agonists. Although agonists at either receptor would be direct-acting like apomorphine, results have shown that D1-like agonists produce apomorphine-like effects whereas D2-like agonists produce amphetamine-like effects! Thus, high rates of lever pressing for food or shock avoidance were decreased by either D1- or D2-like agonists whereas low rates increased following D2-like agonists but decreased following D1-like agonists (see Beninger and Miller, 1998). The reward value of BSR was increased by both D1- and D2-like agonists, as it was with both amphetamine and apomorphine (Ranaldi and Beninger, 1994). Both D1- and D2-like agonists were self-administered (Woolverton, Goldberg and Ginos, 1984; Woolverton, 1986; Wise, Murray and Bozarth, 1990; Self and Stein, 1992; Weed, Vanover and Woolverton, 1993; Chaperon and Thiébot, 1996). There is one study evaluating the effects of the D1-like agonist SKF 38393 on responding for cocaine self-administration; responding was impaired (Katz and Witkin, 1992). Finally, in conditioned reward studies, D1-like agonists impaired, whereas D2-like agonists enhanced, the acquisition of responding (Beninger and Ranaldi, 1992; Beninger and Rolfe, 1995; Ranaldi, Pantalony and Beninger, 1995).

Taken together, these results show that the response-impairing effects of apomorphine in a number of tasks in the specific cue category are related to its action at D1-like receptors. This suggests a critical role for D1-like receptors in the control of behaviour by rewarding stimuli.

### 3.2. Dopamine Antagonists and Studies in the Specific Cue Category

Many studies have evaluated the effects of antagonists relatively specific for D1- or D2-like receptors on the tasks discussed in the previous section. In general, effects consistent with a block of reward are seen with agents affecting either DA receptor subtype family (Beninger, Hoffman and Mazurski, 1989; Miller, Wickens and Beninger, 1990; Beninger, 1991, 1993). However, in recent years a number of studies provide results consistent with a greater effect of D1-like antagonists on reward and a greater effect of D2-like antagonists on motor performance. These differences are discussed in detail elsewhere (Beninger and Nakonechny, 1996; Beninger and Miller, 1998) and only an overview will be given here.

In one study, rats reached through an aperture in the cage wall and pressed a pressure transducer for food reward. Treatment with either a D1- or D2-like antagonist led to a decrease in responding, suggestive of a block of the usual effects of reward. However, detailed analyses of the strength of the operant response showed that the decline in response strength was related closely to the decrease in responding for food following the D2-like antagonist but not the D1-like antagonist (Fowler and Liou, 1994). Results suggest that the decrease in responding produced by the D2-like antagonist may have been related to the effects of the drug on motor performance, whereas the effects of the D1-like antagonist were more clearly on reward.

As discussed in the previous section, the acquisition of responding for conditioned reward is enhanced by amphetamine. Treatment with a D1-like antagonist shifted the amphetamine dose-response curve to the right, indicative of a decrease in reward. Of two D2-like antagonists tested, one also shifted the curve to the right but the maximum level of responding seen following treatment with the D1-like antagonist was never seen; the other decreased the amphetamine enhancement of responding in a dose-dependent manner but failed to shift the amphetamine dose-response curve to the right (Ranaldi and Beninger, 1993). These results implicate both D1- and D2-like receptors in the control of responding by conditioned rewards. The results also suggest that D1-like antagonists may produce effects somewhat specific to reward whereas D2-like antagonists affect reward and motor responding, as also suggested by data reviewed above from studies of operant responding for food.

In summary, D1- and D2-like antagonists appear to reduce the effects of reward in a number of tasks in the specific cue category. However, finer analyses suggest that the actions of D2-like antagonists may be related more closely to their motoric effects, whereas D1-like DA receptor antagonists may reduce reward. (For a further discussion of this dissociation and its possible underlying mechanisms see: Miller, Wickens and Beninger, 1990).

### 3.3. Dopamine Agonists and Studies in the Contextual Cue Category

In place conditioning experiments, one compartment of a test chamber with two or more compartments is paired with a putative reward over several sessions. In the test, animals are left with access to the entire apparatus and if they spend more time in the compartment paired with a drug or other stimulus, that stimulus is said to be rewarding (Carr, Fibiger and Phillips, 1989; Hoffman, 1989). In this paradigm, both amphetamine and apomorphine have been found to be rewarding (see Hoffman, 1989). Differential effects of these two drugs would not be expected because there are not specific stimuli in the drug-paired environment that must come to control responding. Apomorphine, but not amphetamine, might be expected to impair expression of the conditioned place preference if it was administered on the test day (perhaps after pairing one compartment with food during conditioning) because it might disrupt the selective control of responding by the stimuli from the reward-paired chamber. To our knowledge, this experiment has not been done.

In conditioned activity experiments, a single chamber is paired with a drug or other stimulus (e.g. food) over several days and then a test is given without the drug or other stimulus. The dependent variable is activity level, usually in comparison to a control group that has not received the drug in the test environment. Some have argued that this is an example of classical conditioning, the unconditioned stimulus being the drug, the conditioned stimulus being the drug-paired environment, and activity being the unconditioned response that becomes the conditioned response (Stewart and Eikelboom, 1987). However, Martin-Iverson and Fawcett (1996) have carried out detailed studies of the elements of activity that are produced by the drug *versus* those seen on the drug-free test day and shown that they differ. These results argue against a classical conditioning interpretation of conditioned activity. We favour an interpretation in terms of reward-related learning. Environmental stimuli associated with the drug acquire the ability to elicit approach and other responses that are manifested as increased activity on the test day.

Both amphetamine (Pickens and Crowder, 1967) and apomorphine (Möller, Nowak and Kuschinsky, 1987) have been reported to produce conditioned activity. If the drugs were given on the test day, activity should be seen because of the unconditioned effects of the drug, and since there are no other cues to compete with those paired with the drug, they too may influence activity. The additive effects of the influence of the drug itself and the cues associated with it may augment the drug response, a phenomenon termed sensitization (Stewart, 1992). The possible role of conditioning in sensitization will not be discussed further here.

The effects of D1- and D2-like agonists in these paradigms have been assessed. Like amphetamine and apomorphine, they produce similar effects, in contrast to their differential action in tasks in the specific cue category. Thus, place preference learning has been reported following pairing of one compartment with the D1-like agonist SKF 82958 (Abrahams *et al.*, 1998) or with a number of D2-like agonists (Morency and Beninger,

1986; Hoffman, Dickson and Beninger, 1988; Hoffman and Beninger, 1988, 1989; White, Packard and Hiroi, 1991). Place preference conditioning also was found following the weakly D3-selective agonist 7-OH-DPAT (Mallet and Beninger, 1994; Kling-Petersen *et al.*, 1995b; Chaperon and Thiébot, 1996). However, some researchers failed to find this effect (DeFonseca *et al.*, 1995; Khroyan, Baker and Neisewander, 1995). Moreover, D3 receptor *antagonists* also have been reported to produce a place preference (Richardson *et al.*, 1993; Kling-Petersen *et al.*, 1995a,b). Clearly further studies are needed to resolve the role of D3 receptors in this paradigm. Conditioned activity has been reported following pairing of a test environment with injections of D1- or D2-like agonists (Mazurski and Beninger, 1991).

In summary, D1- and D2-like agonists have similar actions in tasks in the contextual cue category. Thus, agonists acting at either receptor subtype family produce place conditioning and conditioned activity.

### 3.4. Dopamine Antagonists and Studies in the Contextual Cue Category

Place preference conditioning has been reported for places associated with food, water, psychostimulants or opiates (Carr, Fibiger and Phillips, 1989; Hoffman, 1989). Place preferences based on water or amphetamine were blocked by D1- or D2-like antagonists (Leone and Di Chiara, 1987; Hoffman and Beninger, 1989; Hiroi and White, 1991; Agmo *et al.*, 1993) and place conditioning with morphine was blocked by acute treatment with D1-like antagonists (Bechara and van der Kooy, 1985; Leone and Di Chiara, 1987). These results implicate both DA receptor families in place learning. However, some studies show differential effects. Thus, morphine place conditioning was blocked by chronic systemic or intra-accumbens injections of a D1-like antagonist but not a D2-like antagonist (Shippenberg and Hertz, 1987, 1988; Shippenberg, Bals-Kubik and Herz, 1993) and place conditioning based on cocaine was blocked by a D1- but not by a D2-like antagonist (Cervo and Samanin, 1995). These latter findings suggest that, at least in the case of place conditioning with morphine or cocaine, D1-like receptors may play a more critical role than D2-like receptors.

It should be noted here that the role of DA in opiate reward measured in place conditioning tasks is not straightforward. Van der Kooy and his co-workers have shown that DA antagonists acting at either both receptor families or at the D2-like receptor block morphine or heroin place preferences in opiate-dependent but not in non-dependent animals (Mackey and van der Kooy, 1985; Bechara *et al.*, 1992; Nader *et al.*, 1994). These important observations that there are at least two independent motivational systems and that DA is involved in one but not the other eventually will have to be integrated into our understanding of the role of DA in reward-related learning.

In the above studies reporting that a D1-like antagonist blocked place conditioning, control experiments showed that the same doses of SCH 23390 given alone did not produce a place aversion. However, a number of studies have found that D1-like antagonists, at some doses, produce a place aversion when given systemically (Shippenberg and Herz, 1988; Shippenberg *et al.*, 1991; Acquas and DiChiara, 1994) or into the nucleus accumbens (Shippenberg *et al.*, 1991; Shippenberg, Bals-Kubik and Herz, 1993). In contrast, D2-like antagonists, given alone, failed to produce a place aversion (Shippenberg and Herz, 1988; Shippenberg *et al.*, 1991). Perhaps these results also indicate a more important role for D1- than D2-like receptors in reward.

## 4. INCENTIVE LEARNING

Stimuli that immediately precede the presentation of a rewarding stimulus acquire the ability to elicit approach and other responses. This form of learning is termed incentive motivational learning, or, more simply, incentive learning, and the stimuli that have acquired response-eliciting properties are termed conditioned incentive motivational stimuli or simply conditioned incentive stimuli (Bolles, 1972; Bindra, 1974; Beninger, 1983). Conditioned incentive stimuli also acquire the ability themselves to act as rewarding stimuli, changing the ability of stimuli that signal them to elicit approach and other responses. For this reason, conditioned incentive stimuli sometimes are referred to as conditioned rewards.

Incentive learning is reward-related learning. Thus, DA plays an important and perhaps critical role in incentive learning. In a number of previous papers, and in the arguments presented in the previous section, we and others have suggested that it is DA acting at D1-like receptors that is critical for incentive learning (Miller, Wickens and Beninger, 1990; Wickens, 1990, 1993; Beninger, 1991, 1992, 1993; Beninger and Ranaldi, 1994; Kötter, 1994; Wickens and Kötter, 1995; Beninger and Nakonechny, 1996; Beninger and Miller, 1998; Josselyn, Miller and Beninger, 1997). Details of the mechanism can be found in these references. For the purposes of the present discussions, we will focus on some of the elements of learning, but not the possible intracellular biochemical mechanisms that may mediate the plastic changes in synaptic connections underlying incentive learning. This latter material is covered in the above references.

In this section we will begin by reviewing the possible interaction of dopaminergic and glutamatergic afferents to the striatum (including the nucleus accumbens) that underlies incentive learning. We then will review some recent studies from animal learning laboratories that provide further details about the role of motivation in incentive learning. Finally, we will assess the implications of these findings for our neuronal model of incentive learning.

### 4.1. Dopaminergic-Glutamatergic Interactions in Incentive Learning

The heaviest projection area of the cortex is to the striatum, and the corticostriatal neurones are glutamatergic (Graybiel *et al.*, 1994). They co-terminate on the dendrites of medium spiny striatal neurons along with dopaminergic afferents (Smith and Bolam, 1990). It is at these co-terminal regions that the synaptic interactions underlying incentive learning are thought to occur (Beninger, 1993; Kötter, 1994; Wickens and Kötter, 1995).

The striatum can be viewed as an interface between sensory/perceptual regions of the brain and motor output regions. From this point of view, various stimuli in the environment of an animal are thought to be represented by the activation of a specific set of neurones in the cortex and therefore a specific set of corticostriatal afferents. In the striatum, these afferents synapse on output neurones which influence motor behaviour via projections to the thalamus (projecting in turn to the cortex) and via projections to brainstem areas (that project in turn to motor regions) (Graybiel *et al.*, 1994).

Dopaminergic neurones branch extensively in the striatum (see Beninger, 1993) and contact medium spiny neurones on the same dendrites that have glutamatergic terminals (Smith and Bolam, 1990). With this arrangement of connections, the dopaminergic system is situated perfectly to modulate the strength of glutamatergic corticostriatal projections. Since striatal output influences motor behaviour, DA may modulate the ability of various stimuli to elicit responses. It now is well established that biologically important stimuli, including reward, activate dopaminergic neurones (Schultz, Dayan and Montague, 1997). When reward occurs, the release of DA could lead to a change in the strength of corticostriatal synapses that were activated recently by the stimuli present in the environment associated with reward.

This model of synaptic events associated with incentive learning is presented in Figure 2.1. A single line is used to represent the corticostriatal afferents activated by a particular stimulus although in reality many axons would be involved. Similarly, only a single dopaminergic axon is shown. In panel A, the synapse between afferents activated by the stimulus aspects of food and striatal medium spiny neurones is blackened to indicate that it has been strengthened by previous conditioning. In panel B, a lever is being pressed, activating striatal projections of cortical neurones that were stimulated by presentation of that lever stimulus. In panel C, food is presented immediately following the lever press; striatal projections of cortical neurones activated by the stimulus aspects of food are active as are dopaminergic neurones activated by reward. We suggest that DA acting at D1-like receptors alters the strength of the most recently activated corticostriatal glutamatergic synapses. The mechanism for storing information about recent activity and for altering synaptic effectiveness has been discussed in detail elsewhere (Beninger, 1993; Kötter, 1994; Wickens and Kötter, 1995; Wickens, chapter 4 of this volume). In panel D, the lever now is a conditioned incentive stimulus with an enhanced ability to influence striatal output.

This model forms the basis of a putative neuronal mechanism that underlies the ability of rewarding stimuli to modify behaviour, i.e. to produce learning. In recent years, researchers who study animal learning have identified a number of motivational influences on incentive learning. These studies have focused on the incentive value of the primary reward itself rather than the incentive value of stimuli associated with reward. In the next section, this work will be reviewed briefly. Following that, these new findings will be considered from the point of view of the above model in an attempt to begin to bring together two current lines of study of incentive learning that at present have not been integrated.

### 4.2. Motivational Influences on Incentive Learning

Dickinson and Balleine (1994) have proposed that motivational states control instrumental responding indirectly through their effects on the primary incentive value of rewarding stimuli. They refer to this process as incentive learning and relate it to Tolman's (1949) formulation of cathexis. Both accounts propose that the primary incentive value of a reward itself in a given motivational state must be learned through experience or contact with the reward in the relevant motivational state.

One study presents experimental evidence that alterations in motivational state, which decrease the value of food reward, change the incentive value of stimuli controlling responding (Balleine and Dickinson, 1991). In this experiment, animals were trained to lever press for a sucrose solution, which then was devalued by a lithium chloride injection that induced nausea. Subsequent performance of the instrumental action in an extinction test was reduced; however, *this effect occurred only if animals were re-exposed to the*

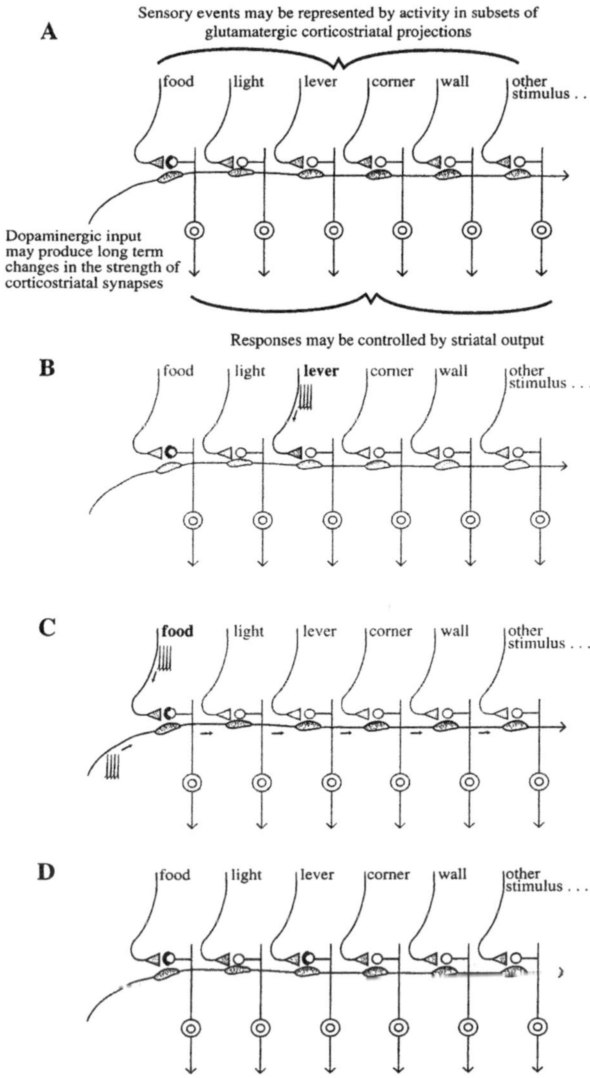


**Figure 2.1.** Possible synaptic events associated with incentive learning. **A**: Descending projections represent corticostriatal glutamatergic neurones that are activated by particular environmental stimuli; although each stimulus is represented by one neurone, presumably many neurones are activated by a particular stimulus. Medium spiny output neurones of the striatum are thought to control responses. Dopaminergic projections to the input-output interface are represented by a single neurone. It is proposed that dopaminergic inputs activated by rewarding stimuli produce long term changes in the strength of corticostriatal synapses. The synaptic darkening on the terminal from the cortical efferent activated by the stimulus properties of food represents synaptic strengthening previously produced by incentive learning. **B**: When the lever is pressed, the stimulus aspects of the lever activate cortical neurones that project to the striatum. **C**: If food follows the lever press response, the stimulus aspects of food activate the associated cortical efferents, and the reward properties of food activate the dopaminergic striatal afferents. Dopamine is thought to produce a long term change in the strength of the most recently active corticostriatal synapses (Beninger, 1993). **D**: Incentive learning, involving an increase in the ability of the stimulus aspects of the lever to control responding, is represented by the synaptic darkening in the corticostriatal synapse activated by the lever.

*sucrose solution prior to testing*. Furthermore, the effect was specific to the outcome associated with the instrumental action. Thus, re-exposure to the devalued outcome selectively attenuated performance of the action trained with that outcome, relative to other actions trained with different outcomes.

Using similar experimental procedures, Dickinson and colleagues have shown that alterations of other motivational variables similarly change the incentive value of stimuli controlling responding. This effect has been seen following changes in food deprivation (Balleine, 1992; Balleine, Ball, and Dickinson, 1994), a decrement in water deprivation (Lopez, Balleine, and Dickinson, 1992), a shift between rather than within motivational states (Dickinson and Dawson, 1988, 1989), and the induction of drug states (Balleine, Ball, and Dickinson, 1994). It is important to note that *pre-exposure* to a primary rewarding stimulus while in a non-deprived state, for example, followed by training while deprived, and then testing in extinction while again non-deprived, similarly led to reduced responding compared to non-pre-exposed animals (Balleine, 1992). Thus, experience with the primary reward while at one level of deprivation can influence responding in extinction following training at another level of deprivation whether that experience preceded or followed training.

The finding that shifts in food deprivation lead to changes in responding only if animals are pre- or re-exposed to the primary incentive stimulus while in the changed motivational state is of particular interest. This is so because these findings indicate that motivational states do not determine directly the primary incentive value of stimuli that fulfil basic biological needs. Thus, the incentive value of primary rewards such as food, as they relate to a particular motivational state, must be learned through direct contact with the stimulus in that state.

Further studies examined the physiological mechanisms by which hunger controls the assignment of incentive value to a food reward. Consistent with their previous findings, Balleine and Dickinson (1994) found that animals that were trained concurrently with two actions and two outcomes in a deprived state and then re-exposed to one of the outcomes in a non-deprived state, exhibited reduced levels of the action associated with the re-exposed outcome during an extinction test while in a non-deprived state. They then evaluated the effects of blockade of endogenous cholecystokinin (CCK), a peptide thought to modulate the effects of food deprivation on motivation to eat (Beinfield, 1995). Administration of devazepide (a CCK receptor antagonist specific for a subtype known as $CCK_A$ receptors), during re-exposure reduced the usual effects of re-exposure on responding. Conversely, the increased incentive value of a food reward produced by a shift from a non-deprived to a deprived state was blocked by injections of exogenous CCK during the re-exposure phase in a deprived condition (Balleine, Davies and Dickinson, 1995). Thus, the ability of a change in food deprivation to alter the primary incentive value of food is mediated by changes in CCK levels.

In related experiments, devaluation, produced by re-exposure to a stimulus previously paired with a lithium chloride injection, was blocked by administration of the anti-emetic ondansetron (Balleine, Garner, and Dickinson, 1995). This finding suggests that feedback consequences of the re-exposed stimulus are critical for incentive learning.

Balleine, Ball and Dickinson (1994) argue that the feedback signals about the animal's motivational state determine the assignment of primary incentive value to rewarding stimuli by modulating affective or hedonic reactions to the stimulus. This hypothesis is based on their finding that the incentive value of a food reward is increased during incentive learning when animals experience the food under the influence of the benzodiazepine agonist midazolam (Balleine, Ball, and Dickinson, 1994). Benzodiazepines enhance positive ingestive reactions to food (Berridge and Treit, 1986), and midazolam also increases the incentive value of a non-nutritive saccharin solution. The authors of the Balleine *et al.* study argue that, since the animals in that study were not food deprived, the incentive learning effect was probably not mediated by feedback from post-ingestional properties of the food. Rather, they propose that incentive learning involves the formation of an association between the animal's current motivational state and the affective properties of the goal object, which themselves are determined by that motivational state.

In summary, according to the work of Dickinson and colleagues, when motivational states control instrumental responding indirectly through their modulation of the incentive value of primary rewarding stimuli, incentive learning is defined to occur. This effect is seen when animals are given exposure to a rewarding stimulus while in one motivational state, trained to respond for the same reward when in another state, and then tested in extinction in the original motivational state. Some of the physiological mechanisms subserving these effects have been identified.

### 4.3. Understanding Motivational Influences on Incentive Learning from the Point of View of the Neuronal Model of Incentive Learning

The recent studies of Dickinson and his colleagues (reviewed above) provide valuable insights into the relationships that control incentive learning. At present it is not possible to integrate all of these findings into the neuronal model for incentive learning that has been proposed (see Figure 2.1). However, some links can be found, and speculation about others is possible. These links will be discussed in this section.

We begin by considering the pre- and re-exposure experiments, involving a change in the level of food deprivation. If an animal is trained to lever press for food while food deprived, and is then tested in extinction while no longer food deprived, it will press at the same level as a food-deprived animal unless it has been re-exposed to the food while in the non-deprived state. On the one hand, if the animal has not experienced the food in the non-deprived state and is tested in extinction while non-deprived, previous incentive learning should be intact and the lever and related stimuli will control responding. Hence, no difference is seen between deprived and non-deprived animals. On the other hand, if food is experienced while in the non-deprived state, less responding is seen subsequently in extinction.

To understand the possible neuronal connections underlying this incentive learning effect, the role of DA in incentive learning *versus* stimulus-stimulus associative learning is important to note. Thus, there is plenty of evidence that animals treated with DA receptor blocking drugs, or depleted of DA, can learn relationships between or among stimuli (Beninger, 1983), although in these animals, stimuli associated with reward do not come to control responding. The results of the motivational shift experiments, involving training while food-deprived and testing in extinction while no longer food deprived, might be understood as follows. A number of assumptions are necessary. First of all, the cells activated by the stimulus properties of food when food is experienced while food-deprived and those activated by food when it is experienced in a non-deprived state must be different. Those cells activated by food, experienced while food-deprived, must acquire stronger response-eliciting (incentive) properties. The two different sets of cells activated by food experienced in different states of deprivation are assumed to be connected by stimulus-stimulus associative learning.

When training occurs while food-deprived, the stimulus properties of food are strong incentive stimuli and the lever stimulus acquires strong incentive properties (see Figure 2.1). By stimulus-stimulus learning the lever stimulus and stimulus properties of food (while in a food-deprived state) would become associated. The responding in extinction of non-food-deprived animals, without previous experience of food while not food-deprived, would be controlled by the incentive properties of the lever. Responding may also be influenced by the incentive properties of the cells responding to the stimulus properties of food when food-deprived, these being associatively activated. For the animals with previous experience of food while non-deprived, the cells associatively activated most strongly in response to stimulus aspects of the lever might be those normally responding to the stimulus aspects of food while in a non-deprived state. These cells would have weaker incentive properties and as a result less responding may occur.

We acknowledge the speculative nature of the above explanation. However, the studies of Dickinson and Balleine and others make it clear that motivational states do not influence directly the level of responding, but rather do so indirectly by modulating the incentive value of rewarding and reward-related stimuli. It is challenging to explain the finding that *pre-exposure* to food while non-deprived influences responding in extinction while not deprived, after training while deprived. This is because the lever-related stimuli have no association with a non-food-deprived state prior to extinction testing. Only the primary rewarding stimulus–food–was associated previously with non-deprivation. Thus, there must be a means for prior information about the incentive quality of food while non-deprived to influence the ability of the lever to control responding. The proposed relationship provides one possibility. This explanation would apply similarly to studies in which other motivational systems were altered or where animals went from being less to more food-deprived.

The finding that CCK may play a role in the effects of food deprivation on incentive learning provides valuable details of the possible physiological and neuronal mechanisms that may underlie these effects. Based on their finding that midazolam, a benzodiazepine that would augment GABAergic neurotransmission, increased the incentive value of food reward, Balleine *et al.* (1994) suggested that incentive learning involves the formation of an association between the animal's current motivational state and the affective properties of the goal object, which themselves are determined by that motivational state. From the present point of view, we would speculate that food in the presence of midazolam leads to a larger DA signal, and therefore to stronger incentive learning represented by the strength of glutamatergic cortico-striatal synapses. To our knowledge, *in vivo* studies have not assessed the possible interactions of benzodiazepines with DA release produced by food.

## 5. SUMMARY AND CONCLUSIONS

In this chapter we have provided an overview of some of the recent studies of the role of DA in the control of locomotor activity and reward-related incentive learning. In locomotor control, both D1- and D2-like receptors are involved, and the nucleus accumbens appears to be a region of considerable importance; the caudate-putamen and frontal cortex also may play a role. There is a co-operative interaction between D1- and D2-like receptors in normosensitive animals that is lost in animals with supersensitive DA receptors following chronic DA depletion. Preliminary studies of D3 receptors suggest an inhibitory influence on locomotor activity possibly in the nucleus accumbens, and possibly at pre-synaptic DA receptors. Further studies with agents that are relatively selective for DA receptor subtypes are awaited eagerly.

For reward-related learning, tasks were divided into the "specific cue" and "contextual" categories. Lever pressing tasks were included in the former category. In tasks of this category, amphetamine generally increased responding whereas apomorphine impaired responding. The response-impairing effects of apomorphine appeared to be related to its action at D1-like receptors. D1- and D2-like antagonists appeared to reduce the effects of reward in a number of these tasks. However, finer analyses suggest that the actions of D2-like antagonists may be related more closely to their motoric effects whereas D1-like DA receptor antagonists may reduce reward. These results suggested a critical role for D1-like receptors in the control of behaviour by rewarding stimuli.

Tasks in the contextual cue category include place conditioning and conditioned activity. Like amphetamine and apomorphine, D1- and D2-like agonists have similar actions in these tasks, producing rewarding effects in place conditioning, and producing conditioned activity. Antagonists at either receptor family block the effects of reward in place conditioning tasks and block conditioned activity. Some findings suggest that, at least in the case of place conditioning with morphine or cocaine, D1-like receptors may play a more critical role than D2-like receptors. These results implicate both DA receptor families in place learning and conditioned activity, and some results suggest a more important role for D1-like receptors.

The second major section of this chapter reviewed theories of incentive learning. A neuronal model involving dopaminergic-glutamatergic interactions in the striatum was reviewed as a basis for understanding the role of DA in reward-related learning and the contribution of D1-like receptors.

The studies of Dickinson and his colleagues were reviewed, focusing on their idea that motivational states control instrumental responding indirectly through their effects on the primary incentive value of rewarding stimuli. For example, devaluation studies revealed that responding in extinction was reduced following a reduction in food deprivation only if food was experienced in the deprived state prior to the extinction test. Further studies identified some of the physiological variables mediating these effects.

We attempted to integrate the animal learning studies of incentive learning with the psychopharmacological ones that suggested the dopamine-glutamate interaction model. A role for both incentive and stimulus-stimulus associative learning was identified. At present, there continues to be a need to seek a common language and mechanism for understanding the variety of results that these lines of study have produced.

## ACKNOWLEDGMENTS

This chapter is dedicated to Claire. Funding was provided by grants from the Natural Sciences and Engineering Research Council of Canada to RJB and MCO.

## REFERENCES


Abrahams, B.S., Rutherford, J.D., Mallet, P.E. and Beninger, R.J. (1998) Place conditioning with the dopamine D1-like agonist SKF 82958 but not SKF 81297 or SKF 77434. *European Journal of Pharmacology*, **343**, 111–118.

Acquas, E. and DiChiara, G. (1994) D1 receptor blockade stereospecifically impairs the acquisition of drug-conditioned place preference and place aversion. *Behavioural Pharmacology*, **5**, 555–569.

Agmo, A., Federman, I., Navarro, V., Padua, M. and Velazquez, G. (1993) Reward and reinforcement produced by drinking water: Role of opioids and dopamine receptor subtypes. *Pharmacology Biochemistry and Behavior*, **46**, 183–194.

Ahlenius, S. and Salmi, P. (1994) Behavioral and biochemical effects of the dopamine D–3 receptor-selective ligand, 7-OH-DPAT, in the normal and the reserpine-treated rat. *European Journal of Pharmacology*, **260**, 177–182.

Balleine, E. (1992) Instrumental performance following a shift in primary motivation depends on incentive learning. *Journal of Experimental Psychology: Animal Behavior Processes*, **18**, 236–250.

Balleine, B., Ball, J., and Dickinson, A. (1994). Benzodiazepine-induced outcome revaluation and the motivational control of instrumental action in rats. *Behavioral Neuroscience*, **108**, 573–589.

Balleine, B., Davies, A. and Dickinson, A. (1995) Cholecystokinin attenuates incentive learning in rats. *Behavioral Neuroscience*, **109**, 312–319.

Balleine, B. and Dickinson, A. (1991). Instrumental performance following reinforcer devaluation depends upon incentive learning. *The Quarterly Journal of Experimental Psychology*, **43**, 279–296.

Balleine, B. and Dickinson, A. (1994). Role of cholecystokinin in the motivational control of instrumental action in rats. *Behavioral Neuroscience*, **108**, 590–605.

Balleine, B., Garner, C. and Dickinson, A. (1995). Instrumental outcome devaluation is attenuated by the anti-emetic ondansetron. *The Quarterly Journal of Experimental Psychology*, **48B**, 235–251.

Baxter, B.L., Gluckman, M.I., Srein, L. and Scerni, R. (1974) Self-injection of apomorphine in the rat: Positive reinforcement by a dopamine receptor stimulant. *Pharmacology Biochemistry and Behavior*, **2**, 387–391.

Bechara, A., Harrington, F., Nader, K. and van der Kooy, D. (1992) Neurobiology of motivation: Double dissociation of two motivational mechanisms mediating opiate reward in drug-naive versus drug-dependent animals. *Behavioral Neuroscience*, **106**, 798–807.

Bechara, A. and van der Kooy, D. (1985) Opposite motivational effects of endogenous opioids in brain and periphery. *Nature*, London, **314**, 533–534.

Beinfield, M.C. (1995). Cholecystokinin/Gastrin. In *Psychopharmacology: The Fourth Generation of Progress*, F.E. Bloom and D.J. Kupfer (eds), New York: Raven Press, pp. 585–594.

Beninger, R.J. (1983) The role of dopamine in locomotor activity and learning. *Brain Research Reviews*, **6**, 173–196.

Beninger, R.J. (1991) Receptor subtype-specific dopamine angonists and antagonists and conditioned behaviour. In *The mesolimbic dopamine system: From motivation to action*, edited by P. Willner and J. Scheel-Krüger. Chichester: John Wiley and Sons, pp. 273–299.

Beninger, R.J. (1992) D–1 receptor involvement in reward-related learning. *Journal of Psychopharmacology*, **6**, 34–42.

Beninger, R.J. (1993) Role of D1 and D2 receptors in learning. In *D1:D2 Dopamine Receptor Interactions: Neuroscience and Pharmacology*, edited by J. Waddington, London: Academic Press, pp. 115–157.

Beninger, R.J., Cooper, T.A. and Mazurski, E.J. (1985). Automating the measurement of locomotor activity. *Neurobehavioral Toxicology and Terratology*, **7**, 79–85.

Beninger, R.J., Hoffman, D.C. and Mazurski, E.J. (1989) Receptor\ subtype-specific dopaminergic agents and conditioned behavior. *Neuroscience and Biobehavioral Reviews*, **13**, 113–122.

Beninger, R.J., Mazurski, E.J. and Hoffman, D.C. (1991) Receptor subtype-specific dopaminergic agents and unconditioned behavior. *Polish Journal of Pharmacology and Pharmacy*, **43**, 507–528.

Beninger, R.J. and Miller, R. (1998) Dopamine D1-like receptors and reward-related incentive learning. *Neuroscience and Biobehavioral Reviews*, **22**,335–345.

Beninger, R.J., Musgrave, M.A. and Dickson, P.R. (1990) Unilateral injections of a D2 but not D1 agonist into the frontal cortex of rats produce a contralateral directional bias. *Pharmacology Biochemistry and Behavior*, **37**, 387–392.

Beninger, R.J. and Nakonechny, P.L. (1996) Dopamine $D_1$-like receptors and molecular mechanisms of incentive learning. In *Dopamine Disease States*, R.J. Beninger, T. Palomo and T. Archer (eds), Madrid: CYM Press, pp. 407–431.

Beninger, R.J. and Ranaldi, R. (1992) The effects of amphetamine, apomorphine, SKF 38393, quinpirole and bromocriptine on responding for conditioned reward in rats. *Behavioural Pharmacology*, **3**, 155–163.

Beninger, R.J. and Ranaldi, R. (1994) Dopaminergic agents with different mechanisms of action differentially affect responding for conditioned reward. In *Strategies for studying brain disorders: Vol 1. Depressive, anxiety and drug abuse disorders*, T. Palomo and T. Archer (eds), London: Farrand Press, pp. 411–428.

Beninger, R.J. and Rolfe, N.G. (1995) Dopamine D1-like receptor agonists impair responding for conditioned reward in rats. *Behavioural Pharmacology*, **6**, 785–793.

Berridge, K.C., and Treit, D. (1986). Chlordiazepoxide directly enhances positive ingestive reactions. *Pharmacology, Biochemistry and Behavior*, **24**, 217–221.

Bindra, D. (1974). A motovational view of learning, performance and behavior modification. *Psychological Review*, **81**, 199–213.

Bolles, R.C. (1972) Reinforcement, expectancy, and learning. *Psychological Review*, **79**, 394–407.

Carr, G.D., Fibiger, H.C. and Phillips, A.G. (1989). Conditioned place preference as a measure of drug reward. In *The Neuropharmacological Basis of Reward*, J.M. Liebman and S.J. Cooper (eds), Oxford: Clarendon Press, pp. 264–319.

Cervo, L. and Samanin, R. (1995) Effects of dopaminergic and glutamatergic receptor antagonists on the acquisition of cocaine conditioned place preference. *Brain Research*, **673**, 242–250.

Chaperon, F. and Thiébot, M.-H. (1996) Effects of dopaminergic $D_3$-receptor-preferring ligands on the acquisition of place conditioning in rats. *Behavioural Pharmacology*, **7**, 105–109.

Civelli, O., Bunzow, J.R. and Grandy, D.K. (1993) Molecular diversity of the dopamine receptors. *Annual Reviews of Pharmacology and Toxicology*, **32**, 281–307.

Clark, D. and White, F.J. (1987) Review: D1 dopamine receptor–the search for a function: A critical evaluation of the D1/D2 dopamine receptor classification and its functional implications. *Synapse*, **1**, 347–388.

Colle, L.M. and Wise, R.A. (1991) Circling induced by intra-accumbens amphetamine injections. *Psychopharmacology*, **105**, 157–161.

Colpaert, F.C., vanBevan, W.F.M. and Leysen, J.E.M.P. (1976) Apomorphine: Chemistry, pharmacology, biochemistry. *International Review of Neurobiology*, **19**, 225–268.

Daly, S.A. and Waddington, J.L. (1993) Behavioural effects of the putative D–3 dopamine receptor agonist 7-OH-DPAT in relation to other 'D–2-like' agonists. *Neuropharmacology*, **32**, 509–510.

DeFonseca, F.R., Rubio, P., MartinCalderon, J.L., Caine, S.B., Koob, G.F. and Navarro, M. (1995) The dopamine receptor agonist 7-OH-DPAT modulates the acquisition and expression of morphine-induced place preference. *European Journal of Pharmacology*, **274**, 47–56.

Dickinson, A. and Balleine, B. (1994) Motivational control of goal-directed action. *Animal Learning and Behavior*, **22**, 1–18.

Dickinson, A., and Dawson, G.R. (1988). Motivational control of instrumental performance: The role of prior experience of the reinforcer. *The Quarterly Journal of Experimental Psychology*, **40B**, 113–134.

Dickenson, A. and Dawson, G.R. (1989) Incentive learning and the motivational control of instrumental performance. *Quarterly Journal of Experimental Psychology*, **41B**, 99–112

Dreher, J.K. and Jackson, D.M. (1989) Role of $D_1$ and $D_2$ dopamine receptors in mediating locomotor activity elicited from the nucleus accumbens of rats. *Brain Research*, **487**, 267–278.

Ellinwood, E.H. and Balster, R.L. (1974) Rating the behavioral effects of amphetamine. *European Journal of Pharmacology*, **28**, 35–41.

Fouriezos, G. and Francis, S. (1992) Apomorphine and electrical self-stimulation of rat brain. *Behavioural Brain Research*, **52**, 72–80.

Fowler, S.C. and Liou, J.-R. (1994) Microcatalepsy and disruption of forelimb usage during operant behavior: Differences between dopamine $D_1$ (SCH–23390) and $D_2$ (raclopride) antagonists. *Psychopharmacology*, **115**, 24–30.

Gallistel, C.R. and Karras, D. (1984) Pimozide and amphetamine have opposing effects on the reward summation function. *Pharmacology Biochemistry and Behavior*, **20**,73–77.

Gilbert, D.B. and Cooper, S.J. (1995) 7-OH-DPAT injected into the accumbens reduces locomotion and sucrose ingestion: $D_3$ autoreceptors-mediated effects? *Pharmacology Biochemistry and Behavior*, **52**, 275–280.

Grace, A.A. (1991) Phasic versus tonic dopamine release and the modulation of dopamine system responsivity: A hypothesis for the etiology of schizophrenia. *Neuroscience*, **41**, 1–24.

Graybiel, A.M., Aosaki, T., Flaherty, A.W. and Kimura, M. (1994) The basal ganglia and adaptive motor control. *Science*, **265**, 1826–1831.

Harris, R.A., Snell, D. and Loh, H.H. (1978). Effects of stimulants, anorectics, and related drugs on schedule-controlled behavior. *Psychopharmacology*, **56**, 49–55.

Hiroi, N. and White, N.M. (1991) The amphetamine conditioned place preference–Differential involvement of dopamine receptor subtypes and two dopaminergic terminal areas. *Brain Research*, **552**, 141–152.

Hoffman, D.C. (1989) The use of place conditioning in studying the neuropharmacology of drug reinforcement. *Brain Research Bulletin*, **24**, 373–387.

Hoffman, D.C. and Beninger, R.J. (1988) Selective D1 and D2 dopamine agonists produce opposing effects in place conditioning but not in conditioned taste aversion learning. *Pharmacology Biochemistry and Behavior*, **31**, 1–8.

Hoffman, D.C. and Beninger, R.J. (1989) The effects of selective dopamine D1 and D2 receptor antagonists on the establishment of agonist-induced place conditioning in rats. *Pharmacology Biochemistry and Behavior*, **33**, 273–279.

Hoffman, D.C., Dickson, P.R. and Beninger, R.J. (1988) The dopamine D2 receptor agonists, quinpirole and bromocriptine produce conditioned place preferences. *Progress in Neuro-Psychopharmacology and Biological Psychiatry*, **12**, 315–322.

Hyland, B.I. (1999) Relationships of substantia nigra dopamine neurone activity to behaviour. In: R. Miller and J.R. Wickens (eds), *Brain dynamics and the striatal complex.* Conceptual Advances in Brain Research (series,) Reading: Gordon and Breach, in press.

Jackson, D.M., Jenkins, D.F. and Ross, S.B. (1988) The motor effects of bromocriptine: A review. *Psychopharmacology*, **95**, 433–447.

Josselyn, S.A., Miller, R. and Beninger, R.J. (1997) Behavioral effects of clozapine and dopamine receptor subtypes. *Neuroscience and Biobehavioral Reviews*, **21**,531–558.

Katz, J.L. and Witkin, J.M. (1992) Selective effects of the $D_1$ dopamine receptor agonist, SKF 38393, on behavior maintained by cocaine injection in squirrel monkeys. *Psychopharmacology*, **109**, 241–244.

Kebabian, J.W. and Calne, D.B. (1979) Multiple receptors for dopamine. *Nature*, **277**, 93–96.

Khroyan, T.V., Baker, D.A. and Neisewander, J.L. (1995) Dose-dependent effects of the $D_3$-preferring agonist 7-OH-DPAT on motor behaviors and place conditioning. *Psychopharmacology*, **122**, 351–357.

Kiyatkin, E.A. (1995) Functional significance of mesolimbic dopamine. *Neuroscience and Biobehavioral Reviews*, **19**, 573–598.

Kiyatkin, E.A. and Gratton, A. (1994) Electrochemical monitoring of extracellular dopamine in nucleus accumbens of rats lever-pressing for food. *Brain Research*, **652**, 225–234.

Kling-Petersen, T., Ljung, E., Wollter, L. and Svensson, K. (1995a) Effects of the dopamine D3- and autoreceptor preferring antagonist (-)-DS121 on locomotor activity, conditioned place preference and intracranial self-stimulation in the rat. *Behavioural Pharmacology*, **6**, 107–115.

Kling-Petersen, T., Ljung, E., Wollter, L. and Svensson, K. (1995b) Effects of dopamine $D_3$ preferring compounds on conditioned place preference and intracranial self-stimulation in the rat. *Journal of Neural Transmission*, **101**, 27–40.

Kötter, R. (1994) Postsynaptic integration of glutamatergic and dopaminergic signals in the striatum. *Progress in Neurobiology*, **44**, 63–196.

LeMoal, M. and Simon, H. (1991) Mesocortical dopaminergic network: Functional and regulatory roles. *Physiological Reviews*, **71**, 155–234.

Leone, P. and Di Chiara, G. (1987) Blockade of D-1 receptors by SCH 23390 antagonizes morphine- and amphetamine-induced place preference conditioning. *European Journal of Pharmacology*, **135**, 251–254.

Lindvall, O. (1979) Dopamine pathways in the rat brain. In *The Neurobiology of Dopamine*, edited by A.S. Horn, J. Korf and B.H.C. Westerink, London: Academic Press, pp. 319–342.

Lopez, M., Balleine, B., and Dickinson, A. (1992). Incentive learning and the motivational control of instrumental performance by thirst. *Animal Learning and Behavior*, **20**, 322–328.

Lucki, I. (1983). Rate-dependent effects of amphetamine on responding under random-interval schedules of reinforcement. *Pharmacology, Biochemistry and Behavior*, **18**, 195–201.

Mackey, W.B. and van der Kooy, D. (1985) Neuroleptics block the positive reinforcing effects of amphetamine but not of morphine as measured by place conditioning. *Pharmacology Biochemistry and Behavior*, **22**, 101–106.

Mallet, P.E. and Beninger, R.J. (1994) 7-OH-DPAT produces place conditioning in rats. *European Journal of Pharmacology*, **261**, R5–R6.

Martin-Iverson, M.T. and Fawcett, S.L. (1996) Pavlovian conditioning of psychomotor stimulant-induced behaviours: Has convenience led us astray? *Behavioural Pharmacology*, **7**, 24–41.

Mas, M., Fumero, B. and Gonzalez-Mora, J.L. (1995) Voltammetric and microdialysis monitoring of brain neurotransmitter release during sociosexual interactions. *Behavioural Brain Research*, **71**, 69–80.

Mazurski, E.J. and Beninger, R.J. (1986) The effects of (+)-amphetamine and apomorphine on responding for a conditioned reinforcer. *Psychopharmacology*, **90**, 239–243.

Mazurski, E.J. and Beninger, R.J. (1991) Effects of selective drugs for dopaminergic D1 and D2 receptors on conditioned locomotion in rats. *Psychopharmacology*, **105**, 107–112.

Messier, C., Mrabet, O. and Destrade, C. (1991a) Locomotor bias produced by intra-accumbens injection of dopamine agonists and antagonists. *Pharmacology Biochemistry and Behavior*, **41**, 177–182.

Messier, C., Mrabet, O., Durkin, T.P. and Destrade, C. (1991b) Bidirectional potentiation between D1 and D2 dopamine agonists–Effects of unilateral intra-accumbens injections on locomotor activity in mice. *Life Sciences*, **49**, 43–40.

Meyer, M.E. (1996) Mesolimbic 7-OH-DPAT affects locomotor activities in rats. *Pharmacology Biochemistry and Behavior*, **55**, 209–214.

Miller, R. and Beninger, R.J. (1991) On the interpretation of asymmetries of posture and locomotion produced with dopamine agonists in animals with unulateral depletion of striatal dopamine. *Progress in Neurobiology*, **36**, 229–256.

Miller, R., Wickens, J.R. and Beninger, R.J. (1990) Dopamine D–1 and D–2 receptors in relation to reward and performance: A case for the D–1 receptor as a primary site of therapeutic action of neuroleptic drugs. *Progress in Neurobiology*, **34**,1 43–183.

Möller, H.-G., Nowak, K. and Kuschinsky, K. (1987) Conditioning of pre- and post-synaptic behavioural responses to the dopamine receptor agonist apomorphine in rats. *Psychopharmacology*, **91**, 50–55.

Molloy, A.G. and Waddington, J.L. (1987) Assessment of grooming and other behavioural responses to the D–1 dopamine receptor agonist SK and F 38393 and its R- and S-enantiomers in the intact adult rat. *Psychopharmacology*, **92**, 164–168.

Morency, M.A. and Beninger, R.J. (1986) Dopaminergic substrates of cocaine-induced place conditioning. *Brain Research*, **399**, 33–41.

Nader, K., Bechara, A., Roberts, D.C.S. and van der Kooy, D. (1994) Neuroleptics block high- but not low-dose heroin place preferences: Further evidence for a two-system model of motivation. *Behavioral Neuroscience*, **108**, 1128–1138.

Neve, K.A. and Neve, R.L. (1997) Molecular biology of dopamine receptors. In *The Dopamine Receptors*, K.A. Neve and R.L. Neve (eds) , pp. 27–76. Totowa, NJ: Humana Press.

Niemegeers, C.E. and Janssen, P.A.J. (1979) A systematic study of the pharmacological activities of dopamine antagonists. *Life Sciences*, **24**, 2201–2216.

Niznik, H.B. and Van Tol, H.H.M. (1992) Dopamine receptor genes: New tools for molecular psychiatry. *Journal of Psychiatry and Neuroscience*, **17**, 158–180.

Phillips, A.G., Blaha, C.D. and Fibiger, H.C. (1989) Neurochemical correlates of brain-stimulation reward measured by ex vivo and in vivo analyses. *Neuroscience and Biobehavioral Reviews*, **13**, 99–104.

Pickens, R.W. and Crowder, W.G. (1967) Effects of CS-US interval on conditioning of drug response with assessment of speed of conditioning. *Psychopharmacology*, **11**, 88–94.

Pickens, R. and Harris, W.C. (1968) Self-administration of d-amphetamine by rats. *Psychopharmacology*, **12**, 158–163.

Pugh, M.T., O'Boyle, K.M., Molloy, A.G. and Waddington, J.L. (1985) Effects of the putative D–1 antagonist SCH23390 on stereotyped behavior induced by the D–2 agonist RU 24213. *Psychopharmacology*, **87**, 308–312.

Pycock, C.J. (1980) Turning behavior in animals. *Neuroscience*, **5**, 461–514.

Ranaldi, R. and Beninger, R.J. (1993) Dopamine $D_1$ and $D_2$ antagonists attenuate amphetamine-produced enhancement of responding for conditioned reward in rats. *Psychopharmacology*, **113**, 110–118.

Ranaldi, R. and Beninger, R.J. (1994) The effects of systemic and intracerebral injections of D1 and D2 agonists on brain stimulation reward. *Brain Research*, **651**, 283–292.

Ranaldi, R., Pantalony, D. and Beninger, R.J. (1995) The D1 agonist SKF 38393 attenuates amphetamine-produced enhancement of responding for conditioned reward in rats. *Pharmacology Biochemistry and Behavior*, **52**, 131–137.

Richardson, N.R., Piercey, M.F., Svensson, K., Collins, R.J., Myers, J.E. and Roberts, D.C.S. (1993) Antagonism of cocaine self-administration by the preferential dopamine autoreceptor antagonist, (+)-AJ 76. *Brain Research*, **619**, 15–21.

Robbins, T.W. (1977) A critique of the methods available for the measurement of spontaneous motor activity. In *Handbook of Psychopharmacology, Volume 7*, edited by L.L. Iversen, S.D. Iversen and S.H. Snyder, New York: Plenum Press, pp 37–82.

Robbins, T.W. (1981). Behavioural determinants of drug action: rate-dependency revisited. In *Theory in Psychopharmacology Volume 1*, edited by S.J. Cooper, London: Academic Press, pp. 1–63.

Robbins, T.W., Watson, B.A., Gaskin, M. and Ennis, C. (1983) Contrasting interactions of pipradol, d-amphetamine, cocaine, cocaine analogues, apomorphine and other drugs with conditioned reinforcement. *Psychopharmacology*, **80**, 113–119.

Scheel-Krüger, J. (1971) Comparative studies of various amphetamine analogues demonstrating different interactions with the metabolism of the catecholamines in the brain. *European Journal of Pharmacology*, **14**, 47–59.

Schultz, W., Dayan, P. and Montague, P.R. (1997) A neural substrate of prediction and reward. *Science*, New York, **275**, 1593–1599.

Self, D.W. and Stein, L. (1992) The D1 agonists SKF–82958 and SKF–77434 are self-administered by rats. *Brain Research*, **582**, 349–352.

Shippenberg, T.S., Bals-Kubik, R. and Herz, A. (1993) Examination of the neurochemical substrates mediating the motivational effects of opioids: Role of the mesolimbic dopamine system and D–1 vs. D–2 dopamine receptors. *Journal of Pharmacology and Experimental Therapeutics*, **265**, 53–59.

Shippenberg, T.S., Bals-Kubik, R., Huber, A. and Herz, A. (1991) Neuroanatomical substrates mediating the aversive effects of D–1 dopamine receptor antagonists. *Psychopharmacology*, **103**, 209–200.

Shippenberg, T.S. and Hertz, A. (1987) Place preference conditioning reveals the involvement of $D_1$-dopamine receptors in the motivational properties of m- and k-opioid agonists. *Brain Research*, **436**, 169–172.

Shippenberg, T.S. and Herz, A. (1988) Motivational effects of opioids: Influence of D–1 versus D–2 receptor antagonists. *European Journal of Pharmacology*, **151**, 233–243.

Sibley, D.R., Monsma, F.J., Jr. and Shen, Y. (1993) Molecular neurobiology of $D_1$ and $D_2$ dopamine receptors. In *$D_1$: $D_2$ Dopamine Receptor Interactions*, edited by J. Waddington, .London: Academic Press Limited, pp 1–17.

Skinner, B.F. (1938) *The Behavior of Organisms*, New York: Appleton-Century-Crofts.

Smith, A.D. and Bolam, J.P. (1990) The neural network of the basal ganglia as revealed by the study of synaptic connections of identified neurons. *Trends in Neurosciences*, **13**, 259–265.

Smith, I.D., Sutton, M.A. and Beninger, R.J. (1997) Rotational bias in intact rats following intrastriatal injections of dopaminergic drugs. *Pharmacology, Biochemistry and Behavior*, **58**, 431–441.

Stewart, J. (1992) Conditioned stimulus control of expression of sensitization of the behavioral activating effects of opiate and stimulant drugs. In *Learning and Memory: Behavioral and Biological Substrates*, edited by I. Gormezano and E.A. Wasserman, Hillsdale, N.J.: Lawrence Erlbaum Publishers, pp. 129–151.

Stewart, J. and Eikelboom, R. (1987) Conditioned drug effects. In *Handbook of psychopharmacology, Vol. 19: New directions in behavioral pharmacology*, edited by L.L. Iversen, S.D. Iversen and S.H. Snyder, New York: Plenum Press, pp 1–57.

Stewart, R.J., Morency, M.A. and Beninger, R.J. (1985) Differential effects of intrafrontocortical microinjections of dopamine agonists and antagonists on circling behaviour in rats. *Behavioural Brain Research*, **17**, 67–72.

Stoof, J.C. and Kebabian, J.W. (1981) Opposing roles for D–1 and D–2 DA receptors in efflux of cyclic AMP from rat neostriatum. *Nature*, London, **294**, 366–368.

Svensson, K., Carlsson, A., Huff, R.M., Kling-Petersen, T. and Waters, N. (1994a) Behavioral and neurochemical data suggest functional differences between dopamine D–2 and D–3 receptors. *European Journal of Pharmacology*, **263**, 235–244.

Svensson, K., Carlsson, A. and Waters, N. (1994b) Locomotor inhibition by the $D_3$ ligand R(+)–7-OH-DPAT is independent of changes in dopamine release. *Journal of Neural Transmission*, **95**, 71–74.

Tolman, E.C. (1949). The nature and functions of wants. *Psychological Review*, **56**, 357–369.

Ungerstedt, U. (1979) Central dopamine mechanisms and unconditioned behavoiur. In *The Neurobiology of Dopamine*, A.S. Horn, J. Korf and B.H.C. Westerink (eds), London: Academic Press, pp 577–596.

Vezina, P., Blanc, G., Glowinski, J. and Tassin, J.P. (1991) Opposed behavioural outputs of increased dopamine transmission in prefrontal and subcortical areas–A role for cortical D–1 dopamine receptor. *European Journal of Pharmacology*, **3**, 1001–1007.

Waddington, J.L. and O'Boyle, K.M. (1989) Drugs acting on brain dopamine receptors: A conceptual re-evaluation five years after the first selective D–1 antagonist. *Pharmacology and Therapeutics*, **43**, 1–52.

Weed, M.R., Vanover, K.E. and Woolverton, W.L. (1993) Reinforcing effect of the $D_1$ dopamine agonist SKF 81297 in rhesus monkeys. *Psychopharmacology*, **113**, 51–52.

Westerink, B.H.C. (1979) The effects of drugs on dopamine biosynthesis and metabolism in the brain. In *The Neurobiology of Dopamine*, A.S. Horn, J. Korf and B.H.C. Westerink (eds). London: Academic Press, pp 255–291.

Westerink, B.H.C. (1995) Brain microdialysis and its application for the study of animal behaviour. *Behavioural Brain Research*, **70**, 103–124.

White, N.M., Packard, M.G. and Hiroi, N. (1991) Place conditioning with dopamine–D1 and D2 agonists induced peripherally or into nucleus accumbens. *Psychopharmacology*, **103**, 271–270.

Wickens, J. (1990) Striatal dopamine in motor activation and reward-mediated learning: Steps towards a unifying model. *Journal of Neural Transmission*, **80**, 9–31.

Wickens, J. (1993) *A Theory of the Striatum*, Oxford: Pergamon Press.

Wickens, J. and Kötter, R. (1995) Cellular models of reinforcement. In *Models of information processing in the basal ganglia*, J.C. Houk, J Davis and D.G. Beiser (eds), Cambridge: MIT, pp 187–214.

Wise, R.A., Murray, A. and Bozarth, M.A. (1990) Bromocriptine self-administration and bromocriptine-reinstatement of cocaine-trained and heroin-trained lever pressing in rats. *Psychopharmacology*, **100**, 355–361.

Wise, R.A. and Rompré, P.-R. (1989) Brain dopamine and reward. *Annual Review of Psychology*, **40**, 191–227.

Woolverton, W.L. (1986) Effects of a $D_1$ and a $D_2$ dopamine antagonist on the self-administration of cocaine and piribedil by rhesus monkeys. *Pharmacology Biochemistry and Behavior*, **24**, 531–536.

Woolverton, W.L., Goldberg, L.I. and Ginos, J.Z. (1984) Intravenous self-administration of dopamine receptor agonists by rhesus monkeys. *Journal of Pharmacology and Experimental Therapeutics*, **230**, 678–683.

Young, S.D. and Michael, A.C. (1993) Voltammetry of extracellular dopamine in rat striatum during ICSS-like electrical stimulation of the medial forebrain bundle. *Brain Research*, **600**, 305–307.

# 3 Stimulants and Motor-Related Striatal Neuronal Activity

**George V. Rebec**

*Program in Neural Science, Department of Psychology, Indiana University*
*Bloomington, IN 47405 USA*
*Telephone: (812) 855-4832; FAX: (812) 855-4520*
*e-mail: rebec@othello.ucs.indiana.edu*

In awake, unrestrained rats, most striatal neurones discharge at relatively low rates (<6 spikes/s) during quiet rest, but increase spike activity during movement. At doses that elicit nonfocused motor activation, psychomotor stimulants like amphetamine activate these motor-related units, while suppressing the activity of nonmotor-related neurones. Behavioural clamping and other procedures, based on assessment of videotape records, suggest that these neuronal changes represent a direct effect of amphetamine rather than a secondary effect of movement. The transition to focused stereotyped behaviour, which occurs at relatively high amphetamine doses, is accompanied by a more complex neuronal response, involving either a further activation or a reduction in the activity of individual motor-related units depending on their behavioural response characteristics. Although dopamine has been implicated in both the excitatory and inhibitory responses of striatal neurones to amphetamine, the activation of motor-related units requires corticostriatal input, suggesting a role for glutamate. In a more direct assessment of how dopamine and glutamate influence striatal activity in awake animals, these substances were applied iontophoretically during periods of quiet rest. Whereas glutamate alone had strongly excitatory effects on all neurones tested, the relative magnitude of these effects was enhanced still further when glutamate was applied during low-dose dopamine iontophoresis. By itself, dopamine caused only modest changes in basal firing rate. It appears, therefore, that dopamine modulates striatal activity by adjusting the signal-to-noise ratio of the glutamate response. Thus, by facilitating dopamine transmission, amphetamine has the net effect of enhancing the gain on glutamate to increase information flow through striatal circuits.

KEYWORDS: amphetamine; cerebral cortex; dopamine; glutamate; motor-related neurone; striatum

## 1. INTRODUCTION

Amphetamine and other psychomotor stimulants, including cocaine, induce characteristic patterns of repetitive or stereotyped movement (Randrup and Munkvad, 1974). In humans, these motor-activation patterns are often accompanied by perseverative thought processes that closely resemble some forms of psychosis (Ellinwood, Sudilovsky and Nelson, 1973; Schiorring, 1981). With repeated stimulant use, certain aspects of both the motor stereotypy and the psychosis are enhanced, suggesting a common neurobiological substrate (Robinson and Becker, 1986; Segal and Janowsky, 1978). Research on stimulant-induced motor-activation patterns, therefore, may have profound psychiatric implications (Rebec and Bashore, 1984; Ridley and Baker, 1982).

Although many forebrain and brain stem systems participate in the behavioural response to psychomotor stimulants, most attention centres on the striatum because it processes a wide range of relevant afferent information for behavioural output, including cognitive, motivational, and sensorimotor aspects of movement (Calabresi, DeMurtas and Bernard, 1997; Mink, 1996). An important component of this processing function is the dense dopamine (DA) innervation that striatal neurones receive from midbrain nuclei. Both amphetamine and cocaine increase striatal DA transmission, and this action is widely believed to shape the motor-activation patterns associated with these drugs (Fischman and Johanson, 1996; Segal and Kuczenski, 1994; Seiden, Sabol and Ricaurte, 1993).

To determine how the striatum participates in stimulant-induced motor activation, it is necessary to assess striatal function when subjects are in an awake, behaving state. Although recordings of striatal activity obtained from *in vitro* and anaesthetized or immobilized preparations can have important predictive value, they are no substitute for neuronal data obtained from normally functioning, freely moving subjects. In fact, the effects of amphetamine on the activity of striatal neurones in behaving animals are reversed when the same animals are tested under conditions of general anaesthesia (Rebec *et al.*, 1991; Warenycia and McKenzie, 1984). Neurobehavioural relationships, therefore, are best established when neuronal activity is monitored in a behavioural context.

Although many striatal recordings have been obtained from behaving primates, rats are the subjects of choice in stimulant pharmacology and thus are most relevant for establishing the neuronal correlates of stimulant-induced behavioural effects. In this line of work, drugs and other neuroactive substances are injected systemically or applied directly to the recording site via intrastriatal infusion or iontophoresis. These procedures not only allow for new insight into the neuronal mechanisms underlying stimulant-induced behavioural activation, but also reveal principles of striatal operation that may not be readily apparent, or that simply may not exist in other recording preparations. An especially intriguing concept, now emerging from work with ambulant rats, is the role of DA not as a transmitter with classical excitatory or inhibitory functions but as a modulator of afferent input, adjusting the gain by which other transmitters, most notably glutamate (GLU), influence striatal activity. The following sections develop this and related concepts within the context of stimulant-induced changes in striatal function.

## 2. FUNCTIONAL ACTIVITY OF STRIATAL NEURONES

One of the most striking features of striatal neurones is their relatively low level of spontaneous activity. Discharge rates of <6 spikes/s interrupted by periods of complete silence are typically reported for both *in vitro* and *in vivo* preparations (Chang and Wilson, 1990). Awake, unrestrained animals display a similarly low level of striatal activity, or silence, during quiet rest (Evarts *et al.*, 1984). Because 90-95% of striatal neurones are classified as medium spiny cells (Groves, 1983), it seems likely that they comprise this slow-firing or silent population. This assumption has been confirmed by intracellular staining of neurones recorded both intra- and extracellularly (Wilson and Groves, 1981). Subsequent intracellular-labelling studies have shown that a small group of tonically active striatal neurones, many of which have firing rates of >15 spikes/s, correspond to a small population of aspiny interneurones (Chang and Wilson, 1990).

### 2.1. Membrane Properties

The low level of medium spiny activity reflects the dominance of an inwardly rectifying potassium current, which keeps the membrane hyperpolarized (Calabresi, Misgeld and Dodt, 1987). Episodes of firing are the result of maintained plateau depolarizations, which are driven by afferent input (Wilson, 1993). Interestingly, however, inward rectification is present on the dendrites, making these cells relatively insensitive to individual excitatory events, or uncoordinated input activity. Only when afferent excitation arrives in a coordinated fashion, causing depolarization over a relatively large area of the dendritic tree, can sufficient charge reach the cell body to overcome the effect of inward rectification. At this point, individual inputs, arriving within the context of a coordinated afferent excitation, become more likely to generate spike activity. This model of medium spiny physiology suggests that the temporal pattern of spikes in these units is not a faithful representation of spike activity in a particular afferent fiber, but rather reflects an envelope of afferent activity integrated over periods of perhaps 100 ms or longer (Wilson, 1993). A further implication is that a change in the level of inhibitory afferent input plays a relatively small role in the transition to a depolarized state. Thus, intrastriatal inhibitory networks, which include GABAergic collaterals from neighboring medium spiny units, may not be critical for setting the level of spontaneous activity, but could be important for coordinating interactions among adjacent neurones during episodes of firing.

### 2.2. Neurobehavioural Relationships

Relative to resting baseline, most striatal neurones increase discharge rate during movement (Alexander, Crutcher and Delong, 1990). This finding is largely based on unit recordings from primates, trained to make discrete movements in response to a sensory cue. Detailed analysis of the firing patterns of these neurones indicates that despite the correlation with movement, the actual role of these units in behaviour seems relatively complex. Some striatal neurones, for example, fire preferentially in relation to a limb movement triggered by a light, whereas others respond only with respect to spontaneous movements (Kimura, 1990; Schultz and Romo, 1988). Similarly, saccade-related activity differs when saccades are made under memory-guided *vs* visually guided conditions (Hikosaka *et al.*, 1989a). A sensory stimulus, moreover, may initiate a change in firing rate only when the stimulus is behaviourally significant (Apicella, Scarnati and Schultz, 1991; Hikosaka, Sakamoto and Usui, 1989b; Schultz and Romo, 1988). In the ventral striatum, many neurones also have place- and reward-related firing patterns (Apicella, Scarnati and Schultz, 1991). Thus, rather than signal simple motor or sensory events, striatal neurones appear to convey context-dependent information, representing processes related to selective attention, reinforcement, or other complex functions. Changes in striatal activity, therefore, may represent a process by which such functions gain access to motor output.

Although relatively few studies have assessed the behavioural response properties of striatal neurones in rats, the available data suggest that these neurones, like those recorded from primates, typically increase activity during movement (Aldridge *et al.*, 1993; Gardiner, Iverson and Rebec, 1988; West *et al.*, 1987). This motor-related response, which has been reported for ~80% of the neuronal population of the striatum (Haracz *et al.*, 1993; Rebec, 1998), typically occurs in relation to gross body movements (e.g., locomotion and rearing), but some units respond specifically during discrete movements of the head or neck. In either case, the unit response usually consists of spike bursts closely coincident with behaviour onset. Although such cells are classified as motor-related, the changes in firing rate may convey complex, context-dependent information (see above). Other striatal neurones (~20%) maintain a relatively high basal rate (>15 spikes/s), that either fails to change or changes only inconsistently during movement. Such tonically active units, which may include striatal interneurones, are difficult to classify on the basis of motor responsiveness, but they are termed nonmotor-related to distinguish them from motor-related neurones.

Standard histological analysis of recording sites indicates that both motor and nonmotor-related neurones are widely distributed throughout dorsal and ventral striatum (Haracz *et al.,* 1993). Although responses to discrete head or limb movements are common in extreme dorsolateral areas (Carelli and West, 1991; West *et al.,* 1990), such responses have been identified in other striatal locations (White *et al.,* 1995). The use of calbindin immunohistochemistry to reveal the striosome (patch)-matrix organization of the striatum also has failed to reveal regional differences in the location of motor- and nonmotor-related units (Trytek *et al.*, 1996). Both types of cells, for example, are found predominantly in matrix or along the patch-matrix border. Thus, although striatal neurones can be distinguished on the basis of their responsiveness to spontaneous movement patterns, such distinctions are not clearly linked to histological location.

Rat striatal neurones have also been reported to alter firing rate in association with sensory-triggered movements much like that reported for primates (Carelli and West, 1991; Gardiner and Kitai, 1992; West *et al.*, 1990). Many of these neuronal responses, moreover, depend on the behavioural context in which they occur (e.g., a change in firing rate to a sensory-triggered movement may not occur when the movement occurs outside the task). Interestingly, the proportion of striatal neurones having both sensory- and motor-related components appears to be higher in rats than in primates, suggesting a greater convergence of sensory and motor signals in rat striatum (Gardiner and Katai, 1992).

Most assessments of how striatal neurones participate in the motor-activating effects of stimulant drugs have focused on amphetamine, though preliminary evidence suggests that in many respects these results also apply to cocaine (White, Doubles, and Rebec, 1998). They are summarized in the following section.

## 3. STRIATAL RESPONSES TO AMPHETAMINE

The motor-activating effects of amphetamine include species-specific aspects of investigative behaviour, which in the rat are manifest as forward locomotion and rearing as well as head bobbing and sniffing (Randrup and Munkvad, 1974; Rebec and Bashore, 1984; Segal and Janowsky, 1978). Expression of these behaviours follows a complex, dose-dependent pattern. At relatively low doses, amphetamine-induced behaviours occur at mild to moderate intensity over a relatively large area of the test chamber. As doses increase, a constricted form of stereotypy emerges in which an early phase of locomotion and rearing gives way to a phase of intense, highly focused head bobbing, sniffing, and occasional oral behavior.

### 3.1. Dose-Dependent Changes in Striatal Unit Activity

Recordings of striatal unit activity indicate that the behavioural activation associated with a relatively low dose (1.0 mg/kg *d*-amphetamine, sc) is accompanied by a clear divergence in firing rate between motor- and nonmotor-related units (Haracz *et al.*, 1989; Wang, Haracz and Rebec, 1992). Relative to the quiet resting rate, motor-related neurones increase and nonmotor-related neurones decrease activity. Both neuronal responses emerge within 5–15 min after injection, peak shortly thereafter, and remain above or below the resting baseline rate for the duration of the behavioural response.

Because amphetamine increases movement, the activation of motor-related cells may represent a secondary or feedback effect of behaviour rather than a direct action of the drug in striatum. This interpretation, however, is not supported by evidence that an intrastriatal infusion of amphetamine, which also activates and inhibits motor- and non-motor-related neurones, respectively, causes a neuronal change ~5–10 min before the onset of behavioural activation (Wang and Rebec, 1993). In addition, when behaviour is clamped or held constant by analyzing neuronal activity only during movements matched in both form and intensity, post-amphetamine discharge activity exceeds the pre-amphetamine level in every case (Haracz *et al.*, 1993). A similar analysis based on a computerized videotape system supports these findings (West *et al.*, 1997). Thus, it seems unlikely that behavioural feedback alone is responsible for amphetamine-induced changes in motor-related neuronal activity.

If motor-related neurones are activated nonselectively in response to low doses of amphetamine, then the transition to a more focused behavioural pattern, which occurs at high doses, may reflect a corresponding focus of neuronal activity. It is possible, for example, that during focused stereotypy, neurones normally responsive to head bobbing continue to increase activity, while locomotor- or rearing-related cells decline in rate. According to this model, the behaviours that emerge during focused stereotypy are a reflection of the pattern of neuronal activation in the striatum. Consistent with this view, rats tested with 5.0 mg/kg *d*-amphetamine (sc) entered a period of focused stereotypy in which striatal neurones active during early periods of locomotion and rearing returned to baseline levels, whereas head-bobbing units reached peak activity (Rebec, White and Puotz, 1997). This study also showed that some motor-related units progressively increased firing rate after a high dose of amphetamine, regardless of the behavioural response, and nonmotor-related cells showed mirror-image inhibitions. It appears, therefore, that while motor- and nonmotor-related neurones simply respond in different directions during amphetamine-induced periods of nonfocused behavioural activation, the emergence of focused stereotypy is associated with a more complex neuronal pattern in which the inhibition of some motor-related neurones results in a more focused striatal activation.

### 3.2. Mechanisms Underlying Amphetamine-Induced Changes in Striatal Activity

The effects of amphetamine on striatal neuronal activity and behaviour are blocked by the subsequent injection of DA antagonists, confirming a DA mechanism (Haracz *et al.*, 1993; Rosa-Kenig, Puotz and Rebec, 1993). This research has shown that both non-selective DA antagonists (e.g., clozapine and haloperidol) and relatively selective D1

(SCH-23390) and D2 (eticlopride) antagonists reverse the amphetamine-induced neuronal changes. DA, however, does not act alone in mediating the excitatory effects of amphetamine on motor-related units. This neuronal response, for example, is also blocked by damage to sensorimotor and frontal cortex (Tschanz *et al.,* 1991, 1994). Thus, amphetamine appears to enhance the level of excitatory drive that corticostriatal afferents exert on motor-related neurones. This conclusion supports the view that DA acts as a gain-enhancing neuromodulator by facilitating the activity of neurones receiving substantial excitatory input (Servan-Schreiber, Printz and Cohen, 1990; Williams and Millar, 1990). According to this model, cerebrocortical ablation, which removes excitatory input originating from GLU afferents, should selectively attenuate the excitatory, but not inhibitory, effects of amphetamine on striatal neurones. Haracz *et al.* (1998) tested this hypothesis in rats that sustained bilateral cerebrocortical damage. Data were obtained from both motor- and nonmotor-related units, and behavioural clamping analysis was performed on both neuronal populations. Compared to sham-lesioned controls, the excitatory effects of amphetamine on motor-related units in cortically lesioned rats were significantly attenuated, whereas the inhibitory responses of nonmotor-related units were unaffected. Thus, the differential responses of motor- and nonmotor-related neurones to amphetamine appear to reflect different underlying mechanisms. Although DA is likely to play a key role in both types of drug-induced responses, these results implicate corticostriatal (i.e., GLU) mechanisms in the excitatory actions of amphetamine on striatal motor-related neurones.

To begin an assessment of how DA and GLU influence striatal activity in awake, unrestrained rats, Pierce and Rebec (1995) first identified motor- and nonmotor-related neurones and then applied these substances iontophoretically when the animals resumed a resting posture. GLU caused a frank, dose-dependent activation of all recorded units. DA, in contrast, had weakly excitatory and weakly inhibitory effects on motor- and nonmotor-related neurones, respectively. In some cases, it was also possible to test motor-related units with simultaneous applications of both DA and GLU. Under these conditions, DA and GLU had supra-additive effects (i.e., both substances together caused a greater increase in firing rate than either substance alone). It appears, therefore, that a major action of DA, which by itself has relatively modest effects on striatal activity, is to enhance the strength of the GLU signal on motor-related units. This conclusion fits nicely with evidence that, without cerebrocortical input, the activating effects of amphetamine in the striatum are significantly attenuated (see above).

For a more direct assessment of how amphetamine alters striatal neuronal activity, Kiyatkin and Rebec (1997) applied the drug iontophoretically in quietly resting rats. Without inducing behavioural activation, iontophoresis of amphetamine (5–40 nA; 20–30 s) caused clear, dose-dependent inhibitions of striatal activity, and this effect was blocked by DA antagonists. An example of the amphetamine-induced inhibition is shown in Figure 3.1. These results suggest that although DA may be responsible for the inhibitory effects of iontophoretic amphetamine, DA alone cannot account for the neuronal activation that predominates after systemic injection of the drug. With systemic administration, amphetamine appears to cause sufficient cerebrocortical activation that the drug-induced increase in striatal DA transmission occurs in conjunction with a certain level of GLU release. Thus, for those neurones receiving substantial DA and GLU input, the result is a synergistic acceleration of firing rate.

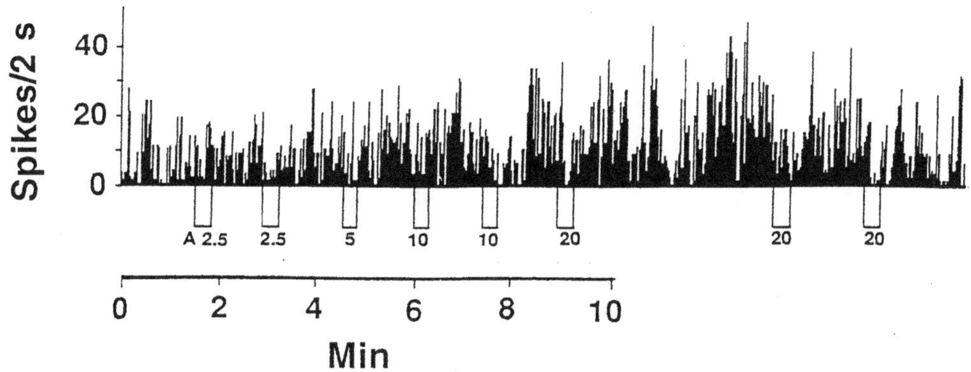


**Figure 3.1.** Rate-meter histogram illustrating the response of a single striatal neurone recorded from an awake, unrestrained rat at rest to iontophoretic application of amphetamine. The drug (A) is applied for 15–20 s intervals (open boxes) at increasing ejection currents (indicated below each box). Note the inhibitory effect of A.

## 4. NEUROCHEMICAL MECHANISMS OF STRIATAL NEURONAL PROCESSING

Evidence for DA-GLU interactions in the response of striatal neurones to amphetamine has led to further use of iontophoresis in the freely moving preparation to obtain detailed profiles of DA and GLU action unencumbered by anaesthesia or immobilization (Kiyatkin and Rebec, 1996a). For these experiments, animals were thoroughly habituated to the recording chamber. This allowed for prolonged periods of quiet rest, which provided a stable baseline for iontophoretic testing. Each recorded neurone, moreover, received multiple iontophoretic tests. Initially, DA and GLU were applied briefly (15–30 s) over a wide range of ejection currents (5–80 nA). GLU activated all tested units with brief onset and offset response latencies (0.5–4.0 s); the magnitude of the response was highly correlated with basal activity such that the slowest firing cells showed the largest percentage increase. DA, in contrast, had relatively weak effects, which in most cases were manifest as an inhibition, though some neurones either failed to respond or showed mild excitations. DA onset-offset latencies were longer (2–20 s) than those for GLU, but shorter than those typically reported for anaesthetized animals (Siggins, 1978). To assess DA-GLU interactions, DA (5–80 nA) was applied for prolonged periods (2–3 min) and brief applications of GLU (5–40 nA; 15 s) were performed before, during, and after DA iontophoresis. The result was an overall enhancement of the GLU response relative to the DA-induced change in basal activity. An example of this effect for a single striatal neurone is shown in Figure 3.2. Note the increase in the signal-to-noise ratio of the GLU response during prolonged DA iontophoresis. For all neurones tested, the relative increase in the GLU signal occurred with low current applications of DA (10–30 nA); high DA ejection currents tended to decrease the GLU response. Single-unit recording and iontophoresis in behaving monkeys also suggest a role for DA in modulating the relative strength of the GLU signal (Rolls *et al.*, 1984).

These results extend evidence for DA as a modulator of striatal activity. They not only confirm a relatively weak action of DA on basal levels of activity but also indicate that,

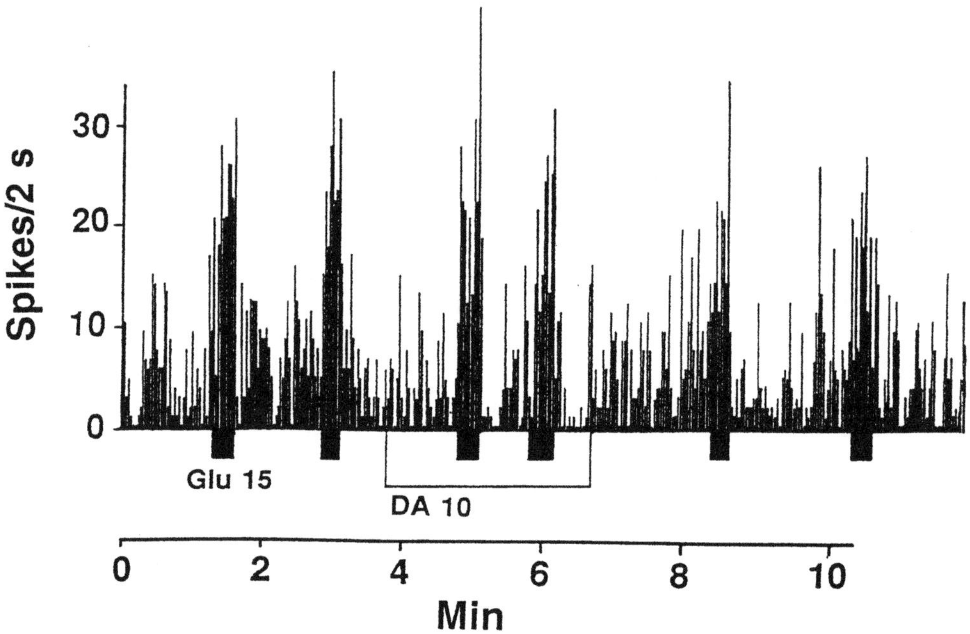


**Figure 3.2.** Rate-meter histogram of a single striatal neurone recorded from the freely moving preparation during a period of quiet rest. GLU iontophoresis at 15 nA (filled black bars) increases neuronal activity over background firing. This effect is enhanced during prolonged co-iontophoresis of 10 nA DA.

within limits, DA has the net effect of amplifying the phasic activation induced by GLU. Thus, DA release, which is known to be triggered by behaviourally important stimuli (Blackburn, Pfaus and Phillips, 1992; Le Moal and Simon, 1991; Martel and Fantino, 1996; Richardson and Gratton, 1996; Salamone *et al.*, 1994; Schultz, Dayan and Montague, 1997) as well as amphetamine (Segal and Kuczenski, 1994; Seiden, Sabol and Ricaurte 1993; see also Chapter 1 by Hyland), may represent a key mechanism for enhancing the flow of GLU-mediated information through striatal circuits.

Separate experiments tested possible interactions between DA and acetylcholine (ACh), another presumed modulator of striatal function (Kiyatkin and Rebec, 1996b). Iontophoresis of ACh alone (5–30 nA; 15–20 s) typically excited striatal neurones, though inhibitions were encountered among fast-firing units. Prolonged applications of DA had no net effect on the relative magnitude of the ACh response. Thus, although ACh may modulate striatal activity, DA does not regulate this effect in the same way that it regulates neuronal responsiveness to GLU.

## 5. CONCLUSIONS

Efforts to record the activity of individual neurones and to assess their responsiveness to drugs and other neuroactive substances in awake, unrestrained animals represent an important step toward identifying the role of striatal mechanisms in stimulant-induced motor activation. It is already apparent that neuronal information processing in the striatum of behaving animals represents a complex interaction of afferent inputs. In fact, medium spiny neurones require a certain level of afferent signalling just to switch from an inactive, hyperpolarized state to a state in which subsequent afferent activity can elicit extracellular spikes. Recent work with the behaving preparation indicates that most of this excitatory input is supplied by GLU, whereas DA functions as a gate or filter, adjusting the level of background activity to strengthen the relative magnitude of the GLU response. A presumed model of this interaction is summarized in Figure 3.3. It is consistent with recent views of striatal function, in which DA release is thought to be timed to coincide with the arrival of GLU signals from cerebrocortex (Mink, 1996; Wickens, Begg and Arbuthnott, 1996; see also Wickens, Chapter 4 this volume). DA fibers, moreover, typically form *en passant* contacts with the same medium spiny neurones that receive GLU input. In fact, the DA projection contacts dendritic spine shafts in an area <2 $\mu$m away from GLU terminals on spine heads (Mink, 1996). In view of neuroanatomical (Nirenberg *et al.*, 1996) and neurochemical evidence (Garris, Ciolkowski and Wightman, 1994; Gonon and Sundstrom, 1996) that DA is likely to diffuse several microns beyond its release site before reuptake, medium spiny afferents seem ideally suited to express DA-GLU interactions.

Under resting conditions, DA release appears to be relatively low. Midbrain DA neurones, for example, are known to fire at a slow rate, and striatal dialysates indicate an extracellular level of <10 nM (Parsons and Justice, 1992). This level appears to increase dramatically, however, in response to behaviourally meaningful stimuli (e.g., Fiorino, Coury and Phillips, 1997; Rebec *et al.*, 1997), and some evidence suggests that DA release

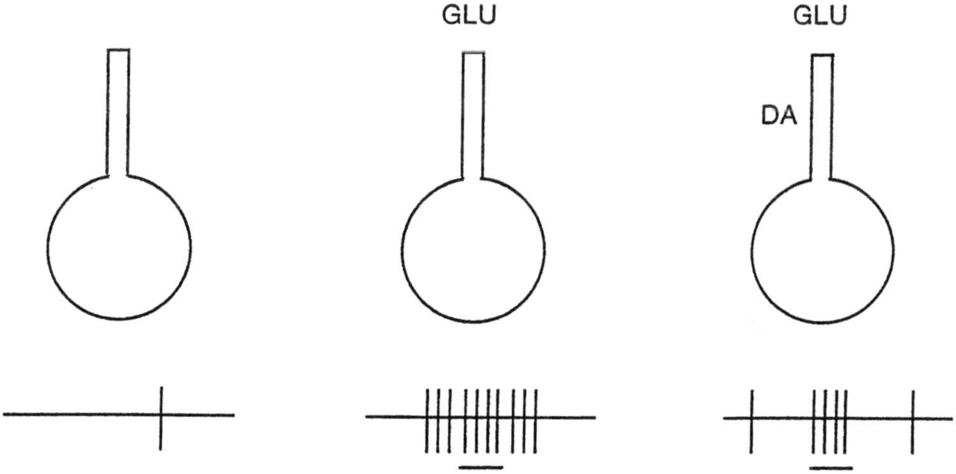


**Figure 3.3.** Schematic representation of presumed DA-GLU interactions on striatal medium spiny neurones. At left, with no GLU input, the neurone maintains a low or "silent" level of activity. With some level of endogenous GLU input, centre, basal activity is enhanced, making it relatively difficult to detect a further increase in the GLU signal (bar underlining a portion of spike activity). At right, the presence of DA acts as a gate or filter to enhance the strength (signal-to-noise ratio) of the GLU signal relative to background activity.

can occur selectively in specific striatal regions (Rebec *et al.*, 1997). To the extent that striatal neurones form functionally distinct clusters, as increasing evidence indicates (Graybiel *et al.*, 1994), targeted DA release may ensure that activity in specific clusters is relatively enhanced, thus increasing the "weight" or influence of such activity on downstream nuclei. By nonselectively increasing DA transmission throughout the striatum as well as activating corticostriatal afferents, amphetamine and related stimulants cause an overactivation of striatal circuits that may form the basis for the nonfocused behavioural activation patterns induced by relatively low doses. Indeed, striatal motor-related units are routinely activated by just such a dose of amphetamine (see above). More problematic, however, is the explanation for the more complex neuronal response to a high dose capable of inducing behavioural stereotypy. It is conceivable that DA plays some role in this effect as well, given that a DA antagonist can reverse both the excitatory and inhibitory responses in motor-related neurones to 5.0 mg/kg *d*-amphetamine (Rebec *et al.*, 1997). Also noteworthy in this regard is recent evidence that iontophoretic amphetamine not only inhibits striatal neurones but also inhibits the neuronal response to GLU (Kiyatkin and Rebec, 1997). High dose applications of DA also attenuate GLU-induced excitations (see above). Thus, a high dose of amphetamine may cause sufficient DA release to override the GLU response and begin shutting down some striatal neurones. Why this effect should occur relatively selectively on neurones with activity linked to locomotion and rearing (rather than head bobbing) to create a highly focused behavioural response remains to be determined. Of course, it also is possible that high doses of amphetamine may engage other neurochemical mechanisms, including serotonin, which ample evidence implicates in amphetamine-induced stereotyped behaviour (see Kuczenski and Segal, 1994). Changes in GABA, which may regulate lateral interactions within the striatum (see Chang and Wilson, 1990), are also possible. Further work with the freely moving preparation is required to assess these and other influences on amphetamine-induced changes in striatal neuronal activity.

Remarkable progress has been made at the molecular level in identifying both the mechanisms by which amphetamine increases DA transmission (Giros *et al.*, 1996) and the large number of receptor subtypes at which DA acts (Schwartz *et al.*, 1992; Seeman and Van Tol, 1994). Corresponding anatomical advances have begun to disentangle the connections and pathways that link the striatum with other nuclei (Chesselet and Delfs, 1996; Parent, Cote and Lavoie, 1995). Relating these and other recent developments in neurobiology to behaviour requires further use of the behavioural preparation, not only to characterize neuronal responses to amphetamine and other psychomotor stimulants, but also to assess how endogenous transmitter systems influence individual striatal neurones in a behavioural context. Such information is essential to reveal how a drug-induced change in striatal neuronal operations leads to a change in the pattern of motor activation.

## ACKNOWLEDGEMENTS

Preparation of this report and data collection by the Rebec laboratory were supported, in part, by DA 02451. Eugene Kiyatkin prepared the rate-meter histograms. Faye Caylor provided editorial assistance.

## REFERENCES


Aldridge, J.W., Berridge, K.C., Herman, M. and Zimmer, L. (1993) Neuronal coding of serial order: syntax of grooming in the neostriatum. *Psychological Science* **4**, 391–395.

Alexander, G.E., Crutcher, M.D. and Delong, M.R. (1990) Basal ganglia-thalamocortical circuits: parallel substrates for motor, oculomotor, "prefrontal" and "limbic" functions. In *The Prefrontal cortex: its structure, function and pathology (Progress in Brain Research vol. 85),* edited by H.B.M. Uylings, C.G. van Eden, J.P.C. de Bruin, M.A. Corner and M.G.P. Feenstra Elsevier, Amsterdam, pp. 119–146.

Apicella, P., Scarnati, E. and Schultz, W. (1991) Tonically discharging neurons of monkey striatum respond to preparatory and rewarding stimuli. *Experimental Brain Research* **84**, 672–675.

Blackburn, J.R., Pfaus, J.G. and Phillips, A.G. (1992). Dopamine functions in appetitive and defensive behaviours. *Progress in Neurobiology* **39**, 247–279.

Calabresi, P., Misgeld, U. and Dodt, H.U. (1987). Intrinsic membrane properties of neostriatal neurons can account for their low level of spontaneous activity. *Neuroscience* **20**, 293–303.

Calabresi, P., DeMurtas, M. and Bernard, G. (1997) The neostriatum beyond the motor function: Experimental and clinical evidence. *Neuroscience* **78**, 39–60.

Carelli, R.M. and West, M.O. (1991) Representation of the body by single neurons in the dorsolateral striatum of the awake, unrestrained rat. *Journal of Comparative Neurology* **309**, 231–249.

Chang, H.T. and Wilson, C.J. (1990) Anatomical analysis of electrophysiologically characterized neurons in the rat strio-pallidal system. In *Handbook of Chemical Neuroanatomy, Vol 8, Analysis of Neuronal Microcircuits and Synaptic Interactions,* edited by A. Bjorklund, T. Hokfelt, F.G. Wouterlood, A.N. van den Pol, pp. 351–402, Amsterdam: Elsevier.

Chesselet, M.F. and Delfs, J.M. (1996) Basal ganglia and movement disorders: An update. *Trends in Neurosciences* **19**, 417–422.

Ellinwood, E.H., Sudilovsky, A. and Nelson, L.M. (1973) Evolving behavior in the clinical and experimental amphetamine (model) psychosis. *American Journal Psychiatry* **130**, 1088–1093.

Evarts, E.V., Kimura, M., Wurtz, R.H. and Hikosaka, O. (1984) Behavioral correlates of activity in basal ganglia neurons. *Trends in Neuroscience* **7**, 447–453.

Fischman, M.W. and Johanson, C.E. (1996) Cocaine. In:*Pharmacological Aspects of Drug Dependence: Towards an Integrated Neurobehavioral Approach,* edited by C.R. Schuster and M. J. Kuhar, pp. 159–195, New York, Springer.

Fiorino, D.F., Coury, A. and Phillips, A.G. (1997) Dynamic changes in nucleus accumbens dopamine efflux during the Coolidge effect in male rats. *Journal of Neuroscience* **17**, 4849–4855.

Gardiner, T.W., Iverson, D.A. and Rebec, G.V. (1988) Heterogeneous responses of neostriatal neurons to amphetamine in freely moving rats. *Brain Research* **463**,–268–274.

Gardiner, T.W. and Kitai, S.T. (1992) Single-unit activity in the globus pallidus and neostriatum of the rat during performance of a trained head movement. *Experimental Brain Research* **88**, 517–530.

Garris, P.A., Ciolkowski, E.L., and Wightman, R.M. (1994) Heterogeneity of evoked dopamine overflow within the striatal and striato-amygdaloid regions. *Neuroscience* **59**, 417–427.

Giros, B., Jaber, M., Jones, S.R., Wightman, R.M. and Caron, M.G. (1996) Hyperlocomotion and indifference to cocaine and amphetamine in mice lacking the dopamine transporter. *Nature* **379**, 606–612.

Gonon, F. and Sundstrom, L. (1996) Excitatory effects of dopamine released by impulse flow in the rat nucleus accumbens in vivo. *Neuroscience* **75**, 13–18.

Graybiel, A.M., Aosaki, T., Flaherty, A.W. and Kimura, M. (1994) The basal ganglia and adaptive motor control. *Science*, New York, **205**, 1826–1831.

Groves, P.M. (1983) A theory of the functional organization of the neostriatum and the neostriatal control of voluntary movement. *Brain Research Review* **5**, 109–132

Haracz, J.L., Tschanz, J.T., Greenberg, J. and Rebec, G.V. (1989) Amphetamine-induced excitations predominate in single neostriatal neurons showing motor-related activity. *Brain Research* **489**, 363–368.

Haracz, J.L., Tschanz, J.T., Wang, Z., Griffith, K.E. and Rebec, G.V. (1998) Amphetamine effects on striatal neurons: implications for models of dopamine function. *Neuroscience and Biobehavioral Reviews* **22**, 613–622.

Haracz, J.L., Tschanz, J.T., Wang, Z., White, I.M. and Rebec, G.V. (1993) Striatal single-unit responses to amphetamine and neuroleptics in freely moving rats. *Neuroscience Biobehavioral Reviews* **17**, 1–12.

Hikosaka, O., Sakamoto, M. and Usui, S. (1989a) Functional properties of monkey caudate neurons I. Activities related to saccadic eye movement. *Journal of Neurophysiology* **61**, 780–798.

Hikosaka, O., Sakamoto, M. and Usui, S. (1989b) Functional properties of monkey caudate neurons III. Activities related to expectation of target and reward. *Journal of Neurophysiology* **61**, 814–832.

Kimura, M. (1990) Behaviorally contingent property of movement-related activity of the primate putamen. *Journal of Neurophysiology* **63**, 1277–1296.

Kiyatkin, E.A. and Rebec, G.V. (1996a) Dopaminergic modulation of glutamate-induced excitations of neurons in the neostriatum and nucleus accumbens of awake, unrestrained rats. *Journal of Neurophysiology* **75**, 142–153.

Kiyatkin, E.A. and Rebec, G.V. (1996b) Modulatory action of dopamine on acetylcholine-responsive striatal and accumbal neurons in awake, unrestrained rats. *Brain Research* **713**, 70–78.

Kiyatkin, E.A. and Rebec, G.V. (1997) Iontophoresis of amphetamine in the neostriatum and nucleus accumbens of awake, unrestrained rats. *Brain Research* **771**, 14–24.

Kuczenski, R. and Segal, D.S. (1994) Neurochemistry of amphetamine. In *Amphetamine and Its Analogs,* edited by A. Cho and D. S. Segal, pp. 81–113. San Diego, CA: Academic Press.

Le Moal, M. and Simon, H. (1991). Mesocorticolimbic dopaminergic network—functional and regulatory roles. *Physiological Reviews* **71**, 155–234.

Martel, P. and Fantino, M. (1996) Mesolimbic dopaminergic system activity as a function of food reward: A microdialysis study. *Pharmacology Biochemistry and Behavior* **53**,–221–226.

Mink, J.W. (1996) The basal ganglia: Focused selection and inhibition of competing motor programs. *Progress in Neurobiology* **50**, 381–425.

Nirenberg, M.J., Vaughan, R.A., Uhl, G.R., Kuhar, M.J. and Pickel, V.M. (1996) The dopamine transporter is localized to dendritic and axonal plasma membranes of nigrostriatal dopaminergic neurons. *Journal of Neuroscience* **16**, 436–447.

Parent, A., Cote, P.Y. and Lavoie, B. (1995) Chemical anatomy of primate basal ganglia. *Progress in Neurobiology* **46**, 131–197.

Parsons, L.H. and Justice, Jr., J.B. (1992) Extracellular concentration and in vivo recovery of dopamine in the nucleus accumbens using microdialysis. *Journal of Neurochemistry* **58**, 212–218.

Pierce, R.C. and Rebec, G.V. (1995) Iontophoresis in the neostriatum of awake, unrestrained rats: differential effects of dopamine, glutamate, and ascorbate on motor- and nonmotor-related neurons. *Neuroscience* **67**, 313–324.

Randrup, A. and Munkvad, I. (1974) Pharmacology and physiology of stereotyped behavior. *Journal of Psychiatric Research* **11**, 1–10

Rebec, G.V. (1998) Real-time assessments of dopamine function during behavior: Single-unit recording, iontophoresis, and fast-scan cyclic voltammetry in awake, unrestrained rats. *Alcoholism: Clinical and Experimental Research* **22**, 32–40.

Rebec, G.V. and Bashore, T.R. (1984) Critical issues in assessing the behavioral effects of amphetamine. *Neuroscience and Biobehavioral Reviews* **8**, 153–159.

Rebec, G.V., Christensen, J.R.C., Guerra, C. and Bardo, M.T. (1997) Regional and temporal differences in dopamine efflux in the nucleus accumbens during free-choice novelty. *Brain Research* **776**, 61–67.

Rebec, G.V., Haracz, J.L., Tschanz, J.T., Wang, Z. and White, I. (1991) Responses of motor- and nonmotor-related neostriatal neurons to amphetamine and neuroleptic drugs. In *Basal Ganglia III,* edited by G. Bernardi, pp. 463–470, New York: Plenum.

Rebec, G.V., White, I.M. and Puotz, J.K. (1997) Responses of neurons in dorsal striatum during amphetamine-induced focused stereotypy. *Psychopharmacology* **130**, 343–351.

Richardson, N.R. and Gratton, A. (1996) Behavior-relevant changes in nucleus accumbens dopamine transmission elicited by food reinforcement: An electrochemical study in rat. *Journal of Neuroscience,* **16**, 8160–8169.

Ridley, R.M. and Baker, H.F. (1982) Stereotypy in monkeys and humans. *Psychological Medicine* **12**, 61–72.

Robinson, T.E. and Becker, J.B. (1986) Enduring changes in brain and behavior produced by chronic amphetamine administration: a review and evaluation of animal models of amphetamine psychosis. *Brain Research Reviews* **11**, 157–198.

Rolls, E.T., Thorpe, S.J., Boytim, M., Szabo, I. and Perrett, D.I. (1984) Responses of striatal neurons in the behaving monkey. 3. Effects of iontophoretically applied dopamine on normal responsiveness. *Neuroscience* **12**, 1201–1212.

Rosa-Kenig, A., Puotz, J.K. and Rebec, G.V. (1993) Involvement of D1 and D2 dopamine receptors in amphetamine-induced changes in striatal activity in behaving rats. *Brain Research* **619**, 347–351.

Salamone, J.D., Cousins, M.S., McCullough, L.D., Carriero, D.L. and Berkowitz, R.J. (1994) Nucleus accumbens dopamine release increases during instrumental lever pressing for food but not free food consumption. *Pharmacology Biochemistry and Behavior* **49**, 25–31.

Schiorring, E. (1981) Psychopathology induced by "speed drugs." *Pharmacology Biochemistry and Behavior* **1**, 109–122.

Schultz, W., Dayan, P. and Montague, P.R. (1997) A neural substrate of prediction and reward. *Science*, New York, **275**, 1593–1599.

Schultz,W. and Romo, R. (1988) Neuronal activity in the monkey striatum during the initiation of movements. *Experimental Brain Research* **71**, 431–436.

Schwartz, J -C., Giros, B., Martres, M.-P. and Sokoloff, P. (1992) The dopamine receptor family: molecular biology and pahrmacology. *Seminars in the Neurosciences* **4**, 99–108.

Seeman, P. and Van Tol, H.H.M. (1994) Dopamine receptor pharmacology. *Trends in Pharmacological Sciences* **15**, 264–270.

Segal, D.S. and Kuczenski, R. (1994) Behavioral pharmacology of mphetamine. (Eds.), In *Amphetamine and Its Analogs,* A. Cho and D.S. Segal (eds) San Diego, CA: Academic Press, pp. 115–150.

Segal, D.S. and Janowsky, D.S. (1978) Psychostimulant-induced behavioral effects: Possible models of schizophrenia. In *Psychopharmacology: A Generation of Progress*, edited by M.A. Lipton, A. DiMascio, and K.F. Killam,. New York: Raven Press, pp. 1113–1123.

Seiden, L.S., Sabol, K.E. and Ricaurte, G.A. (1993) Amphetamine: Effects on catecholamine systems and behavior. *Annual Reviews in Pharmacological Toxicology* **33**, 639–677.

Servan-Schreiber, D., Printz, H. and Cohen, J.D. (1990) A network model of catecholamine effects: Gain, signal-to-noise ratio, and behavior. *Science*, New York, **249**, 892–895.

Siggins, G.R. (1978) Electrophysiological role of dopamine in striatum: Excitatory or inhibitory? In *Psychopharmacology: A Generation of Progress*, edited by M.A. Lipton, A. DiMascio, and K.F. Killam, pp. 143–157, New York: Raven Press.

Trytek, E.S., White, I.W., Schroeder, D.M., Heidenreich, B.A. and Rebec, G.V. (1996) Localization of motor- and nonmotor-related neurons within the matrix-striosome organization of rat striatum. *Brain Research* **707**, 221–227

Tschanz, J.T., Griffith, K.E., Haracz, J.L. and Rebec, G.V. (1994) Cortical lesions attenuate the opposing effects of amphetamine and haloperidol on neostriatal neurons in freely moving rats. *European Journal of Pharmacology* **257**, 161–167.

Tschanz, J.T., Haracz, J.L., Griffith, K.E. and Rebec, G.V. (1991) Bilateral cortical ablations attenuate amphetamine-induced excitations of neostriatal motor-related neurons in freely moving rats. *Neuroscience Letters* **134**, 127–130.

Wang, Z. and Rebec, G.V., (1993), Neuronal and behavioral correlates of intrastriatal infusions of amphetamine in freely moving rats, *Brain Research* **627**, 79–88.

Wang, Z., Haracz, J.L. and Rebec, G.V. (1992) BMY–14802, a sigma ligand and potential antipsychotic drug, reverses amphetamine-induced changes in neostriatal single-unit activity in freely moving rats. *Synapse* **12**, 312–321.

Warenycia, M.W. and McKenzie, G.M. (1984) Immobilization of rats modifies the response of striatal neurons to dexamphetamine. *Pharmacology Biochemistry and Behavior* **21**, 53–59.

West, M.O., Carelli, R.M., Pomerantz, M., Cohen, S.M., Gardner, J.P., Chapin, J.K. and Woodward, D.J. (1990) A region in the dorsolateral striatum of the rat exhibiting single-unit correlations with specific locomotor limb movements. *Journal of Neurophysiology* **64**, 1233–1246.

West, M.O., Michael, A.J., Knowles, S.E., Chapin., J.K. and Woodward, D.J. (1987) Striatal unit activity and the linkage between sensory and motor events. In *Basal Ganglia and Behavior: Sensory Aspects of Motor Functioning*, edited by J.S. Schneider and T.I. Lidsky, pp. 27–35, Toronto: Huber.

West, M.O., Peoples, L.L., Michael, A.J., Chapin, J.K. and Woodward, D.J. (1997) Low-dose amphetamine elevates movement-related firing of rat striatal neurons. *Brain Research* **745**, 331–335.

White, I.M., Doubles, L. and Rebec, G.V. (1998) Cocaine-induced activation of striatal neurons during focused stereotypy in rats. *Brain Research* **810**, 146–152.

White, I.M., Flory, G.S., Hooper, K.C., Speciale, J., Banks, D.A. and Rebec, G.V. (1995) Phencyclidine-induced excitation of striatal neurons in behaving rats: reversal by haloperidol and clozapine. *Journal of Neural Transmission* **102**, 99–112.

Wickens, J.R., Begg, A.J. and Arbuthnott, G.W. (1996) Dopamine reverses the depression of rat corticostriatal synapses which normally follows high-frequency stimulation of cortex in vitro. *Neuroscience* **70**, 1–5.

Williams, G.V. and Millar, J. (1990) Concentration-dependent actions of stimulated dopamine release on neuronal activity in rat striatum. *Neuroscience* **39**, 1–16.

Wilson C.J. (1993) The generation of natural firing patterns in neostriatal neurons. In: *Chemical Signalling in the Basal Ganglia, (Progress in Brain Research, Vol. 99),* G. W. Arbuthnott and P. C. Emson (eds), Amsterdam: Elsevier, pp. 277–297.

Wilson, C.J. and Groves P.M. (1981) Spontaneous firing patterns of identified spiny neurons in the rat neostriatum. *Brain Research* **220**, 67–80.

# 4 Dopamine Regulation of Synaptic Plasticity in the Neostriatum: A Cellular Model of Reinforcement

**Jeffery R. Wickens**

*Department of Anatomy and Structural Biology, School of Medical Sciences*
*University of Otago, P.O. Box 913, Dunedin, New Zealand.*
*tel: (0064) 3 479-7373; FAX: (0064) 3 479-7254;*
*e-mail: jeff.wickens@stonebow.otago.ac.nz*

Synaptic plasticity is a possible mechanism for learning and memory functions of the neostriatum. A three-factor rule for synaptic plasticity in the corticostriatal pathway has been proposed, in which dopamine inputs selectively strengthen synapses at which a conjunction of presynaptic and postsynaptic activity has occurred. Experimental evidence concerning the role of dopamine in synaptic plasticity is reviewed. It is argued that a conjunction of all three factors produces long-term potentiation. Preliminary evidence for a strict temporal coincidence of these factors is presented. This proposal implies that the timing of dopamine inputs required for induction of long-term potentiation is such that the dopamine cells must fire in advance of behavioural consequences of corticostriatal output activity. It is argued that this is compatible with recent evidence concerning the timing of the dopamine signal during learning.

KEYWORDS: Dopamine, Long-term potentiation, Long-term depression, corticostriatal pathway.

## 1. INTRODUCTION

The idea that learning and memory mechanisms of the brain involve activity-dependent synaptic plasticity gained widespread acceptance after the discovery of long-term potentiation (LTP) in the hippocampus (Bliss and Lomo, 1973), and its subsequent experimental analysis in the hippocampus and other brain regions (Bear and Kirkwood, 1993; Castro-Alamancos, Donoghue and Connors, 1995; Kirkwood *et al.*, 1993; Smith 1987). The induction of hippocampal LTP requires repeated conjunctions of presynaptic and postsynaptic activity. This requirement for both presynaptic and postsynaptic activity can be termed a "two-factor rule" for synaptic modification.

Synaptic modification according to a two-factor rule is a possible mechanism for the formation of neural assemblies on the basis of correlated firing of neurones representing external sensory events (Braitenberg, 1978; Hebb, 1949). By forming such assemblies a coherent representation of the external world is built from the fragments of information impinging on the senses (Palm, 1982). A similar mechanism may be involved in the formation of motor programmes by establishing strengthened synaptic linkages among groups of neurones representing combinations or sequences of motor actions (Wickens, Hyland and Anson, 1994). The wide range of potential uses for a two-factor rule for synaptic modification do not, however, exclude other possible rules for synaptic modification.

More complicated rules for synaptic modification could in principle be implemented using combinations of neurones connected by synapses individually governed by two-factor rules (Brindley, 1967). There are, however, certain operations for which a three-factor rule has advantages. The following sections discuss these advantages in relation to reward-related learning. A proposal that reward-related learning involves synaptic modification in the neostriatum, and that this operates according to a three-factor rule (Miller, 1981), is considered. Evidence that dopamine mediates strengthening of synaptic connections in the striatum is reviewed, focusing on long-term changes in synaptic efficacy brought about by interactions among the three factors represented by presynaptic corticostriatal inputs, postsynaptic striatal output neurones, and dopamine inputs from the midbrain. A refinement of the proposed rule is suggested on the basis of experimentally determined requirements for synaptic plasticity in the neostriatum.

## 2. NEURAL MECHANISMS OF REWARD-RELATED LEARNING

Reward, more properly referred to as positive reinforcement, is a key factor in the acquisition of new behaviours. The formal psychological concept of positive reinforcement learning probably originated with Thorndike (1911), who wrote: "Any act which in a given situation produces satisfaction becomes associated with that situation so that when the situation recurs the act is more likely than before to recur also." The effect of positive reinforcement (satisfaction) is to strengthen the association between the situation and the act. This formulation implicitly requires a mechanism capable of integrating three factors: situation, action and reinforcement.

The neural systems mediating reinforcement are non-specific systems that project diffusely within the brain (Crow, 1968; Donahoe and Palmer, 1988). A reinforcement signal widely propagating in the brain can be integrated into a three-factor rule for synaptic modification by plausible biological mechanisms. A precedent for such a mechanism is given by heterosynaptic plasticity mechanisms in invertebrates. In heterosynaptic plasticity, the effectiveness of synaptic transmission in one pathway is modified by activity in another pathway. For example, in *Aplysia* synapses between sensory and motor neurones can be strengthened by their being active in association with serotonin released, in response to noxious stimuli, from a different set of neurones (Kandel and Tauc, 1965). In principle, such activity-dependent heterosynaptic plasticity has the potential to combine three factors (presynaptic and postsynaptic activity, and neurochemical reinforcement).

Synaptic plasticity models for reward-related learning in the mammalian brain have developed along similar lines. The facilitation of simple reflexes described in invertebrates is obviously different from the reinforcement learning which occurs in mammals. In the mammalian brain, the behavioural effects of reinforcement cannot be reduced to changes in the strength of connections between sensory and motor neurones. It is necessary to understand the effects of heterosynaptic plasticity in the context of the neural circuits within which the synapses are embedded.

A number of authors have proposed that different types of learning are dealt with by anatomically separable parts of the brain, which employ different rules for synaptic modification (Hirsh, 1974; Miller, 1988; Mishkin, Malamut and Bachevalier, 1984). In particular, Miller (1981) argued that the neostriatum is a substrate for the strengthening of connections by the action of a three-factor rule for synaptic modification. The third factor in this formulation is dopamine, which is proposed to exert a heterosynaptic effect on recently active synapses (Beninger, 1983; Miller, 1988; Miller and Wickens, 1991; Wickens, 1992; Wickens and Kotter, 1995). Dopamine was proposed to be a neural mediator of reward, acting diffusely over the striatum, and producing strengthening at specific synapses selected on the basis of recent presynaptic and postsynaptic activity. The basic outline of the scheme is shown in Figure 4.1. According to this model the dopaminergic input to the striatum conveys a reward signal, as discussed in Chapters 1 and 2 of this book. When a sequence of activity in the corticostriatal system has led to a reward, the release of a pulse of dopamine in the striatum is supposed to strengthen the synapses that were active immediately before reward was obtained. The specific rules for synaptic modification are considered below.

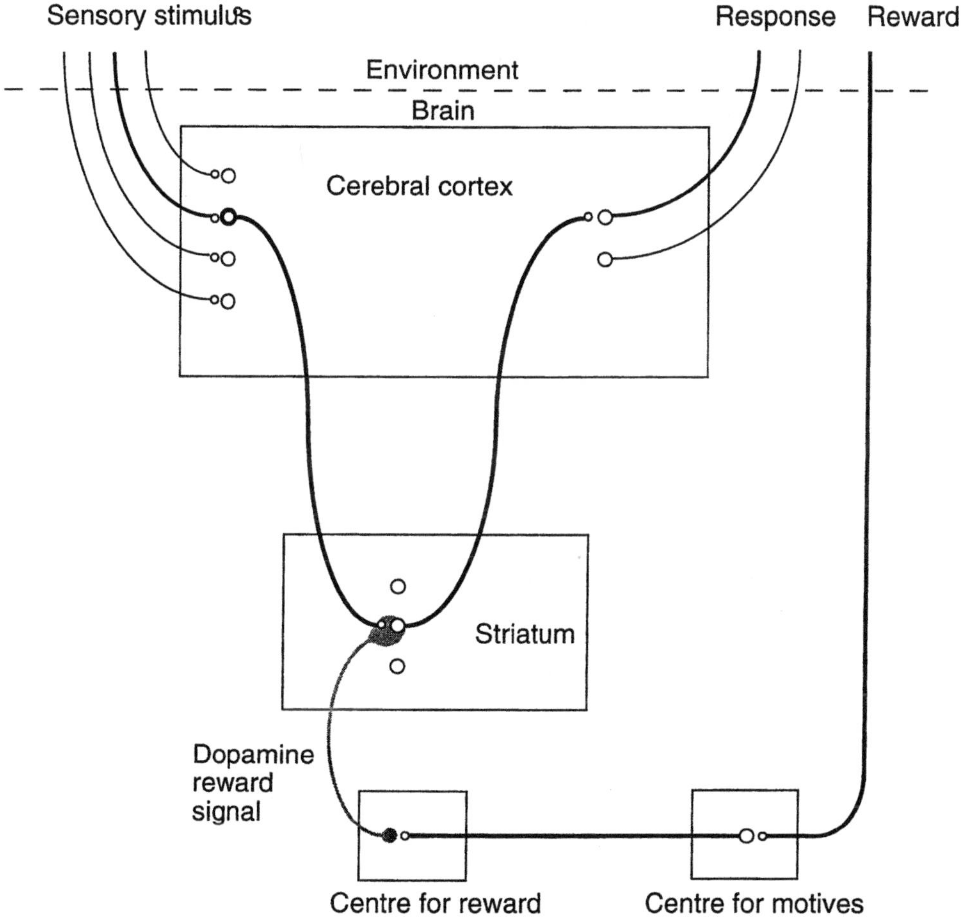


**Figure 4.1.** Scheme of striatal involvement in reward-related learning.

## 3. THREE FACTOR RULE FOR SYNAPTIC MODIFICATION

The three-factor rule for synaptic modification originally proposed for the corticostriatal synapses is as follows:

1. A conjunction of presynaptic activity in corticostriatal afferents and postsynaptic activity in striatal projection neurones puts the synapses in question into a "state of readiness" for synaptic modification.
2. Activation of a reward signal within the duration of the "state of readiness" leads to strengthening of those synapses.
3. Repeated conjunctions of presynaptic and postsynaptic activations in the absence of the reward signal leads to weakening of synaptic connections.
4. Synapses that are not in a state of readiness do not change.

There are a number of similarities between these rules and a set of functions proposed quite independently to solve difficult control problems in the context of machine learning (Barto, Sutton and Anderson, 1983; Barto, Sutton and Brouwer, 1981). Related ideas have been proposed by Donahoe (1988) from more psychological considerations. In recent years these different lines have converged to the extent that more advanced reinforcement learning algorithms are now considered in relation to basal ganglia function (Barto, 1994; Wickens and Kotter, 1995). These concepts have considerable explanatory appeal, but direct evidence to support them has been limited. The existing evidence will now be considered.

A number of indirect arguments can be made for the proposed three-factors rule. The neostriatum is a site of convergence of dopamine and glutamate projections. The glutamate afferents to the neostriatum from the cerebral cortex (McGeorge and Faull, 1989) and the dopaminergic afferents from the midbrain terminate in close proximity to each other on individual striatal neurones, frequently converging on individual dendritic spines (Smith *et al.,* 1994). The close juxtaposition of three processes of cortical, striatal and nigral neurones makes it possible for interactions to occur between them. Intracellular signalling pathways in the dendritic spines of the postsynaptic spiny projection neurones are likely to mediate such interactions, with the involvement of dopamine and glutamate receptors and postsynaptic calcium ions (Kotter, 1994). At the molecular level dopamine and glutamate produce co-operative effects on gene expression in a subset of striatal neurones (Berretta, Robertson and Graybiel, 1992), suggesting that they may affect neuronal activity over extended periods of time. These considerations suggest that the machinery necessary to implement a three-factor rule for synaptic modification exists in the neostriatum.

In order to test the hypothesis that the proposed three-factor rule for synaptic modification actually operates on the corticostriatal synapses, each of the three factors needs to be under the direct control of the experimenter. The development of a corticostriatal slice preparation (Arbuthnott, MacLeod and Rutherford, 1985; Kawaguchi, Wilson and Emson, 1989) enabled intracellular recordings to be made from striatal neurones during repeated test stimulation of the cerebral cortex over relatively long periods of time. In this situation the effect of presynaptic activity and postsynaptic activity can be directly controlled. Control over dopamine receptor activation by bath application, or by pressure ejection of dopamine agonists and antagonists, in conjunction with presynaptic and postsynaptic control, makes it possible to test the hypothesis directly. The evidence obtained from these experiments is reviewed below. However, there is also a body of evidence derived from experiments which were not designed to test this particular hypothesis. The results from these experiments are described first, before considering more direct tests.

## 4. EFFECTS OF DOPAMINE ON STRIATAL PROJECTION NEURONES

Dopamine agonists appear to produce an overall increase in neural output from the striatum, but at the single neurone level the effects are non-uniform with a mixture of increases and decreases. Striatal output activity measured using regional cerebral glucose utilisation (a measure of the overall output activity) is increased by dopamine agonist drugs (Sirinathsinghji *et al.*, 1988; Trugman and Wooten, 1986; Wooten and Collins, 1983). Cells in the substantia nigra *pars reticulata* which receive inhibitory input from striatal output neurones display highly variable responses to the systemic administration of the dopamine agonist apomorphine: After a dose sufficient to stimulate postsynaptic dopamine receptors, many cells exhibit increased firing rates, some cells are markedly inhibited, and a large group show only modest or fluctuating changes in rate (Waszczak *et al.*, 1984). The variability is reduced by lesions of the neostriatum, which is compatible with a nonuniform effect of dopamine on nigrostriatal neurones. Cell-to-cell variability in the effects of dopamine agonists on functional output activity may be due to an activity-dependent action of dopamine. West *et al.*, (1986) and Haracz *et al.*, (1993) found that amphetamine increased the activity of striatal units that were active in association with movements on the part of the animal, but inhibited units with little or no associated activity. (See Chapter 3 by G. Rebec; this volume).

Since the action potential firing of striatal output neurones is largely due to excitatory input from the cerebral cortex, the changes in firing rates brought about by dopamine are probably due to changes in the effectiveness of the corticostriatal inputs. Consistent with this, Warenycia (1987) found that systemic administration of amphetamine produced excitation in control animals, but inhibition in animals subjected to bilateral removal of the fronto-parietal cortex. The authors concluded that the excitatory response depended on an intact cerebral cortex, and required intact corticostriatal afferents.

Overall, the evidence may indicate that dopamine increases output from the most active cells (which are in the minority) and decreases output from the less active cells. This is compatible with the operation of the proposed three-factor rule on excitatory corticostriatal inputs. However, only one of the three factors (dopamine) was been under direct control in the above experiments; the presynaptic input and the postsynaptic output occurred "spontaneously". Thus, the contribution of these two factors to the results is not known.

Experimental control over presynaptic activity can be achieved, in part, by implanting a stimulating electrode in the cortex. Hirata, Yim and Mogenson (1984) measured the effect of substantia nigra stimulation on extracellular single unit responses of striatal neurones to cortical stimulation. In these experiments they applied a train of 10 pulses at 10 Hz to the substantia nigra. They observed both increases and decreases in the responses to cortical stimulation applied 100 msec after the last train. Although they were not searching for long-term changes, they noted in the discussion section of their paper that some of the increases and decreases in responses to cortical stimulation persisted for several minutes after substantia nigra stimulation. Schneider *et al.*, (1984) investigated the effects of amphetamine on the responses of cat neostriatal neurones recorded intracellularly. Systemic administration of amphetamine resulted in long-lasting changes in the responses to afferent inputs (Schneider *et al.*, 1984), with excitatory responses showing increased amplitudes. Conversely, dopamine depletion reduced the responses of striatal neurones to peripheral sensory stimulation (Schneider, 1991).

## 5. EFFECTS OF DOPAMINE ON STRIATAL INTERNEURONES

In a series of studies by Kimura and collaborators, long-lasting changes in responses of tonically active striatal neurons (TANs) related to the acquisition and performance of learnt behaviour have been demonstrated in awake, behaving animals. In these experiments the proportion of TANs responding to sensory cues predicting reward increases in parallel with learning of behavioural responses. The higher proportion of responding TANs persists for as long as performance is maintained (Aosaki *et al.*, 1994). The acquisition of both behavioural and neuronal responses studied in these experiments is dependent on the nigrostriatal dopamine system (Aosaki, Graybiel and Kimura, 1994).

It is important to note that the TANs are probably cholinergic interneurones rather than spiny projection neurones (Kawaguchi *et al.*, 1995; Wilson, Chang and Kitai, 1990). Also, the long-lasting changes in responses of TANs involved the development of inhibitory responses expressed as transient reductions in firing rate. The afferent activity underlying the inhibitory responses is not known. The inhibitory responses are unlikely to be direct responses to dopaminergic inputs. Although cholinergic interneurones do receive some dopaminergic inputs (Kubota *et al.*, 1987) which probably have inhibitory effects on the cholinergic interneurones, the changes in firing of dopamine neurones associated with learning have a different time course from the TANs (Kimura, 1995).

The increase in inhibitory responses in TANs could be secondary to increased responses in spiny projection neurones. The spiny projection neurones show a phasic pattern of firing in relation to sensory stimuli which has an acquired association with movement (Kimura, 1986, 1990, 1992; Kimura *et al.*, 1990). Synaptic inputs to cholinergic interneurones from the collaterals of spiny projection neurones are probable, because substance P synapses onto cholinergic interneurones are frequent (Martone *et al.*, 1992). An increase in the responses of the spiny projection neurones might lead to a suppression of firing in TANs receiving potentially inhibitory collaterals of the spiny projection neurones. Alternatively, facilitation of inhibitory inputs from inhibitory interneurones might occur in response to dopamine (Yan and Surmeier, 1997).

## 6. SYNAPTIC PLASTICITY IN THE CORTICOSTRIATAL PATHWAY

Synaptic plasticity in the corticostriatal pathway is a possible basis for the long-lasting changes in neuronal responses in the striatum. Experimental study of synaptic plasticity in the striatum has advanced rapidly over recent years. Both long-term potentiation (LTP) and long-term depression (LTD) of synaptic responses have been described in the striatum.

Long-term depression can be induced in the synapses connecting the cerebral cortex to the striatum by high-frequency stimulation of the cerebral cortex (Calabresi *et al.*,

1992a,b,c; Calabresi *et al.*, 1994; Lovinger, Tyler and Marritt, 1993; Walsh, 1993). Striatal LTD is a depolarisation-dependent process that requires activation of voltage-sensitive calcium channels in the postsynaptic cell during the conditioning tetanus (Calabresi *et al.*, 1992b; Calabresi *et al.*, 1994). This finding is compatible with the third part of the synaptic modification rule proposed above: Repeated conjunctions of pre-synaptic and postsynaptic activations in the absence of the reward signal leads to weakening of synaptic connections. Activation of glutamate metabotropic receptors is also a requirement for induction of LTD (Calabresi *et al.*, 1992b) which supports the proposed requirement for presynaptic activity in conjunction with postsynaptic activity. However, another requirement complicates this interpretation.

Activation of dopamine receptors is also a requirement for LTD induction (Calabresi *et al.*, 1992a). LTD cannot be induced in slices prepared from dopamine-depleted animals, but can be restored by bath application of exogenous dopamine, or coapplication of both D-1 and D-2 dopamine receptor agonists. LTD can be prevented from occurring in normal slices by pretreatment with either D-1 or D-2 antagonists (Calabresi *et al.*, 1992a,b). These results seem to imply that part 3 of the synaptic modification rule requires revision. However, the residual dopamine level in slices is apparently enough to support LTD. Thus, the tonic level of firing activity of the dopamine cells that would normally occur *in vivo* would be sufficient to support LTD. The phasic firing activity that occurs in association with reward (see Chapter 1 by Hyland; this volume) does not appear to be necessary.

It should be noted that the kind of stimulation which induces LTD in the neostriatum would be expected to induce LTP in the hippocampus. It is possible that the bias toward LTD in the neostriatum reflects cellular properties of striatal neurones. Striatal neurones display very low levels of excitability under normal conditions because they express a high density of $K^+$ channels. These $K^+$ channels oppose $Ca^{2+}$ channel activation (Bargas, Galarraga and Aceves, 1988, 1989; Calabresi, Misgeld and Dodt, 1987; Kita, Kita and Kitai, 1985) thus limiting $Ca^{2+}$ influx produced by HFS. The lower levels of $Ca^{2+}$ brought about by plasticity-inducing stimulation might favour LTD over LTP. Trains of stimuli that would normally induce LTD lead to LTP in slices exposed to the $K^+$ channel blocker, TEA (Walsh, 1991).

The finding that LTP can be induced in the presence of extracellular $K^+$ channel blockers might indicate that striatal LTP could follow a two-factor rule, but the level of postsynaptic activation required is higher than that normally achieved by plasticity-inducing stimulation. However, LTP is not facilitated in striatal cells loaded with intra-cellular $K^+$ channel blockers (Wickens *et al.*, 1998). These results suggest that the facilitating effect of extracellular TEA on the induction of LTP is not due to blockade of $K^+$ channels in the postsynaptic neurone. An alternative possibility is that extracellular TEA facilitates LTP by presynaptic effects on dopamine terminals. In squid nerve terminals TEA causes prolonged action potentials and increased release of neurotransmitter by presynaptic mechanisms (Kusano, Livengood and Werman, 1967). Extracellular TEA increases the amounts of dopamine released in response to stimulation in the striatum (Boireau *et al.*, 1991). Thus, the experimental manipulations that facilitate the induction of LTP in the striatum may act by increasing the phasic release of dopamine that occurs in association with HFS. This possibility is compatible with the proposed three-factor rule, but experiments with dopamine antagonists are needed to test it.

Similarly, HFS results in LTP in slices perfused with magnesium-free solution (Calabresi *et al.*, 1992c; Walsh, 1991). This form of LTP is blocked by antagonists of the

N-methyl-D-aspartate (NMDA) receptors. This finding again suggest that the bias towards LTD in the neostriatum might reflect the limited influx of $Ca^{2+}$ brought about by HFS, and that treatments which increase $Ca^{2+}$ influx favour LTP. However, LTP in magnesium-free fluid is blocked by the dopamine D-1 receptor antagonist SCH23390 (Kerr and Wickens, 1996; 1998). These findings suggest that NMDA receptor activation is not sufficient for induction of LTP in the neostriatum, and that dopamine receptor activation is also necessary.

A possible interpretation of the foregoing results is that both magnesium-free and TEA-containing solutions act by presynaptic actions on neurotransmitter release. A transient increase in dopamine release in the striatum has been measured after cortical HFS (Calabresi *et al.*, 1995). It is very likely that the release of dopamine from striatal slices is facilitated in magnesium-free or TEA-containing solutions, and that this in turn facilitates LTP.

We have recently tested the idea that phasic release of dopamine may facilitate LTP, by directly applying dopamine as a pulse, timed to coincide with HFS. In these experiments we employed a conditioning protocol in which presynaptic corticostriatal fibres are stimulated in conjunction with activation of the postsynaptic neostriatal neurone. In the control group this causes LTD of synaptic responses, as previously reported. When dopamine is applied in brief pulses coinciding with the pre- and postsynaptic conjunction of activity, LTP of responses is seen (Arbuthnott, Ingham and Wickens, 1998; Begg, 1993; Wickens, Begg and Arbuthnott, 1996). Thus, *pulsatile application* of dopamine acts like a switch determining whether HFS leads to LTD or LTP.

Further studies have aimed to analyse the effect on synaptic plasticity of the timing of dopamine pulses in relation to HFS. These experiments were conducted in superfused rat brain slices containing neocortex and neostriatum. Intracellular records were made from neostriatal cells using sharp microelectrodes filled with potassium acetate (2 M). Postsynaptic potentials were evoked by stimulating the neocortex. After a 20 min control period, HFS (6 trains of 20 pulses at 100 Hz) was applied to the cortex in conjunction with depolarisation of the postsynaptic cell by current injection. Dopamine was pressure-ejected in pulses (10–20 ms, 50 psi) from one barrel of a double-barrel micropipette positioned adjacent to the recorded neurone. Groups of neurones were compared, which had dopamine pulses applied at different times in relation to HFS onset. The extracellular dilution of dopamine was estimated before and after the experiment, using the depolarisation produced by a matched pulse of KCl from the second barrel of the pipette. Application of dopamine 200 or 500 ms before HFS, frequently resulted in long-term potentiation. Delay of dopamine application 200 or 500 ms after onset of HFS was associated with long-term depression of synaptic responses. Figure 4.2 shows examples of the results obtained using dopamine (10 $\mu$M inside the pipette).

These results suggest that dopamine modulation of synaptic plasticity produces increased synaptic efficacy if dopamine pulses reach the postsynaptic striatal neurones in advance of the presynaptic and postsynaptic conjunction of activity. The temporal requirements suggested are consistent with the advance activation of dopamine neurones by stimuli predicting reward. The firing of the dopamine neurones, while initially following the reinforcing events, may come to occur earlier in trials in which rewards have become predictable (Ljungberg, Apicella and Schultz, 1992).

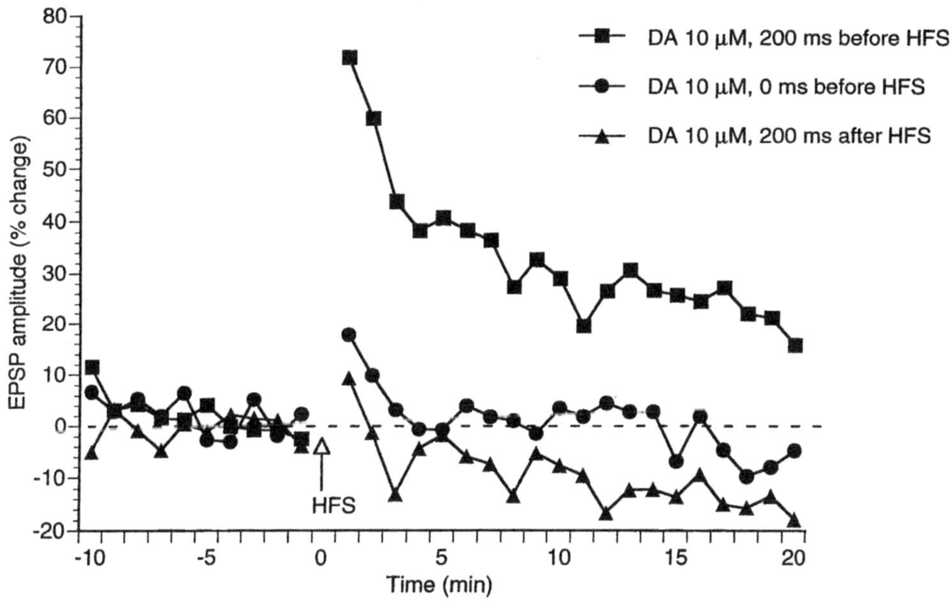


**Figure 4.2.** Temporal requirements for dopamine facilitation of LTP in the striatum. Graph shows changes in response peak amplitude relative to a 10 min control period. HFS (arrow) was applied to the cortex in conjunction with depolarisation of the postsynaptic cell by current injection. Data is from three different cells which had dopamine pulses applied at different times in relation to HFS onset. Unpublished data from Bushby and Wickens.

## 7. CONCLUSIONS

Evidence to date strongly supports the premise that a presynaptic and postsynaptic conjunction of activity brings about LTD of corticostriatal synaptic connections. While there is some evidence that dopamine is required for such LTD, the phasic activation of the dopamine neurones is not. Pulsatile application of dopamine is associated with LTP of corticostriatal connections. The temporal requirements for such an effect are strict. If the dopamine pulse is delayed, LTD is more likely to occur. This is compatible with the advance timing of the dopamine signal that is acquired during learning of new behaviours. Further careful parametric studies of the requirements for inducing synaptic plasticity in the striatum are needed to test all features of the three-factor rules for synaptic modification proposed.

## REFERENCES


Aosaki T., Graybiel A.M. and Kimura M. (1994) Effect of the nigrostriatal dopamine system on acquired neural responses in the striatum of behaving monkeys. *Science, New York* **265**, 412–415

Aosaki T., Tsubokawa H., Ishida A., Watanabe K., Graybiel A.M. and Kimura, M. (1994) Responses of tonically active neurons in the primate's striatum undergo systematic changes during behavioral sensorimotor conditioning. *Journal of Neuroscience* **14**, 3969–3984

Arbuthnott G.W., Ingham C.A. and Wickens J.R. (1998) Modulation by dopamine of rat corticostriatal input. *Advances in Pharmacology* **42**, 733–736

Arbuthnott G.W., MacLeod N, Rutherford A (1985) The rat cortico-striatal pathway *in vitro*. *Journal of Physiology (London)* **367**, 102P

Bargas J., Galarraga E., and Aceves J. (1988) Electrotonic properties of neostriatal neurons are modulated by extracellular potassium. *Experimental Brain Research* **71**, 390–398

Bargas J., Galarraga E. and Aceves J. (1989) An early outward conductance modulates the firing latency and frequency of neostriatal neurons of the rat brain. *Experimental Brain Research* **75**, 146–156

Barto A.G. (1994) Adaptive critics and the basal ganglia. In *Models of information processing in the basal ganglia*, edited by J.C. Houk, J.L. Davis, D.G. Beiser, pp. 215–232, MIT Press, Cambridge.

Barto A.G., Sutton R.S. and Anderson C.W. (1983) Neuronlike elements that can solve difficult learning control problems. *IEEE Transactions on Systems Man and Cybernetics* **15**, 835–846

Barto A.G., Sutton R.S. and Brouwer P.S. (1981) Associative search network: A reinforcement learning associative memory. *Biological Cybernetics* **40**, 201–211

Bear M.F. and Kirkwood A. (1993) Neocortical long-term potentiation. *Current Opinion in Neurobiology* **3**, 197–202

Begg A.J. (1993) *A long lasting effect of dopamine on synaptic transmission in the corticostriatal pathway* in vitro. B. Med. Sci. Thesis, University of Otago.

Beninger R.J. (1983) The role of dopamine in locomotor activity and learning. *Brain Research* **287**, 173–196

Berretta S., Roberston H.A. and Graybiel A.M. (1992) Dopamine and glutamate agonists stimulate neuron-specific expression of Fos-like protein in the striatum. *Journal of Neurophysiology* **68**, 767–777

Bliss T.V.P. and Lomo T. (1973) Long-lasting potentiation of synaptic transmission in the dentate area of the unanaesthetized rabbit following stimulation of the perforant path. *Journal of Physiology (London)* **232**, 331–356.

Boireau A., Richard F., Olivier V., Aubeneau M., Miquet J.M., Dubedat P., Laduron P., Doble A. and Blanchard J.C. (1991) Differential effects of potassium channel blockers on dopamine release from rat striatal slices. *Journal of Pharmacy and Pharmacology* **43**, 798–801

Braitenberg V. (1978) Cell assemblies in the cerebral cortex. In: R. Heim, G. Palm (eds), *Theoretical approaches to complex systems* Springer, Berlin. pp. 171–188

Brindley G.S. (1967) The classification of modifiable synapses and their use in models for conditioning. *Proceedings of the Royal Society, London [Biology]* **168**, 361–376

Calabresi P., Fedele E., Pisani A., Fontana G., Mercuri N.B., Bernardi G. and Raiteri M. (1995) Transmitter release associated with long-term synaptic depression in rat corticostriatal slices. *European Journal of Neuroscience* **7**, 1889–1894

Calabresi P., Maj R., Mercuri N.B. and Bernardi G. (1992a) Coactivation of D1 and D2 dopamine receptors is required for long-term synaptic depression in the striatum. *Neuroscience Letters* **142**, 95–9

Calabresi P., Maj R., Pisani A., Mercuri N.B. and Bernardi G. (1992b) Long-term synaptic depression in the striatum: physiological and pharmacological characterization. *Journal of Neuroscience* **12**, 4224–33

Calabresi P., Misgeld U. and Dodt H.U. (1987) Intrinsic membrane properties of neostriatal neurons can account for their low level of spontaneous activity. *Neuroscience* **20**, 293–303

Calabresi P., Pisani A., Mercuri N.B. and Bernardi G. (1992c) Long-term potentiation in the striatum is unmasked by removing the voltage-dependent magnesium block of NMDA receptor channels. *European Journal of Neuroscience* **4**, 929–935

Calabresi P., Pisani A., Mercuri N.B. and Bernardi G. (1994) Post-receptor mechanisms underlying striatal long-term depression. *Journal of Neuroscience* **14**, 4871–81

Castro-Alamancos M.A., Donoghue J.P. and Connors B.W. (1995) Different forms of synaptic plasticity in somatosensory and motor areas of the neocortex. *Journal of Neuroscience* **15**, 5324–33

Crow T.J. (1968) Cortical synapses and reinforcement, a hypothesis. *Nature, London* **219**, 736–737

Donahoe J.W. and Palmer D.C. (1988) The interpretation of complex human behavior: Some reactions to Parallel Distributed Processing. J.L. McClelland, D.E. Rumlhart and the PDP Research Group. *Journal of the Experimental Analysis of Behavior* **51**, 399–416

Haracz J.L., Tschanz J.T., Wang Z., White I.M. and Rebec G.V. (1993) Striatal single-unit responses to amphetamine and neuroleptics in freely moving animals. *Neurosciences and Biobehavioral Reviews* **17**, 1–12

Hebb D.O. (1949) *The organization of behaviour: A neuropsychological theory*. Wiley, New York

Hirata K., Yim C.Y. and Mogenson G.J. (1984) Excitatory input from sensory motor cortex to neostriatum and its modification by conditioning stimulation of the substantia nigra. *Brain Research* **321**, 1–8

Hirsh R. (1974) The hippocampus and contextual retrieval of information from memory: a theory. *Behavioral Biology* **12**, 421–444

Kandel E.R. and Tauc L. (1965) Mechanisms of heterosynaptic facilitation in the giant cell of the abdominal ganglion of *Aplysia delipans*. *Journal of Physiology (London)* **181**, 28–47

Kawaguchi Y., Wilson C.J., Augood S.J. and Emson P.C. (1995) Striatal interneurones: chemical, physiological and morphological characterization. *Trends in Neuroscience* **18**, 527–35

Kawaguchi Y., Wilson C.J. and Emson P.C. (1989) Intracellular recording of identified neostriatal patch and matrix spiny cells in a slice preparation preserving cortical inputs. *Journal of Neurophysiology* **62**, 1052–68

Kerr J.N.D. and Wickens J.R. (1996) Dopamine receptor activation is necessary for induction of neostriatal long-term potentiation in low magnesium solutions. *Proceedings of the Otago Medical Research Society* **74**, 22

Kerr J.N.D. and Wickens J.R. (1998) Dopamine D–1 but not D–2 receptor activation is required for magnesium-free LTP in the rat neostriatum, in vitro. *Submitted*

Kimura M. (1986) The role of primate putamen neurons in the association of sensory stimuli with movement. *Neuroscience Research* **3**, 436–443

Kimura M. (1990) Behaviorally contingent property of movement-related activity of the primate putamen. *Journal of Neurophysiology* **63**, 1277–1296

Kimura M. (1992) Behavioral modulation of sensory responses of primate putamen neurons. *Brain Research* **578**, 204–214

Kimura M. (1995) Role of basal ganglia in behavioral learning. *Neuroscience Research* **22**, 353–358

Kimura M., Kato M. and Shimazaki H. (1990) Physiological properties of projection neurons in the monkey striatum to the globus pallidus. *Experimental Brain Research* **82**, 672–6

Kirkwood A., Dudek S.M., Gold J.T., Aizenman C.D. and Bear M.F. (1993) Common forms of synaptic plasticity in the hippocampus and neocortex in vitro. *Science, New York* **260**, 1518–21

Kita H., Kita T. and Kitai S.T. (1985) Regenerative potentials in rat neostriatal neurons in an in vitro slice preparation. *Experimental Brain Research* **60**, 63–70

Kötter R. (1994) Postsynaptic integration of glutamatergic and dopaminergic signals in the striatum. *Progress in Neurobiology* **44**, 163–196

Kubota Y., Inagaki S., Shimada S., Kito S., Eckenstein F. and Tohyama M. (1987) Neostriatal cholinergic neurons receive direct synaptic inputs from dopaminergic axons. *Brain Research* **413**, 179–184

Kusano K., Livengood D.R. and Werman R. (1967) Tetraethylammonium ions: effect of presynaptic injection on synaptic transmission. *Science, New York* **155**, 1257–9

Ljungberg T., Apicella P. and Schultz W. (1992) Responses of monkey dopamine neurons during learning of behavioral reactions. *Journal of Neurophysiology* **67**, 145–163

Lovinger D.M., Tyler E.C. and Marritt A. (1993) Short- and long-term depression in the rat neostriatum. *Journal of Neurophysiology* **70**, 1937–1949

Martone M.E., Armstrong D.M., Young S.J. and Groves P.M. (1992) Ultrastructural examination of enkephalin and substance P input to cholinergic neurons within the rat neostriatum. *Brain Research* **594**, 253–262

McGeorge A.J. and Faull R.L. (1989) The organisation of the projections from the cerebral cortex to the striatum in the rat. *Neuroscience* **29**, 503–537

Miller R. (1981) *Meaning and purpose in the intact brain*. Oxford University Press, Oxford

Miller R, (1988) Cortico-striatal and cortico-limbic circuits: A two tiered model of learning and memory function. In: H. Markowitsch (ed), *Information Processing by the Brain: Views and Hypotheses from a Cognitive-Physiological Perspective*. Hans Huber Press, Bern. pp. 179–198

Miller R., Wickens J.R. (1991) Corticostriatal cell assemblies in selective attention and in representation of predictable and controllable events. *Concepts in Neuroscience* **2**, 65–95

Mishkin M., Malamut B. and Bachevalier J. (1984) Memories and habits: Two neural systems. In: G, Lynch, J.L. McGaugh, and N.M. Weinberger, (eds) *The Neurobiology of Learning and Memory* The Guilford Press, New York, pp. 65–77.

Palm G. (1982) *Neural Assemblies*. Springer, Berlin, Heidelberg and New York

Schneider J.S. (1991) Responses of striatal neurons to peripheral sensory stimulation in symptomatic MPTP-exposed cats. *Brain Research* **544**, 297–302

Schneider J.S., Levine M.S., Hull C.D. and Buchwald N.A. (1984) Effects of amphetamine on intracellular responses of caudate neurones in the cat. *Journal of Neuroscience* **4**, 930–937

Sirinathsinghji D.J.S., Dunnett S.B., Isacson O., Clarke D.J., Kendrick K. and Bjorklund A. (1988) Striatal grafts in rats with unilateral neostriatal lesions. II In vivo monitoring of GABA release in globus pallidus and substantia nigra. *Neuroscience* **24**, 803–811

Smith S.J. (1987) Progress on LTP at hippocampal synapses: A post-synaptic $Ca^{2+}$ trigger for memory storage? *Trends in Neuroscience* **10**, 142–144

Smith Y., Bennett B.D., Bolam J.P., Parent A. and Sadikot A.F. (1994) Synaptic relationships between dopaminergic afferents and cortical or thalamic input in the sensorimotor territory of the striatum in monkey. *Journal of Comparative Neurology* **344**, 1–19

Thorndike E.L. (1911) *Animal Intelligence*. Macmillan, New York

Trugman J.M. and Wooten G.F. (1986) The effects of L-DOPA on regional cerebral glucose utilization in rats with unilateral lesions of the substantia nigra. *Brain Research* **379**, 264–274

Walsh J.P. (1991) Long-term potentiation (LTP) of the excitatory synaptic input to medium spiny neurons of the rat striatum. *Society for Neuroscience Abstracts* **17**, 852

Walsh J.P. (1993) Depression of excitatory synaptic input in rat striatal neurons. *Brain Research* **608**, 123–128

Warenycia M.W., McKenzie G.M., Murphy M. and Szerb J.C. (1987) The effects of cortical ablation on multiple unit activity in the striatum following dexamphetamine. *Neuropharmacology* **26**, 1107–1114

Waszczak B.L., Lee E.K., Ferraro T., Hare T.A. and Walters J.R. (1984) Single unit responses of substantia nigra pars reticulata neurons to apomorphine: Effects of striatal lesions and anesthesia. *Brain Research* **306**, 307–318

West M.O., Micheal A.J., Knowles S.F., Chapin J.K. and Woodward D.J. (1986) Striatal unit activity and the linkage between sensory and motor events. In: J.S. Schneider, T.I. Lidsky (eds), *Basal Ganglia and Behaviour: Sensory aspects of motor functioning,* Hans Huber, Stuttgart. pp. 27–35

Wickens J.R. (1992) The contribution of the striatum to cortical function. In *Information Processing in the Cortex,* edited by A. Aertsen, V. Braitenberg, pp. 271–284, Berlin, Springer.

Wickens J.R., Begg A.J. and Arbuthnott G.W. (1996) Dopamine reverses the depression of rat cortico-striatal synapses which normally follows high frequency stimulation of cortex in vitro. *Neuroscience* ,**70**, 1–5

Wickens J.R., Hyland B. and Anson G. (1994) Cortical cell assemblies: A possible mechanism for motor programs. *Journal of Motor Behaviour* **26**, 66–82

Wickens J.R. and Kötter R. (1995) Cellular models of reinforcement. In *Models of Information Processing in the Basal Ganglia.*, edited by J.C. Houk, J.L. Davis, D.G. Beiser, pp. 187–214, Cambridge, M.I.T. Press.

Wickens J.R., McKenzie D., Constanzo E. and Arbuthnott G.W. (1998) Effects of potassium channel blockers on LTP in the corticostriatal pathway. *Neuropharmacology* **37**, 523–533

Wilson C.J., Chang H.T. and Kitai S.T. (1990) Firing patterns and synaptic potentials of identified giant aspiny interneurons in the rat neostriatum. *Journal of Neuroscience* **10**, 508–19

Wooten G.F. and Collins R.C. (1983) Effects of dopaminergic stimulation on functional brain metabolism in rats with substantia nigra lesions. *Brain Research* **263**, 267–275

Yan Z. and Surmeier D.J. (1997) D5 dopamine receptors enhance Zn2+-sensitive GABA(A) currents in striatal cholinergic interneurons through a PKA/PP1 cascade. *Neuron* **19**, 1115–1126

# 5 The Amygdaloid Complex: Input Processor for the Midbrain Dopaminergic Nuclei and the Striatum

**Robert Miller**

*Department of Anatomy and Structural Biology, School of Medical Science, University of Otago, P.O. Box 913, Dunedin, New Zealand*
*Tel: (0064) 3-479-7357; FAX (0064) 3-479-7254; e-mail: anat09@rivendell.otago.ac.nz*

This chapter explores the information processing which takes place in structures afferent to the midbrain dopaminergic cell groups. In particular it discusses the processing of motivationally significant cues which are "upstream" from the reward neuronal groups of the midbrain, as well as their relation to stimuli which can become associated with these cues. Psychological evidence suggests that Pavlovian association between motivational cues and related stimuli occurs in structures afferent to the reward system. In addition, these structures build up expectancies of the magnitude of a motivational cue, and a signal is derived (by a process akin to mathematical differentiation) representing the difference between expected and actual magnitude. This signal, (rather than the absolute magnitude of the motivational cue) is often used to control the reward system. Anatomical evidence is reviewed about afferents to the midbrain dopamine neurones. It is concluded that cell groups in the amygdala are a major source of input from centres of motivational processing to the midbrain dopaminergic nuclei. Studies of the effect of lesions of amygdala on various instrumental paradigms are reviewed. This data supports the view that the amygdala is a major input processor of motivational signals used by the dopaminergic nuclei. Specifically, the amygdala can form Pavlovian associations between neutral stimuli and motivational cues; and the amygdala also appears to be capable of computing the derivative of the magnitude of these cues, representing the difference between expected and actual magnitude. Electrophysiological evidence relating to these two aspects of information processing in the amygdala is also briefly reviewed. Further careful scrutiny of published evidence is required before the assumptions necessary for construction of a neuronal model of the amygdala can be defined.

KEYWORDS: instrumental conditioning; Pavlovian conditioning; amygdala; substantia nigra; reward; dopamine neurones; contrast effects; secondary reinforcers.

## 1. INTRODUCTION

In chapter 2 of this book, Beninger and Olmstead review behavioural experiments indicating that the dopaminergic input to the striatum conveys a reward or incentive signal, which either reinforces behaviour with motivationally favourable consequences, or makes approach responses more likely towards stimuli associated with fulfilment of motivational demands. In chapter 4, Wickens, presents data obtained from intracellular studies in cortico-striatal slice preparations, which suggest that such reinforcement seen at the behavioural level is mediated by dopamine-dependent synaptic change in cortico-striatal synapses. Chapter 1 by Hyland complements these accounts, by describing single unit physiological experiments which investigate the circumstances in which midbrain dopamine neurones fire in the free-moving animal: specifically, when a stimulus with incentive value occurs unpredictably, it generates in midbrain dopamine neurones a brief high-frequency burst of action potentials. Putting all these lines of evidence together one can build a scheme for reward- or incentive-mediated learning in which the midbrain dopamine neurones discharge a burst of impulses when they receive signals about motivationally favourable changes in the environment; activity in such neurones causes a brief pulsatile release of dopamine in the striatum; this strengthens synapses which have been active immediately preceding the motivationally favourable change; and this effect, in turn, is expressed as reinforcement of the behaviour which led to the motivationally favourable change (see Figure 5.1).

Other evidence, including some presented in chapters 1 and 2, suggests that this model is an oversimplification. In particular, there appears to be a variety of processes taking place in the input pathways to the dopamine neurones, which are important determinants

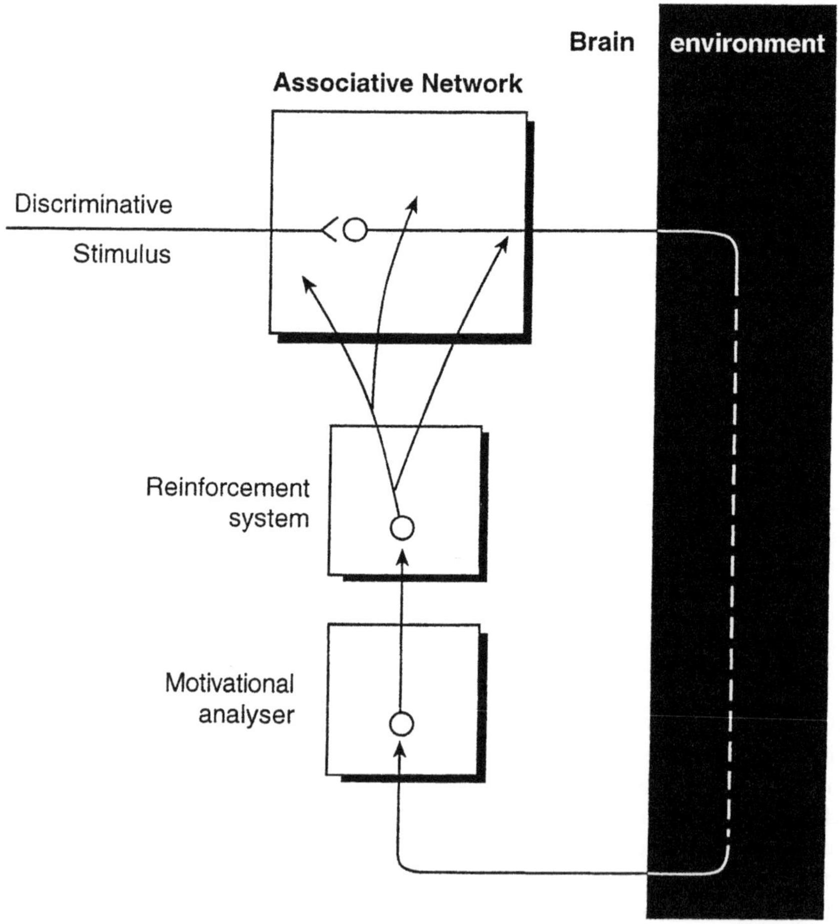


**Figure 5.1:** Block diagram of classical scheme for instrumental conditioning. Circles may represent single neurones (in which case rule-dependent synaptic modification may apply directly—see chapter 4 by Wickens), or they may represent larger groups of neurones.

of the relation between motivationally significant events and the reinforcement or incentive effects produced by dopamine neurones. The present chapter reviews such processes.

The chapter starts (section 2) with evidence which is entirely psychological, about the relation of reinforcement processes, both to stimuli which may activate these processes, and to the motivational context in which this activation may occur. These relationships turn out to be more complex than might be expected.

The sections which follow this develop the idea that a particular brain structure—the amygdaloid complex—is of special relevance for such psychological processes, modulating the input to the midbrain regions which generate the reinforcement signal. This thesis arises from several bodies of evidence. First, there is substantial evidence that certain nuclei of the amygdala and the hypothalamus contain neurones which project directly to the midbrain dopamine neurones. While there are other significant inputs to the latter neurones, the amygdala and hypothalamus appear to be the structures afferent to midbrain dopamine neurones which are most closely related to motivational processes. The amygdala itself can also provide input to the midbrain dopamine neurones indirectly via the hypothalamus. On connectional grounds therefore, the amygdala and hypothalamus, especially the former, are the most likely regions to mediate the various aspects of information processing identified in psychological experiments as taking place on the afferent side of the reinforcement/incentive signal. The anatomical evidence leading to this conclusion is reviewed in section 3.

To confirm the inference that the information processing identified in section 2 actually takes place in motivational centres such as the amygdala and hypothalamus, further bodies of information are relevant. In fact, these implicate the amygdala more than the hypothalamus. There are many studies which have investigated the effect of lesions in the amygdaloid complex both on instrumental learning itself, and on the variations of such learning described in section 2, from which inferences were drawn about information processing in the input pathways to the reinforcement/incentive system. These are reviewed in section 4.

Beyond this, single unit electrophysiological studies of the amygdala provide further evidence for the modulation of motivational signals, which may influence the reward/incentive and other related systems. This topic is reviewed in section 5. In fact, the combined conclusions of the psychological and lesion studies lead to important predictions for electrophysiological experiments, on which there is virtually no evidence currently available. Thus, one objective of section 5 is to define as precisely as possible such predictions, to be addressed in the future by electrophysiological studies.

The development of an actual neuronal model of the information processing achieved in the amygdala is some way in the future, but some of the requirements for such a model are discussed briefly in the concluding section 6.

## 2. PSYCHOLOGICAL EVIDENCE OF INFORMATION PROCESSING IN THE INPUT PATHWAYS TO THE REWARD/INCENTIVE SYSTEM

### 2.1. Rewarding Stimuli, Motivational States, and Pavlovian Association

Since the earliest studies of reinforcement-mediated learning, it has been clear that stimuli used as primary reinforcers can readily be replaced by secondary reinforcers, if these are presented in regular fashion as predictors of the so-called primary reinforcing stimulus. In principal, this process appears to be a special variety of Pavlovian conditioning. The electrophysiological studies of Schultz, reviewed by Hyland (chapter 1), which identify the midbrain dopamine neurones as ones which carry reward-related signals, confirm the relationship between primary and secondary rewarding stimuli, defined according to the protocol of Schultz's experiments: When the rewarding stimulus can be reliably predicted by a preceding cue (i.e. when there is a "secondary reinforcer"), the primary reward loses its ability to activate the dopamine neurones, but the activation is transferred to the earliest cue which permits prediction of the primary reward. This finding already has significant implications for information processing in the pathways afferent to the midbrain dopamine neurones: Processes must exist which not only give the secondary stimulus control of the dopamine neurones, but also inhibit the responsiveness of these neurones to the "primary" rewarding stimuli, when these become predictable.

Beyond this however, more fundamental questions arise about what might be the nature of the most fundamental primary rewarding stimuli. In typical appetitive conditioning experiments, the reinforcer is the immediate signal (such as a light or tone) indicating that food is available. However, such a signal is not really a primary reinforcer, because it is itself a predictor of subsequent signals, more directly related to fulfilment of motivational goal. These could include, first of all, the sight of food. However, this also is not primary, because it is a predictor of subsequent, more immediately-relevant stimuli, such as the taste and smell, indicating the quality of food which has actually been ingested. However, arguably, these too are not primary in that they also are predictors of goal fulfilment, rather than fulfilment itself. Ultimately one might conclude that the primary rewarding stimulus is an interoceptive and/or humoral one, such as the experience that blood sugar has been replenished in a hungry animal. But is such a humoral event a stimulus, as usually understood?

The chapter by Beninger and Olmstead (chapter 2) reviews recent psychological evidence which helps to clarify this issue. This evidence suggests that, if stimuli are defined as those which can be directly manipulated by the experimenter, *any* stimulus can acquire reinforcing properties, provided that it is presented in such a way that it can be used to predict favourable changes in the animal's motivational state. The fundamental and primary rewarding "stimulus" thus turns out not to be a stimulus, but a change of motivational state. The difference is that a motivational state is a relatively enduring internal condition (slow changes in which might be detected or generated by chemosensors or other mechanisms, actually within the brain) rather than a rapid phasic sensory pattern transmitted to the brain by identifiable sense organs in the periphery.

This viewpoint has been elaborated from experiments involving appetitive conditioning. Under the circumstances of most such experiments, the stimuli used are ones which have already become associated with favourable changes in motivational state. Of course, by appropriate conditioning procedures, the power to control reinforcement processes may be transferred from one such stimulus to another. Nevertheless, the various stimuli are really all equivalent instances of "secondary reinforcers". It requires more specialised experiments, such as those reviewed by Beninger and Olmstead (1999) to explicitly demonstrate the association of specific exteroceptive stimuli with changes in motivational states (as opposed to other stimuli).

A number of questions are left unanswered. It is not yet clear whether specific stimuli in the senses of taste or smell may directly control reinforcement processes, regardless of the postingestive changes produced by the relevant food items. There is, however, evidence that taste representation in the brain includes not only the quality of taste (as motivationally neutral information), but also, and in separate neurones, its hedonic value (Yamamoto *et al.*, 1989). It is also unresolved how far the viewpoint described above applies to other types of reinforcing stimulus: With regard to somatic rather than chemical stimuli, it is unclear whether painful stimuli need to be associated with a favourable change in some motivational state before that can act as reinforcers, rather than themselves being in direct control of the reinforcement systems.

Despite these uncertainties, with regard to the structure of psychological processes, one must conclude that two sorts of association can be made in the pathways afferent to the reward/incentive system: those between motivational states and correlated neutral stimuli, to give the latter motivational significance; and those between one stimulus which has, in this way already acquired motivational significance, and another neutral stimulus which predicts it. The clear implication is that both of these associative processes must occur in one of the brain structures which send axonal projections to the midbrain dopamine ("reward") neurones.

### 2.2. Modification of the Potency of Rewarding Stimuli by Failure of Expectancies

Apart from such basic associative processes, a line of psychological evidence built up over many years, indicates that there are other, more complex types of information processing which modulate the effectiveness of motivational signals in control of the reinforcement system. In particular, the original "law of effect" has required significant amendment to cope with a variety of evidence showing that the rewarding potency of a motivationally significant stimulus depends not only on the stimulus intensity *per se*, but also on *expectations* of what that intensity might be, based on earlier experience. Usually, failure of expectations shows up as a *contrast effect*: A given stimulus will have decreased potency as a reward if it turns out to be of lower intensity than expected, compared with a stimulus of the same intensity presented in circumstances where expectations are fulfilled; and a given stimulus will have increased potency if it exceeds expectations than if it is presented under circumstances where expectations are met. These two effects are nowadays referred to (respectively) as "negative" and "positive" contrast. Sometimes however expectancy can apparently have the opposite effect: Failure of an expected reward can enhance or energise behaviour, in excess of that predicted from the stimulus intensity in itself. This is generally referred to as a *frustration* effect.

Contrast effects have been shown in experiments with several different designs. The first demonstration of contrast effects was achieved by Crespi (1942) in experiments in which rats ran an alleyway for food on a series of trials, with runway speeds in successive trials as the dependent measure. Both positive and negative contrast effects were shown, when unexpected reward sizes were presented, following, respectively, upward or downward shifts in reward magnitude. Bower (1961) showed contrast effects in a different design, in which rats alternated between two easily distinguishable alleyways. One group found a large reward in one alley, and a small one in the other, while the control group received a small reward in either. The experimental group produced slower runway speeds than the control group for equivalent small rewards. A third design to show contrast effects involves anticipation. Flaherty and Checke (1982) utilised the fact that rats drinking each day from 0.15% saccharin-flavoured water gradually increase their water consumption. If however each daily exposure to the fluid is followed 5 minutes later by a much more attractive fluid reward—32% sucrose—the acquisition curve for the saccharin solution is retarded. In controls in which the second solution on each day was less attractive—2% sucrose—no such lowering of the acquisition curve took place.

In studies with each of the above designs, negative contrast has generally been a more reliable effect than positive contrast. The latter effect can be shown, but requires special procedures to avoid a "ceiling" of performance obscuring positive contrast when rewarding stimuli exceed expectations. Experiments with such designs are reviewed by Mackintosh (1974) dealing with literature up to 1973, and by Flaherty (1982) dealing with more recent literature.

A small number of papers shows that similar effects of failed expectancy can be obtained with aversive stimuli. Miller (1960) and Holmes and Brookshire (1968) showed that rats which had acquired an asymptotic running speed for food in an alleyway suppressed their responding much more severely if an intense shock in the goal box was introduced suddenly, than if the same intense shock was introduced gradually as a series of shocks increasing in intensity over a number of trials. Nation, Wrather and Mellgren (1974) trained three groups of rats to escape from shocks of 0.2, 0.4 or 0.8 mA. Then all three groups were shifted to a 0.4 mA aversive stimulus for a further 20 trials. Both positive and negative contrast effects were observed in the post-shift trials. In a second experiment the contrast group experienced either 0.2 or 0.4 mA aversive stimuli on alternate trials, and was compared with a control group receiving consistently one or other of these two values of aversive stimulus. In this design both positive and negative contrast effects were again shown (see also Nation, Mellgren and Wrather, 1975).

The "frustration" effect was first shown by Amsel and Roussel (1952). Rats ran for food in a double alleyway, in which there were two goal boxes, the first one of which also served as the start box for the second alley. Initially subjects (rats) were rewarded in both goal boxes. Subsequently, reward in the first goal box was given in a randomly intermittent manner, while that in the second was maintained at the initial 100% schedule. It was found that, on those trials in which reward was omitted in the first goal box, runway speed in the second component was increased compared with trials where reward was present in both goal boxes. The frustration effect has also been shown in lever-pressing tasks (Scull, Davies and Amsel, 1970; Staddon and Innis, 1969; Wookey and Strongman, 1971). Amsel's term "frustration" referred to a supposed motivating effect shown only with unexpected omission of reward, not with unexpected presentation of it. However, one experiment using a lever pressing task (Zimmerman, 1971) showed not only a frustration effect after unexpected omission of reward, but also the opposite—a decrease of responding after unscheduled presentations.

It is easy to envisage how expectations of a motivationally significant stimulus arise, as a result of Pavlovian-type associations between an initially neutral stimulus and a regularly-following one which already has its motivational significance. It is more difficult to understand why the efficacy of a reference stimulus acting as a reward should change when expectations fail to materialise (compared with when the same stimulus occurs in a situation which fulfils expectations). The variety of such effects is a major factor making their explanation more difficult. However, one way of describing contrast effects, is that in the input to the reinforcement system, there is an element of *differentiation* (in the mathematical sense) of the elementary motivational cues available to the animal. Thus, the signal controlling the reinforcement system responds to *change* of motivational cues over trials rather than to their absolute value. The "change score" could then have either positive or negative values. In support of this model, in some behavioural experiments (e.g. that of Salinas, Packard and McGaugh [1993] in which rats ran an alleyway for food) learning curves are identical for animals consistently given large or small rewards. Here changes across trials are zero in both groups, so there is no differential reinforcement between groups, despite differences in absolute magnitude of reward.

The model of "mathematical differentiation" of motivational value in the input to the reinforcement system may have broader relevance. In active avoidance, learning appears to be mediated to a large degree because a rewarding signal is generated when an expected aversive stimulus fails to materialise, even though there is no explicit reward. Conversely, in extinction, unlearning may occur because absence of a reward is computed as a punishment, even though there is no explicit punishment. Amsel's frustration effect may have a similar explanation. According to Adelman and Maatsch (1955) extinction is much slower if animals are allowed to escape from the environment in which non-reward occurs, than if they are confined in the empty goal box (see also McAllister *et al.*, 1972). Thus, escape from the situation of non-reward, such as occurs in the Amsel-type experiment, may be itself rewarding.

## 2.3. The Relationship Between Associations Made with Neutral Stimuli, and the Effects of Failed Expectation

It has already been made clear that neutral secondary stimuli may become linked with stimuli which already have motivational significance, or with motivational states themselves, by a process comparable with Pavlovian conditioning. Such a process must take place in the afferent pathways of the reward and punishment systems. Presumably this is the way in which expectations of reward are built up. Any effects of failed expectancy depend on the prior build up of expectancy, and so the above associations must occur before the envisaged differentiation process. Some of the available evidence clearly points to such a relationship. In the study of Amsel and Hancock (1957) using the double alley design of Amsel and Roussel, it was arranged that the first alley should have either similar or different floor characteristics to the first goal box. After 54 preliminary trials of continuous reinforcement in goal box one, the shift to 50% reinforcement was made. The frustration effect in alley two was much greater when alley-one cues were similar to goal box-one cues than when they were different. In other words, the more closely alley cues in alley one are associated with reward, the stronger are the expectancies which are built up and the more dramatic the effect when expectancies are disconfirmed. In a second experiment Amsel and Hancock carried out the same procedure as far as post-shift conditions were concerned, but omitted the pretraining trials. In this case the frustration effect did not appear immediately but grew progressively over the 36 daily sessions. In other words alternation of reward and non-reward is by itself insufficient to produce the frustration effect. Expectancies need to be built up before they can be disconfirmed, and this can only be by association of secondary stimuli with more primary motivational ones.

In the experiments of Miller (1960) and Holmes and Brookshire (1968) it was shown that appetitive responding became relatively invulnerable to punishing shock if the latter was introduced as a gradual build up of intensity over a number of trials. Miller (1960) found that this effect was seen only if the gradual build up occurred in the apparatus where the eventual combination of intense shock and reward was to be tested. Holmes and Brookshire

(1968) showed that repeated fixed level intense shocks had no effect in reducing subsequent vulnerability to punishment unless it is delivered in the test environment. Thus there is some agreement with the studies mentioned in the previous paragraph. When punishment is expected in combination with reward, the suppressant effects of punishment are less severe than when punishment occurs unexpectedly; and this effect of failed expectancy depends on similarity of secondary cues (present during establishment of the expectation) to those present during testing.

Other evidence tells a different story however, in which secondary stimuli associate not with the absolute level of a motivational cue, in order to build up the expectancy, but with the derivative of this level obtained by comparison of the contrasted values of the motivational cues. Such evidence implies that secondary stimuli, as well as being involved upstream from the site where the effects of failed expectancy are generated, can also link with the downstream consequences of this process. Shanab and co-workers have performed the most clear-cut demonstrations of this role for secondary stimuli, in a Crespi type experiment. Shanab, Domino and Steinhauer (1982) allowed two groups of rats access to an 18% sucrose reward, for either 2 or 20 sec. After the shift, all animals received the 2-sec. access. Half of each group received a tone stimulus in post-shift trials. There followed a two-month rest period, and animals were again tested under the same conditions as immediately after the shift. Strong negative contrast was retained only in the groups given the tone as a signal for reward shift. Shanab, Moyalem and Steinhauer (1982b) gave rats pre-shift access to water reward for 2 or 8 sec., and then all animals were shifted to the longer access time. As the shift was introduced, half of all groups were exposed to light and tone just as the reward was presented, half throughout the post-shift trials. Positive contrast was seen only when the CS was present, and more significantly so if it was present throughout the trial than just at the time of reward. These findings leave some uncertainty as to whether the stimulus really acted as a secondary reinforcer, or rather as a discriminative stimulus "setting the scene" for responding. Both roles for stimuli are real phenomena, though they are often indistinguishable. However, if the stimuli had a discriminative role, responding by the group with consistently large reward should also have been enhanced when the stimulus was introduced at the shift. That any such effect did not obscure the positive contrast effect indicates that the stimuli were linked with the derivatives of motivational cues as secondary reinforcers, rather than just with responding.

The frustration effect also provides some indication that secondary stimuli can link with the signal resulting from failed expectancy of a motivationally significant stimulus, as well as with such stimuli themselves. Assuming that the frustration effect is a consequence of escape from a situation of failed expectancy of an appetitive stimulus, that situation is defined by the association of secondary cues with the apparently aversive signal produced by failure of the expected appetitive stimuli.

In summary, the evidence indicates that the main role of sensory stimuli in expectancy effects is in forming associations with the motivational cues prior to their modification by differentiation or some similar process, later to be revealed when expectancy is disconfirmed. However, in addition, significant evidence indicates that signals can also link specifically with the modified version of motivational cues following the disconfirmation. Therefore, in instrumental learning there are three possible roles of sensory signals: the role as discriminative signals (which become part of stimulus-response links), as well as the two discussed here which are subdivisions of the role of signals as secondary reinforcers (see Figure 5.2).

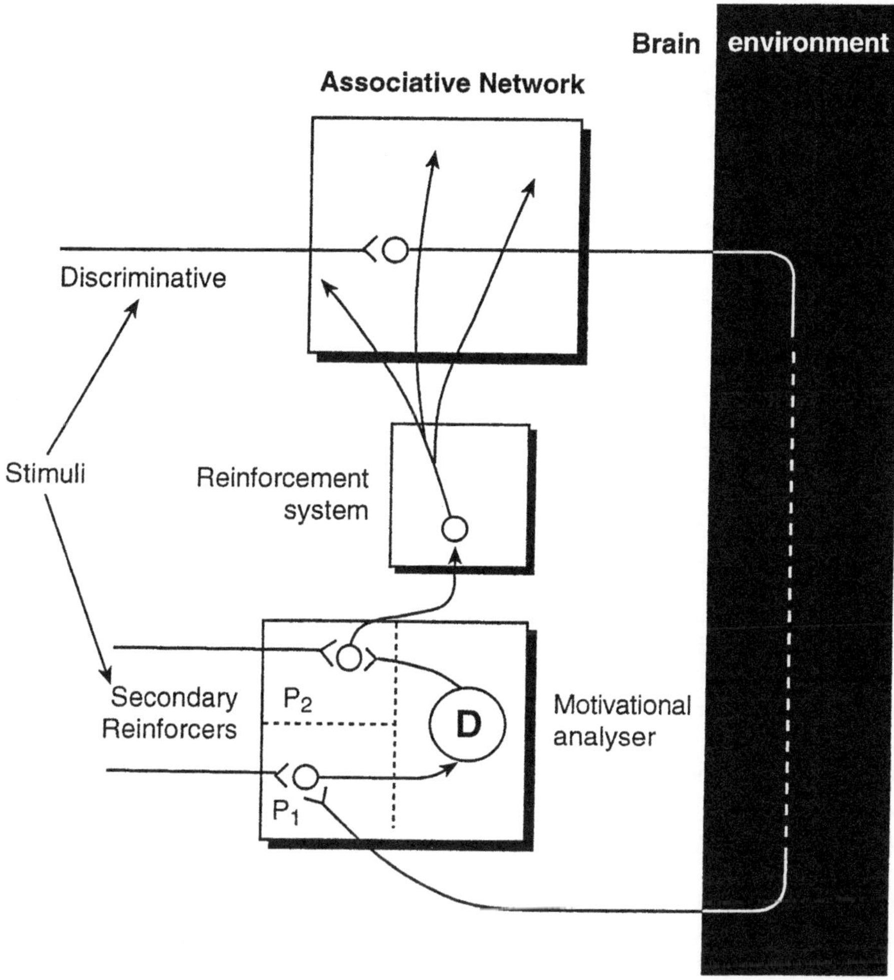


**Figure 5.2:** Modified scheme for instrumental conditioning, incorporating psychological evidence such as that discussed in section 2 of this chapter. **D** represents a component of the "Motivational analyser" which differentiates motivational states with respect to trials on which they have been associated with other stimuli. The derivative produced here (the "change score") represents the difference between actual size of a motivational cue and that expected on the basis of the other stimuli. **P1** represents Pavlovian association between a neutral stimulus and a motivationally significant consequence of behaviour, prior to the differentiation process. **P2** represents a similar process of association with the signal derived after the differentiatiuon process. Compare with Figures 1 and 3.

## 3. ANATOMY: THE AMYGDALOID COMPLEX AS A PROVIDER OF INPUT TO THE MIDBRAIN DOPAMINERGIC REWARD STRUCTURES

As mentioned above there is considerable evidence for the view that the striatum is the primary site in the forebrain for acquisition of input/output links based on reward and punishment; and that the midbrain dopaminergic cell groups provide the reward signal. This group of dopaminergic neurones is however, rather enigmatic by comparison with the major motor or sensory pathways. It is a neurochemical or histochemical concept in the first instance, rather than a component of a connected pathway defined by the type of information it is known to transmit. Information on the afferent pathways to this cell group, and the corresponding influences that control them is rather scanty.

The best known afferent pathway to the substantia nigra originates in the striatum itself, but this is known to terminate mainly in the *pars reticulata* rather than the *pars compacta* (SNC), where dopaminergic cell bodies are located. Evidence for the existence of connections from the striatum to the SNC (i.e the dopaminergic cell bodies) is harder to find. However, such evidence does exist (e.g. Somogyi *et al.,* 1981). Most recent evidence (Gerfen, 1984; 1985) suggests that the projection from striatum to SNC originates not within the mass of striatal tissue (the "matrix") but from numerous smaller regions embedded within this tissue (the "patches"). The latter patches appear to receive their own afferents, not from the broad mass of sensory/motor/associational cortex, but from a more specialised zone of cortex—the prefrontal cortex (Gerfen, 1984; 1985).

Neither the striatum nor the prefrontal cortex are recognised as major structures for representation of motivationally significant events. Thus, one is left searching for inputs to midbrain dopamine neurones which could convey signals of motivational significance to the reward pathway. However, if dopaminergic neurones of the SNC and ventral tegmental area (VTA) do indeed serve the role of a reinforcement system, they should have direct inputs from those parts of the nervous system which act as analysers of the motivational significance of the sensory input. The most obvious regions in the forebrain which could serve this role are: (a) the hypothalamus; and (b) the amygdala. The paragraphs below review evidence about connections from these regions to the midbrain dopaminergic neurones.

A number of papers report on the existence of efferents from the hypothalamus terminating in the neighbourhood of the midbrain dopaminergic neurones. In the older literature, there is better agreement on their distribution to VTA than to SNC. Hypothalamic afferents to VTA have been shown by Wolf and Sutin (1966) using silver stains for degenerating fibres, after large lateral hypothalamic lesions in rat. Using autoradiographic tracing methods, similar connections from all parts of the lateral hypothalamus have been demonstrated in rat by Nauta and Domesick (1978), Saper, Swanson and Cowan (1979) and by Hosoya and Matsushita (1981); and in cat by Trojiano and Siegel (1975). Similar autoradiographic methods have shown afferents to the VTA from anterior, ventromedial and paraventricular hypothalamic nuclei (Conrad and Pfaff, 1976). Retrograde transport methods have shown labelling of many neurones in lateral hypothalamus, as well as some in medial and lateral pre-optic areas after injection of horseradish peroxidase into VTA of the rat (Phillipson, 1979).

From VTA some of these projections can be traced further in a caudo-lateral direction, in a thin sheet of tissue lying between the subthalamic nucleus and the dorsal aspect of SNC (Conrad and Pfaff, 1976; Nauta and Domesick, 1978; Hosoya and Matsushita, 1981). The view has been expressed that these connections do not enter SNC. However, this probably reflects the fact that stains for degenerating fibres do not easily demonstrate fine unmyelinated connections. More recent studies, with tracer injections do show the existence of hypothalamo-nigral connections. Swanson (1976) traces connections from the lateral pre-optic area and neighbourhood to substantia nigra. Nauta and Domesick (1978) and Arbuthnott *et al.* (1976) report labelling throughout most of SNC in rat (albeit less densely than in the tissue dorsal to it) after injection of tritiated amino acids into the lateral hypothalamus. Saper, Swanson and Cowan (1979) describe one rat brain with label injected into the posterior part of the lateral hypothalamus whose projections passed through, and probably terminated in SNC. Panksepp (1981) illustrates prominent labelling of the entire SNC and VTA following injections of tritiated proline into the basomedial hypothalamus. Veazey, Amaral and Cowan (1982) illustrate a small degree of labelling of SNC after injections into the lateral mammillary regions of the hypothalamus. Finally Chiba and Murata (1985) using the horseradish peroxidase-wheat germ antigen method traced efferents from the medial pre-optic area of the hypothalamus to both SNC and VTA.

In summary there is substantial evidence for the existence of pathways from various hypothalamic nuclei to the midbrain regions thought to be involved in reinforcement. Not all hypothalamic nuclei give such a projection. Almost all of this evidence has been obtained using the techniques dependant upon axonal transport, rather than the older degeneration methods. This may be an indication that these connections consist of extremely fine unmyelinated axons.

Evidence for the existence of amygdaloid efferents to the neighbourhood of the midbrain dopamine neurones has been obtained quite recently using methods based on axonal transport, rather than classical degeneration techniques. Indeed Price and Amaral (1981) remark that "prior to 1975 subcortical amygdalofugal projections were thought to extend no further caudally than the posterior hyopthalamus". Dealing first with anterograde transport studies, Hopkins and Holstege (1978) injected tritiated amino acids into the central nucleus of the amygdala in cats, and traced labelled projections to the dorsolateral *pars reticulata*, and in one case to the most caudal part of SNC. The VTA also contained labelled projections. Krettek and Price (1978) using rat and cat found that the central nucleus was the only part of the amygdala projecting to substantia nigra, label being detectable in the lateral edge of SNC, *pars lateralis* (which also contains dopaminergic neurones) and extending into the *pars reticulata*. Post and Mai (1980) confirmed in rats that the central amygdaloid nucleus (but not other amygdaloid nuclei) projects to the substantia nigra, label being densest in *pars lateralis* and lateral SNC. Price and Amaral (1981) report on an autoradiographic tracing study in monkey after injections into the central nucleus of the amygdala. Heavy labelling was reported in this species throughout SNC, and labelling was also found in VTA. It correlated well with the distribution of neuromelanin-containing cells (a presumed marker for the dopaminergic cells). Gonzales and Chesselet (1990) report that, after anterograde tracer injections into the central nucleus, "a sparse projection of bouton-like swellings" was seen in rostral and medial SNC and VTA from all subregions of the central nucleus, and a more dense projection of labelled axons and bouton-like swellings was seen in the lateral SNC and *pars lateralis* from some parts of central nucleus. No projections were found in *pars reticulata*. Wallace, Magnuson and Gray (1992) conducted a more comprehensive study of projections from the amygdala to brainstem monoaminergic cell groups. The central nucleus of the amygdala projects to the lateral part of the A9 group, where "amygdaloid terminals appeared to contact the majority of tyrosine hydroxylase immunoreactive cells".

Retrograde transport studies also indicate that the central nucleus is an important origin of amygdalo-nigral fibres. Bunney and Aghajanian (1976) found labelled neurones in this nucleus in rat after HRP injections into the lateral substantia nigra at all levels. Phillipson (1978) also using rat, found labelled neurones in central, medial and basolateral nuclei after HRP injections into the VTA. Cassell, Gray and Kiss (1986) have also traced connections from HRP injections in lateral SNC or medial SNC/VTA to cell bodies in the central nucleus of the amygdala.

In addition to the direct connections from amygdala to midbrain one must bear in mind the possibility of indirect connections. Many of the amygdaloid nuclei are known to project to large areas of the hypothalamus including the regions of origin of hypothalamo-nigral connections discussed above. For instance, the central, basolateral, basomedial and anterior amygdaloid nuclei project to the lateral hypothalamus in rat and cat (McBride and Sutin, 1977; Krettek and Price, 1978). Parts of the basal amygdaloid nucleus in cat also project to the medial and lateral preoptic areas of the hypothalamus (McBride and Sutin, 1977). Cortical, medial and parts of the basal nuclei also project to the ventromedial hypothalamic nuclei (see also Zaborsky, 1982).

With respect to the overall functioning of the striatum, it is also relevant to mention evidence that the amygdala projects directly to certain parts of the striatum. This was first documented by Royce (1978) with retrograde tracers injected into the caudate nucleus of cats, which labelled cell bodies in the basolateral nucleus of the amygdala. The projection from the basolateral nucleus was confirmed with anterograde tracers, for rat as well as cat, by Krettek and Price (1978), Kelley, Domesick and Nauta (1982) and Russchen and Price (1984), who emphasised that the connection terminated mainly in the ventral striatum. However, Kita and Kitai (1990), while supporting the existence of a strong projection to the ventral striatum (nucleus accumbens) in rats, with anterograde tracing methods, also confirmed Royce's evidence for projections to the dorsal striatum. In most recent studies the main issue has been which compartment —the striosomes or the matrix—is the predominant site of termination of the amygdalostriatal pathway. Ragsdale and Graybiel (1988) provided evidence with anterograde tracers in the cat, that the striosomes were the preferred target. However, in the rat, Kita and Kitai (1990) were unable to find a clear segregation of projections between compartments in the dorsal striatum, and found a preference for the matrix compartment in the ventral striatum. Projections from the amygdala to the striatum are commonly collaterals of projections which also project to the prefrontal cortex (McDonald, 1991; Shinonaga, Takada and Mizuno, 1994). Robinson and Peart (1988) present evidence that, like the corticostriatal projections, this projection is an excitatory pathway, utilising amino acids as transmitter. The functional role of amygdalostriatal pathways is discussed in section 4.2.4.

## 4. LESIONS OF THE AMYGDALID COMPLEX

### 4.1. Amygdaloid Lesions and Basic Forms of Instrumental Conditioning

Two structures were mentioned above—the hypothalamus and amygdala—as motivational centres giving projections to the midbrain dopamine neurones. Lesions of the hypothalamus have very severe consequences for behaviour, which preclude any sophisticated behavioural analysis. However, the general concept we have of the hypothalamus is that it is the region organising a wide variety of endocrine, autonomic and motor programs for motivated behaviour. As such we can regard it as a final common pathway for motivated behaviour, rather than a region involved in the preceding information processing, such as balancing the conflicting demands of various motives in both the long and short term. The amygdala is a more promising region in which to identify the latter functions.

Obviously, if the amygdaloid complex really is the structure in which the various transformations of motivationally-significant stimuli are produced, before they impinge on the midbrain dopamine neurones, lesions of the amygdala should interfere with these transformations. However, before one can determine the role of lesions of the amygdala on the various transformations of rewarding stimuli reviewed in section 2, it is necessary to consider the effect of such lesions on basic paradigms of instrumental conditioning in themselves. In the paragraphs below, consideration is therefore given first to basic appetitive, escape, and passive avoidance learning. As mentioned above, active avoidance learning itself involves control of responding by an expectancy, and therefore comes in a later section.

In appetitive tasks, provided that one avoids the immediate post-operative week or two in which hypophagia is prominent, experimental animals with large amygdaloid lesions appear to be able to acquire appetitively motivated responses with little impairment. This has been found in monkeys, in various types of appetitive task, by Weiskrantz (1956), Schwartzbaum (1960) and Aggleton and Passingham (1982), in cats by Ursin (1965), and in rats by Pellegrino (1968), Henke, Allen and Davidson (1972), Henke (1973), and McDonough and Manning (1979). No doubt many other investigators have made similar observations, but the finding, though quite significant, may not have been reported, since it is essentially negative.

Two papers report on the effects of amygdaloid lesions on escape learning. Kemble and Beckman (1969) using rats found that the amygdala-lesioned animals actually had faster escape performance compared to non-lesioned control. The design of the escape task used by these workers was a two-way arrangement so that "escape" involved return to a chamber where the animal had previously been shocked. Since amygdaloid-lesioned animals are considerably impaired in passive avoidance learning (see below), it is not surprising that, with this particular experimental design, the effect of lesions was to enhance escape performance. The other paper is that of Werka (1980) using cats, some of whom had amygdaloid lesions restricted to the dorsolateral part of the central nucleus. Lesions had a relatively small effect on response latency.

The effect of amygdaloid lesions on passive avoidance learning has been studied using a variety of passive avoidance methods. The general finding is that response suppression mechanisms are markedly impaired in the lesioned animals. In various types of passive avoidance conditioning most studies (Russo *et al.*, 1976; Nagel and Kemble, 1976 [experiments 1 and 4]) find an impairment, in some cases quite severe. Impairment is also seen in tests involving suppression of pretrained responses such as licking water from a spout (Pellegrino, 1968; Kemble and Tapp, 1968; Coover, Ursin and Levine, 1973; Grossman, Grossman and Walsh, 1975) or suppression of approach responses to food (Horvath, 1963; Ursin, 1965). Other examples of impaired response suppression after amygdaloid lesions include suppression of a previously acquired one-way avoidance response (Werka, Skar and Ursin, 1978) and suppression of rearing (an innate response) (Nagel and Kemble, 1976, experiment 2). Pellegrino (1968) also showed that amygdala-lesioned animals were impaired in the "differential reinforcement of low rates" schedule.

The passive avoidance deficit seems somewhat enigmatic, since the other simple instrumental paradigms discussed above are not much affected by amygdaloid lesions. However, punishment-mediated learning, of all the so-called simple instrumental paradigms, is in fact usually contaminated to a considerable degree by implicit Pavlovian contingencies. In almost all of the experiments described it is possible for the aversive stimulus to produce an unconditioned response (usually "freezing") which can, in test trials, then be evoked as a conditioned response elicited by the concurrent apparatus cues acting as CS's. There is substantial evidence (next subsection), that the amygdala is important for Pavlovian processes linking neutral stimuli to motivationally significant ones. The deficit in passive avoidance tasks after amygdaloid lesions is likely to arise from an impairment in such Pavlovian conditioning, rather than from a basic impairment in instrumental processes.

## 4.2. Amygdaloid Lesions and the Association of Neutral Stimuli to Motivationally Significant Ones

### *4.2.1. Experiments with appetitive stimuli*

Some of the earliest studies of amygdaloid lesions produced evidence indirectly suggesting an inability to link neutral stimuli with appetitive stimuli: Pribram and Bagshaw (1953) reported that their amygdalectomised monkeys could not recognise food from non-food objects by vision alone. The oral tendencies of amygdalectomised animals (part of the original Kluver-Bucy syndrome seen after amygdalohippoocampal lesions) may be a secondary result of such a deficit, since animals incapable of linking secondary stimuli derived from distance cues (vision and hearing) may be forced to make excessive use of those senses (especially taste and smell) by which food is recognised. Schreiner and Kling (1953) noted that their amygdalectomised cats were delicate eaters, even after recovery from the transient post-operative hypophagia. They also showed an unusual attitude to their food. For instance, they would repeatedly take into their mouths meat containing methyl salicylate (an aversive taste) and on each occasion would then eject it, without ever seeming to learn the link between the noxious taste and the visual properties of the food it was associated with.

Jones and Mishkin (1972) performed an explicit test of the role of the amygdala in linking neutral stimuli to motivationally significant ones derived from food. The test involved learning, over a series of trials, which of two objects was used to hide a food reward. After learning the initial discrimination, several reversals of the task followed. The subjects were monkeys, including groups with lesions in orbitofrontal cortex, "temporal pole plus amygdala", and hippocampal region. Only the animals with lesions in the "temporal pole plus amygdala" showed a severe impairment in acquisition and reversal of this task. The authors comment that there are interpretations other than that of an impairment in linking neutral to motivationally significant stimuli. For instance serial reversal learning involves the ability to produce opposite responses to the same object at different times, and it might be this, rather than cue-reward associations *per se* which is impaired by the lesion

To overcome such objections, Spiegler and Mishkin (1981) used an experimental design involving single trial association of objects and food reward. In each trial a half-peanut was first hidden under the object, or the object was unbaited. In the test which followed 10 sec. later, the animals (monkeys) had to choose the object (if it had been baited) or a grey card (if the object had not been unbaited), in order to retrieve the food reward. The objects themselves were chosen from amongst about 500 objects, so that a particular object was not presented to a particular animal more than once in three or four weeks of testing. In such a test it is possible that animals learn to make the correct choice merely because they have seen the reward being placed, rather than because they had learned to link it with a particular object. A more difficult version of the task was therefore also applied in which sets of two objects were presented in acquisition before either was presented in test. For the two objects one was usually positive, the other negative. A third strategy was to present sets of 4, 6, 8 or 10 objects before any testing was done. After acquiring these tasks to criterion, lesions were made, with different groups having lesions in the anterior part of the infero-temporal cortex, the amygdala, the posterior inferotemporal cortex or the fusiform-hippocampal gyrus plus hippocampus. Subsequently animals were retrained to criterion. Only animals with lesions in anterior inferotemporal cortex or amygdala were impaired during the retraining. Moreover, other results were cited in which the test was only a simple object recognition rather than an object-reward association. In this only anterior inferotemporal-lesioned animals were impaired, amygdalectomised ones performing normally. The authors suggest that these results support a sequential processing hypothesis, in which object recognition is performed by association areas of the temporal lobe, and subsequently the amygdala (to which these areas project) makes the linkage with motivationally significant cues. Since the second of these two processes subsumes the first, object-reward association is impaired by both lesions, although these association are elaborated in the amygdala rather than the association cortical areas.

Gaffan, Gaffan and Harrison (1988) conducted a similar experiment involving training in one of two situations: (a) The monkey was offered a choice of two objects presented together, one of which, when touched, gave food reward. (b) Only one of two objects was presented at once, and one of them gave a reward, the other did not, in either case regardless of response. In the subsequent test trial, both objects were presented together side by side, and the animal had to choose one rather than the other. Disconnection of the amygdala from the inferotemporal cortex (by lesioning one on one side, the other on the other side) produced a profound impairment in both tasks. This confirms that the amygdala has a specific role in learning associations between visual stimuli and reward, and this applies even when the association is not response-contingent (as in the second of the two tasks).

Peinado-Manzano (1988) showed that the role of the amygdala in association of visual stimuli with reward applied also to rats. Lesions of the central or lateral amygdaloid nuclei were not impaired in a basic instrumental pretraining task, but when a successive discrimination (GO signalled by bright light and NO-GO signalled by a dim light [or vice versa]) was introduced they were impaired, especially in reversal learning. Preoperatively-trained tasks were performed as well as in unlesioned animals.

The above studies point to a role for the amygdala in associating a neutral stimulus with the sight of food, which already has motivational significance. A number of other studies explore whether the amygdala plays a part in forming associations more remotely related to motivational significance. Murray and Mishkin (1985) showed that amygdalectomy in monkeys also impairs association between two neutral stimuli (cross-modal matching). The basic test was delayed non-matching to sample. After acquiring this task in full light, animals were tested on performance using tactile cues alone (i.e. in darkness). Amygdalectomized animals were severely impaired. Gaffan and Harrison (1987) used a task where an auditory stimulus (click or white noise) had been given significance as a secondary reinforcer, indicating which of two novel objects was to be touched. The actual food reward came only after four successful choices of positive objects, which individually received only secondary reinforcement. Bilateral amygdalectomy severely retarded the use of the secondary reinforcers within each problem. Disconnection of the amygdala from the auditory association cortex had a similar effect, but disconnection from the visual association cortex produced no impairment. Thus, intact connections were needed in this task for the input modality of the secondary reinforcer, but not for that of the discriminative stimuli. A subsequent study (Gaffan, Gaffan and Harrison, 1989) had the same basic design, but used visual rather than auditory secondary reinforcers. In this case, bilateral amygdala lesions did not impair the role of the secondary reinforcers in learning. Thus, the amygdala may not be necessary when the secondary reinforcer is in same modality as the discriminative cue. Hatfield *et al.* (1996) carried out experiments similar to those of Gaffan's group in rats. Animals with lesions in the basolateral amygdaloid nucleus could not form a second order conditioned association between a light stimulus (previously paired with food) and a tone. Rats with lesions of the central amygdaloid nucleus were unimpaired.

### *4.2.2. Experiments with aversive stimuli*

Several papers show the role of the amygdala in linking neutral stimuli to painful shocks. Most of them employ Pavlovian procedures, and so do not specifically prove the role of the amygdala in formation of secondary rewards and punishments. The evidence is nevertheless highly relevant, since the formation of secondary reinforcers is in principle an example of Pavlovian association. Brady *et al.* (1954) in the course of avoidance conditioning in cats noticed that control animals produced large emotional responses to the clicker CS, at least in the early stages of training. Amygdalectomised cats on the other hand showed almost no such responses to the clicker. Weiskrantz (1956) trained monkeys to lever press for food, and measured the degree of suppression produced by two levels of shocks, or by a CS (automobile horn previously paired with shock). Acquisition of the conditioned suppression was slower in the amygdalectomised animals, or those with inferotemporal cortical lesions. Bagshaw and Coppock (1968) used the galvanic skin response produced in monkeys by painful shocks. In normal monkeys this response could be conditioned to a neutral stimulus ("lights off") if the latter was paired with the shocks. In amygdalectomised monkeys such conditioning was completely eliminated. Blanchard and Blanchard (1972) found that large amygdaloid lesions, or smaller ones confined to the corticomedial nuclei in rats reduced the "freezing" response of the animals produced by shock. Moreover, in normal animals, the "freezing" response could be conditioned to a previously neutral stimulus, whereas this process was much reduced in the lesioned animals. Spevack, Campbell and Drake (1975) used the conditioned suppression of responding in rats. Large amygdaloid lesions reduced the linking of CS to UCS, so that suppression of a licking response by shocks could not be linked to the CS.

While most of the evidence about the role of the amygdala in association with aversive stimuli has been based on a Pavlovian paradigm, one study used instrumental techniques. Shibata *et al.* (1986) employed a test of conflict behaviour involving a session consisting of six cycles of unpunished reinforcement by food, alternating with a period in which every response was punished, this schedule being signalled by a distinctive tone. Rats with lesions of the central amygdala showed a significant and long-lasting (at least 21 days post-lesion) increase in punished responding.

This study (Shibata *et al.* 1986) also indicates the role of the amygdala in determining how behaviour should be directed to one amongst competing motives. Results of

Simonov (1999), using a different sort of conflict task, show this more clearly. Rats had to learn to respond to a tone to obtain food, in morning sessions, and to respond to the same tone to avoid noxious shocks in evening sessions. Normal rats could acquire this task, but, impairments were produced in rats, for certain parameter values, with lesions of the amygdala.

### *4.2.3. Experiments using social stimuli*

In Schreiner and Kling's (1953) experiments, amygdalectomised monkeys, though generally more docile than normal, had an explosive temperament when provoked directly by noxious stimuli. Weiskrantz (1956) describes how normal monkeys showed excitement and avoidance when the experimenter approached, whereas amygdalectomised monkeys showed almost none of this. It is not clear whether the normal response is to a primary aversive stimulus or to a secondary conditioned one. However, the operated animals also showed less fear and hostility to objects such as sticks and gloves, which are almost certainly secondary aversive stimulus objects. Dicks, Myers and Kling (1969) studied free-ranging monkeys in their natural habitat. Animals with uncus/amygdala lesions quickly became social outcasts, and died, for reasons other than their transient aphagia. The authors comment that the lesioned animals responded appropriately when social interaction was forced upon them, but nevertheless were retarded in their ability to foresee and avoid dangerous confrontations. For instance they did not respond with submissiveness to the group leader, and so easily became involved in fights with him. Much of this impairment is interpretable as a loss of recognition of a variety of acquired (secondary) social triggering stimuli.

More recently Fonberg and associates have investigated the effect of amygdaloid lesions on social behaviour of cats and dogs. Fonberg and Kostarczyk (1980) found that lesions of the dorsomedial amygdala markedly impaired pre-trained social responses of the dog to the experimenter. These responses included commands such as "sit", "paw", "lay", appropriate responses being reinforced by verbal responses ("good dog") and petting (stroking the back of the head of the dog). Other responses included running from the technician to the experimenter, or jumping a barrier between the two. The experimenter never fed the animals, so responses were not reinforced secondarily by the presence of the food-giver. Lesioned animals displayed both aphagia and lack of social responses, but the latter impairment was more permanent than the former, and so cannot be a consequence of the former. The impairment here seems to be in social responding generally, but a part of this is the failure to recognise either primary or secondary social reinforcers. The hypersexuality seen after amygdaloid lesions (Schreiner and Kling, 1953) includes not only an increase in frequency of sexual behaviour but also a widening of the range of sexual objects. This may also be a reflection of a failure to link neutral stimuli to social stimuli with a primary motivational significance.

Campbell (1978) made lesions in cats in either the posterior temporal cortex or the basolateral amygdala. The latter but not the former produced impairment on a task involving the recognition of a silhouette of a cat. Zagrodzka, Brudnias-Stepowska and Fonberg (1983) produced damage to the dorsal part of the amygdala in cats. Operated cats, when not in the company of other cats showed normal predatory behaviour towards mice. However, in the presence of other cats they lost their social rank and did not show predatory behaviour.

### *4.2.4. Lesions and direct amygdalo-striatal communication*

In section 3 evidence was reviewed that, amongst the direct efferent projections of the amygdala, are not only the midbrain dopamine neurones, but also their target, that is various parts of the striatum itself. These are similar to the corticostriatal projections, in utilising excitatory amino acids as transmitters. It is therefore probable that signals relayed to the striatum from the amygdala (like those from the cortex) serve the role of discriminative stimuli in instrumental conditioning, this being in addition to their role as secondary reinforcers of stimuli which already have motivational significance, or of changes of motivational state. The existence of such connections and their possible modifiability would, in the adult, if not in the infant animal, predispose the animal to respond to such secondary motivational cues, rather than to any neutral stimulus which might be linked to goal fulfilment. For instance, because of the existence of such connections the animal would be more likely to approach food objects than entirely neutral objects to fulfil a hunger drive.

Two papers give some evidence of this: Cador, Robbins and Everitt (1989) trained thirsty rats to choose between two levers, the choice being made on the basis that one produced a light-noise compound stimulus, which had previously been associated with water delivery. Preference for the lever producing the secondary reinforcers was increased by microinjections of amphetamine into the ventral striatum. This is a typical rewarding effects of an indirect dopamine agonist, but acts directly on dopamine terminals and neurones in the ventral striatum, rather than by controlling firing rate of the midbrain dopamine neurones. In animals with excitotoxic lesions of the basolateral amygdala, the amphetamine microinjection produced a much smaller enhancement of responding to the secondary reinforcer. However, since amphetamine is providing a dopaminergic signal directly in the ventral striatum, the effect of amygdaloid lesions cannot be achieved by reducing the effect of the secondary signal on control of dopamine neurone firing. What is referred to as a "secondary reinforcer" is therefore probably a discriminative signal (albeit one which already has motivational associations), setting the scene in which responding is reinforced, rather than controlling reinforcement *per se*. More detail of these experiments is provided by Everitt and Robbins (1992).

The second paper, that of Hiroi and White (1991) uses place-preference conditioning, induced by two pairings of an amphetamine injection (2.0 mg/kg s.c.) with a distinct compartment of the animal's cage. Again such acute injections of the indirect dopamine agonist are likely to act predominantly on the terminals of dopamine neurones within the striatum, rather than on firing rate of dopamine neurones in the midbrain. Excitotoxic lesions of the lateral amygdaloid nucleus prevented the expression of conditioned place preference acquired in this way. Lesions elsewhere in the amygdala, or in related structures had no such effect. Again, one must conclude that signals relayed in an amygdalostriatal pathway, rather than those controlling dopamine neurone firing rate, maintain the place preference.

A discriminative stimulus relayed to the striatum from the amygdala (rather than from the neocortex) is likely to be an unusual one, in that it may already contain an association of neutral and motivational information. Therefore, assuming that the dopaminergic reinforcement signal has its essential actions in some part of the striatal complex, the target of reinforcement may be a signal to approach a stimulus which is already coded as being associated with motivational fulfilment. We see here the possibility of higher level programming of purposeful behaviour. In regular examples of instrumental conditioning, a link between a neutral stimulus and a response strategy is reinforced if their joint activation is followed by rewarding stimuli. In the above experiments, however, the link is forged instead between a valued goal and a response strategy, if that link turns out to lead to further enhancement of reward. In chapter 12 of this book by Joel and Weiner, a detailed exposition of this process is presented, as part of an account of the hierarchical relations between different cortico-striatal loops, set in action by critical instances of dopaminergic reinforcement in different parts of the striatum. Reinforcement in the ventral striatum, with which we are particularly concerned here, appears to control the long-term strategic selection of goals, rather than immediate objectives or motor programs.

### 4.3. The Amygdala and Contrast Effects

A number of papers show the role of the amygdala in contrast effects. Some have been based on the experimental design of Crespi (1942). Schwartzbaum (1960) used monkeys, some of which had large lesions of the amygdala. In tests involving bar pressing for food, the lesioned animals showed a smaller change in response rate than did controls when a much larger reward was given for each response; and if a smaller reward was given, the lesioned animals had a smaller reduction in bar pressing than the control animals. In other words, there was an impairment in both positive and negative contrast effects. Henke and Bunnell (1971) trained rats to lever press for food on a continuous reinforcement (CRF) schedule for 14 daily sessions, followed by 5 daily extinction sessions, and then by six more days of the CRF schedule. In normal animals, return of the CRF was accompanied by an increase in lever pressing rates to a value above that established in the initial CRF period. In animals with lesions of the amygdala this did not happen (i.e. positive contrast was impaired). Becker *et al.* (1984) provide further specific anatomical information on the parts of the amygdala involved in negative contrast. One group of rats was shifted from 32% sucrose reward to 4% sucrose and was compared with another group continuously receiving 4% sucrose. Sham-operated animals showed a large negative contrast effects in the first situation, compared with the second. In animals with lesions confined to the cortico-medial and central nuclei of the amygdala, negative contrast was completely abolished. In animals with lesions of basolateral, lateral and basomedial nuclei negative contrast was significantly attenuated, though still quite clearly in existence. Most of the above papers report rather large amygdaloid lesions (usually the medial nucleus is spared). It is impossible to draw any firm conclusion about the most important nuclei involved in mediating contrast effects, though it may be noted that in the last two cited papers, where more limited lesions were made, the central nucleus was a common target of the most potent lesions, and this nucleus is a major origin of fibres descending as far as the midbrain dopaminergic neurones.

Other studies are based on the experimental design of Bower: Henke (1972) trained rats on a variable interval (VI) schedule for water reward, and then transferred them to a related schedule in which 5 minute periods of VI-20sec. alternated with 5 minute periods of unsignalled extinction. In this test, control animals gradually increased their responding in the VI component over the 25 sessions, so that eventually they responded more vigorously in this component than they had to a similar component before the transfer. Such a positive contrast effect was not seen in animals with large lesions of the amygdala: a steady level of lever pressing during the VI component was seen throughout the 25 sessions. Henke, Allen and Davidson (1972) carried out similar experiments with food-rewarded schedules, the principal difference being that the VI and the extinction schedules were signalled by turning lights on or off. Again amygdaloid lesions abolished positive contrast. Henke (1979) confirmed the finding, and showed that it was restricted to lesions of the amygdala, while those of the hippocampus, septum, stria terminalis or fornix all showed significant positive contrast. Goomas, Hamm and Skinner (1980; see also Goomas and Steele, 1980) compared groups of rats running an alleyway for 2 *vs* 12 pellets of food, with each group subdivided into those with or without amygdalectomy. In the unlesioned group, the learning curve defined by runway speed was initially steeper in the large-reward groups, but by trials 50–100 the small-reward group had caught up. In the lesioned groups, the small-reward group caught up with the large-reward group much sooner, within 20 trials. When reward was delayed progressively in 15 sec steps up to 45 sec, the lesioned animals did not decrease speed significantly, while the intact animals were significantly slower than lesioned ones for 30 and 45 sec delays. These results imply that negative contrast is attenuated after lesions of the amygdala. A recent investigation, used reversible inactivation of the amygdala, with injections of local anaesthetic (Salinas, Packard and McGaugh, 1993). Inactivation of the amygdala was produced after a downward shift in reward magnitude, and after a negative contrast had been exhibited in six trials on the post-shift day. Nevertheless, expression of the contrast was subsequently abolished. The amygdala appears to be involved in storage of memory for reward reduction as well as in initial production of the contrast effect. Similar effects were recently described using GABA agonists injected into the amygdala (Salinas and McGaugh, 1996).

Two papers deal with the contrast effects using aversive stimuli in relation to the effects of amygdaloid lesions, but the results are in disagreement. Kemble and Beckman (1969) ran rats in an escape test, using three levels of shock strength (0.1, 0.5 and 1.0 mA). Rats were shifted from the 0.1 mA shock level (over 7 initial ten-trial sessions) to the 1.0 mA shock (for a following 7 ten-trial sessions). Both controls and lesioned animals accelerated their escape responses considerably after the shift. However, the amygdala lesioned animals did so less than the controls, i.e. there was an attenuation of positive contrast. The second paper is that of Werka (1980). As already mentioned, this study showed that amygdaloid lesions had relatively small effects on escape learning, using shocks of about 0.5–0.7 mA. However, in post-training test trials a very significant lesion effect was found in the latency of escape responses at lower shock intensities than those used in training. The test sessions used a sequence of shocks of gradually increasing intensity, and were interspersed at various times during the more protracted training sessions. Thus the start of each test session would have involved a sudden shift from 0.5–0.7 mA to a much lower level (about 0.05 mA). The lesioned animals had much *longer* escape latencies than controls with these low intensity shocks, though their latencies increased to match controls when the intensity was increased back to 0.5 mA and above. It is difficult to provide a satisfactory explanation of this result within the bounds of the theoretical framework advanced here. One would have expected, from other evidence, that when the shock strength was suddenly reduced, this should have caused a more profound reduction in vigour of escape in the intact animals than in those with amygdaloid lesions, in whom the absolute rather than the "contrasted" value of the aversive stimulus should have determined responding. In fact the evidence demonstrated the opposite pattern.

Despite this inconsistency, another experiment, using taste aversion showed a loss of contrast effect after amygdaloid lesions. Hatfield *et al.* (1996) devalued a food reinforcer with lithium chloride, and produced a reduction of responding (i.e. negative contrast) in normal rats. Rats with basolateral lesions were insensitive to such changes in the value of a reinforcer.

Several studies investigate the role of the amygdala in Amsel's frustration effect: Henke and Maxwell (1973) and Henke (1977) used the double alley design of Amsel and Rousell (1952) to observe frustration effects after omission of reward in the first alley. Acceleration of second alley runway speed after the intermittent omission of first alley reward (frustration effect) was not seen in animals with large amygdaloid lesions. Henke (1973) reported on two further experiments using operant analogues of the Amsel and Roussell double alley design. In the first, rats bar-pressed for water on a fixed interval (FI) 2 mins schedule, delivery of reward being accompanied by audible "clicks". The animals were then shifted to a variant of the schedule in which there were two alternating components. The second was consistently maintained at FI 2 mins. The first was a randomly alternated sequence of "reward plus click" and "no reward or click". Control animals accelerated their response rate in the 2 min interval which followed no reward. However, this did not happen in animals with amygdaloid lesions. Henke and Maxwell (1973) also report on an experiment where frustration effects are produced in a lever pressing task, in which there were two FR-25 components, separated by a ten second time-out period, which was signalled by a distinctive stimulus. If reward in the first component was given in a semi-random 50% basis, responses during the time out period (i.e. after non-reward) accelerated. Once more, this effect was greatly diminished in amygdaloid lesioned animals. Goomas and Steele (1980) used a runway design in which immediate reward was shifted to reward with a 60 sec delay. In intact rats this led to an initial increase in runway speed (a typical frustration effect) followed by a decline in speed in later days of testing. In amygdalectomised rats, the initial increase was delayed later, and the subsequent decline was much less than in intact rats.

### 4.4. The Amygdala and Active Avoidance

In section 2.2. it was suggested that in active avoidance behaviour, rewarding effects are produced by absence of expected aversive stimuli. This could then be regarded as a variety of positive contrast effect. It is therefore, highly relevant that many studies have shown that lesions of the amygdala impair active avoidance behaviour. The majority of these find that significant impairment of acquisition of this task is produced by amygdaloid lesions. The degree of impairment varies considerably between studies. This variation, as well as the minority of studies showing no impairment, depends on technical details of the task in each study, such as the time during which it is possible for an animal to give an avoidance response, or the intensity of the aversive stimulus used. In rats, Coover, Ursin and Levine (1973) found that control animals readily reached an asymptote of 95% avoidance after ten sessions, whereas amygdaloid lesioned animals never responded above 80% avoidance. Similar findings were reported by Bush, Lovely and Pagano (1973) and by Molino (1975). In these studies avoidance learning, though impaired can still occur. On the other hand Yeudall and Walley (1977) using a more demanding avoidance task, found that amygdala-lesioned animals could not learn significantly (only one or two avoidances occurred per 25 trials, whereas controls reached

18/25 avoidances after 200 trials. Similarly Eclancher and Karli (1980) found that 50% of amygdalectomised rats failed to reach a criterion of 10 successive avoidances, even after 900 trials, compared with only 20% in the control group. These workers also found that in animals, lesioned at seven days of age and tested for avoidance learning as adults, there was no deficit. In cats, two early papers (Brady *et al.*, 1954; Horvath, 1963) found that amygdaloid lesions greatly increased the trials required to reach avoidance criterion. In monkeys, Weiskrantz (1956) found that two amygdalectomised animals took 250 and 350 trials to reach avoidance criterion as opposed to 150 and 200 trials in two unoperated monkeys.

Several papers also report the effects of amygdaloid lesions on one-way active avoidance learning. Once again, an impairment is generally found, but with some inconsistencies, and variations in the degree of impairment. Robinson (1963), and McNew and Thompson (1966) found, with rats, that control animals reached a criterion with fewer trials than amygdalectomised ones. Werka, Skar and Ursin (1978) conducted a study with the most detailed anatomical analysis. Lesions of the central nucleus produced the severest impairment of acquisition. Other lesions (basolateral nucleus, or cortex lateral to amygdala) produced lesser degrees of impairment. In cats, Horvath (1963) and Ursin (1965) found that trials-to-criterion was increased above control levels in animals with lateral amygdaloid lesions.

A few of the early studies have dealt with retention of a two-way avoidance task acquired pre-operatively (Brady *et al.*, 1954; Horvath, 1963; Thatcher and Kimble, 1966; Werka, Skar and Ursin, 1978). Amygdaloid lesions do impair retention of two-way avoidance performance provided that the task has not been overtrained.

The general conclusion to be drawn from these studies is that amygdalectomy seriously impairs 2-way or 1-way avoidance learning, but there is nevertheless some other machinery which can accomplish the same function, albeit less effectively when the amygdala is destroyed bilaterally, or can even take over completely when the operation is performed neonatally.

### 4.5. Extinction and Amygdaloid Lesions

If the amygdala has an important role in mediating the effects of failed expectancy of appetitive or aversive stimuli, then amygdaloid lesions should result in impairment not only of contrast effects, frustration effects and active avoidance: Such lesions should also retard extinction, since an extinction schedule is also, initially at least, a disconfirmation of expectations. The evidence on this is in fact rather equivocal. For this issue the most relevant type of extinction is extinction of a previously established appetitive response. Two studies in this category find that extinction is retarded (i.e. there is perseveration). Weiskrantz (1956) obtained such a result in monkeys trained pre- or post-operatively to lever press for food reward on a VI–1min. schedule. Kemble and Beckman (1970) made such observations in rats traversing a runway for food. Against these studies are several others where there was no significant effect of amygdaloid lesions on extinction (Pellegrino, 1968; Henke and Bunnell, 1971; Henke, 1972; Henke, Allen and Davidson, 1972; McDonough and Manning, 1979)

It is difficult to reconcile the inconsistencies in this evidence. Extinction may be retarded by amygdaloid lesions under some circumstances, but it would be wrong to conclude that effects due to disconfirmed expectancies, mediated by the amygdala, are anything like a complete account of extinction. There must be some other important process carried out elsewhere in the brain which can mediate such extinction when amygdaloid mechanisms fail.

## 5. ELECTROPHYSIOLOGICAL STUDIES

### 5.1. Amygdaloid Neuronal Responses and Associations Between Neutral Stimuli and those with Motivational Significance

A number of single unit electrophysiological studies have sought neuronal responses in the amygdala to sensory stimuli which have (or have acquired) motivational significance. In some of these experiments, motivational significance has been acquired before the acute experiments, either in explicit training, or implicitly in the normal activities of the animals before the experiment. In such studies, it is regularly found that a proportion (generally a small proportion) of neurones respond to motivationally significant stimuli. These responses may be made to the explicit CSs of a conditioning regime (Fuster and Uyeda, 1971). Alternatively they may be to incidental stimuli which have acquired significance. In the visual sense these are such things as the sight of a syringe used for delivering hypertonic saline into the mouth of the animal (Sanghera, Rolls and Roper-Hall, 1979) or the sight of a rat (Jacobs and McGinty, 1972), while in the auditory sense they include sounds such as "kitty", "meiaow", a cat howl or rustling of wood chips (Jacobs and McGinty, 1972). Some neurones may also respond to neutral stimuli (Sanghera, Rolls and Roper-Hall, 1979). There are nevertheless neurones in the amygdala whose responsiveness is to motivationally significant stimuli alone. For instance in the study of Ono *et al.* (1983) neurones were found responding to the sight of a glove, a syringe, cotton wool, tape, pipette or sponge. These same neurones gave no response to simpler stimuli, such as a cube, a column, or a triangle. One study (Jacobs and McGinty, 1972) also documents selective responsiveness to complex stimuli involving more than one modality. This is compatible with the results of the lesion study of Murray and Mishkin (1985), discussed in section 4.2.1., that association between initially neutral stimuli may be disrupted by amygdaloid lesions.

Some neurones respond only to attractive stimuli, some only to aversive stimuli, but many respond to both. The majority of studies find that most neurones responding to motivationally significant stimuli do so with the same sign to both attractive and aversive stimuli (Fuster and Uyeda, 1971; Sanghera, Rolls and Roper-Hall, 1979) and, in the first of these two studies, there was no reversal of the sign of the response when the significance of a former CS+ and CS- was reversed. However, Pascoe and Kapp (1985) found differential responses to CS+ vs CS- (greater *vs* lesser degrees of slowing or acceleration of firing) in a Pavlovian conditioning paradigm with paraorbital shock as UCS. Yasoshima, Shimura and Yamamoto (1995) in a taste aversion paradigm, found that while attractive and aversive tastes produced similar responses in neurones in the corticomedial nuclei, in neurones from the basolateral nucleus, responses to attractive stimuli tended to be of the opposite sign to those to aversive stimuli. According to Muramoto *et al.* (1993) the sign of response to an attractive UCS was the same as to an auditory CS, suggesting that the response to the UCS is transferred to the CS in Pavlovian fashion.

Spontaneous firing rates of amygdaloid neurones are often low, which makes it difficult to say whether any of the stimulus objects inhibit firing. However, in the study of Ono *et al.* (1983) inhibition could be tested in another way, by combining presentation of food objects with the excitatory non-food stimuli (e.g. raisin plus syringe). When such tests were applied to neurones in the anterolateral amygdala, it was generally found that the presence of food objects blocked the activation produced by the non-food objects. Such inhibitory effects were never seen when combining two non-food objects. Nishijo *et al.* (1986) found that neurones in the dorsal amygdala responded, usually with inhibition, to the visual appearance of food, and others in the same region responded, usually with excitation, to the appearance of non-food or aversion-related stimuli. Muramoto *et al.* (1993) report that responses to taste stimuli, to which an aversion has been conditioned, generally produce excitation of neurones in the basolateral amygdala, but generally produce inhibition in the central nucleus. Such findings suggest that there are processes in the amygdala which differentiate between attractive and aversive stimuli. The findings imply that some complex and sophisticated information processing may be occurring within the amygdala. However, at present there is no consensus about the details of this differential activity, however intriguing the available evidence may be.

In addition to these studies giving details of already-acquired responsiveness to motivationally significant sensory stimuli, a few recent studies investigate the actual process by which links between neutral and motivationally significant stimuli are acquired. Applegate *et al.* (1982) recorded from amygdaloid neurones during Pavlovian heart-rate conditioning with auditory (tone) CSs and an aversive UCS, in rabbits. Ten out of the 34 neurones developed short latency (about 100 msec.) excitatory responses to the CS within about 3 five-trial blocks of pairing. Repeated exposure to the CS alone ("orienting") was not accompanied by such a change. Some of these 10 neurones also showed rapid abatement of the firing increase if the CS was presented unpaired to the UCS (i.e. in extinction). Similar results were obtained by Richardson and Thompson (1984). Yajeya *et al.* (1991) found that basolateral amygdaloid neurones initially showed regular (14 Hz mean) firing, but after one trial passive avoidance conditioning, this fell almost to zero, and recovered slowly over 24–48 hrs, in parallel with extinction of the avoidance response.

An important question is whether the associative changes documented above arise by plastic changes occurring actually within the amygdala, or are produced elsewhere, and are relayed to the amygdala via its afferent pathways. Several results suggest that the latter possibility must be taken seriously: For instance, in conditioned taste aversion, although neurones in the amygdala can be shown to respond to the acquired aversiveness of a taste stimulus, the same is true for neurones lower in the taste pathway (such as nucleus tractus solitarius, or the primary gustatory cortex [Chang and Scott, 1984; Yamamoto *et al.*, 1989]). Likewise, in classical conditioning of rabbits' heart rate, with an aversive UCS and auditory CS, differential conditioning of neuronal responses to CS+ *vs* CS- could be shown not only in the central nucleus of the amygdala, but also in the medial geniculate nucleus of the thalamus, which is afferent to the amygdala (McEchron *et al.*, 1995). Indirect arguments are however available to suggest that plastic associative processes occurring actually within the amygdala make a major contribution to the response patterns described above. In the study of Nishijo *et al.* (1988) several amygdaloid neurones were analysed which initially responded to vision plus oral sensation of an attractive food, and later, when the food was serruptitiously salted, neuronal responses were attenuated. In discussing the results the authors point out that the inferotemporal cortex projects massively to the amygdala; cooling of this region limits visual responsiveness in the amygdala; responses in the amygdala have a longer latency than those in the inferotemporal cortex; and the inferotemporal cortex itself responds to objects in terms of their visual properties, but does not associate them with motivational value.

### 5.2. Electrophysiological Implications of Contrast Effects

In the discussion above of psychological and lesion studies of the contrast and related effects (section 2.2.) the implication was made that the amygdaloid complex somehow achieves a mathematical differentiation of the motivational value of particular events, so that changes rather than the absolute value of a motivational state can come to control reinforcement systems. This hypothesis clearly has implications for single unit studies. The hypothetical process does not appear to differentiate motivational value with respect to time. If it did, a suitable neuronal model would involve habituation of a motivational signal in the time after a change, the habituated motivational signal then being used for association with secondary stimuli. However, the evidence considered so far suggests rather that motivational value is differentiated with respect to the *number of occasions* (e.g trials in a conditioning experiment) on which a given event with a given value has occurred. Therefore, one would expect that a neurone which responds vigorously on the first occasion when an event with given motivational value occurs, should respond less vigorously the more the event recurs. For a thoroughly familiar reward, neuronal response should drop to a minimum, or even disappear. Rewards of different magnitude which do not change over many trials may then have quite similar effects on the reinforcement system.

The only electrophysiological study of which the author is aware which is relevant to this prediction is that of Ben-Ari (1972) and Ben-Ari and Le Gal La Salle (1974). Amygdaloid neurones were activated by somatic stimuli (i.e. 8v shocks applied to the paws via needle electrodes). Most of the responses were inhibitory in sign, a smaller proportion excitatory. Over the course of about 50 presentations of the stimulus the response waned until it was negligible. This was observed both in the neurones with excitatory and in those with inhibitory responses. This process of "habituating" was accompanied by progressive changes in spontaneous activity levels. In the habituating units with inhibitory responses, spontaneous activity gradually increased, while it decreased in those few with excitatory responses (Ben-Ari,1972; Ben-Ari and Le Gal La Salle, 1974). Ben-Ari and coworkers also describe other more complex effects: In some neurones, after an initial series of presentations of an auditory or visual stimulus, resulting in almost complete habituation, this first stimulus was then paired with a somatic one 500 msec. later. In some cases this brought about a sudden reactivation of the response to the visual or auditory stimulus ("dishabituation"). Once the response had reappeared it would then often continue to make its appearance, even when the paw shock was omitted. Ben-Ari and Le Gal La Salle (1972) also describe a single unit in which a sequence of clicker stimuli produced no response, and a subsequent sequence of pairings of the clicker with a forepaw shock also produced no response; but when the clicker was then presented alone again a vigorous excitatory response to the clicker was elicited. Such a response profile may be an indication of a special responsiveness to disconfirmed expectancy. There is of course much work needed to prove this and to evaluate fully the predictions of the previous paragraph.

There is however, a lesser prediction to be made, where a little evidence is available. If the amygdala achieves a mathematical differentiation of motivational value, the information from which this computation must start should give a rather accurate measurement of the absolute reward value. Without this, the change scores across trials could not be computed. If however no differentiation is performed a simple categorical representation of whether reward had occurred or not would suffice. A few studies do appear to indicate that neurones in the amygdala respond in a manner giving a rather systematic representation of magnitude of reward (Nishijo *et al.*, 1988; Peinado-Manzano, 1989; Kesner, Walser and Winsenreid, 1989; Fukuda and Ono, 1993).

### 5.3. Synaptic Plasticity and the Information Processing in the Amygdala

Before leaving the electrophysiological evidence about associative processes in the amygdala, the actual processes of synaptic plasticity should be briefly mentioned. Chapman *et al.* 1990) have observed long-term synaptic potentiation in single neurones in the amygdala, following high frequency stimulation of axons in the external capsule, afferent to the amygdala. This was observed in 80% of cells in which suitable recordings could be made, and thus appears to be more robust than in some other forebrain structures (e.g. the cerebral neocortex). Otherwise, it is not a very unusual finding. However, the review of the psychological and lesion studies presented above suggests that information processing of some sophistication may be occurring within the amygdala. One possible explanation of such processing might implicate forms of synaptic plasticity rather different from those identified in cortical structures. In this context, the very recent paper of Li *et al.* (1998) should be mentioned. In the basolateral nucleus of the amygdala, low frequency stimulation induced enduring synaptic potentiation, but if the low frequency train was preceded by a high frequency train, synaptic depression was produced instead.

## 6. SEPARATION OF ROLES FOR DIFFERENT NUCLEI IN THE AMYGDALOID COMPLEX?

In anatomical terms the amygdala is a complex and heterogeneous structure. It has a number of distinct nuclei, each with their own cytology and their own afferent and efferent connections. Even within an individual nucleus there may be regional variations in cytology and connections. There may also be significant species variations. In this respect it differs from most other forebrain structures—such as neocortex, hippocampus, striatum or thalamus—each of which seems to be built up by accumulation of repeated processing units, which, despite some differences within a macrostructure, are recognisable as "variations on a theme". Amongst the evidence so far reviewed, little attention has been devoted to the difficult subject of the possible separate roles of different nuclei in the amygdala. However, a few suggestions are appropriate at this stage.

Behaviour will become completely disorganised unless it is directed mainly towards one goal at a time, and this has implications for the control of the reinforcement system as well as other systems. However, computation of which single motive (amongst a variety) should be the focus for behavioural control systems at any one time, is a complex operation. This may be reflected in the heterogeneous nuclear composition of the amygdala, because such complex operations require a detailed genetic blueprint, closely related to the ecology in which a species has evolved, rather than a versatile "all-purpose" structure such as the cerebral cortex, capable of information processing in a far more generalised sense. It seems unlikely that the concept of repeated processing units applies to the amygdaloid complex.

One possibility is that the different nuclei of which the amygdala is composed may each be concerned with different motives. However, a brief look at some of the detail of the lesion and single unit studies referred to above shows that this is not the case. Individual nuclei may, in different experiments, be involved in appetitive tasks using either the sight of food or an attractive taste, and aversive tasks using painful shocks in a variety of paradigms, as well as taste aversion. An alternative view, is that motivated behaviour can be grouped into two broad categories, and it is these which have separate representation in the different nuclei. This view has support from older researchers, such as Kaada (1971). His review led to the conclusions that some nuclear groups are concerned with facilitating feeding and sexual activity, flight and defence, while other areas exerted an overall inhibitory effect on a wide range of behaviours. This approach has not received much emphasis in more recent reviews of the amygdala (Ben-Ari, 1981; Aggleton, 1992).

A full exploration of this topic, including modern research as well as the older evidence is well beyond the scope of the present chapter. However, a few suggestions arising from modern studies will be made. Kilcross, Robbins and Everitt (1997) showed that lesions in different parts of the amygdala cause differential impairment in tasks using aversive stimuli with different experimental designs. Rats with lesions of the central nucleus of the amygdala showed a reduction in their ability to suppress behaviour elicited by a conditioned fear stimulus (a Pavlovian design), but were simultaneously able to direct their actions to avoid further presentations of the aversive stimuli (an instrumental design). In other words, the positive affect produced by avoidance of the aversive stimulus was intact, but the negative affect from its actual occurrence was not. In contrast, animals with lesions of basolateral amygdala could not avoid the conditioned aversive stimuli, but showed normal conditioned suppression.

In the single unit study of Yasoshima, Shimamura and Yamamoto (1995) different patterns of response to taste stimuli were seen in basolateral and central nuclei. In the basolateral nucleus, neurones were not often sensitive to taste stimuli in the preconditioning period. Moreover they did not respond to aversive gustatory stimuli such as lithium chloride or quinine. However, they acquired facilitatory responses to taste stimuli when the same taste acquired aversive properties after its pairing with lithium chloride. In contrast, in the central nucleus of the amygdala, neurones have a baseline responsiveness to aversive taste stimuli, and acquire a response to other taste stimuli as conditioned taste aversion developed. In discussion the authors suggest that the basolateral nucleus communicates qualitative and cognitive aspects of taste to the hypothalamus, while the central nucleus communicates innate aversiveness of certain tastes.

From such studies the possibility is raised that different nuclear divisions have different information processing roles in the overall computation of which motivational signals should prevail. The complex connectivity within the amygdala between its various nuclear subdivisions would become significant if this were the case. Much work is needed to clarify the details of these differences between nuclei and their interactions.

## 7. SUMMARY: TOWARDS A NEURONAL MODEL

The concept of the function of the amygdala which emerges from the evidence reviewed above is that it is concerned with central decisions about the motivational goals which an animal should pursue. This involves balancing the competing claims of various motives, on the basis not only of motivational cues immediately available, but also with respect to a wide variety of other sensory stimuli which offer predictions of how favourable changes in motivational state may be achieved. In addition, the amygdala estimates the motivational value of all these cues in a semi-quantitative sense. As a result it is possible to compare the motivational value of particular cues from one occasion to the next when a similar event occurs. From this it appears that *changes* in motivational state may sometimes have greater importance than absolute values. Hence the amygdala can compute the best motives to follow in the long as well as the short term. The objective of this complex process is to reach a clear decision about which motive should dominate behaviour at any time. When this decision is taken, it has major implications for control of the reinforcement system discussed in other chapters of this book. Such decisions also affect control of other systems involved in behaviour, such as the autonomic nervous system, the endocrine system, as well as certain fixed motor strategies by which motivated behaviour may sometimes be expressed. Since a focus on one motive takes away the focus on concomitantly active motives, the different values of competing motives must somehow be reduced to a single "currency" as regards behavioural control. One may thus regard the amygdala, figuratively, as a central accountant or "Minister of Finance" deciding priorities for expenditure of the organisms' resources.

A significant part of the above computation is an operation akin to mathematical differentiation, so that favourable or unfavourable changes are brought to prominence. This implies quantitative coding of the magnitude of rewards and punishments, before the "change score" can be made explicit. Such differentiation may be a very fundamental part of the operations of the amygdala. In fact, in chapter 2 Beninger and Olmstead imply that reinforcement is *not* controlled by particular stimuli which innately have motivational significance. *Any* stimulus can acquire such significance, provided it occurs in such a way as to predict a change in the motivational state. This formulation already implies the process of differentiation.

The concept of amygdala function developed in this chapter, and the overall relationship of the amygdala to the nigro-striatal system of information processing is summarised in Figure 5.3. The parallels between this figure and those presented above (Figures 5.1 and 5.2) purely on the basis of psychological evidence, should be noted.

The amygdala is complex. We are not yet in sight of a neuronal model. Conclusions such as those summarised above are derived fairly well from empirical evidence. Nevertheless, the evidence is in some ways insufficiently precise to permit effective neuronal modelling. Major empirical and theoretical questions remain before such modelling becomes a realistic objective for research. These include: Is a "motivational state" the right concept to deal with here? What is the neurobiological substrate of a "motivational state? Are changes in this state sufficient to explain the normal control of reinforcement systems? If a differentiation process occurs, what is the underlying neural machinery? What is the role of different nuclei with respect to different motives and different aspects of the computational process referred to? Does one need to invoke a radically new form of synaptic plasticity in the amygdala? Alternatively, is the seemingly complex and inscrutable computation accomplished therein

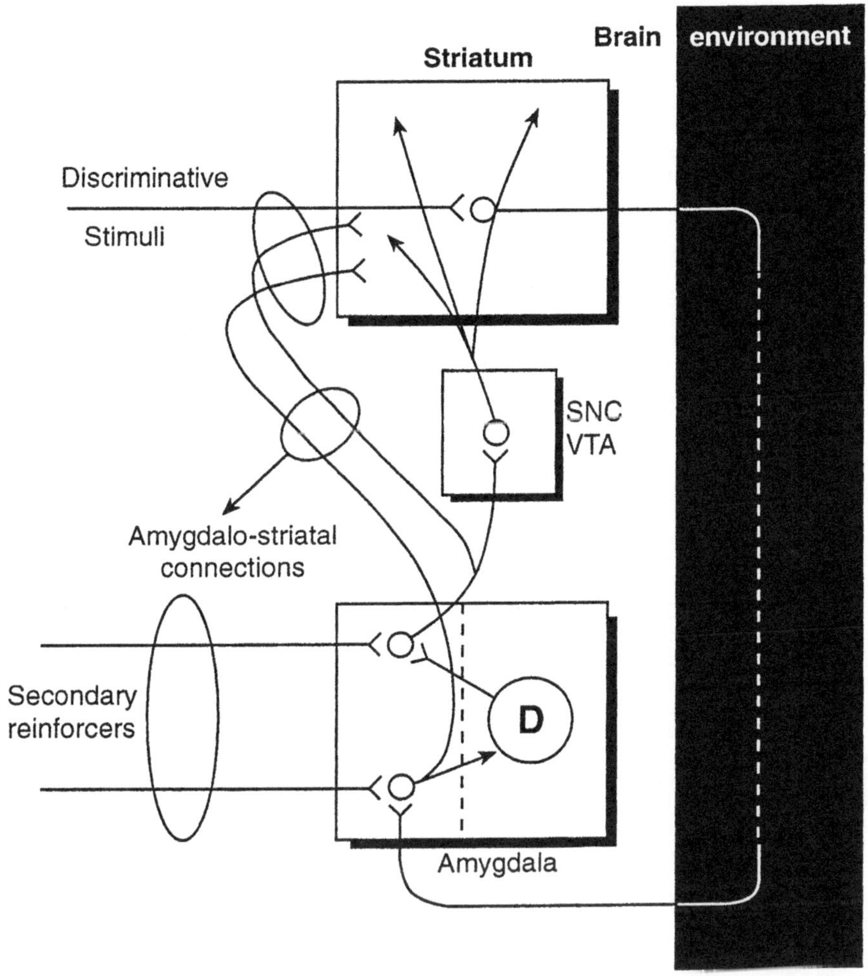


**Figure 5.3.** Scheme for instrumental conditioning in which specific brain structures are identified with psychological processes. The Figure includes direct links from amygdala to striatum as well as the indirect ones controlling dopamine neuronal firing. Compare with Figures 1 and 2.

achieved by more familiar principles of synaptic plasticity, which are made complex only by the interaction of several such familiar processes in several nuclei. Elucidation of these questions will require exhaustive scholarship of existing literature before the critical questions for new experiments can be adequately defined.

## REFERENCES.

Adelman, H.M. and Maatsch, J.L. (1955) Resistance to extinction as a function of the type of response elicited by frustration. *Journal of Experimental Psychology* **50**, 61–65

Aggleton, J.P. and Passingham, R.E. (1982) An assessment of the reinforcing properties of foods after amygdaloid lesions in rhesus monkeys. *Journal of Comparative and Physiological Psychology* **96**, 71–77.

Aggleton, J.P. (1992) *The amygdala: neurobiological aspects of emotion, memory, and mental dysfunction.* Wiley, New York.

Amsel, A. and Hancock, W. (1957) Motivational properties of frustration: III Relation of frustration effect to antedating goal factors. *Journal of Experimental Psychology* **53**, 126–131.

Amsel, A. and Roussel, J. (1952) Motivational properties of frustration: I Effect on a running response of addition of frustration to the motivational complex. *Journal of Experimental Psychology* **43**, 363–368

Applegate, C.D., Frysingher, R.C., Kapp, B.S. and Gallagher, M. (1982) Multiple unit activity recorded from amygdala central nucleus during Pavlovian heart rate conditioning in rabbit. *Brain Research* **238**, 457–462

Arbuthnott, G.W., Mitchell, M.J., Tulloch, I.F. and Wright, A.K. (1076) Efferent pathways from lateral hypothalamic neurones. *Journal of Physiology, London* **263**, 131P

Bagshaw, M.H. and Coppock, H.W. (1968) Galvanic skin response conditioning deficit in amygdalectomised monkeys. *Experimental Neurology* **20**, 188–196

Becker, H.C., Jarvis, M.F., Wagner, G.C. and Flaherty, C.F. (1984) Medial and lateral amygdalectomy differentially influences consummatory negative contrast. *Physiology and Behavior* **33**, 707–712.

Ben-Ari, Y. (1972) Plasticity at Unitary level. 1. An experimental design. *Electroencephalography and Clinical Neurophysiology* **32**, 655–665.

Ben-Ari, Y. (1981) *The Amygdaloid complex. INSERM Symposium no. 20.* Elsevier, N.Holland.

Ben-Ari, Y. and Le Gal La Salle, G. (1972) Plasticity at unitary level. 2 Modifications during sensory-sensory association procedures. *Electroencephalography and Clinical Neurophysiology* **32**, 667–679

Ben-Ari, Y. and Le Gal La Salle, G. (1974) Lateral amygdala unit activity: 2 Habituating and non-habituating neurons. *Electroencephalography and Clinical Neurophysiology* **37**, 463–472

Blanchard, D.C. and Blanchard, R.J. (1972) Innate and conditioned reactions to threat in rats with amygdaloid lesions. *Journal of Comparative and Physiological Psychology* **81**, 281–290.

Bower, G.H. (1961) A contrast effect in differential conditioning. *Journal of Experimental Psychology* **62**, 196–199

Brady, J.V., Schreiner, L., Geller, I. and Kling, A. (1954) Subcortical mechanisms in emotional behaviour: the effect of rhinencephalic injury. *Journal of Comparative and Physiological Psychology* **47**, 179–186

Bunney, B.S. and Aghajanian, G.K. (1976) The precise localization of nigral afferents in the rat as determined by a retrograde tracing technique. *Brain Research* **117**, 423–435

Bush, D.F., Lovely, R.H. and Pagano, R.R. (1973) Injections of ACTH induces recovery from shuttle-box avoidance deficits in rats with amygdaloid lesions. *Journal of Comparative and Physiological Psychology* **83**, 168–172

Cador, M., Robbins, T.W. and Everitt, B.J. (1989) Involvement of the amygdala in stimulus-reward associations: interaction with the ventral striatum. *Neuroscience* **30**, 77–86

Campbell, A. (1978) Deficits in visual learning produced by posterior temporal lesions in cats. *Journal of Comparative and Physiological Psychology* **92**, 45–57

Cassell, M.D., Gray, T.S. and Kiss, J.Z. (1986) Neuronal architecture in the rat central nucleus of the amygdala: a cytological, hodological and immunocyotchemical study. *Journal of Comparative Neurology* **246**, 478–499.

Chang, F.-C.T. and Scott, T.R. (1984) Conditioned taste aversions modify neural responses in the rat nucleus tractus solitarius. *Journal of Neuroscience* **4**, 1850–1862

Chapman, P.F., Kairiss, E.W., Keenan, C.L. and Brown, T.H. (1990) Long-term synaptic potentiation in the amygdala. *Synapse* **6**, 271–278

Chiba, T and Murata, Y. (1985) Afferent and efferent connections of the medial preoptic area in the rat: a WGA-HRP study. *Brain Research Bulletin* **14**, 261–272

Conrad, L.C.A. and Pfaff, D.W. (1976) Efferents from the medial basal forebrain and hypothalamus in the rat II An autoradiographic study of the anterior hypothalamus. *Journal of Comparative Neurology* **169**, 221–62

Coover, G., Ursin, H. and Levine, S. (1973) Corticosterone and avoidance in rats with basolateral amygdala lesions. *Journal of Comparative and Physiological Psychology* **85**, 111–122

Crespi, L.P. (1942) Quantitative variation of incentive and performance in the white rat. *American Journal of Psychology* **55**, 467–517

Dicks, D., Myers, R.E., and Kling, A. (1969) Uncus and amygdala lesions: effects on social behaviour in free range monkeys. *Science, New York* **165**, 69–71

Eclancher, F. and Karli, P. (1980) Effects of infant and adult amygdaloid lesions upon acquisition of two-way active avoidance by the adult rat: influence of rearing conditions. *Physiology and Behavior* **24**, 887–893

Everitt, B.J. and Robbins, T.W. (1992) Amygdala ventral-striatal interactions and reward-related processes. In: *The amygdala: neurobiological aspects of emotion, memory, and mental dysfunction.* edited by J.P. Aggleton., Wiley, New York, pp. 401–429

Flaherty, C.F. (1982) Incentive contrast: a review of behavioural changes following shifts in reward. *Animal Learning and Behavior.* **10**, 409–440

Flaherty, C.F. and Checke, S. (1982) Anticipation of incentive gain. *Animal Learning and Behavior* **10**, 177–182

Fonberg, E. and Kostarczyk, E. (1980) Motivational role of social reinforcement in dog-man relations *Acta Neurobiologiae Experimentalis* **40**, 117–136.

Fukuda, M. and Ono, T. (1993) Amygdala-hypothalamic control of feeding behaviour in monkey. Single cell responses before and after reversible blockade of temporal cortex or amygdala projections. *Behavioral Brain Research* **55**, 233–241

Fuster, J.M. and Uyeda, A.A. (1971) Reactivity of limbic neurons of the monkey to appetitive and aversive signals. *Electroencephalography and Clinical Neurophysiology* **30**, 281–293

Gaffan, D. and Harrison, S. 1987 Amygdalectomy and disconnection in visual learning for auditory secondary reinforcement by monkeys. *Journal of Neuroscience* **7**, 2285–2292.

Gaffan, E.A., Gaffan, D. and Harrison, S. (1988) Disconnection of the amygdala from visual association cortex impairs visual reward-association learning in monkeys. *Journal of Neuroscience* **8**, 3144–3150.

Gaffan, D., Gaffan, E.A. and Harrison, S. (1989) Visual-visual associative learning and reward-association learning in monkeys: The role of the amygdala. *Journal of Neuroscience* **9**, 558–564

Gerfen, C.R. (1984) The neostriatal mosaic: Compartmentalization of cortico-striatal input and striatonigral output systems. *Nature,* London, **311**, 461–464

Gerfen, C.R. (1985) The neostriatal mosaic. 1 Compartmental organization of projections from the striatum to the substantia nigra in the rat. *Journal of Comparative Neurology*, **236**, 454–476.

Gonzales, C. and Chesselet, M.-F. (1990) Amygdalonigral pathway: an anterograde study in the rat with Phaseolus vulgaris leucoagglutinin (PHA-L) *Journal of Comparative Neurology* **297**, 182–200.

Goomas, D.T., Hamm, C. and Skinner, J. (1980) Runway performance of amygdalectomised rats: magnitude of reinforcement and delay of food reward. *Physiological Psychology* **8**, 97–100

Goomas, D.T. and Steele, M.K. (1980) The collapse effect and delay of reinforcement with amygdalectomized rats. *Physiological Psychology* **8**, 463–466

Grossman, S.P., Grossman, L. and Walsh, L. (1975) Functional organization of the rat amygdala with respect to avoidance behaviour. *Journal of Comparative and Physiological Psychology* **88**, 829–850

Hatfield, T., Han, J.-S., Conley, M., Gallgher, M. and Holland, P. (1996) Neurotoxic NMDA (presumably axon-sparing) lesions of basolateral, but not central amygdala interfere with Pavolvian second-order conditioning and reinforcer devaluation effects. *Journal of Neuroscience* **16**, 5256–5265.

Henke, P.G. (1972) Amygdalectomy and mixed reinforcement schedule contrast effects. *Psychonomic Science* **28**, 301–302

Henke, P.G. (1973) Effects of reinforcement omission on rats with lesions in the amygdala *Journal of Comparative and Physiological Psychology* **84**, 760–767

Henke, P.G. (1977) Dissociation of the frustration effect and the partial reinforcement extinction effect after limbic lesions in rats.*Journal of Comparative and Physiological Psychology* **91**, 1032–1038

Henke, P.G. (1979) Limbic lesions and the energizing, aversive and inhibitory effects of non-reward in rats. *Canadian Journal of Psychology* **33**, 133–140

Henke, P.G. and Bunnell, B.N. (1971) Reinforcement and extinction interactions after limbic lesions in rats. *Communications in Behavioral Biology* **6**, 329–333.

Henke, P.G., Allen, J.D. and Davidson, C. (1972) Effect of lesions in the amygdala on behavioural contrast. *Physiology and Behavior* **8**, 173–176.

Henke, P.G. and Maxwell, D. (1973) Lesions in the amygdala and the frustration effect. *Physiology and Behavior* **10**, 647–650.

Hiroi, N. and White, N.M. (1991) The lateral nucleus of the amygdala mediates expression of the amphetamine-produced conditioned place preference. *Journal of Neuroscience* **11**, 2107–2116.

Holmes, P.A. and Brookshire, K.H. (1968) Exposure to electric shock and performance in an approach-avoidance conflict situation. *Journal of Comparative and Physiological Psychology* **65**, 531–534

Hopkins, D.A. and Holstege, G. (1978) Amygdaloid projections to the mesencephalon, pons and medulla oblongata in the cat. *Experimental Brain Research* **32**, 529–547

Horvath, F.E. (1963) Effects of basolateral amygdalectomy on three types of avoidance behaviour in cats. *Journal of Comparative and Physiological Psychology* **56**, 380–389

Hosoya, Y. and Matsushita, M. (1981) Brainstem projections from the lateral hypothalamic area in the rat as studied with autoradiography. *Neuroscience Letters* **24**, 111–116

Jacobs, B.L. and McGinty, D.J. (1972) Participation of the amygdala in complex stimulus recognition and behavioural inhibition: Evidence from unit studies. *Brain Research* 36, 431–436

Jones, B. and Mishkin, M. (1972) Limbic lesions and the problem of stimulus-reinforcement associations. *Experimental Neurology* **36**, 362–377

Kaada, B.R. (1971) Stimulation and regonal ablation of the amygdalid complex with reference to functional representations. In *The neurobiology of the amygdala (Adv.Behav.Biol.2)* edited by B.Eleftheriou, pp. 205–264, New York, Plenum Press

Kelly, E.A., Domesick, V.B. and Nauta, W.J.H. (1982) The amygdalostriatal projection in the rat. An anatomical study by anterograde and retrograde tracing methods. *Neuroscience* **7**, 615–630

Kemble, E.D. and Tapp, J.T. (1968) Passive and active avoidance performance following small amygdaloid lesions in rats. *Physiology and Behavior* **3**, 713–718

Kemble, E.D. and Beckman, G.J. (1969) Escape latencies at three levels of electric shock in rats with amygdala lesions. *Psychonomic Science* **14**, 205–206.

Kemble, E.D. and Beckman, G.J. (1970) Runway performance of rats after amygdaloid lesions. *Physiology and Behavior* **5**, 45–47

Kesner, R.P., Walser, R.D. and Winzenried, G. (1989) Central but not basolateral amygdala mediates memory for positive affective experiences. *Behavioral Brain Research* **33**, 189–195

Kilcross, S., Robbins, T.W. and Everitt, B.J. (1997) Different types of fear-conditioned behaviour mediated by separate nuclei within amygdala. *Nature, London* **388**, 377–380

Kita, H. and Kitai, S.T. (1990) Amygdaloid projections to the frontal cortex and the striatum in the rat. *Journal of Comparative Neurology* **298**, 40–49.

Krettek, J.E. and Price, L. (1978) Amygdaloid projections to subcortical structures within the basal forebrain and brainstem in the rat and cat. *Journal of Comparative Neurology* **178**, 225–254

Li, H., Weiss, S.R.B., Chuang, D.-M., Post, R.M. and Rogawski, M.A. (1998) Bidirectional synaptic plasticity in the rat basolateral amygdala: Characterization of an activity-dependent switch sensitive to the presynaptic metabotropic glutamate receptor anatagonist 2S-alpha-ethylglutamic acid. *Journal of Neuroscience* **18**, 1662–1670.

McAllister, D.E., McAllister, W.R., Brooks, C.I., and Goldman, J.A. (1972) Magnitude and shift of reward in instrumental aversive learning in rats *Journal of Comparative and Physiological Psychology* **80**, 490–501

McBride, R.L. and Sutin, J. (1977) Amygdaloid and pontine projections to the ventromedial nucleus of the hypothalamus. *Journal of Comparative Neurology* **174**, 377–396.

McDonald, A.J. (1991) Projection neurons of the basolateral amygdala: a correlative Golgi and retrograde tract tracing study. *Brain Research Bulletin* **28**, 179–185.

McDonough, J.H. and Manning, F.J. (1979) The effect of lesions in the amygdala or dorsomedial frontal cortex on reinforcement omission and non-contingent reinforcement in rats. *Physiological Psychology* **7**, 167–172

McEchron, M.D., McCabe, P.M., Green, E.J., Llabre, M.M. and Schneiderman, N. (1995) Simultaneous single unit recording in the medial nucleus of the medial geniculate nucleus and amygdaloid central nucleus throughout habituation, acquisition and extinction of the rabbit's classically conditioned heart rate. *Brain Research* **682**, 157–166

MacKintosh, N.J. (1974) *The psychology of animal learning*. London, New York, San Francisco: Academic Press.

McNew, J.J. and Thompson, R. (1966) Role of the limbic system in active and passive avoidance conditioning in the rat. *Journal of Comparative and Physiological Psychology* **61**, 173–180

Miller, N.E. (1960) Learning resistance to pain and fear: effects of overlearning, exposure and rewarded exposure in context. *Journal of Experimental Psychology* **60**, 137–145.

Molino, A. (1975) Sparing of functions after infant lesions of selected limbic structures in the rat. *Journal of Comparative and Physiological Psychology* **89**, 868–881

Muramoto, K., Ono, T., Nishijo, H. and Fukuda, M. (1993) Rat amygdaloid neuron responses during auditory discrimination. *Neuroscience* **52**, 621–636.

Murray, E.A. and Mishkin, M. (1985) Amygdelectomy impairs crossmodal association in monkeys. *Science, New York* **228**, 604–606.

Nagel, J.A. and Kemble, E.D. (1976) Effects of amygdaloid lesions on the performance of rats in four passive avoidance tasks. *Physiology and Behavior* **17**, 245–250.

Nation, J.R., Wrather, W.R. and Mellgren, R.L, (1974) Contrast effects in escape conditioning of rats. *Journal of Comparative and Physiological Psychology* **86**, 69–73

Nation, J.R., Mellgren, R.L. and Wrather (1975) Contrast effects with shifts in punishment levels. *Bulletin of the Psychonomic Society* **5**, 167–169.

Nauta, W.J.H. and Domesick, V.B. (1978) Crossroads of limbic and striatal circuitry: hypothalamo-nigral connections. In *Limbic Mechanism*, edited by K.E. Livingston and O. Hornykiewicz, New York, Plenum, pp. 75–93.

Nishijo, H., Ono, T., Nakamura, K., Kawabata, M. and Yamatani, K. (1986) Neuron activity in and adjacent to the dorsal amygdala of monkey during operant feeding behavior. *Brain Research Bulletin* **17**, 847–854.

Nishijo, H., Ono, T. and Nishino, H. (1988) Single neuron responses in amygdala of alert monkey during complex sensory stimulation with affective significance. *Journal of Neuroscience* **8**, 3570–3583

Ono, T., Fukuda, M., Nishino, H., Sasaki, K. and Muramoto, K.-I. (1983) Amygdaloid neuronal response to complex visual stimuli in an operant feeding situation in the monkey. *Brain Research Bulletin* **11**, 115–18.

Panksepp, J. (1981) Hypothalamic integration of behaviour: rewards punishments and related psychological processes. In *Handbook of the Hypothalamus Vol 3B (Behavioural studies of the hypothalamus)*, edited by P.J. Morgane and J. Panksepp, Marcel Dekker Inc. New York, pp. 289–431.

Pascoe, J.P. and Kapp, B.S. (1985) Electrophysiological characteristics of amygdaloid central nucleus neurons during Pavlovian fear conditioning in the rabbit. *Behavioral Brain Research* **16**, 117–133.

Pellegrino, L (1968) Amygdaloid lesions and behavioural inhibition in the rat. *Journal of Comparative and Physiological Psychology* **65**, 483–491

Phillipson, O.T. (1978) Afferetn projections to A10 dopaminergic neurons in the rat as shown by the retrograde transport of horseradish peroxidase. *Neuroscience Letters* **9**, 353–359

Phillipson, O.T. (1979) Afferent projections to the ventral tegmental area of Tsai and interfascicular nucleus: a horseradish peroxidase study in the rat. *Journal of Comparative Neurology* **187**, 117–144

Peinado-Manzano, A. (1988) Effects of bilateral lesions of the central and lateral amygdala on free operant successive discrimination. *Behavioral Brain Research* **29**, 61–71

Peinado-Manzano, A. (1989) Intervention of the lateral and central amygdala on the association of visual stimuli with different magnitudes of reinforcement. *Behavioral Brain Research* **32**, 289–295.

Post, S. and Mai, J.K. (1980) Contribution to the amygdaloid projection field in the rat. A quantitative autoradiographic study. *Journal fur Hirnforschung* **21**, 199–225.

Pribram, K.H. and Bagshaw, M. (1953) Further analysis of the temporal lobe syndrome utilizing fronto-temporal ablations. *Journal of Comparative Neurology* **99**, 347–375

Price, J.L. and Amaral, D.G. (1981) An autoradiographic study of the projections of the central nucleus of the monkey amygdala. *Journal of Neuroscience* **1**, 1242–59

Ragsdale, C.W. and Graybiel, A.M. (1988) Fibers from the basolateral nucelus of the amygdala selectively innervate striosomes in the caudate nucleus of the cat. *Journal of Comparative Neurology* **269**, 506–522

Richardson, R.T. and Thompson, R.F. (1984) Amygdaloid unit activity during classical conditioning of the nictitating membrane response in rabbit. *Physiology and Behavior* **32**, 527–539.

Robinson, E. (1963) Effect of amygdalectomy on fear motivated behaviour in rats. *Journal of Comparative and Physiological Psychology* **56**, 814–820.

Robinson, T.G. and Beart, P.M. (1988) Excitant amino acid projections from rat amgdala and thalamus to nucleus accumbens. *Brain Research Bulletin* **20**, 467–471.

Royce, G.J. (1978) Cells of orgiin of subcortical afferehnts to the caudate nucleus: a horseradish peroxidase syudy in the cat. *Brain Research* **153**, 465–475.

Russchen, F.T. and Price, J.L. (1984) Amygdalostriatal projections in the rat, topographical organization and fiber morphology shown using the lectin PHA-L as an anterograde tracer. *Neuroscience Letters* **47**, 15–22.

Russo, N.J. II, Kapp, B.S., Holmqvist, B.K. and Musty, R.E. (1976) Passive avoidance and amygdala lesions: relationship with pituitary-adrenal system. *Physiology and Behavior* **16**, 191–199.

Salinas, J.A., Packard, M.G. and McGaugh, J.L. (1993) Amygdala modulates memory for changes in reward magnitude: reversible post-training inactivation with lidocaine attenuates the response to a reduction in reward. *Behavioral Brain Research* **59**, 153–159.

Salinas, J.A. and McGaugh, J.L. (1996) The amygdala modulates memory for changes in reward magnitude: involvement of the amygdaloid GABAergic system. *Behavioral Brain Research* **80**, 87–98.

Sanghera, M.K., Rolls, E.T. and Roper-Hall, A. (1979) Visual responses of neurons in the dorsolateral amygdala of the alert monkey. *Experimental Neurology* **63**, 610–626.

Saper, C.B., Swanson, L.W. and Cowan, W.M. (1979a) Some efferent connections of the rostral hypothalamus in the squirrel monkey (Saimiri sciurans) and cat. Journal of Comparative Neurology, **184**, 205–242.

Schreiner, L. and Kling, A. (1953) Behavioral changes following rhinencephalic injury in cat. *Journal of Neurophysiology* **16**, 643–659.

Schwartzbaum, J.S. (1960) Changes in reinforcing properties of stimuli following ablation of the amygdaloid complex in monkeys. *Journal of Comparative and Physiological Psychology* **53**, 388–395

Scull, J., Davies, K. and Amsel, A. (1970) Behavioural contrast and frustration effect in multiple and mixed fixed-interval schedules in the rat. *Journal of Comparative and Physiological Psychology* **71**, 478–483

Shanab, M.E., Domino, J. and Steinhauer, G. (1982) Sustained negative contrast obtained following signalled shifts in sucrose reinforcement. *Bulletin of the Psychonomic Society* **19**, 237–240

Shanab, M.E., Moyalem, O. and Steinhauer, G. (1982) Effects of signalled shifts in water reinforcement. *Bulletin of the Psychonomic Society* **19**, 237–240

Shibata, K., Kataoka, Y., Yamashita, K. and Ueki, S. (1986) An important role of the central amygdaloid nucleus and mammilary body in the mediation of conflict behaviour in rats. *Brain Research* **372**, 159–162.

Shinonaga, Y., Takada, M. and Mizuno, N. (1994) Topographic organization of collateral projections from the basolateral amygdaloid nucelus to both the prefrontal cortex and nucleus accumbens in the rat. *Neuroscience* **58**, 389–397

Simonov, P. (1998) Brain mechanisms of emotion. In: *Conceptual Advances in Russian neuroscience: Complex brain Function.* CABR series, Amsterdam: Harwood Academic Publishers (In press).

Somogyi, P., Bolam, J.P., Totterdell, S. and Smith, A.D. (1981) Monosynaptic input from the nucleus accumbens-ventral striatal region to retrogradely labelled nigrostriatal neurones. *Brain Research* **217**, 245–263.

Spevack, A.A., Campbell, C.T. and Drake, L. (1975) Effecy of amygdalectomy and CER in rats. *Physiology and Behavior* **15**, 199–207.

Spiegler, B.J. and Mishkin, M. (1981) Evidence for the sequential participation of inferior temporal cortex and amygdala in the acquisition of stimulus-reward associations. *Behavioral Brain Research* **3**, 303–317

Staddon, J.E.R. and Innis, N.K. (1969) Reinforcement omission on fized interval schedules. *Journal of the Experimental Analysis of Behavior* **12**, 689–670

Swanson, L.W. (1976) An Autoradiographic study of the efferent connections of the preoptic region in the rat. *Journal of Comparative Neurology* **167**, 227–256.

Thatcher, R.W. and Kimble, D.P. (1966) Effect of amygdaloid lesions on retention of an avoidance response in overtrained and non-overtrained rats. *Psychonomic Science* **6**, 9–10

Troiano, R. and Siegel, A. (1975) The ascending and descending connections of the hypothalamus in the cat. *Experimental Neurology* **49**, 161–173

Ursin, H. (1965) Effect of amygdaloid lesions on avoidance behaviour and visual discrimination in cats. Experimental Neurology, **11**, 298–317

Veazey, R.B., Amaral, D.G. and Cowan, W.M. (1982) The morphology and connections of the posterior hypothalamus in the cynomolgus monkey (Macac fascicularis) II Efferent connections. *Journal of Comparative Neurology* **207**, 135–156

Wallace, D.M., Magnuson, D.J. and gray, T.S. (1992) Organization of amygdaloid projections to brainstem dopaminergic, noradrenergic and adrenergic cell groups in the rat. *Brain Research Bulletin* **28**, 447–454.

Weiskrantz, L. (1956) Behavioural changes associated with ablation of the amygdaloid complex in monkeys. *Journal of Comparative and Physiological Psychology* **49**, 381–391

Werka, T. (1980) Acquisition of the escape reflex in cats after the nucleus centralis of the amygdala lesions. *Acta Neurobiologiae Experimentalis* **40**, 433–449

Werka, T., Skar, J. and Ursin, H. (1978) Exploration and avoidance in rats with lesions in amygdala and piriform cortex. *Journal of Comparative and Physiological Psychology* **92**, 672–681

Wolf, G. and Sutin, J. (1966) Fibre degeneration after lateral hypothalamic lesions in the rat. *Journal of Comparative Neurology* **127**, 137–156.

Wookey, P.E. and Strongman, K.T. (1971) Emotional and instrumental effects of reward shift. *Psychological Record* **21**, 181–189

Wookey, P.E. & Strongman, K.T. (1974) Frustration and eleation effects in operant analogues of the double runway. *British Journal of Psychology* **65**, 305–313

Yajeya, J, Patino, A., Riolobos, A.S., Criado, J.M. and De La Fuente, A. (1991) Passive avoidance conditioning and unitary activity in the basolateral amygdaloid nucleus of the rat. *Acta Physiologica Scandinavica* **141**, 549–553.

Yamamoto, T., Matsuo, R., Kiyomitsu, Y. and Kitamura, R. (1989) Taste responses of cortical neurons in freely ingesting rats. *Journal of Neurophysiology* **61**, 1244–1258.

Yasoshima, Y., Shimura, T. and Yamamoto, T. (1995) Single unit responses of the amygdala after conditioned taste aversion in conscious rats. *NeuroReport* **6**, 2424–2428.

Yeudall, L.T. and Walley, R.E. (1977) Methylphenidate, amygdalectomy and active avoidance performance in the rat. *Journal of Comparative and Physiological Psychology* **91**, 1207–1219

Zaborsky, L. (1982) Afferent connections of the medial basal hypothalamus. *Advances in Anatomy, Embryolology and Cell Biology* **69**, 1–107.

Zagrozka, J., Brudnias-Stepowska, Z. and Fonberg, E. (1983) Impairment of social behaviour in amygdalar cats. *Acta Neurobiologiae Experimentalis* **43**, 63–77.

Zimmerman, D.W. (1971) Rate changes after unscheduled omission and presentation of reinforcement. *Journal of the Experimental Analysis of Behavior* **15**, 261–270.

# 6 Synaptology and Physiology of Neostriatal Neurones


## Ben D. Bennett and Charles J. Wilson

*Department of Anatomy and Neurobiology, University of Tennessee, Memphis*
*855 Monroe Avenue, Memphis, TN 38163, USA*
*tel: (901) 448-4045; FAX (901) 448-7193; e-mail ben@marlin.utmem.edu*



The neostriatum contains two populations of cells, principal neurones and interneurones. The principal cells are the spiny projection neurones which provide synaptic input to other basal ganglia nuclei and, through local axon collaterals, contact interneurones and other spiny cells. The interneurones are local circuit cells providing synaptic input to both spiny neurones and interneurones. Detailed anatomical and physiological data on the spiny cells have provided the framework which underpins the interpretation that the electrical behaviour of these neurones *in vivo* is a result of a dynamic interaction between their intrinsic membrane properties and their synaptic input. Information concerning the intrinsic electrical characteristics of, and physiological properties of synaptic input to interneurones is beginning to emerge and is anticipated to be combined with existing and forthcoming anatomical data to explain the behaviour of these cells. The purpose of this chapter is to provide a synthesis of the currently available information concerning the synaptic organisation and physiological properties of the different populations of neostriatal neurones.



KEYWORDS: spiny neurone, interneurone, intrinsic properties, synaptic input.


## 1. INTRODUCTION

The neostriatum is one of the largest subcortical nuclei in the brain and is considered to represent the first stage of neural computation in the basal ganglia. The spiny projection neurones, which constitute the overwhelming majority of cells in this nucleus, are the basic neural substrate for information processing within this nucleus. The spiny cells have been the subject of intensive study during the last twenty years, and many features of their electrophysiological characteristics and anatomical organisation have been elucidated. Physiological data have shed light on the mechanisms underlying the characteristic electrical behaviour of spiny neurones observed *in vivo*. Specifically, these data have demonstrated how incoming excitatory inputs interact with the intrinsic membrane properties to influence membrane potential state transitions and action potential generation. The combination of both physiological and anatomical data are therefore providing a functional role for synaptic inputs identified in the electron microscope. Although there is a great deal of information on the physiological characteristics of specific types of synaptic inputs to spiny neurones, there are many inputs for which there is as yet no identified physiological function.

A comparison of the available physiological and anatomical data for the different classes of neostriatal interneurones reveals a large discrepancy. There is an abundance of information concerning the synaptology of the interneurones, largely provided through ultrastructural immunocytochemical studies, but there are very few data on the intrinsic and synaptic physiology of interneurones. This is a result of their sparsity, rendering them for all practical purposes almost inaccessible to conventional intracellular recording. However, the recent application of visualised recording is beginning to provide previously unattainable information on neostriatal interneurones and, through intracellular labelling, detailed information on the synaptic organisation of individual cells. The purpose of this text is to summarise the known data on the synaptic organisation of the neostriatum, and the electrical properties of the underlying neurones.

## 2. GENERAL ORGANISATION OF THE NEOSTRIATUM

The neostriatum is a heterogeneous structure, exhibiting at least three levels of anatomical and neurochemical organisation. At one level the neostriatum can be considered to contain two compartments, patch/striosome and matrix, which were initially identified on the basis of heterogeneous staining produced by $\mu$-opiate receptor binding (Pert, Kuhar and Snyder, 1976; Herkenham and Pert, 1981) and acetylcholinesterase (AChE) histochemistry (Graybiel and Ragsdale, 1978). Superimposed on this level of complexity is that of the organisation of inputs to the neostriatum which, in the case of the corticostriatal projection, exhibit patterns of innervation which are both topographically organised and preferentially targeted to a specific compartment (Künzle, 1975; Goldman and Nauta, 1977; Ragsdale and Graybiel, 1981; Gerfen, 1984; Donoghue and Herkenham, 1986; Goldman-Rakic and Selemon 1986; Gerfen, 1989). Furthermore, the projections arising from the neostriatum are heterogeneous: Neurones within the matrix compartment project either to the globus pallidus (GP), or to the globus pallidus and the entopeduncular nucleus (EP) and/or the substantia nigra *pars reticulata* (SNr); neurones of the patch compartment project to the substantia nigra *pars compacta* (SNc) (Chang, Wilson and Kitai, 1981; van der Kooy and Carter, 1981; Gerfen, 1985; Fink-Jensen and Mikkelsen, 1989; Parent, 1990; Smith and Bolam, 1990; Bolam and Smith, 1992; Gerfen, 1992; Flaherty and Graybiel, 1993; see chapter 10 by Gerfen). The striatoentopeduncular and striatonigral projections are referred to as the direct pathway and the striatopallidal projection is referred to as one of the indirect pathways. In addition, there are organisational principles which do not conform to this scheme, for instance, the projection neurones of the matrix exhibit multiple projection subtypes with some classes of spiny neurone innervating the GP, EP and SNr (Kawaguchi, Wilson and Emson, 1990).

## 3. NEOSTRIATAL CELL TYPES: MORPHOLOGY AND NEUROCHEMISTRY

### 3.1. Spiny Neurones

In the neostriatum the spiny projection neurones predominate (Kemp and Powell, 1971a) and account for approximately 77% and 95% of the total cell population in primates and rats, respectively (Graveland and DiFiglia, 1985). These neurones possess a perikaryon 10–20 $\mu$m in diameter in rat and primate, which gives rise to 3–5 smooth dendritic trunks that branch, usually within 10–30 $\mu$m, to produce secondary and higher order dendrites which are densely laden with spines (Figure 6.1A; Kemp and Powell 1971a; DiFiglia, Pasik and Pasik, 1976; Dimova, Vuillet and Seite, 1980; Wilson and Groves, 1980; Chang, Wilson and Kitai, 1982). As a consequence the majority of the surface area of a spiny cell is attributable to the spines themselves (see Wilson, 1990). Spiny cells usually possess 25–30 terminal dendrites which arborise in all directions from the perikaryon to fill an approximately spherical volume of 300–500 $\mu$m in diameter (see Wilson, 1990). The main axon arising from spiny neurones emits several collaterals, which arborise profusely, usually within the dendritic domain of the parent cell, before leaving the neostriatum (Kemp and Powell 1971a; DiFiglia, Pasik and Pasik, 1976; Kitai *et al.*, 1976; Preston, Bishop and Kitai, 1980; Wilson and Groves, 1980; Chang, Wilson and Kitai, 1981, 1982; Kawaguchi, Wilson and Emson, 1989, 1990). Although less commonly observed, a population of spiny neurones possessing widespread intrastriatal axon collaterals have also been described (Bishop, Chang and Kitai, 1982; Kawaguchi, Wilson and Emson, 1989, 1990). The volume occupied by these collaterals can be very large, with the arborisations extending over a 1 mm diameter, producing perhaps the most extended axonal field of any neostriatal neurone.

Spiny neurones contain glutamic acid decarboxylase (GAD; Fisher *et al.* 1986; Penny, Afsharpour and Kitai, 1986; Kita and Kitai, 1988), GABA (Smith *et al.*, 1987) and either enkephalin or substance P and dynorphin (Vincent *et al.*, 1982; Bolam *et al.*, 1983b; Haber and Nauta, 1983; Izzo, Graybiel and Bolam, 1987; Alheid and Heimer, 1988). The expression of these neuropeptides is related to the target nuclei of the spiny cell. Thus, the majority of spiny neurones which project to the output nuclei of the basal ganglia, specifically the EP and SNr, contain substance P and dynorphin mRNA and express the D1 dopamine receptor (Gerfen and Young, 1988; Gerfen *et al.*, 1990). In contrast, the striatopallidal neurones contain enkephalin and D2 dopamine receptors (Gerfen and Young 1988; Gerfen *et al.*, 1990). There are a few reports of co-localisation of these peptides in single neurones (Penny, Afsharpour and Kitai, 1986; Besson, Graybiel and Quinn, 1990) and that D1 and D2 receptors are co-localised (c.g. Surmeier *et al.*, 1993). However, more recent data suggest less co-localisation of D1 and D2 receptors than was originally suggested (Surmeier, Song and Yan, 1996). One of the most striking compartmental differences between spiny neurones is the presence of the calcium-binding protein, calbindin D–28k in projection cells of the matrix (Gerfen, Baimbridge and Miller, 1985).

### 3.2. Interneurones

The remaining 5% or 23% of neostriatal cells (in the rat and primate, respectively) are aspiny or sparsely spiny interneurones. The aspiny interneurones possess variably-sized somata, which give rise to aspiny or sparsely spiny dendrites (Kemp and Powell, 1971a; DiFiglia, Pasik and Pasik, 1976; Dimova, Vuillet and Seite, 1980; Chang, Wilson and Kitai, 1982; Bolam 1984; Takagi, Somogyi and Smith, 1984; Graveland and DiFiglia, 1985) and are further distinguishable on the basis of neurochemical content. Three major classes have been identified and include neurones which contain choline acetyltransferase (ChAT; Armstrong *et al.*, 1983; Levey *et al.*, 1983; Bolam, Wainer and Smith, 1984; Phelps, Houser and Vaughn, 1985; Graybiel, Baughman and Eckenstein, 1986; DiFiglia, 1987), GABAergic interneurones (Ribak, Vaughn and Roberts, 1979; Bolam *et al.*,

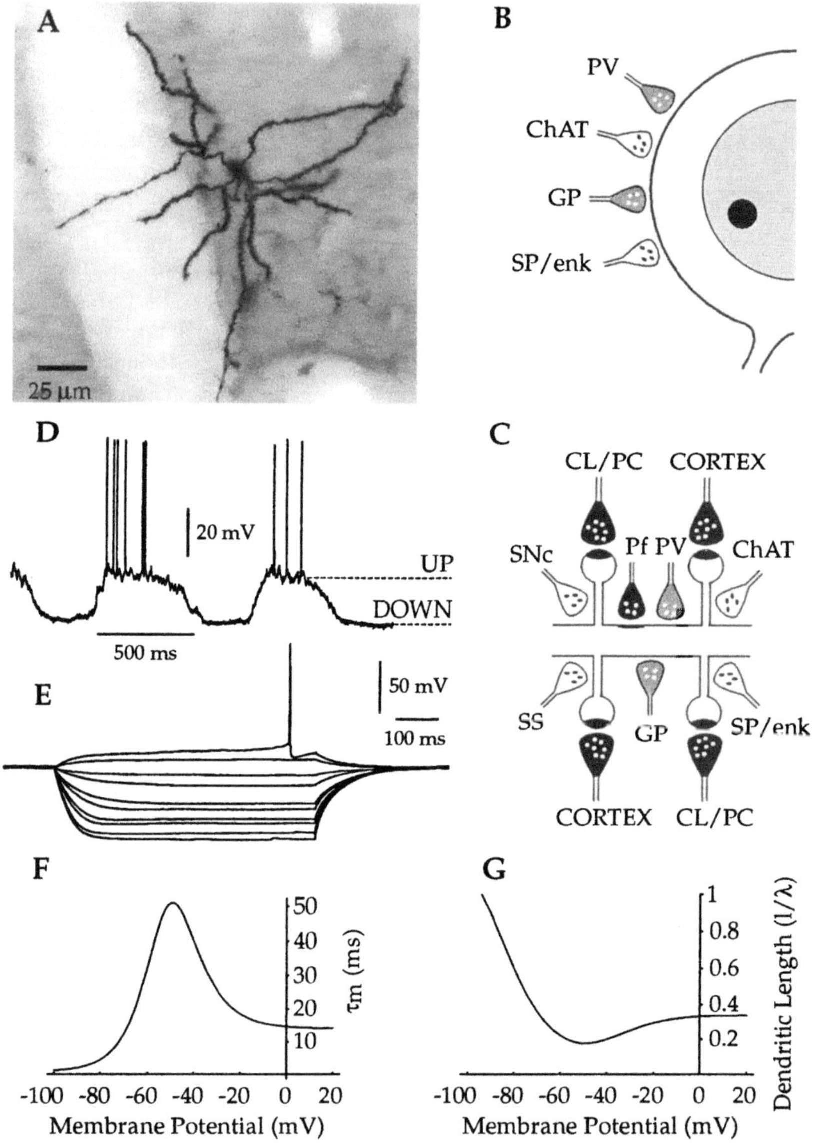


**Figure 6.1. A.** Neostriatal spiny projection neurone filled *in vivo*. Cell image in this, and all subsequent figures prepared using the synthetic projection microscopy technique of Agard *et al.* (1989). **B** and **C:** Schematic representations of the electron microscopically-verified synaptic inputs to the proximal (B) and distal (C) regions of spiny neurones (see Appendix for abbreviations) Excitatory and inhibitory inputs are boutons filled black and gray, respectively. Inputs with undefined functions are white. **D:** Spiny neurone recorded *in vivo* exhibiting spontaneous membrane potential state transitions and action potential generation (same cell as A). **E:** Spiny neurone *in vitro* displaying non-linear responses to intrasomatic current injection. Activation of $I_{Kir}$ (see 5.1.1) produces the non-linearity in the hyperpolarizing direction, whereas the slowly developing ramp potential is produced by time- and voltage-dependent inactivation of $I_{Kir}$. **F** and **G**: Voltage-dependent alteration in the membrane time constant, t (F) and dendritic electrotonic length, l (G). F and G were drawn assuming an inward rectifier with maximal conductance of 1 mS/cm$^2$, half activation voltage of –90 mV, a noninactivating outward rectification of 0.06 mS/cm$^2$ with a half inactivation voltage of –30 mV, and a leak conductance of 0.005 mS/cm$^2$.

1983a, 1985; Oertel and Mugnaini 1984; Kita and Kitai, 1988), a proportion of which contain parvalbumin (Gerfen, Baimbridge and Miller, 1985; Cowan *et al.*, 1990; Kita, Kosaka and Heizman, 1990) and a population in which neuropeptide Y (NPY), somatostatin (SS) and nicotinamide adenine dinucleotide phosphate-diaphorase (NADPH-d) are co-localised (Vincent *et al.*, 1983; Chesselet and Graybiel, 1986; Sandell, Graybiel and Chesselet, 1986; Smith and Parent 1986; Mizukawa *et al.*, 1988).

### *3.2.1. Cholinergic interneurones*

The ChAT-positive neurones are the giant cells of the neostriatum, and were correctly identified by Kölliker (1896), as a population of interneurones. Numerous Golgi-impregnation studies have provided further descriptions of the morphology of these neurones (e.g. Fox *et al.*, 1971; Kemp and Powell 1971a; DiFiglia, Pasik and Pasik, 1976; Dimova, Vuillet and Seite, 1980; Chang, Wilson and Kitai, 1982) and more recent investigations, utilising immunocytochemistry (Bolam, Wainer and Smith, 1984; Phelps, Houser and Vaughn, 1985; Graybiel, Baughman and Eckenstein, 1986; DiFiglia, 1987), have established the cholinergic nature of the giant neurones. These cells possess large spherical, oval or elongated perikarya (approximately 20–35 $\mu$m in diameter in rat and primate) from which smooth or sparsely spiny dendrites radiate (Figure 6.2A; Bolam, Wainer and Smith, 1984; Phelps, Houser and Vaughn, 1985; Graybiel, Baughman and Eckenstein, 1986; DiFiglia, 1987), bifurcating, often at great distances from the parent soma (Wilson, Chang and Kitai, 1990; Kawaguchi, 1992, 1993). The dendrites branch to occupy a volume which extends about 500 $\mu$m in the dorsovental and mediolateral axes and 750–1000 $\mu$m in the rostrocaudal plane (Wilson, Chang and Kitai, 1990). The distal dendrites display many fine branches and are sparsely spiny, forming dendritic tufts (Wilson, Chang and Kitai, 1990; Kawaguchi, 1992). The axon of ChAT-immunoreactive cells usually arises from the soma or from a large primary dendrite and then arborises profusely to give rise to intrastriatal axon collaterals which are usually more widespread than the dendritic field (Wilson, Chang and Kitai, 1990; Kawaguchi, 1992, 1993). The ChAT-positive axons are distributed preferentially in the matrix compartment in primate (Graybiel, Baughman and Eckenstein, 1986) and in the matrix in medial areas of the neostriatum in rats (Kubota and Kawaguchi, 1993). The cell bodies of cholinergic interneurones are found in both compartments and extend their dendrites across compartmental boundaries (Graybiel, Baughman and Eckenstein, 1986; Kawaguchi, 1992; Kubota and Kawaguchi, 1993). The ChAT-positive neurones also exhibit a preferential distribution at a macroscopic level, with a higher density of immunoreactive structures located in the dorsolateral neostriatum (Kubota and Kawaguchi, 1993).

### *3.2.2. Parvalbumin-immunoreactive interneurones*

GABAergic interneurones were identified initially in studies of GAD-immunoreactivity (Ribak, Vaughn and Roberts, 1979) and more recently through further investigations of GAD- (Bolam *et al.*, 1985; Aronin, Chase and DiFiglia, 1986; Kita and Kitai, 1988), GABA- (Bolam *et al.*, 1983a) and parvalbumin-containing (Cowan *et al.*, 1990; Kita, Kosaka and Heizmann, 1990; Côté, Sadikot and Parent, 1991; Lapper *et al.*, 1992) neurones. These neurones were first described in Golgi studies (e.g. Fox *et al.*, 1971; Kemp and Powell 1971a; DiFiglia, Pasik and Pasik, 1976; Dimova *et al.*, 1980; Chang, Wilson and Kitai, 1982) but their neurochemical nature was unclear in these original experiments. The interneurones in which parvalbumin, GAD and GABA are co-localised possess fusiform or polygonal somata (10–30 $\mu$m in diameter in rat and monkey) which give rise to dendrites that ramify extensively, close to the parent soma (Figure 6.3A; Cowan *et al.*, 1990; Kita, Kosaka and Heizmann, 1990; Côté, Sadikot and Parent, 1991; Lapper *et al.*, 1992). The dendrites are usually smooth in their proximal regions, but become varicose more distally. Two morphological types of parvalbumin-immunopositive neurones, which possess either local or extended dendritic fields, have recently been described (Kawaguchi, 1993). Both types give rise to axons which arborise most densely within or near their dendritic fields (Kawaguchi, 1993). Parvalbumin-immunoreactive fibres are preferentially distributed in the matrix compartment in primate neostriatum (Waldvogel and Faull, 1993; Bennett and Bolam, 1994a) and in the matrix in certain areas of the neostriatum in rats (Kubota and Kawaguchi, 1993). However, this apparent preferential distribution of parvalbumin-containing structures in primate should be viewed with caution. The centromedian/parafascicular thalamic nuclei provide a compartmentally organised projection to the neostriatum in monkeys, and contain parvalbumin-immunoreactive neurones (Jones and Hendry, 1989) and, in both rat and primate, the GP provides input to the neostriatum (see below), of which at least a proportion is likely to be parvalbumin-immunoreactive (Celio, 1990; Hontanilla, Parent, and Gimenez-Amaya, 1994; Kita, 1994; Rajakumar *et al.*, 1994; Bevan, Eaton and Bolam, 1997). Therefore, the apparent preponderance of parvalbumin-immunoreactive structures in the matrix compartment of the neostriatum is likely, at least partially, to result from the compartmental organisation of parvalbumin-containing afferent fibers.

The parvalbumin-immunoreactive perikarya are distributed in both patch and matrix compartments and extend their dendrites across compartmental boundaries (Cowan *et al.*, 1990; Kubota and Kawaguchi, 1993; Waldvogel and Faull, 1993; Bennett and Bolam, 1994a). Parvalbumin-immunoreactive structures are unevenly distributed within the neostriatum with the caudolateral regions of the neostriatum exhibiting the highest density.

### *3.2.3. SS/NPY/NADPH-d-positive neurones*

The interneurones in which SS, NPY and NADPH-d are co-localised (Vincent *et al.*, 1983; Chesselet and Graybiel, 1986; Sandell, Graybiel and Chesselet, 1986; Smith and Parent 1986; Mizukawa *et al.*, 1988) also contain nitric oxide synthase (Dawson *et al.*, 1991) and were initially identified on the basis of morphological features (Kemp and Powell 1971a; DiFiglia, Pasik and Pasik, 1976; Dimova *et al.*, 1980; Chang, Wilson and Kitai, 1982; Bolam 1984; Takagi, Somogyi and Smith, 1984). These neurones are typically bipolar and have somata that are approximately 10–20 $\mu$m in diameter in rat (see Figure 6.4A). The dendrites are smooth and aspiny, branching infrequently (DiFiglia and Aronin, 1982; Takagi *et al.*, 1983; Vincent *et al.*, 1983; Chesselet and Graybiel, 1986; Sandell, Graybiel and Chesselet, 1986; Smith and Parent, 1986; Mizukawa *et al.*, 1988) and their axonal arborisations are the most widespread of the three major classes of interneurones, occasionally appearing to have two axonal origins (Takagi, Somogyi and Smith, 1984; Kawaguchi, 1993). NADPH-d-positive fibres have been reported to be localised preferentially in the matrix compartment in cat caudate-putamen (Chesselet and Graybiel, 1986; Sandell, Graybiel and Chesselet, 1986). Similarly, the distribution of SS-immunoreactive fibres has also been reported to observe compartmental ordering in the neostriatum of the rat (Gerfen, 1984). The cell bodies of this class of interneurone are located in both compartments and extend their dendrites across compartmental boundaries (Chesselet and Graybiel, 1986; Sandell, Graybiel and Chesselet, 1986; Kubota and Kawaguchi, 1993).

#### *3.2.4. Other population of interneurones*

A population of neurones containing the calcium-binding protein, calretinin has been described (Jacobowitz and Winsky, 1991; Résibois and Rogers 1992) and appears to represent a distinct class of interneurones (Bennett and Bolam, 1993a; Figueredo-Cardenas *et al.* 1996). These neurones have perikarya 9–17 μm in diameter in rat, which give rise to aspiny dendrites which taper to become thin, varicose structures (Bennett and Bolam, 1993a). Little is known about the organisation of axonal arborisations of these neurones at the light microscopic level as there are no anatomical data from individually filled calretinin-positive neurones, and immunostained tissue is difficult to interpret because some of the inputs to the neostriatum contain calretinin (Bennett and Bolam, 1993a). Presently, there are also no data on the distribution of the neurones with respect to the compartmental organisation of the neostriatum, but it is known that the somata of these neurones are preferentially located in rostral regions (Bennett and Bolam, 1993a).

The remaining classes of neurones are even less clearly defined and include aspiny calbindin interneurones (Kiyama, Seto-Ohshima and Emson, 1990; Bennett and Bolam, 1993b), vasoactive intestinal polypeptide-immunoreactive cells (Theirault and Landis, 1987; Vincent and Reiner 1988) and cholecystokinin-immunopositive neurones (Takagi *et al.*, 1984; Adams and Fisher, 1990). In addition, there are descriptions of neurones in which some neurochemical markers are co-localised, including calbindin with NADPH-d (Bennett and Bolam, 1993b) and parvalbumin with calretinin (Figueredo-Cardenas *et al.* 1996), which may represent further subclasses of neurones. Nevertheless, the vast majority of interneurones are encompassed by the three major populations described above, namely cholinergic, parvalbumin- and SS/NPY/NADPH-d-positive cells. Hence, for the purposes of this chapter there will be no further discussion of the minor populations of interneurones.

## 4. SYNAPTOLOGY OF NEOSTRIATAL NEURONES

The description that follows will focus on the synaptic organisation of the neurones in the neostriatum that have been confirmed at the electron microscopic level. A description of the macroscopic organisation of inputs to, outputs from and compartmental organisation of, the neostriatum is beyond the scope of this chapter.

### 4.1. Spiny Neurones

#### *4.1.1. Afferents*

The neocortex provides a significant input to the neostriatum (Webster, 1961; Carman *et al.*, 1965; Kemp and Powell, 1970, 1971b; Heimer and Wilson, 1975), which is

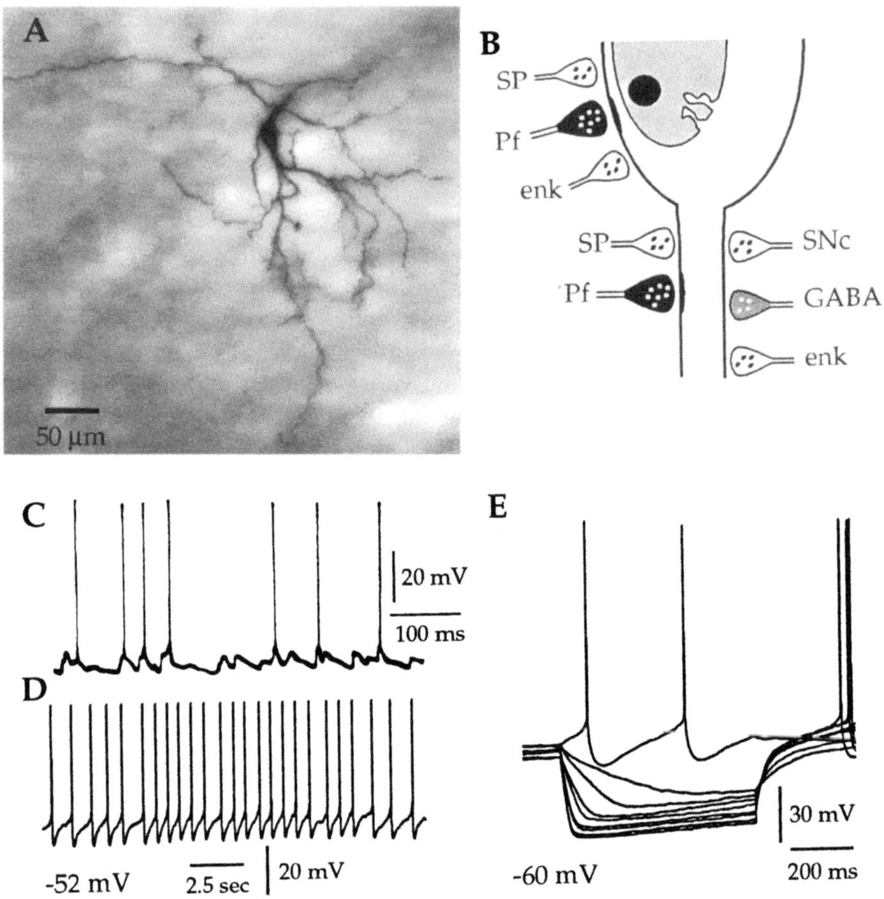


**Figure 6.2. A**. Giant neostriatal cholinergic interneurone filled *in vitro*. Note the thick primary dendrites and thin, widely distributed second and higher order dendritic branches. **B:** Schematic illustration of the synaptic inputs to cholinergic interneurones which have been confirmed at the ultrastructural level (see Appendix for abbreviations). **C** and **D:** Spontaneous action potential generation observed in cholinergic interneurones recorded *in vivo* (C) and *in vitro* (D). *E:* Cholinergic cells are physiologically characterised by their stereotypical responses to somatic current injection, including the hyperpolarisation-induced sag of the membrane potential and the large amplitude, long-duration afterhyperpolarisation.

organised with respect to the cortical area (Künzle 1975, 1977, 1978; Goldman and Nauta, 1977; Jones *et al.*, 1977; Ragsdale and Graybiel, 1981; Crutcher and DeLong, 1984; Gerfen, 1984; Alexander and DeLong, 1985; Selemon and Goldman-Rakic, 1985; Donoghue and Herkenham, 1986; Goldman-Rakic and Selemon, 1986; Malach and Graybiel, 1986; McGeorge and Faull 1989; Cavada and Goldman-Rakic 1991; Flaherty and Graybiel, 1991; Ebrahimi, Pochet and Roger, 1992; Flaherty and Graybiel, 1993; Yeterian and Pandya, 1993; Cowan and Wilson, 1994; Kincaid and Wilson, 1996) and laminar origin of the projection (Gerfen, 1989; Kincaid and Wilson, 1996). The corticostriatal input terminates almost exclusively on dendritic spines (Kemp, 1968; Kemp and Powell, 1970, 1971b; Hassler *et al.*, 1978; Frotscher *et al.*, 1981; Wictorin *et al.*, 1989; Xu, Wilson and Emson, 1989) including spines of identified striatonigral projection neurones (Somogyi, Bolam and Smith, 1981). The projection from the cortex is excitatory (Buchwald *et al.*, 1973; Kitai *et al.*, 1976; Spencer, 1976; Kocsis, Sugimori and Kitai, 1977; Wilson, Chang and Kitai, 1982; Wilson, 1986; Kawaguchi, Wilson and Emson, 1989) and utilises an amino acid transmitter which is probably glutamate (Divac, Fonnum and Storm-Mathisen, 1977; Kim *et al.*, 1977; McGeer *et al.*, 1977; Reubi and Cuénod, 1979; Godukhin, Zharikova and Novoselov, 1980; Fonnum, Storm-Mathisen and Divac, 1981; Hassler *et al.*, 1982). Cortical boutons are packed with round synaptic vesicles, usually contain one or two mitochondria and form asymmetric synaptic specialisations, predominantly with the heads of dendritic spines (Kemp, 1968; Kemp and Powell, 1970, 1971b; Hassler *et al.*, 1978; Frotscher *et al.*, 1981; Somogyi, Bolam and Smith, 1981; Wictorin *et al.*, 1989; Xu, Wilson and Emson, 1989).

As virtually every dendritic spine receives one asymmetric input it is possible to make estimates of the total number of excitatory inputs to an individual spiny neurone by determination of the total number of dendritic spines. The most accurate estimates have been made using high-voltage electron microscopy and indicate that on spiny neurones with the densest investment of spines there are approximately 500 spines per spiny cell dendrite (Wilson *et al.*, 1983). Assuming that the spiny neurones possess 25–30 dendrites (see Wilson, 1990) then the total number of spines, and therefore of excitatory inputs to an individual spiny neurone can be as high as 12,500–15,000 (see Wilson, 1990). Roughly half of these will arise from the corticostriatal projection and the remainder are contributed by the thalamostriatal input (see below). Electron microscopic analyses of corticostriatal axons have demonstrated that synaptic junctions are only formed by the varicose portions of the arborisation (Wouterlood and Groenewegen 1985; Kincaid, Zheng and Wilson, 1995). Hence, using light microscopic analyses it is possible to estimate the number of synaptic contacts formed, and the volume of neostriatal neuropil innervated by an individual corticostriatal axon. The geometry of the dendritic field of an individual spiny cell and the organisation of a single corticostriatal axon indicate that a spiny neurone receives between 1 and 10 synaptic inputs from a single cortical neurone (Wilson, 1995). Hence, an individual spiny neurone receives synaptic input from roughly 750–7500 corticostriatal neurones. There are two broad conclusions which can be drawn from these data. Firstly, there is a great deal of convergence in the corticostriatal projection at the cellular level, with many cortical neurones providing synaptic input to a single spiny cell. Secondly, the spiny neurones that are recipients of input from the same individual corticostriatal axon are not nearest neighbours, but are spatially spread out within the neostriatum in an as yet unidentified manner.

The intralaminar thalamic nuclei provide input to the neostriatum (Powell and Cowan, 1956) which is compartmentally organised in rats (Herkenham and Pert 1981), cats (Ragsdale and Graybiel, 1991) and primates (Sadikot *et al.*, 1992). The target of thalamostriatal boutons on individual neurones appears to be related to the nucleus from which the projection originates (Xu, Wilson and Emson, 1991). Hence, projections from the parafascicular nucleus terminate primarily on dendritic shafts both in rats (Dubé, Smith and Bolam, 1988; Xu, Wilson and Emson, 1991; 89%) and primates (Sadikot *et al.*, 1992; 81%). In primates, the projection arising from the centromedian thalamic nucleus also terminates preferentially on dendritic shafts (66%) but relatively more terminals provide synaptic input to spines compared to terminals originating from the parafascicular nucleus (Sadikot *et al.*, 1992). However, striatal inputs which originate from the central medial and paracentral thalamic nuclei terminate almost entirely on dendritic spines (Chung, Hassler and Wagner,

1977; Xu, Wilson and Emson, 1991). The thalamostriatal projection arising from the intralaminar thalamic nuclei is known to be excitatory (Purpura and Malliani, 1967; Buchwald *et al.*, 1973; Kitai *et al.*, 1976; Kocsis, Sugimori and Kitai, 1977; Wilson, Chang and Kitai, 1983) and probably utilises the excitatory amino acids aspartate and/or glutamate (Streit, 1980; Christie *et al.*, 1987; Fuller, Russchen and Price, 1987). Thalamic boutons are packed with vesicles, usually contain one or two mitochondria and form asymmetric synaptic specialisations (Chung, Hassler and Wagner, 1977; Dubé, Smith and Bolam, 1988; Xu, Wilson and Emson 1991; Lapper and Bolam, 1992). A recent single neurone tracing study indicates that the organisational principles governing the arborisation pattern of thalamostriatal axons may differ markedly from the corticostriatal projection (Deschenes *et al.*, 1996). Individual cells of the parafascicular and ethmoid nuclei give rise to focal, dense clusters of boutons which are spatially widely distributed within the neostriatum (Deschenes *et al.*, 1996). This pattern of innervation is suggestive that individual neurones of the parafascicular/ethmoid region may provide multiple inputs to individual neostriatal spiny neurones and consequently provide a much greater influence upon their firing patterns than individual corticostriatal cells (see below).

The dopaminergic neurones located in the SNc give rise to a compartmentally organised, dopamine-containing projection to the neostriatum (Gerfen, Herkenham and Thibault, 1987; Jiménez-Castellanos and Graybiel, 1987; Langer and Graybiel, 1989). Tyrosine hydroxylase-positive inputs to the neostriatum form symmetric synaptic specialisations, primarily with dendritic spines (59%) and shafts (35%), although neuronal perikarya and axon initial segments also receive this input (Bouyer *et al.*, 1984; Freund, Powell and Smith, 1984; Smith *et al.*, 1993). Both the larger varicose portions of these fibers and the thinner parts of the axon contain vesicles and form *en passant* synapses (Groves, 1980; Freund, Powell and Smith, 1984; Ren and Kita 1996). Analyses of identified striatonigral projection neurones have revealed that approximately two thirds of the dopaminergic input to these neurones are situated on dendritic spines, about one third on dendritic shafts and only a relatively minor input on cell bodies (Freund, Powell and Smith, 1984). However, these numbers are somewhat misleading as they can be incorrectly interpreted to indicate that the majority of spines receives a dopaminergic input. Although all spines are recipients of an asymmetric synaptic input, only about 8% also receive a second, symmetric input (Wilson *et al.*, 1983). There are additional sources of inputs to spines which form symmetric contacts (see below), so assuming that half of the symmetric inputs are dopaminergic, then only 4% of spines receive such input. As approximately two thirds of the dopaminergic input to spiny cells are directed to spines and the remaining one third is directed to dendritic shafts (Freund, Powell and Smith, 1984), then the total number of dopaminergic synapses on a particular spiny neurone is roughly 6% of the total number of spines. Using the numbers described above, this leads to an estimate of about 750–900 dopaminergic synapses on an individual spiny cell. It also should be noted that the distribution of tyrosine hydroxylase-positive synapses on dendritic spines probably does not represent a preferential distribution on dendritic spines, but rather reflects the overall distribution of dendritic membrane (Wilson, 1992).

The neostriatum also receives an ascending input from the GP (Staines, Atmadja and Fibiger, 1981; Arbuthnott *et al.*, 1982; Beckstead, 1983; Jayaraman, 1983; Walker, Arbuthnott and Wright, 1989; Kita, Chang and Fujimoto, 1991; Bolam and Bennett, 1995; Bevan, Eaton and Bolam, 1997) which is likely to be inhibitory, as nearly all GP neurones contain GAD (Oertel and Mugnaini 1984) or GABA (Smith *et al.*, 1987) and pallidal terminals in the EP, SNr and STN are enriched in GABA (Smith and Bolam

1989, 1991; Bolam and Smith 1992; Bevan, Smith and Bolam, 1996). The pallidostriatal projection provides input to the somata and dendritic shafts of spiny neurones (Kita, Chang and Fujimoto, 1991) and the terminals contain one to several mitochondria and form symmetric synaptic specialisations (Kita, Chang and Fujimoto, 1991).

The basolateral nucleus of the amygdala provides an input to the neostriatum (Russchen and Price, 1984; Russchen *et al.*, 1985; Ragsdale and Graybiel, 1988; Kita and Kitai, 1990; McDonald, 1991). The efferent projections of the amygdala are excitatory (Noda, Manohar and Adey, 1968; Dafny, Dauth and Gilman, 1975) although the transmitter or transmitters that mediate this excitation have yet to be elucidated. The amygdalostriatal terminals form asymmetric synaptic specialisations with dendritic spines, are densely packed with round synaptic vesicles and usually contain one or two mitochondria (Kita and Kitai, 1990).

In addition to the inputs described above, spiny neurones may also receive input from extrinsic sources such as the STN (Kita and Kitai, 1987; Carpenter and Jayaraman 1990; Smith, Hazrati and Parent, 1990) and the dorsal raphe nucleus (Soghomonian, Descarries and Watkins, 1989; Lavoie and Parent, 1990; Corvaja, Doucet and Bolam, 1991; Vertes, 1991). However, these inputs are very sparse and largely speculative as only the raphe-striatal projection has been examined at the ultrastructural level (Soghomonian, Descarries and Watkins, 1989; Corvaja, Doucet and Bolam, 1991).

### *4.1.2. Intrinsic connectivity*

Spiny projection neurones provide synaptic input to one another within the neostriatum, providing synaptic input to all regions of spiny neurones within the axon collateral field (Wilson and Groves, 1980; Somogyi, Bolam and Smith, 1981). Indeed, there is inter-connectivity between spiny cells of the direct and indirect pathways and between neurones which are involved in the same pathway (Aronin, Chase and DiFiglia, 1986; Bolam and Izzo, 1988; Yung *et al.*, 1996). Ultrastructural examination of immunostained material has revealed that the substance P- and enkephalin-positive boutons are directed primarily to the somata and dendritic shafts of striatonigral neurones (Aronin, Chase and DiFiglia, 1986; Bolam and Izzo, 1988).

Spiny cells also receive inputs from the three major populations of interneurones including cholinergic cells (Phelps, Houser and Vaughn, 1985; DiFiglia, 1987; Izzo and Bolam, 1988; Pickel and Chan, 1990), GABAergic, parvalbumin-immunoreactive interneurones (Kita, Kosaka and Heizmann, 1990; Bennett and Bolam,1994b) and the population of interneurones in which NADPH-d, SS and NPY are co-localised (DiFiglia and Aronin 1982; Takagi *et al.*, 1983). The ChAT-immunoreactive terminals form symmetric synaptic specialisations with all parts of spiny neurones (Phelps, Houser and Vaughn, 1985; DiFiglia, 1987; Izzo and Bolam, 1988), some of which have been identified as striatonigral projection cells (Izzo and Bolam, 1988). However, proportionately more cholinergic terminals are directed to dendrites and perikarya of spiny cells in rats (Izzo and Bolam, 1988) whereas inputs to spines predominate in primates (DiFiglia, 1987). Parvalbumin-immunoreactive terminals form symmetric synapses with all parts of spiny cells but the majority of inputs are directed to perikarya in rats (Kita, Kosaka and Heizman, 1990) including somata of identified striatonigral neurones (Bennett and Bolam, 1994b). The parvalbumin-positive terminals are enriched in GABA (Bennett and Bolam, 1994b) and are likely, therefore to be inhibitory. The SS-immunoreactive terminals form symmetric synaptic contacts with dendritic shafts and spines in the neostriatum of the rat (DiFiglia and Aronin, 1982; Takagi *et al.*, 1983). The cell bodies of NOS-positive neurones have been demonstrated to contain GAD following colchicine treatment indicating their possible GABAergic nature (Kubota, Mikawa and Kawaguchi, 1993).

Thus, spiny cells are recipients of synaptic input from an extremely diverse collection of axons, arising from both extrinsic and intrinsic sources. Although all of these studies have greatly enhanced our knowledge of the synaptic connectivity of spiny projection neurones in general, it is still unclear the degree to which these inputs converge upon individual neurones, or indeed whether there exists a specificity governing the input to individual projection cells. Nevertheless, some principles governing synaptic input can be gleaned from the available data. The excitatory, glutamatergic input to spiny neurones, arises from extrinsic sources and is directed primarily to the spines and more distal dendritic areas. In contrast, inhibitory, putatively GABAergic input, as well as inputs which are derived from other intrinsic sources, are directed to all regions of the spiny neurones, including the proximal dendrites and somata (Figure 6.1B,C).

## 4.2. Interneurones

### *4.2.1. Afferents*

Cholinergic interneurones have been suggested to be recipients of direct cortical input on the basis of electrophysiological findings (Wilson, Chang and Kitai, 1990), although anatomical investigations indicate that this input is likely to be directed to distal regions of these cells (see Lapper and Bolam 1992). An alternative explanation is that the areas of the cortex that provide axons which innervate cholinergic interneurones may not have been labelled in the study by Lapper and Bolam (1992). These neurones also receive a dopamine-containing input which forms symmetric synaptic contacts with the proximal regions of cholinergic neurones (Kubota *et al.*, 1987b; Dimova *et al.*, 1993). The ChAT-immunopositive cells are innervated by a direct thalamic input from the parafascicular nucleus, which forms asymmetric synapses with the perikarya and dendrites (Lapper and Bolam, 1992). It is not known if cholinergic interneurones are contacted directly by terminals arising from the amygdala, GP, STN or dorsal raphe nucleus. The known afferent inputs to cholinergic neurones arising from extrinsic sources are illustrated schematically in Figure 6.2B.

Parvalbumin-immunoreactive GABAergic interneurones are directly innervated by the cortex in both rats (Kita, 1993; Bennett and Bolam, 1994b) and primates (Lapper *et al.*, 1992). The cortical terminals establish asymmetric synaptic contacts with dendrites and somata of parvalbumin-immunopositive cells (Lapper *et al.*, 1992; Kita, 1993; Bennett and Bolam, 1994b). A population of GABAergic interneurones is known to receive a direct dopaminergic input which forms symmetric synaptic specialisations with the perikarya and dendrites (Kubota *et al.*, 1987a). Presently however, there are no direct demonstrations of dopaminergic input to parvalbumin-containing neurones. Recently, a pallidal input to parvalbumin-immunoreactive neurones has been documented (Bevan, Eaton and Bolam, 1997). Direct synaptic input from the thalamus, amygdala,

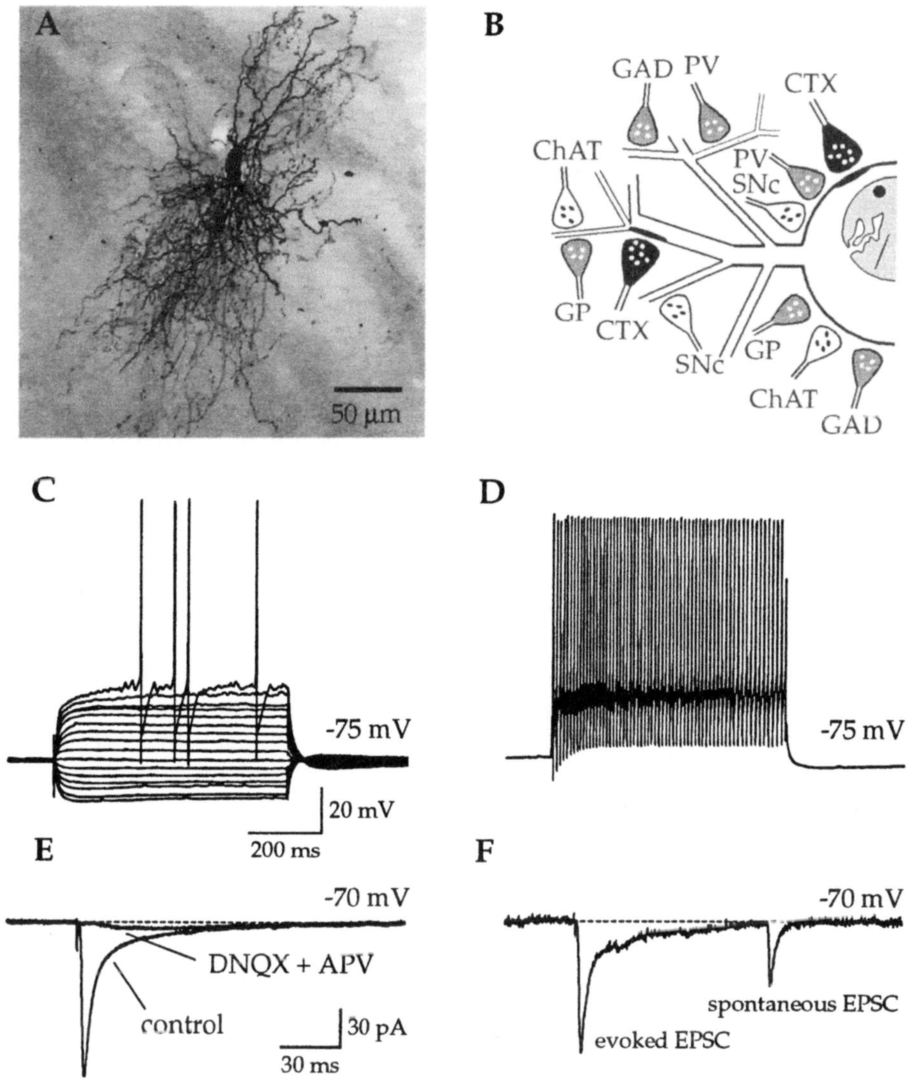


**Figure 6.3. A.** Aspiny neurone filled *in vivo*. The extensively branched dendrites and dense local axon arborisation are characteristic of fast-spiking, parvalbumin-immunopositive neurones. **B:** Schematic illustration of the synaptic inputs to parvalbumin-immunopositive GABAergic interneurones which have been demonstrated at the fine structural level (see Appendix for abbreviations). **C** and **D:** Parvalbumin-containing GABAergic interneurones are fast-spiking cells, exhibiting very brief action potentials followed by large amplitude, short duration afterhyperpolarisations (C), and high frequency, non-accommodating trains of spikes (D) in response to depolarisation. These neurones also exhibit membrane non-linearities in the hyperpolarising direction (C). **E:** Voltage-clamp recordings reveal that electrically-evoked fast, monosynaptic glutamatergic inputs to fast-spiking neurones are blocked by DNQX (20 µM) and APV (50 µM), indicating the involvement of AMPA and NMDA receptors, respectively. **F:** Evoked EPSCs are biphasic whereas spontaneous EPSCs only exhibit the fast decay component.

GP, STN or dorsal raphe nucleus to GABAergic, parvalbumin-containing neurones awaits investigation. Figure 6.3B summarises the known synaptology of parvalbumin-immunoreactive GABAergic interneurones.

Knowledge of the synaptic inputs to the third major class of striatal interneurone, in which SS, NPY and NADPH-d are co-localised, is very limited. These neurones receive a direct dopaminergic input which establishes symmetric synaptic contacts with the perikarya and proximal dendrites (Kubota *et al.*, 1988; Vuillet *et al.*, 1989). Preliminary evidence indicates that at least some of the SS/NPY/NADPH-d neurones receive a direct projection from the GP (Staines, 1983; Bolam and Bennett, 1995; Bevan, Eaton and Bolam, 1997). Although there is a possible cortical input to this class of interneurones, the ultrastructural features of these contacts do not bear the usual hallmarks of cortico-striatal input and await further confirmation (Vuillet *et al.*, 1989). Possible inputs to SS/NPY/NADPH-d-positive neurones from the amygdala, thalamus, STN or dorsal raphe nucleus remain uninvestigated. The synaptic inputs to SS/NPY/NADPH-d-positive neurons are represented in Figure 6.4B.

### *4.2.2. Intrinsic connectivity*

Cholinergic interneurones receive substance P-, enkephalin- and GABA-containing synaptic inputs (Bolam *et al.*, 1986; Bolam, 1989; Martone *et al.*, 1992) which are likely to originate from spiny cells. Substance P- and enkephalin-immunoreactive terminals form symmetric synaptic specialisations with the perikarya and proximal dendrites of ChAT-immunoreactive neurones (Bolam *et al.*, 1986; Martone *et al.*, 1992) although relatively fewer inputs arise from enkephalin-containing inputs (Martone *et al.*, 1992). The GABA-immunopositive terminals form symmetric synapses with the dendrites of cholinergic cells (Bolam, 1989). These neurones do not seem to be recipients of inputs from GABAergic interneurones that contain parvalbumin, at least in proximal regions (Chang and Kita, 1992), and possible relationships between cholinergic interneurones and SS/NPY/NADPH-d interneurones have not been investigated (see Figure 6.2B).

The GABAergic interneurones are recipients of cholinergic (Chang and Kita, 1992) GAD- (Bolam *et al.*, 1985) and parvalbumin-containing synaptic input (Chang and Kita, 1992; Kita, 1993). All three classes of terminal form symmetric synaptic specialisations with the perikarya and dendrites of GABAergic interneurones (Bolam *et al.*, 1985; Chang and Kita, 1992; Kita, 1993). The GAD-immunoreactive input may arise from spiny neurones and/or from GABAergic interneurones (see Figure 6.3B). Inputs to these cells originating from other populations of neostriatal neurones await investigation.

The SS/NPY/NADPH-d neurones are recipients of a GAD-immunoreactive input, which forms symmetric synapses with the perikarya and dendrites of these cells (Vuillet *et al.*, 1990). The GAD-immunopositive terminals may originate from spiny neurones or from GABAergic interneurones. Parvalbumin-immunoreactive terminals have been demonstrated to be in direct synaptic contact with NADPH-d-containing neurones (Bennett and Bolam, 1993c), but whether these terminals arise solely from the GP (see above) or from both GABAergic interneurones and the GP is not known. By similar reasoning, the relative proportion of GAD-positive terminals in contact with NADPH-d-positive cells which are from spiny cells, parvalbumin-immunoreactive neurones or from the GP is not clear, as boutons from all of these sources are likely to contain GAD (see Figure 6.4B). Inputs to the SS/NPY/NADPH-d neurones arising from other populations of striatal neurones have not been investigated.

Clearly, neostriatal interneurones receive a very diverse synaptic input. However, one clear anatomical difference between the interneurones, at least the cholinergic and GABAergic cells, and the spiny projection neurones is the spatial distribution of these inputs. As stated above, excitatory inputs are directed to the more distal regions of spiny projection cells, whereas interneurones receive excitatory input on the proximal dendrites and somata (compare Figures 6.1A,B with 6.2B, 6.3B and 6.4B). This anatomical arrangement, coupled with the profound differences in the electrical properties of interneurones (see below), indicates that the regulation of action potential generation in interneurones is likely to differ dramatically from spiny cells.

## 5. ELECTROPHYSIOLOGICAL PROPERTIES OF NEOSTRIATAL NEURONES

### 5.1. Spiny neurones

Neostriatal spiny neurones, recorded *in vivo* exhibit spontaneous fluctuations in membrane voltage, which consist of transitions between two preferred potentials (Figure 6.1D; Hull *et al.*, 1970; Wilson and Groves, 1981), a relatively depolarised level referred to as the Up state and a more polarised condition called the Down state (see Wilson, 1993, for review). The transition between, and maintenance of, the Up and Down states reflects the interaction between synaptic input and intrinsic voltage-dependent conductances (see Wilson and Kawaguchi, 1996). This interaction governs the pattern of action potential generation in spiny cells, because spikes are only generated from the Up state, and therefore provides a means by which the input-output characteristics of an individual neurone can be dynamically modulated, through mechanisms which influence either intrinsic conductances and/or synaptic input.

#### *5.1.1. Intrinsic conductances*

The spiny projection cells possess four potassium conductances that are largely responsible for the non-linear, behaviour of these neurones (see Figure 6.1E) and include, an inward rectifier ($I_{Kir}$), fast ($I_{Af}$) and slowly ($I_{As}$) inactivating A currents, and a slow persistent current (Nisenbaum, Xu and Wilson, 1994; Nisenbaum and Wilson, 1995; Nisenbaum *et al.*, 1996). Together these conductances can account for both the inward and outward rectification observed in spiny neurones following hyperpolarisation and depolarisation, respectively. However, the outwardly rectifying conductances produce the majority of their effects in the voltage range in which action potentials are generated. Although the characteristics of these conductances will be discussed here, their contribution to voltage behaviour observed *in vivo* will be discussed below.

The characteristics of the inward rectifier are similar to $I_{Kir}$ described previously in other preparations (Katz, 1949; Kandel and Tauc, 1966; Hagiwara and Takahashi, 1974; Constanti and Galvan, 1983) and the time- and voltage-dependence and pharmacological sensitivity are consistent with identification of this current as $I_{Kir}$ (see Nisenbaum and Wilson, 1995). $I_{Kir}$ contributes significantly to the resting potential of the spiny neurones (Nisenbaum and Wilson, 1995). The three outwardly rectifying potassium conductances have distinct effects on the membrane potential in terms of the voltage range and time course of their influence (Nisenbaum, Xu and Wilson, 1994; Nisenbaum and Wilson 1995). $I_{Af}$ makes a transient contribution to outward rectification, which is most prominent at suprathreshold (~–40 mV) membrane potentials. However, $I_{As}$ and the noninactivating $K^+$ conductance provide a more persistent outward rectification at both sub- and supra-threshold potentials. Although the kinetics of $I_{As}$ are slower than $I_{Af}$ they are still sufficiently fast to be involved throughout a depolarising response and to limit the membrane potential. The slow kinetics of activation of the noninactivating $K^+$ conductance indicate that, although this current will also limit depolarisation, this effect will be delayed.

Spiny neurones also contain both $Na^+$ and $Ca^{2+}$ currents, available in the subthreshold range, which interact with outwardly rectifying conductances, particularly $I_{As}$, to influence the delay to action potential initiation (Kita, Kita and Kitai, 1985; Calabresi, Misgeld and Dodt, 1987; Bargas, Galarraga, and Aceves, 1989). Both high-voltage activated $Ca^{2+}$ currents have been described in spiny cells (Bargas, Surmeier and Kitai, 1991; Song and Surmeier 1996), but the relative contribution of the individual currents to the subthreshold voltage behaviour is not yet known.

### *5.1.2. Synaptic input*

Spiny neurones receive convergent monosynaptic excitatory inputs from both the cortex and the thalamus (Wilson, Chang and Kitai, 1982) which are individually rather weak and do not produce significant membrane depolarisation when elicited in isolation (see Wilson, 1995). This input is mediated through the activation of both NMDA and AMPA receptors (Cherubini *et al.*, 1988; Jiang and North, 1991; Calabresi *et al.*, 1992; Kita, 1996). The AMPA receptors in spiny cells, exhibit slow gating kinetics, in terms of the rate of both desensitisation and deactivation, and are relatively impermeable to $Ca^{2+}$, contrasting sharply with the properties of AMPA receptors in cholinergic interneurones (Götz *et al.*, 1997). In contrast, NMDA receptors are relatively homogeneous and exhibit slow gating kinetics in both principal cells and interneurones (Götz *et al.*, 1997).

The dopaminergic input to spiny neurones is somewhat enigmatic, since clearly defined postsynaptic effects of dopamine released from nigrostriatal afferents have yet to be documented. In the study by Wilson, Chang and Kitai (1982), stimulation of the substantia nigra following lesions of both the cortex and ascending fibers from the thalamus could still evoke an EPSP, indicating a possible weak excitatory action for nigrostriatal fibers. Subsequent studies have relied primarily on the application of dopamine or dopamine agonists directly to the neurones. Recently, these have revealed that the highly variable nature of the effects seen in such experiments can be explained by the diversity of modulatory effects of dopamine receptors upon voltage-gated ion channels (e.g. Surmeier *et al.*, 1992; Surmeier and Kitai 1993). In no case did dopamine receptors open classical ligand-gated ion channels. These studies rely upon the assumption that bath application of dopaminergic agonists and antagonists mimic the effects of dopamine released from intact axons, and future studies will reveal whether their results apply to synaptic transmission at dopaminergic synapses. A popular theory for the action of dopamine is that invoked by Freund, Powell and Smith, (1984), which suggests that the dopaminergic inputs may gate the cortical or thalamic input to individual spines by virtue of their placement on spine necks. From a purely numerical standpoint this seems rather unlikely as roughly 4% of spines receive a dopaminergic input (see above) whereas 100% of spines are contacted by a terminal which forms an asymmetric synaptic input (see Wilson *et al.*, 1983). Furthermore, the idea that the dopaminergic inputs could electrically gate the excitatory signals requires that these synapses generate a local synaptic conductance change, which is apparently not the case. In view of the mostly modulatory effects of dopamine, it seems more likely that the nigrostriatal inputs to spines and dendrites may produce a more macroscopic effect by providing a relatively long-lasting alteration in the excitability of a particular segment of dendrite.

### *5.1.3. Interaction between intrinsic properties and synaptic input explain electrical behaviour observed in vivo*

One of the most striking features of spiny neurones recorded *in vivo*, is that the membrane potential exhibits two preferred levels of polarisation and spends little time in transition between these two states (Figure 6.1D; Wilson and Groves, 1981; Wilson, 1992, 1993). The transition to, and maintenance of either state results from the interaction between excitatory synaptic input with the intrinsic membrane conductances. In the absence of excitatory inputs, the membrane potential is maintained at a hyperpolarised level, the Down state, close to the equilibrium potential for potassium, by the inwardly rectifying current $I_{Kir}$ (Wilson, 1993; Wilson and Kawaguchi, 1996). In the Down state the input resistance of the neurone is relatively low, the time constant is short (Figure 6.1F) and consequently the electrotonic structure of the spiny cell is extended (Figure 6.1G). Excitatory synaptic inputs which are not sufficiently synchronous or large enough in number, have little effect on the membrane potential. However, temporally-coincident synaptic inputs produce significant depolarisation, causing voltage-dependent inactivation of $I_{Kir}$, and a subsequent increase in the input resistance and time constant (Figure 6.1F). This leads to an electrotonic collapse of the dendritic tree (Figure 6.1G). The alteration in the electrotonic properties of the cell produce numerous effects, which contribute to further depolarisation. For instance: (i) less temporally-specific inputs can contribute to the depolarisation (due to the increased time constant); (ii) an individual axospinous synaptic input will produce a larger voltage deflection in the dendrite (due to the increased input resistance of the dendrite and subsequent reduction of the impedance mismatch between the spine and dendrite); and, (iii) a larger voltage deflection in the soma will be observed for a given EPSP in the dendrite (due to increased membrane resistivity of the dendrite and consequent shortening of the electrotonic distance between the input and the soma) (Wilson, 1992). These factors produce a dramatic and rapid depolarisation and transition from the Down state but do not explain the maintenance of the membrane potential at a value which is typically just below threshold for action potential generation.

If no other regulating conductances were to become activated following inactivation of $I_{Kir}$, then the neurone would be driven towards the reversal potential for the EPSPs, and would almost certainly be destroyed as a consequence. This does not happen, and there is instead a conspicuous, preferred level of depolarisation referred to as the Up state. The level of depolarisation observed in the Up state is a consequence of the balance between the depolarising influence produced by the barrage of excitatory input, and the opposing, hyperpolarising effect of the outwardly rectifying $K^+$ conductances (Wilson and Kawaguchi, 1996). Two of the three $K^+$ currents described above are primarily responsible for limiting the level of depolarisation. The slow A-current, $I_{As}$ is available immediately and limits the level of depolarisation. $I_{As}$ also exhibits time- and voltage-dependent kinetics which influence the duration to action potential generation as a consequence of

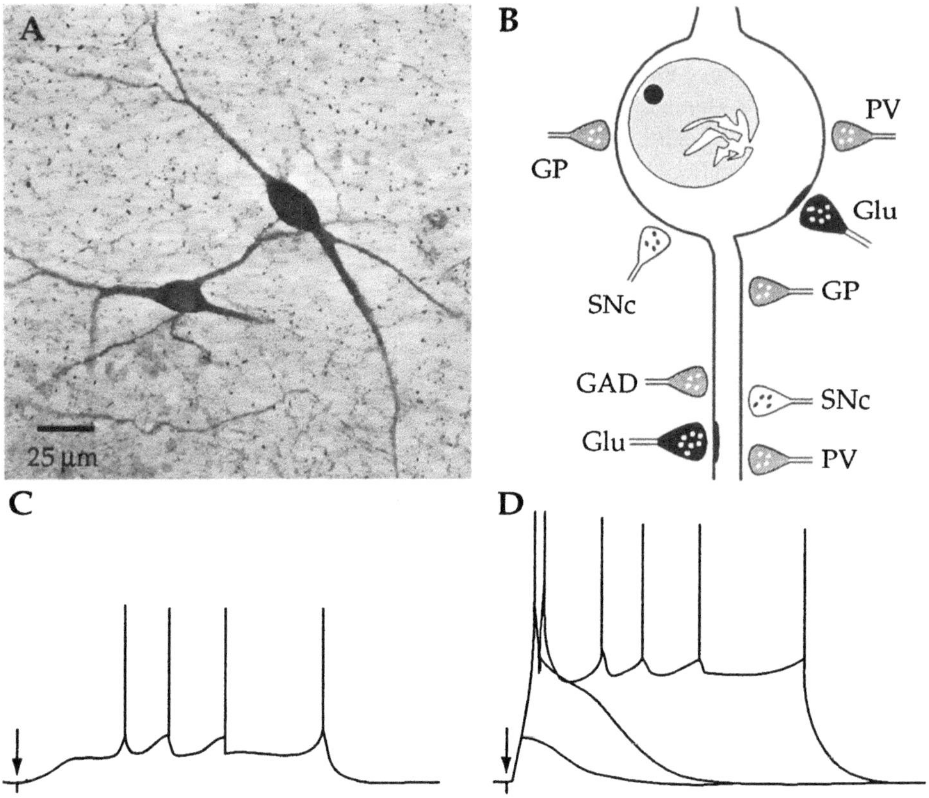


**Figure 6.4. A.** Histochemically stained NADPH-diaphorase neurones in primate putamen, exhibiting bipolar or fusiform somata and long, smooth, infrequently branching dendrites, which are characteristic of this cell class. **B:** Synaptic inputs to SS/NPY-NADPH-d-positive neurones identified in the electron microscope are illustrated shematically (see Appendix for abbreviations). **C** and **D:** Synaptic stimulation during depolarisation (C) and hyperpolarisation (D) induced action potential generation and could elicit persistent, depolarisations (D). C and D are taken with permission from Kawaguchi *et al.* (1995, Striatal interneurones: chemical, physiological and morphological characterization. *Trends in Neuroscience,* 18, 527–535).

both the recent voltage history of the neurone and the intensity of the excitatory input (Nisenbaum, Xu and Wilson, 1994). The slow kinetics of the non-inactivating $K^+$ conductance mean that it will not immediately limit the membrane potential following a transition to the Up state, but will provide a stabilising influence if the depolarising episode is of a sufficient duration. The effects of both $I_{As}$ and the non-inactivating current thus oppose the depolarising influence of the excitatory input. This is produced both by a direct hyperpolarising influence of the current (i.e. $K^+$ leaving the neurone) and by the $K^+$ conductances decreasing the input resistance and the time constant (Figure 6.1F) thereby causing an electrotonic extension of the dendrites (Figure 6.1G). The effects upon the electrotonic structure of the neurone effectively decrease the efficacy of an individual excitatory input. The fast A-current, $I_{Af}$ mainly contributes to action potential repolarisation and to the afterhyperpolarisation (AHP). The time- and voltage-dependence of inactivation of this current are relatively rapid and provide an explanation for both the widening of the action potential and the decrement in the amplitude of the AHP observed when a train of spikes is observed. Hence the more persistent $K^+$ currents (i.e. $I_{As}$ and the non-inactivating current) also contribute to spike repolarisation through voltage-dependent activation during an action potential and/or an increase in driving force.

Following cessation of the excitatory input the membrane potential of the spiny neurone is driven towards the $K^+$ equilibrium potential by the outwardly rectifying $K^+$ currents. As the cell hyperpolarises, these outward currents become inactivated and then $I_{Kir}$ takes over driving the neurone towards $E_K$, and re-establishing the Down state.

## 5.2. Interneurones

### *5.2.1. Intrinsic conductances*

Currently there are very few data concerning the biophysical properties of interneurones which are responsible for their intrinsic membrane characteristics. What data that are available come largely from studies by Kawaguchi (1992, 1993).

The cholinergic interneurones possess a conductance which produces a depolarising sag in the membrane potential in response to persistent hyperpolarising current injection (Figure 6.2E; Kawaguchi, 1992, 1993). The slow inwardly rectifying conductance is likely to be the mixed cation current $I_H$, as it is blocked by millimolar concentrations of caesium (Jiang and North, 1991; Kawaguchi, 1993). Depolarising current injections evoke action potentials which are followed by large, long duration AHPs (Figure 6.2E; Wilson, Chang and Kitai, 1990; Jiang and North, 1991; Kawaguchi, 1992, 1993). The AHP is completely blocked by replacing extracellular calcium with cobalt, indicating the likely involvement of a calcium-activated $K^+$ current (Kawaguchi, 1993). The resting potential of cholinergic cells is depolarised, in comparison to spiny neurones recorded in slices, and is typically in the range of –50 to –60 mV. However, these neurones are never really at rest because all cholinergic neurones observed *in vivo* (Figure 6.2C; Wilson, Chang and Kitai, 1990) and the majority recorded *in vitro* (Figure 6.2D; Bennett and Wilson, 1997) are tonically active. Furthermore, the cholinergic cells possess intrinsic mechanisms which can generate action potentials in the absence of excitatory or inhibitory synaptic input (Wilson, Chang and Kitai, 1990; Bennett and Wilson, 1997).

The GABAergic parvalbumin-immunoreactive neurones are the fast-spiking interneurones of the neostriatum, exhibiting very short duration spikes and a brief, large amplitude AHP (Figure 6.3C,D; Kawaguchi, 1993). A relatively hyperpolarised resting potential is seen in these cells and they typically are polarised to about –80 mV (Kawaguchi, 1993). Although at present there are no pharmacological data available on the intrinsic conductances in parvalbumin-containing interneurones, it is likely that these cells possess an inwardly rectifying, hyperpolarisation-activated conductance, as the membrane potential response to current injection is non-linear (Figure 6.3C; see Figure 6.3B in Kawaguchi, 1993).

The electrical properties of neurones in which SS, NPY and NADPH-d are co-localised exhibit several interesting and unusual features (Figure 6.4C,D). These cells display low-threshold spikes (LTS) in response to synaptic stimulation and depolarising current injection when delivered at hyperpolarised potentials (Figure 6.4D; Kawaguchi, 1993). The LTS can also be evoked by cessation of hyperpolarising current pulses. LTS are sensitive to replacement of calcium with cobalt, providing evidence for a calcium-mediated process (Kawaguchi, 1993). Persistent depolarisations following LTS are also observed and are characterised by a maintained depolarisation that can last hundreds of milliseconds (Figure 6.4D; Kawaguchi, 1993). The persistent depolarisations are prevented by combined application of cobalt and TTX (Kawaguchi, 1993). These neurones also exhibit non-linear membrane properties that are as yet uncharacterised.

### *5.2.2. Synaptic input*

The cholinergic interneurones are recipients of excitatory synaptic input from the cortex and the thalamus (Wilson, Chang and Kitai, 1990). *In vivo* recordings have demonstrated that giant cells are constantly bombarded with excitatory inputs which produce rapid depolarisations of the membrane from which action potentials can be triggered (Figure 6.2C; Wilson, Chang and Kitai, 1990). The relative contributions of AMPA and NMDA receptors *in vivo* are unknown, but it is clear that glutamatergic inputs can activate both classes of receptors, following intrastriatal stimulation *in vitro* (Jiang and North,1991; Kawaguchi, 1992; Bennett and Wilson, 1997).

In addition to the fast amino acid-mediated inputs, slow peptidergic inputs to the cholinergic neurones have also been documented (Aosaki and Kawaguchi, 1996). Repetitive intrastriatal stimulation releases substance P, causing depolarisation and action potential generation in the cholinergic neurones (Aosaki and Kawaguchi, 1996). This effect is a result of activation of neurokinin–1 (NK–1) receptors (Gerfen, 1991), which open a TTX-insensitive cation conductance (Aosaki and Kawaguchi, 1996). Additionally, in some instances blockade of NK–1 receptors unmasks a slow outward, inhibitory current (Aosaki and Kawaguchi, 1996). Direct application of met-enkephalin to these neurones also produces an outward current which is mediated through the activation of $\delta$-opioid receptors (Aosaki and Kawaguchi, 1995).

At present there are very limited data on the synaptic inputs to either GABAergic interneurones or the cells in which SS, NPY and NADPH-d are co-localised. However, it is clear that both of these classes of interneurones are recipients of excitatory input mediated through the activation of glutamate receptors (see Figure 6.3E,F; Kawaguchi, 1993; Kita, 1993.).

### *5.2.3. Interaction between intrinsic properties and synaptic input*

*In vivo*, the cholinergic interneurones exhibit a tonic irregular pattern of action potential generation (Figure 6.2C; Wilson, Chang and Kitai, 1990). Although, as described above, the cholinergic neurones are capable of generating action potentials in the absence of synaptic input (at least fast transmission mediated via GABAergic and glutamatergic inputs), it is clear that the constant barrage of depolarising inputs which they receive influence the timing of individual spikes (Wilson, Chang and Kitai, 1990; Bennett and Wilson, 1997).

## 6. CONCLUSIONS

The available data for spiny neurones and interneurones illustrates the very different ways in which these neurones are controlled. Individual spiny cells receive input from several thousand individual cortical and thalamic neurones. However, probably only 200–300 of these inputs are required to be active at any one time to produce state transitions and action potential generation (see Wilson, 1995). Consequently, different ensembles of cortical or thalamic neurones can effectively control a single spiny neurone at different points in time, but certainly individual excitatory inputs are rather weak and separately have little effect. The neostriatal interneurones are controlled in a manner which is in stark contrast with the spiny cell. Taking the cholinergic neurone as an example, it is clear that individual excitatory inputs are extremely effective in influencing the membrane potential, and therefore a few or even single neurones providing such inputs may be in a position to regulate action potential generation. Consequently, neuromodulators which alter the efficacy of individual synaptic inputs are likely to have profoundly different effects in spiny cells and interneurones.

## ACKNOWLEDGEMENTS

We thank Dr. Edward Stern for providing the data illustrated in Figures 6.1A,D and 6.3A, and also thank Dr. Melburn Park for providing the software (Fourier's Microscope) used in the iterative deconvolution of the cell images.

## APPENDIX

| | |
|---|---|
| APV | 2-amino-5-phosphonopentanoic acid |
| ChAT | choline acetyltransferase |
| CL/PC | central lateral/paracentral thalamic nuclei |
| CTX | cerebral cortex |
| DNQX | 6,7-dinitroquinoxaline-2,3-dione |
| enk | enkephalin |
| EP | entopeduncular nucleus |
| GABA | $\gamma$-aminobutyric acid |
| GAD | glutamic acid decarboxylase |
| Glu | glutamate |
| GP | globus pallidus |
| Pf | parafascicular thalamic nucleus |
| PV | parvalbumin |
| SNc | substantia nigra pars compacta |
| SNr | substantia nigra pars reticulata |
| SP | substance P |
| SS | somatostatin |

## REFERENCES

Adams, C.E. and Fisher, R.S. (1990) Sources of neostriatal cholecystokinin in the cat. *Journal of Comparative Neurology* **292**, 563–574.

Agard, D.A., Hiraoka, Y., Shaw, P. and Sedat, J.W. (1989) Fluorescence microscopy in three dimensions. *Methods in Cell Biol.* **30**: 353–377.

Alexander, G.E. and DeLong, M.R. (1985) Microstimulation of the primate neostriatum. II. Somatotopic organization of striatal microexcitable zones and their relation to neuronal response properties. *Journal of Neurophysiol.* **53**, 1417–1430.

Alheid, G.F. and Heimer, L. (1988) New perspectives in basal forebrain organization of special relevance to neuropsychiatric disorders–the striatopallidal, amygdaloid, and corticopetal components of substantia innominata. *Neuroscience* **27**, 1–39.

Aosaki, T. and Kawaguchi, Y. (1995) Actions of neuropeptides on large aspiny neurons of rat neostriatum *in vitro. Society for Neuroscience Abstracts,* **21**, 913.

Aosaki, T. and Kawaguchi, Y. (1996) Actions of substance P on rat neostriatal neurons *in vitro. Journal of Neuroscience* **16**, 5141–5153.

Arbuthnott, G.W., Wright, A.K., Hamilton, M.H. and Brown, J.R. (1982) Orthograde transport of nuclear yellow: a problem and its solution. *Journal of Neuroscience Methods* **6**, 365–368.

Armstrong D.M., Saper C.B., Levey A.I., Wainer B.H. and Terry R.D. (1983) Distribution of cholinergic neurons in the rat brain demonstrated by the immunocytochemical localisation of choline acetyltransferase. *Journal of Comparative Neurology* **216**, 56–68.

Aronin, N., Chase, K. and DiFiglia, M. (1986) Glutamic acid decarboxylase and enkephalin immunoreactive axon terminals in the rat neostriatum synapse with striatonigral neurons. *Brain Research* **365**, 151–158.

Bargas, J., Galarraga, E. and Aceves, J. (1989) An early outward conductance modulates the firing latency and frequency of neostriatal neurons of the rat brain. *Experimental Brain Research* **75**, 146–156.

Bargas, J., Surmeier, D.J. and Kitai, S.T. (1991) High- and low-voltage activated calcium currents are expressed by neurons cultured from embryonic rat neostriatum. *Brain Research* **541**, 70–74.

Beckstead, R.M. (1983) A pallidostriatal projection in the cat and monkey. *Brain Research Bulletin* **11**, 629–632.

Bennett, B.D. and Bolam, J.P. (1993a) Characterization of calretinin-immunoreactive structures in the striatum of the rat. *Brain Research* **609**, 137–148.

Bennett, B.D. and Bolam, J.P. (1993b) Two populations of calbindin D28k-immunoreactive neurones in the striatum of the rat. *Brain Research* **610**, 305–310.

Bennett, B.D. and Bolam, J.P. (1993c) Identified targets of parvalbumin-immunoreactive terminals in the striatum of the rat. *Brain Research Association Abstracts* **10**, 6.

Bennett, B.D. and Bolam, J.P. (1994a) Localisation of parvalbumin-immunoreactive structures in primate caudate-putamen. *Journal of Comparative Neurology* **347**, 340–356.

Bennett, B.D. and Bolam, J.P. (1994b) Synaptic input and output of parvalbumin-immunoreactive neurons in the neostriatum of the rat. *Neuroscience* **62**, 707–719.

Bennett, B.D. and Wilson, C.J. (1997) Functional microcircuitry of the neostriatum: physiological and anatomical analyses of cholinergic interneurons. *Brain Research Association, Abstracts* **14**, 57.

Besson, M.J., Graybiel, A.M. and Quinn, B. (1990) Co-expression of neuropeptides in the cat's striatum: an immunohistochemical study of substance P, dynorphin B and enkephalin. *Neuroscience* **39**, 33–58.

Bevan, M.D., Smith, A.D. and Bolam, J.P. (1996) The substantia nigra as a site of synaptic integration of functionally diverse information arising from the ventral pallidum and the globus pallidus in the rat. *Neuroscience* **75**, 5–12.

Bevan, M.D., Eaton, S.A. and Bolam, J.P. (1997) Synaptic targets of physiologically, neurochemically and morphologically characterized neurons of the rat globus pallidus. *Society for Neuroscience Abstracts* **27**.

Bishop, G.A., Chang, H.T. and Kitai, S.T. (1982) Morphological and physiological properties of neostriatal neurons: an intracellular horseradish peroxidase study in the rat. *Neuroscience* **7**, 179–191.

Bolam, J.P. (1984) Synapses of identified neurons in the neostriatum. In *CIBA Foundation Symposium* **107**, 30–42.

Bolam, J.P. (1989) Cholinergic neurons in the striatum and basal forebrain receive direct synaptic input from GABA-containing axon terminals. *Neuroscience Letters, Supplement,* **36**, S9.

Bolam J.P., Clarke D.J., Smith A.D. and Somogyi P. (1983a) A type of aspiny neuron in the rat neostriatum accumulates [$^3$H]gamma-aminobutyric acid: Combination of Golgi-staining, autoradiography, and electron microscopy. *Journal of Comparative Neurology* **213**, 121–134.

Bolam, J.P., Somogyi, P., Takagi, H., Fodor, I. and Smith, A.D. (1983b) Localization of substance P-like immunoreactivity in neurons and nerve terminals in the neostriatum of the rat: a correlated light and electron microscopic study. *Journal of Neurocytology* **12**, 325–344.

Bolam, J.P., Wainer, B.H. and Smith, A.D. (1984) Characterization of cholinergic neurons in the rat neostriatum. A combination of choline acetyltransferase immunocytochemistry, Golgi impregnation and electron microscopy. *Neuroscience* **12**, 711–718.

Bolam, J.P., Powell, J.F., Wu, J.-Y. and Smith, A.D. (1985) Glutamate decarboxylase-immunoreactive structures in the rat neostriatum: A correlated light and electron microscopic study including a combination of Golgi impregnation with immunocytochemistry. *Journal of Comparative Neurology* **237**, 1–20.

Bolam, J.P., Ingham, C.A., Izzo, P.N., Levey, A.I., Rye, D.B., Smith, A.D. and Wainer, B.H. (1986) Substance P-containing terminals in synaptic contact with cholinergic neurons in the neostriatum and basal forebrain: a double immunocytochemical study in the rat. *Brain Research* **397**, 279–289.

Bolam, J.P. and Izzo, P.N. (1988) The postsynaptic targets of substance P-immunoreactive terminals in the rat neostriatum with particular reference to identified spiny striatonigral neurons. *Experimental Brain Research* **70**, 361–377.

Bolam, J.P. and Smith, Y. (1992) The striatum and globus pallidus send convergent synaptic inputs onto single cells in the entopeduncular nucleus of the rat: a double anterograde labelling study combined with postembedding immunocytochemistry for GABA. *Journal of Comparative Neurology* **321**, 456–476.

Bolam, J.P. and Bennett, B.D. (1995) Microcircuitry of the neostriatum. In *Molecular and cellular mechanisms of neostriatal function*, M.A. Ariano and D.J. Surmeier (eds). Germany, Springer-Verlag, pp. 1–12.

Bouyer, J.J., Park, D.H., Joh, T.H. and Pickel, V.M. (1984) Chemical and structural analysis of the relation between neocortical inputs and tyrosine hydroxylase-containing terminals in rat neostriatum. *Brain Research* **302**, 267–275.

Buchwald, N.A., Price, D.C., Vernon, L. and Hull, C.D. (1973) Caudate intracellular response to thalamic and cortical inputs. *Experimental Neurology* **38**, 311–323.

Calabresi, P., Misgeld, U. and Dodt, H.U. (1987) Intrinsic membrane properties of neostriatal neurons can account for their low level of spontaneous activity. *Neuroscience* **20**, 293–303.

Calabresi, P., Maj, R. Pisani, A., Mercuri, N.B. and Bernardi, G. (1992) Long-term synaptic depression in the striatum: physiological and pharmacological characterization. *Journal of Neuroscience* **12**, 4224–4233.

Carman, J.B., Cowan, W.M., Powell, T.P.S. and Webster, K.E. (1965) A bilateral cortico-striate projection. *Journal of Neurology Neurosurgery and Psychiatry*, **28**, 71–77.

Carpenter, M.B. and Jayaraman, A. (1990) Subthalamic nucleus of the monkey: connections and immunocytochemical features of afferents. *Journal für Hirnforschung*, **31**, 653–658.

Cavada, C. and Goldman-Rakic, P.S. (1991) Topographic segregation of corticostriatal projections from posterior parietal subdivisions in the Macaque monkey. *Neuroscience* **42**, 683–696.

Celio, M.R. (1990) Calbindin D–28k and parvalbumin in the rat nervous system. *Neuroscience* **35**, 375–475.

Chang, H.T. and Kita, H. (1992) Interneurons in the rat striatum—relationships between parvalbumin neurons and cholinergic neurons. *Brain Research* **574,** 307–311.

Chang, H.T., Wilson, C.J. and Kitai, S.T. (1981) Single neostriatal efferent axons in the globus pallidus: a light and electron microscopic study. *Science, New York* **213**, 915–918.

Chang, H.T., Wilson, C.J. and Kitai, S.T. (1982) A Golgi study of rat neostriatal neurons: light microscopic analysis. *Journal of Comparative Neurology* **208**, 107–126.

Cherubini, E., Herrling, P.L., Lanfumey, L. and Stanzione, P. (1988) Excitatory amino acids in synaptic excitation of rat striatal neurones *in vitro*. *Journal of Physiology (London)* **400**, 677–690.

Chesselet, M.-F. and Graybiel, A.M. (1986) Striatal neurons expressing somatostatin-like immunoreactivity: Evidence for a peptidergic interneuronal system in the cat. *Neuroscience* **17**, 547–571.

Christie, M.J., Summers, R.J., Stephenson, J.A., Cook, C.J. and Beart, P.M. (1987) Excitatory amino acid projections to the nucleus accumbens septi in the rat: a retrograde transport study utilizing D[3H]aspartate and [3H]GABA. *Neuroscience* **22**, 425–439.

Chung, J.W., Hassler, R. and Wagner, A. (1977) Degeneration of two of nine types of synapses in the putamen after center median coagulation in the cat. *Experimental Brain Research* **28**, 345–361.

Constanti, A. and Galvan, M. (1983) Fast inward-rectifying current accounts for anomalous rectification in olfactory cortex neurones. *Journal of Physiology (London )* **335**, 153–178.

Corvaja, M., Doucet, G. and Bolam, J.P. (1991) Characterization of the postsynaptic targets of serotonergic terminals in the substantia nigra and striatum of the rat. *European Journal of Neuroscience, Supplement* **4**, 150.

Côté, P.-V., Sadikot, A.F. and Parent, A. (1991) Complementary distribution of calbindin D–28k and parvalbumin in the basal forebrain and midbrain of the squirrel monkey. *European Journal of Neuroscience,* **3**, 1316–1329.

Cowan, R.L., Wilson, C.J., Emson, P.C. and Heizmann, C.W. (1990) Parvalbumin-containing GABAergic interneurons in the rat neostriatum. *Journal of Comparative Neurology* **302**, 197–205.

Cowan, R.L. and Wilson, C.J. (1994) Spontaneous firing patterns and axonal projections of single corticostriatal neurons in the rat medial agranular cortex. *Journal of Neurophysiology* **71**, 17–32.

Crutcher, M.D. and DeLong, M.R. (1984) Single cell studies of the primate putamen. I. Functional organization. *Experimental Brain Research* **53**, 233–243.

Dafny, N., Dauth, G. and Gilman, S. (1975) A direct input from amygdaloid complex to caudate nucleus of the rat. *Experimental Brain Research* **23**, 203–210

Dawson, T.M., Bredt, D.S., Fotuhi, M., Hwang, P.M. and Snyder, S.H. (1991) Nitric oxide synthase and neuronal NADPH-diaphorase are identical in brain and peripheral tissues. *Proceedings of the National Academy of Sciences, U.S.A.* **88**, 7797–7801.

Deschenes, M., Bourassa, J., Doan, V.D. and Parent, A. (1996) A single-cell study of the axonal projections arising from the posterior intralaminar thalamic nuclei in the rat. *European Journal of Neuroscience,* **8**, 329–343.

DiFiglia, M. (1987) Synaptic organization of cholinergic neurons in the monkey striatum. *Journal of Comparative Neurology* **255**, 245–258.

DiFiglia, M., Pasik, P. and Pasik, T. (1976) A Golgi study of neuronal types in the neostriatum of monkeys. *Brain Research* **114**, 245–256.

DiFiglia, M. and Aronin, N. (1982) Ultrastructural features of immunoreactive somatostatin neurons in the rat caudate nucleus. *Journal of Neuroscience* **2**, 1267–1274.

Dimova, R., Vuillet, J. and Seite, R. (1980) Study of the rat neostriatum using a combined Golgi-electron microscope technique and serial sections. *Neuroscience* **5**, 1581–1596.

Dimova, R., Vuillet, J., Nieoullon, A., Kerkerian-Le Goff, L. (1993) Ultrastructural features of the choline acetyltransferase-containing neurons and relationships with nigral dopaminergic and cortical afferent pathways in the rat striatum. *Neuroscience* **53**, 1059–1071.

Divac, I., Fonnum, F. and Storm-Mathisen, J. (1977) High affinity uptake of glutamate in terminals of corticostriatal axons. *Nature (London)* **266**, 377–378.

Donoghue, J.P. and Herkenham, M. (1986) Neostriatal projections from individual cortical fields conform to histochemically distinct striatal compartments in the rat. *Brain Research* **365**, 397–403.

Dubé, L., Smith, A.D. and Bolam, J.P. (1988) Identification of synaptic terminals of thalamic or cortical origin in contact with distinct medium-size spiny neurons in the rat neostriatum. *Journal of Comparative Neurology* **267**, 455–471.

Ebrahimi, A., Pochet, R. and Roger, M. (1992) Topographical organization of the projections from physiologically identified areas of the motor cortex to the striatum of the rat. *Neuroscience Research* **14**, 39–60.

Figueredo-Cardenas, G., Medina, L. and Reiner, A. (1996) Calretinin is largely localized to a unique population of striatal interneurons in rats. *Brain Research* **709**,145–150.

Fink-Jensen, A. and Mikkelsen, J.D. (1989) The striato-entopeduncular pathway in the rat. A retrograde transport study with wheatgerm-agglutinin-horseradish peroxidase. *Brain Research* **476**, 194–198.

Fisher, R.S., Buchwald, N.A., Hull, C.D. and Levine, M.S. (1986) The GABAergic striatonigral neurons of the cat: demonstration by double peroxidase labeling. *Brain Research* **398**, 148–156.

Flaherty, A.W. and Graybiel, A.M. (1991) Corticostriatal transformations in the primate somatosensory system. Projections from physiologically mapped body-part representations. *Journal of Neurophysiology,* **66**, 1249–1263.

Flaherty, A.W. and Graybiel, A.M. (1993) Two input systems for body representations in the primate striatal matrix: Experimental evidence in the squirrel monkey. *Journal of Neuroscience* **13**, 1120–1137.

Fonnum, F.F., Storm-Mathisen, J. and Divac, J. (1981) Biochemical evidence for glutamate as the neurotransmitter in corticostriatal and corticothalamic fibres in rat brain. *Neuroscience* **6**, 863–873.

Fox, C.A., Andrade, A.N., Schwyn, R.C. and Rafols, J.A. (1971) The aspiny neurons and the glia in the primate striatum: a golgi and electron microscopic study. *Journal für Hirnforschung* **13**, 341–362.

Freund, T.F., Powell, J. and Smith, A.D. (1984) Tyrosine hydroxylase-immunoreactive boutons in synaptic contact with identified striatonigral neurons, with particular reference to dendritic spines. *Neuroscience* **13**, 1189–1215.

Frotscher, M., Rinne, U., Hassler, R. and Wagner, A. (1981) Termination of cortical afferents on identified neurons in the caudate nucleus of the cat. *Experimental Brain Research* **41**, 329–337.

Fuller, T.A., Russchen, F.T. and Price, J.L. (1987) Sources of presumptive glutamergic/aspartergic afferents to the rat ventral striatopallidal region. *Journal of Comparative Neurology* **258**, 317–338.

Gerfen, C.R. (1984) The neostriatal mosaic: compartmentalization of corticostriatal input and striatonigral output systems. *Nature (London)* **311**, 461.

Gerfen, C.R. (1985) The neostriatal mosaic I. Compartmental organization of projections from the striatum to substantia nigra in the rat. *Journal of Comparative Neurology* **236**, 454–476.

Gerfen, C.R. (1989) The neostriatal mosaic: striatal patch-matrix organization is related to cortical lamination. *Science, New York* **246**, 385–388.

Gerfen, C.R. (1991) Substance P (neurokinin–1) receptor mRNA is selectively expressed in cholinergic neurons in the striatum and basal forebrain. *Brain Research* **556**, 165–170.

Gerfen, C.R. (1992) The neostriatal mosaic—multiple levels of compartmental organization. *Trends in Neuroscience,* **15**, 133–139.

Gerfen, C.R., Baimbridge, K.G. and Miller, J.J. (1985) The neostriatal mosaic: Compartmental distribution of calcium-binding protein and parvalbumin in the basal ganglia of the rat and monkey. *Proceedings of the National Academy of Sciences, USA.* **82**, 8780–8784.

Gerfen, C.R., Herkenham, M. and Thibault, J. (1987) The neostriatal mosaic II. Patch- and matrix-directed mesostriatal dopaminergic and non-dopaminergic systems. *Journal of Neuroscience* **7**, 3915–3934.

Gerfen, C.R. and Young, W.S. (1988) Distribution of striatonigral and striatopallidal peptidergic neurons in both patch and matrix compartments: an in situ hybridization histochemistry and fluorescent retrograde tracing study. *Brain Research* **460**, 161–167.

Gerfen, C.R., Engber, T.M., Mahan, L.C., Susel, Z., Chase, T. N., Monsma, F. J. and Sibley, D.R. (1990) D1 and D2 dopamine receptor regulated gene expression of striatonigral and striatopallidal neurons. *Science, New York* **250**, 1429–1432.

Godukhin, O.V., Zharikova, A.D. and Novoselov, V.I. (1980) The release of labelled L-glutamic acid from rat neostriatum in vivo following stimulation of frontal cortex. *Neuroscience* **5**, 2151–2154.

Goldman, P.S. and Nauta, W.J.H. (1977) An intricately patterned prefronto-caudate projection in the Rhesus monkey. *Journal of Comparative Neurology* **171**, 369–386.

Goldman-Rakic, P.S. and Selemon, L.D. (1986) Topography of corticostriatal projections in nonhuman primates and implications for functional parcellation of the neostriatum. In *Cerebral cortex, volume 5, (Sensory-motor areas, and aspects of cortical circuitry)* edited by E.G. Jones and A. Peters, New York, Plenum Press, pp. 447–466.

Götz, T., Kraushaar, U., Geiger, J., Lübke, J., Berger, T. and Jonas, P. (1997) Functional properties of AMPA and NMDA receptors expressed in identified types of basal ganglia neurons. *Journal of Neuroscience* **17**, 204–215.

Graveland, G.A. and DiFiglia, M. (1985) The frequency and distribution of medium-sized neurons with indented nuclei in the primate and rodent neostriatum. *Brain Research* **327**, 307–311.

Graybiel, A.M. and Ragsdale, C.W. (1978) Histochemically distinct compartments in the striatum of human, monkey, and cat demonstrated by acetylcholinesterase staining. *Proceedings of the National Academy of Sciences, USA* **75**, 5723–5726.

Graybiel, A.M., Baughman, R.W. and Eckenstein, F. (1986) Cholinergic neuropil of the striosomal boundaries. *Nature (London)* **323**, 625–627.

Groves, P.M. (1980) Synaptic endings and their postsynaptic targets in neostriatum: synaptic specializations revealed from analysis of serial sections. *Proceedings of the National Academy of Sciences, USA* **77**, 6926–6929.

Haber, S.N. and Nauta, J.H. (1983) Ramifications of the globus pallidus in the rat as indicated by patterns of immunohistochemistry. *Neuroscience* **9**, 245–260.

Hagiwara, S. and Takahashi, K. (1974) The anomalous rectification and cation selectivity of the membrane of a starfish egg cell. *Journal of Membrane Biology.* **18**, 61–80.

Hassler, R., Chung, J.W., Rinne, U. and Wagner, A. (1978) Selective degeneration of two out of the nine types of synapses in cat caudate nucleus after cortical lesions. *Experimental Brain Research* **31**, 67–80.

Hassler, R. Haug, P., Nitsch, C., Kim, J.S. and Paik, K. (1982) Effect of motor and premotor cortex ablation on concentrations of amino acids, monoamines, and acetylcholine and on the ultrastructure in rat striatum. A confirmation of glutamate as the specific cortico-striatal transmitter. *Journal of Neurochemistry* **38**, 1087–1098.

Heimer, L. and Wilson, R.D. (1975) The subcortical projections of the allocortex: similarities in the neural associations of the hippocampus, the piriform cortex, and the neocortex. In *Golgi centennial symposium, perspectives in neurobiology,* edited by M. Santini, New York, Raven Press, pp. 173–193.

Herkenham, M. and Pert, C.B. (1981) Mosaic distribution of opiate receptors, parafascicular projections and acetylcholinesterase in rat striatum. *Nature (London)* **291**, 415–418.

Hontanilla, B., Parent, A. and Gimenez-Amaya, J.M. (1994) Compartmental distribution of parvalbumin and calbindin D–28k in rat globus pallidus. *NeuroReport* **5**, 2269–2272.

Hull, C.D., Bernardi, G. and Buchwald, N.A. (1970) Intracellular responses of caudate neurons to brain stem stimulation. *Brain Research* **22**, 163–179.

Izzo, P.N., Graybiel, A.M. and Bolam, J.P. (1987) Characterization of substance P- and [met]enkephalin-immunoreactive neurons in the caudate nucleus of cat and ferret by a single section Golgi procedure. *Neuroscience* **20**, 577–587.

Izzo, P.N. and Bolam, J.P. (1988) Cholinergic synaptic input to different parts of spiny striatonigral neurons in the rat. *Journal of Comparative Neurology* **269**, 219–234.

Jacobowitz, D.M. and Winsky, L. (1991) Immunocytochemical localization of calretinin in the forebrain of the rat. *Journal of Comparative Neurology* **304**, 198–218.

Jayaraman, A. (1983) Topographic organization and morphology of peripallidal and pallidal cells projecting to the striatum in cats. *Brain Research* **275**, 279–286.

Jiang, Z.G. and North, R.A. (1991) Membrane properties and synaptic responses of rat striatal neurones *in vitro. Journal of Physiology (London)* **443**, 533–553.

Jiménez-Castellanos, J. and Graybiel, A.M. (1987) Subdivisions of the dopamine-containing A8-A9-A10 complex identified by their differential mesostriatal innervation of striosomes and extrastriosomal matrix. *Neuroscience* **23**, 223–242.

Jones, E.G., Coulter, J.D., Burton, H. and Porter, R. (1977) Cells of origin and terminal distribution of corticostriatal fibres arising in the sensory-motor cortex of monkeys. *Journal of Comparative Neurology,* **173**, 53–80.

Jones E.G. and Hendry, S.H.C. (1989) Differential calcium binding protein immunoreactivity distinguishes classes of relay neurons in monkey thalamic nuclei. *European Journal of Neuroscience* **1**, 222–246.

Kandel, E.R. and Tauc, L. (1966) Anomalous rectification in the metacerebral giant cells and its consequences for synaptic transmission. *Journal of Physiology (London)* **183**, 287–304.

Katz, B. (1949) Les constantes electriques de la membrane du muscle. *Archives des Sciences Physiologiques,* **3**, 285.

Kawaguchi, Y. (1992) Large aspiny cells in the matrix of the rat neostriatum *in vitro*: physiological identification, relation to the compartments and excitatory postsynaptic currents. *Journal of Neurophysiology* **67**, 1669–1683.

Kawaguchi, Y. (1993) Physiological, morphological, and histochemical characterization of three classes of interneurons in rat neostriatum. *Journal of Neuroscience* **13**, 4908–4923.

Kawaguchi, Y., Wilson, C.J. and Emson, P.C. (1989) Intracellular recording of identified neostriatal patch and matrix spiny cells in a slice preparation preserving cortical inputs. *Journal of Neurophysiology* **62**, 1052–1069.

Kawaguchi, Y., Wilson, C.J. and Emson, P.C. (1990) Projection subtypes of rat neostriatal matrix cells revealed by intracellular injection of biocytin. *Journal of Neuroscience* **10**, 3421–3438.

Kawaguchi, Y., Wilson, C.J., Augood, S.J. and Emson, P.C. (1995) Striatal interneurones: chemical, physiological and morphological characterization. *Trends in Neuroscience* **18**, 527–535.

Kemp, J.M. (1968) An electron microscopic study of the termination of afferent fibres in the caudate nucleus. *Brain Research* **11**, 464–467.

Kemp, J.M. and Powell, T.P.S. (1970) The cortocostriate projection in the monkey. *Brain* **93**, 525–546.

Kemp, J.M. and Powell, T.P.S. (1971a) The structure of the caudate nucleus of the cat: light and electron microscopy. *Philosophical Transaction of the Royal Society, London,* **262**, 383–401.

Kemp, J.M. and Powell, T.P.S. (1971b) The termination of fibres from the cerebral cortex and thalamus upon dendritic spines in the caudate nucleus: a study with the Golgi method. *Philosophical Transaction of the Royal Society, London,* **262**, 429–439.

Kim, J.S., Hassler, R., Haug, P. and Paik, K.S. (1977) Effect of frontal cortex ablation on striatal glutamic acid level in rat. *Brain Research* **132**, 370–374.

Kincaid, A.E., Zheng, T. and Wilson, C.J. (1995) The spacing of synapses along corticostriatal axons. *Society for Neuroscience Abstracts* **21**, 911.

Kincaid, A.E. and Wilson, C.J. (1996) Corticostriatal innervation of the patch and matrix in the rat neostriatum. *Journal of Comparative Neurology.* **374**, 578–592.

Kita, H. (1993) GABAergic circuits of the striatum. In *Chemical Signalling in the Basal Ganglia, (Progress in Brain Research, vol. 99),* edited by G.W. Arbuthnott and P.C. Emson, *North Holland*, Elsevier, pp. 51–72.

Kita, H. (1994) Parvalbumin-immunopositive neurons in rat globus pallidus, a light and electron microscopic study. *Brain Research* **657**, 31–41.

Kita, H. (1996) Glutamatergic and GABAergic postsynaptic responses of striatal spiny neurons to intrastriatal and cortical stimulation recorded in slice preparations. *Neuroscience* **70**, 925–940.

Kita, H., Kita, T. and Kitai, S.T. (1985) Active membrane properties of rat neostriatal neurons in an *in vitro* slice preparation. *Experimental Brain Research* **60**, 54–62.

Kita, H. and Kitai, S.T. (1987) Efferent projections of the subthalamic nucleus in the rat: Light and electron microscopic analysis with the PHA-L method. *Journal of Comparative Neurology,* **260**, 435–452.

Kita, H. and Kitai, S.T. (1988) GAD-immunoreactive neurons in rat neostriatum: their morphological types and populations. *Brain Research* **447**, 346–352.

Kita, H. and Kitai, S.T. (1990) Amygdaloid projections to the frontal cortex and the striatum in the rat. *Journal of Comparative Neurology* **298**, 40–49.

Kita, H., Kosaka, T. and Heizmann, C.W. (1990) Parvalbumin-immunoreactive neurons in the rat neostriatum: a light and electron microscopic study. *Brain Research* **536**, 1–15.

Kita, H., Chang, H.T. and Fujimoto, K. (1991) Pallido-neostriatal projections of the rat. *Society for Neuroscience Abstracts* **17**, 453.

Kitai, S.T., Kocsis, J.D., Preston, R.J. and Sugimori, M. (1976) Monosynaptic inputs to caudate neurons identified by intracellular injection of horseradish peroxidase. *Brain Research* **109**, 601–606.

Kiyama, H., Seto-Ohshima, A. and Emson, P.C. (1990) Calbindin-D28K as a marker for the degeneration of the striatonigral pathway in Huntington's disease. *Brain Research* **525**, 209–214.

Kocsis, J.D., Sugimori, M. and Kitai, S.T. (1977) Convergence of excitatory synaptic inputs to caudate spiny neurons. *Brain Research* **124**, 403–413.

Kölliker, A. (1896) *Handbuch der Gewebelehre des Menschen.* Bd. II, Englemann, Leipzig.

Kubota, Y., Inagaki, S., Kito, S. and Wu, J.-Y. (1987a) Dopaminergic axons directly make synapses with GABAergic neurons in the rat neostriatum. *Brain Research* **406**, 147–156.

Kubota, Y., Inagaki, S., Shimada, S., Kito, S., Eckenstein, F. and Tohyama, M. (1987b) Neostriatal cholinergic neurons receive direct synaptic inputs from dopaminergic axons. *Brain Research* **413**, 179–184.

Kubota, Y., Inagaki, S., Kito, S., Shimada, S., Okayama, T., Hatanaka, H., Pelletier, G., Takagi, H. and Tohyama, M. (1988) Neuropeptide Y-immunoreactive neurons receive synaptic inputs from dopaminergic axon terminals in the rat neostriatum. *Brain Research* **458**, 389–393.

Kubota, Y. and Kawaguchi, Y. (1993) Spatial distributions of chemically identified neurons in relation to patch and matrix compartments of rat neostriatum. *Journal of Comparative Neurology* **332**, 499–513.

Kubota, Y., Mikawa, S. and Kawaguchi, Y. (1993) Neostriatal GABAergic interneurones contain NOS, calretinin or parvalbumin. *NeuroReport* **5**, 205–208.

Künzle, H. (1975) Bilateral projections from precentral motor cortex to the putamen and other parts of the basal ganglia. An autoradiographic study in *Macaca fascicularis. Brain Research* **88**, 195–209.

Künzle, H. (1977) Projections from the primary somatosensory cortex to basal ganglia and thalamus in the monkey. *Experimental Brain Research* **30**, 481–492.

Künzle, H. (1978) An autoradiographic analysis of the efferent connections from premotor and adjacent prefrontal regions (areas 6 and 9) in *Macaca fascicularis. Brain Behavior and Evolution* **15**, 185–234.

Langer, L.F. and Graybiel, A.M. (1989) Distinct nigrostriatal projection systems innervate striosomes and matrix in the primate striatum. *Brain Research* **498**, 344–350.

Lapper, S.R. and Bolam, J.P. (1992) Input from the frontal cortex and the parafascicular nucleus to cholinergic interneurones in the dorsal striatum of the rat. *Neuroscience* **51**, 533–545.

Lapper, S.R., Smith, Y., Sadikot, A.F., Parent, A. and Bolam, J.P. (1992) Cortical input to parvalbumin-immunoreactive neurones in the putamen of the squirrel monkey. *Brain Research* **580**, 215–224.

Lavoie, B. and Parent, A. (1990) Immunohistochemical study of the serotoninergic innervation of the basal ganglia in the Squirrel monkey. *Journal of Comparative Neurology* **299**, 1–16.

Levey, A.I., Armstrong, D.M., Atweh, S.F., Terry, R.D. and Wainer, B.H. (1983) Monoclonal antibodies to choline acetyltransferase: production, specificity, and immunohistochemistry. *Journal of Neuroscience* **3**, 1–9.

Malach, R. and Graybiel, A.M. (1986) Mosaic architecture of the somatic sensory-recipient sector of the cat's striatum. *Journal of Neuroscience* **6**, 3436–3458.

Martone, M.E., Armstrong, D.M., Young, S.J. and Groves, P.M. (1992) Ultrastructural examination of enkephalin and substance P input to cholinergic neurons within the rat neostriatum. *Brain Research* **594**, 253–262.

McDonald, A.J. (1991) Topographical organization of amygdaloid projections to the caudatoputamen, nucleus accumbens, and related striatal-like areas of the rat brain. *Neuroscience* **44**, 15–33.

McGeer, P.L., McGeer, E.G., Scherer, V. and Singh, K. (1977) A glutamatergic corticostriatal path? *Brain Research* **128**, 369–373.

McGeorge, A.J. and Faull, R.L.M. (1989) The organization of the projection from the cerebral cortex to the striatum in the rat. *Neuroscience* **29**, 503–537.

Mizukawa, K., McGeer, P.L., Vincent, S.R. and McGeer, E.G. (1988) Ultrastructure of reduced nicotinamide dinucleotide phosphate (NADPH) diaphorase-positive neurons in the cat cerebral cortex, amygdala and caudate nucleus. *Brain Research* **452**, 286–292.

Nisenbaum, E.S., Xu, Z.C. and Wilson, C.J. (1994) Contribution of a slowly inactivating potassium current to the transition to firing of neostriatal spiny projection neurons. *Journal of Neurophysiology*, **71**, 1174–1189.

Nisenbaum, E.S. and Wilson, C.J. (1995) Potassium currents responsible for inward and outward rectification in rat neostriatal spiny projection neurons. *Journal of Neurophysiology,* **15**, 4449–4463.

Nisenbaum, E.S., Wilson, C.J., Foehring, R.C. and Surmeier, D.J. (1996) Isolation and characterization of a persistent potassium current in neostriatal neurons *Journal of Neurophysiology*, **76**, 1180–1194.

Noda, H., Manohar, S. and Adey, W.R. (1968) Responses of cat pallidal neurons to cortical and subcortical stimuli. *Experimental Neurology* **20**, 585–610.

Oertel, W.H. and Mugnaini, E. (1984) Immunocytochemical studies of GABAergic neurons in rat basal ganglia and their relations to other neuronal systems. *Neuroscience Letters* **47**, 233–238.

Parent, A. (1990) Extrinsic connections of the basal ganglia. *Trends in Neuroscience* **13**, 254–258.

Penny, G.R., Afsharpour, S. and Kitai, S.T. (1986) The glutamate decarboxylase-, leucine enkephalin-, methionine enkephalin- and substance P-immunoreactive neurons in the neostriatum of the rat and cat: evidence for partial population overlap. *Neuroscience* **17**, 1011–1045.

Pert, C.B., Kuhar, M.J. and Snyder, S.H. (1976) Opiate receptors: autoradiographic localisation in rat brain *Proceedings of the National Academy of Sciences, USA* **73**, 3729–3733.

Phelps, P.E., Houser, C.R. and Vaughn, J.E. (1985) Immunocytochemical localisation of choline acetyltransferase within the rat neostriatum: a correlated light and electron microscopic study of cholinergic neurons and synapses. *Journal of Comparative Neurology* **238**, 286–307.

Pickel, V.M. and Chan, J. (1990) Spiny neurons lacking choline acetyltransferase immunoreactivity are major targets of cholinergic and catecholaminergic terminals in rat striatum. *Journal of Neuroscience Research* **25**, 263–280.

Powell, T.P.S. and Cowan, W.M. (1956) A study of thalamo-striate relations in the monkey. *Brain* **79**, 364–395.

Purpura, D.P. and Malliani, A. (1967) Intracellular studies of the corpus striatum. I. Synaptic potentials and discharge characteristics of caudate neurons activated by thalamic stimulation. *Brain Research* **6**, 324–340.

Preston, R.J., Bishop, G.A. and Kitai, S.T. (1980) Medium spiny neuron projection from the rat striatum: an intracellular horseradish peroxidase study. *Brain Research* **183**, 253–263.

Ragsdale, C.W. and Graybiel, A.M. (1981) The frontostriatal projection in the cat and monkey and its relationship to inhomogeneities established by acetylcholinesterase histochemistry. *Brain Research* **208**, 259–266.

Ragsdale, C.W. and Graybiel, A.M. (1988) Fibres from the basolateral nucleus of the amygdala selectively innervate striosomes in the caudate nucleus of the cat. *Journal of Comparative Neurology* **269**, 506–522.

Ragsdale, C.W. and Graybiel A.M. (1991) Compartmental organization of the thalamostriatal connection in the cat. *Journal of Comparative Neurology* **311**, 134–167.

Rajakumar, N., Rushlow, W., Naus, C.C., Elisevich, K. and Flumerfelt, B.A. (1994) Neurochemical compartmentalization of the globus pallidus in the rat: an immunocytochemical study of calcium-binding proteins. *Journal of Comparative Neurology* **346**, 337–348.

Ren, J.Q. and Kita, H. (1996) An electron microscopic analysis of neostriatal dopamine fibers. *Society for Neuroscience Abstracts* **22**, 1088.

Résibois, A. and Rogers, J.H. (1992) Calretinin in rat brain: an immunohistochemical study. *Neuroscience* **46**, 101–134.

Reubi, J.C. and Cuénod, M. (1979) Glutamate release *in vitro* from corticostriatal terminals. *Brain Research* **176**, 185–188.

Ribak, C.E., Vaughn, J.E. and Roberts, E. (1979) The GABA neurons and their axon terminals in rat corpus striatum as demonstrated by GAD immunocytochemistry. *Journal of Comparative Neurology* **187**, 261–283.

Russchen, F.T. and Price, J.L. (1984) Amygdalostriatal projections in the rat. Topographical organization and fiber morphology shown using the lectin PHA-L as an anterograde tracer. *Neuroscience Letters* **47**, 15–22.

Russchen, F.T., Bakst, I., Amarel, D.G. and Price, J.L. (1985) The amygdalostriatal projections in the monkey. An anterograde tracing study. *Brain Research* **329**, 241–257.

Sadikot, A.F., Parent, A., Smith, Y. and Bolam, J.P. (1992) Efferent connections of the centromedian and parafascicular nuclei in the squirrel monkey- a light and electron microscopic study of the thalamostriatal projection in relation to striatal heterogeneity. *Journal of Comparative Neurology* **320**, 228–242.

Sandell, J.H., Graybiel, A.M. and Chesselet, M.-F. (1986) A new enzyme marker for striatal compartmentalization: NADPH-diaphorase activity in the caudate nucleus and putamen of the cat. *Journal of Comparative Neurology* **243**, 326–334.

Selemon, L.D. and Goldman-Rakic, P.S. (1985) Longitudinal topography and interdigitation of corticostriatal projections in the Rhesus monkey. *Journal of Neuroscience* **5**, 776–794.

Smith, Y. and Parent, A. (1986) Neuropeptide Y-immunoreactive neurons in the striatum of cat and monkey: morphological characteristics, intrinsic organization and co-localization with somatostatin. *Brain Research* **372**, 241–252.

Smith, Y., Parent, A., Seguela, P. and Descarries, L. (1987) Distribution of GABA-immunoreactive neurons in the basal ganglia of the squirrel monkey (*Saimiri sciureus*). *Journal of Comparative Neurology* **259**, 50–64.

Smith, Y. and Bolam, J.P. (1989) Neurons of the substantia nigra reticulata receive a dense GABA-containing input from the globus pallidus in the rat. *Brain Research* **493**, 160–167.

Smith, A.D. and Bolam, J.P. (1990) The neural network of the basal ganglia as revealed by the study of synaptic connections of identified neurones. *Trends in Neuroscience* **13**, 259–266.

Smith, Y., Hazrati, L.-N. and Parent, A. (1990) Efferent projections of the subthalamic nucleus in the squirrel monkey as studied by the PHA-L anterograde tracing method. *Journal of Comparative Neurology* **294**, 306–323.

Smith, Y. and Bolam, J.P. (1991) Convergence of synaptic inputs from the striatum and the globus pallidus onto identified nigrocollicular cells in the rat—a double anterograde labelling study. *Neuroscience* **44**, 45–73.

Smith, Y., Bennett, B.D., Bolam, J.P., Parent, A. and Sadikot, A.F. (1993) Synaptic interactions between the dopaminergic afferents and the cortical or thalamic input at the single cell level in the striatum of monkey. *Society for Neuroscience Abstracts* **19**, 977.

Soghomonian, J.-J., Descarries, L. and Watkins, K.C. (1989) Serotonin innervation in adult rat neostriatum. II. Ultrastructural features: a radioautographic and immunocytochemical study. *Brain Research* **481**, 67–86.

Somogyi, P., Bolam, J.P. and Smith, A.D. (1981) Monosynaptic cortical input and local axon collaterals of identified striatonigral neurons. A light and electron microscopic study using the Golgi-peroxidase transport-degeneration procedure. *Journal of Comparative Neurology* **195**, 567–584.

Song, W.J. and Surmeier, D.J. (1996) Voltage-dependent facilitation of calcium channels in rat neostriatal neurons. *Journal of Neurophysiology*, **76**, 2290–2306.

Spencer, H.J. (1976) Antagonism of cortical excitation of striatal neurons by glutamic acid diethylester: evidence for glutamic acid as an excitatory transmitter in the rat striatum.*Brain Research* **102**, 91–101.

Staines, W.A. (1983) Experiment 8: demonstration of a pallidal innervation of striatal somatostatin containing neurons. *PhD Thesis*, University of British Columbia, 223–230.

Staines, W.A., Atmadja, S. and Fibiger, H.C. (1981) Demonstration of a pallidostriatal pathway by retrograde transport of HRP-labeled lectin. *Brain Research* **206**, 446–450.

Streit, P. (1980) Selective retrograde labelling indicating the transmitter of neuronal pathways. *Journal of Comparative Neurology* **191**, 429–463.

Surmeier, D.J., Eberwine, J., Wilson, C.J., Cao, Y., Stefani, A. and Kitai, S.T. (1992) Dopamine receptor subtypes colocalize in rat striatonigral neurons. *Proceedings of the National Academy of Sciences, USA* **89**, 10178–10182.

Surmeier, D.J. and Kitai, S.T. (1993) D1 and D2 dopamine receptor modulation of sodium and potassium currents in rat neostriatal neurons. In: *Chemical Signalling in the Basal Ganglia, (Progress in Brain Research, Vol. 99),* edited by G. Arbuthnott and P.C. Emson, North Holland, Elsevier, pp. 309–324.

Surmeier, D.J., Reiner, A., Levine, M.S. and Ariano, M.A. (1993) Are neostriatal dopamine receptors co-localized? *Trends in Neuroscience* **16**, 299–305.

Surmeier, D.J., Song, W.J. and Yan, Z. (1996) Coordinated expression of dopamine receptors in neostriatal medium spiny neurons. *Journal of Neuroscience* **16**, 6579–6591.

Takagi, H., Somogyi, P., Somogyi, J. and Smith, A.D. (1983) Fine structural studies on a type of somatostatin-immunoreactive neuron and its synaptic connections in the rat neostriatum: a correlated light and electron microscopic study. *Journal of Comparative Neurology* **214**, 1–16.

Takagi, H., Mizuta, H., Matsuda, T., Inagaki, S., Tateishi, K. and Hamaoka, T. (1984) The occurrence of cholecystokinin-like immunoreactive neurons in the rat neostriatum: light and electron microscopic analysis. *Brain Research* **309**, 346–349.

Takagi, H., Somogyi, P. and Smith, A.D. (1984) Aspiny neurons and their local axons in the neostriatum of the rat: a correlated light and electron microscopic study of Golgi-impregnated material. *Journal of Neurocytology* **13**, 239–265.

Theriault, E. and Landis, D.M.D. (1987) Morphology of striatal neurons containing VIP-like immunoreactivity. *Journal of Comparative Neurology* **256**, 1–13.

van der Kooy, D. and Carter, D.A. (1981) The organization of the efferent projections and striatal afferents of the entopeduncular nucleus and adjacent areas in the rat. *Brain Research* **211**, 15–36.

Vertes, R.P. (1991) A PHA-L analysis of ascending projections of the dorsal raphe nucleus in the rat. *Journal of Comparative Neurology* **313**, 643–668.

Vincent, S., Hökfelt, T., Christensson, I. and Terenius, L. (1982) Immunohistochemical evidence for a dynorphin immunoreactive striato-nigral pathway. *European Journal of Neuroscience* **85**, 251–252.

Vincent, S.R., Johansson, O., Hökfelt, T., Skirboll, L., Elde, R.P., Terenius, L., Kimmel, J. and Goldstein, M. (1983) NADPH-diaphorase: a selective histochemical marker for striatal neurons containing both somatostatin- and avian pancreatic polypeptide (APP)-like immunoreactivities. *Journal of Comparative Neurology* **217**, 252–263.

Vincent, S.R. and Reiner, P.B. (1988) A population of very small striatal neurons in the cat displays vasoactive intestinal polypeptide immunoreactivity. *Neuroscience Letters* **89**, 277–282.

Vuillet, J., Kerkerian, L., Kachidian, P., Bosler, O. and Nieoullon, A. (1989) Ultrastructural correlates of functional relationships between nigral dopaminergic or cortical afferent fibres and neuropeptide Y-containing neurons in the rat striatum. *Neuroscience Letters* **100**, 99–104.

Vuillet, J., Kerkerian-Le Goff, L., Kachidian, P., Dusticier, G., Bosler, O. and Nieoullon, A. (1990) Striatal NPY-containing neurons receive GABAergic afferents and may also contain GABA—an electron microscopic study in the rat. *European Journal of Neuroscience* **2**, 672–681.

Waldvogel, H.J. and Faull, R.L.M. (1993) Compartmentalization of parvalbumin immunoreactivity in the human striatum. *Brain Research* **610**, 311–316.

Walker, R.H., Arbuthnott, G.W. and Wright, A.K. (1989) Electrophysiological and anatomical observations concerning the pallidostriatal pathway in the rat. *Experimental Brain Research* **74**, 303–310.

Webster, K.E. (1961) Cortico-striate interrelations in the albino rat. *Journal of Anatomy* **95**, 532–544.

Wictorin, K., Clarke, D.J., Bolam, J.P. and Björklund, A. (1989) Host corticostriatal fibres establish synaptic connections with grafted striatal neurons in the ibotenic acid lesioned striatum. *European Journal of Neuroscience* **1**, 189–195.

Wilson, C.J. (1986) Postsynaptic potentials evoked in spiny neostriatal projection neurons by stimulation of ipsilateral and contralateral neocortex. *Brain Research* **367**, 201–213.

Wilson, C.J. (1990) Basal Ganglia, In:*The synaptic organization of the brain*, edited by G. M. Shepherd 3rd Edition, Oxford, Oxford University Press, pp. 279–316.

Wilson, C.J. (1992) Dendritic morphology, inward rectification, and the functional properties of neostriatal neurons. In *Single Neuron Computation*, edited by T. McKenna, J. Davis and S.F. Zornetzer, San Diego, Academic Press, pp. 141–171.

Wilson, C.J. (1993) The generation of natural firing patterns in neostriatal neurons. In *Chemical Signalling In The Basal Ganglia, (Progress In Brain Research, vol 99)*, edited by G.W. Arbuthnott and P.C. Emson, Oxford, Elsevier, pp. 277–297.

Wilson, C.J. (1995) The contribution of cortical neurons to the firing pattern of striatal spiny neurons. In *Models of Information Processing in the Basal Ganglia*, edited by J.C. Houk, J.L. Davis and D.G. Beiser, Cambridge, MIT Press, pp.29–50.

Wilson, C.J. and Groves, P.M. (1980) Fine structure and synaptic connections of the common spiny neuron of the rat neostriatum: a study employing intracellular injection of horseradish peroxidase. *Journal of Comparative Neurology* **194**, 599–615.

Wilson, C.J. and Groves, P.M. (1981) Spontaneous firing patterns of identified spiny neurons in the rat neostriatum. *Brain Research* **220**, 67–80.

Wilson, C.J., Chang, H.T. and Kitai, S.T. (1982) Origins of postsynaptic potentials evoked in identified rat neostriatal neurons by stimulation in substantia nigra. *Experimental Brain Research* **45**, 157–167.

Wilson, C.J., Chang, H.T. and Kitai, S.T. (1983a) Origins of postsynaptic potentials evoked in spiny neostriatal neurons by thalamic stimulation in the rat. *Experimental Brain Research* **51**, 217–226.

Wilson, C.J., Groves, P.M., Kitai, S.T. and Linder, J.C. (1983b) Three-dimensional structure of dendritic spines in the rat neostriatum. *Journal of Neuroscience* **3**, 383–388.

Wilson, C.J., Chang, H.T. and Kitai, S.T. (1990) Firing patterns and synaptic potentials of identified giant aspiny interneurons in the rat neostriatum. *Journal of Neuroscience* **10**, 508–519.

Wilson, C.J. and Kawaguchi, Y. (1996) The origins of two-state spontaneous membrane potential fluctuations of neostriatal spiny neurons. *Journal of Neuroscience* **16**, 2397–2410.

Wouterlood, F.G. and Groenewegen, H.J. (1985) Neuroanatomical tracing by use of Phaseolus vulgaris-leucoagglutinin (PHA-L): electron microscopy of PHA-L-filled neuronal somata, dendrites, axons and axon terminals. *Brain Research* **326**, 188–191.

Xu, Z.C., Wilson, C.J. and Emson, P.C. (1989) Restoration of the corticostriatal projection in rat neostriatal grafts: electron microscopic analysis. *Neuroscience* **29**, 539–550.

Xu, Z.C., Wilson, C.J. and Emson, P.C. (1991) Restoration of thalamostriatal projections in rat neostriatal grafts: an electron microscopic analysis. *Journal of Comparative Neurology* **303**, 22–34.

Yeterian, E.H. and Pandya, D.N. (1993) Striatal connections of the parietal association cortices in Rhesus monkeys. *Journal of Comparative Neurology* **332**, 175–197.

Yung, K.K.L., Smith, A.D., Levey, A.I. and Bolam, J.P. (1996) Synaptic connections between spiny neurons of the direct and indirect pathways in the neostriatum of the rat: evidence from dopamine receptor staining and neuropeptide immunostaining. *European Journal of Neuroscience,* **8**, 861–869.

# 7 Neural Dynamics and Surround Inhibition in the Neostriatum: A Possible Connection

**Jeffery R. Wickens and Dorothy E. Oorschot**

*Department of Anatomy and Structural Biology, School of Medical Sciences, University of Otago, P.O. Box 913, Dunedin, New Zealand.*
*tel: (0064) 3 479-7373; FAX: (0064) 3 479-7254;*
*e-mail: jeff.wickens@stonebow.otago.ac.nz; dorothy.oorschot@stonebow.otago.ac.nz*

The domain hypothesis proposes that mutual inhibition among spiny projection neurones is a central principle of neostriatal organization. This hypothesis also proposes that the level of competition among striatal neurones may be regulated by dopamine and acetylcholine levels, with dopamine favouring high levels of competition, and acetylcholine favouring coactivation. It is argued that high competition levels within domains correspond to looseness and movement of the limbs, whereas low competition levels correspond to cocontraction of antagonist groups of muscles and fixation of the limbs, or rigidity. A pivotal assumption of the hypothesis is that under high dopamine conditions inhibitory interactions occur among striatal projection neurones. Existing data for and against the hypothesis are evaluated. It is argued that further experiments are needed to test the hypothesis definitively.

KEYWORDS: Inhibition, competition, dopamine, acetylcholine, rigidity, cocontraction.

## 1. INTRODUCTION: OUTLINE OF THE DOMAIN HYPOTHESIS

In this chapter we consider possible links between the neurodynamics of striatal activity at the network level and the effects of dopamine and acetylcholine on spiny projection neurones at the cellular level. We focus on the idea that different dynamic modes of neural network activity may prevail in the neostriatum depending on the levels of dopamine and acetylcholine.

The domain hypothesis proposes that mutual inhibition among spiny projection neurones is a central principle of neostriatal organization. The key assumption of this hypothesis is that inhibitory interactions occur between the spiny projection neurones that make synaptic contacts with one another. The local network of inhibitory neurones so formed defines a domain of inhibition. Such a network has useful dynamic properties for certain requirements of motor control. In particular, in a network of mutually inhibitory neurones a dynamic of competition should prevail, in which the most active neurones suppress their less active neighbours.

Mutual inhibition among neurones might serve to prevent cocontraction of antagonist muscles. This would be useful for the motor system because the active movement of a limb usually requires relaxation of antagonist muscles in parallel with contraction of agonist groups of muscles. Neural mechanisms for such reciprocal inhibition of antagonistic groups of muscles exist at multiple levels of the motor system. The idea that the neostriatum contains such a mechanism is supported by evidence that during a temporary lesion of the striatum induced by local cooling of the putamen the normal reciprocal inhibition of antagonist muscles is replaced by cocontraction (Hore and Villis, 1980).

Mutual inhibition to prevent cocontraction is unlikely to be useful all of the time. It is often necessary to fix one part of a limb to provide a stable base while moving another part of the limb. For example, in the movements involved in bringing food to the mouth during eating, it is necessary to stabilize the shoulder girdle while moving the elbow and wrist joints. This consideration implies that a mechanism is required to switch mutual inhibition on and off at different joints. We propose that the cholinergic interneurones are well suited to play such a role, on the basis of a model which predicts that the effect of acetylcholine should be to reduce the level of competition in the striatum (Wickens, Alexander and Miller, 1991). Each cholinergic interneurone might control a set of antagonist groups of muscles, allowing cocontraction as and when required.

In Parkinson's disease the normal relationship between antagonists appears to break down, and instead there is cocontraction of muscles that are normally reciprocally inhibited (Hayashi *et al.*, 1988). These observations suggest that the striatal mechanism that regulates the balance of activation of antagonist groups of muscles requires an intact dopamine system to function normally. A further extension of the domain hypothesis is that under conditions of high dopaminergic tone a neural dynamic of competition should prevail in a domain, such that activation of a group of muscles by activity of a striatal neurone is accompanied by suppression of activity in neurones representing antagonist groups of muscles. This may correspond to the conditions required for free movement or looseness in the limbs. Under conditions of low dopaminergic tone and high cholinergic tone mutual inhibition is less effective and a dynamic of coactivation prevails. These effects of dopamine may be mediated by inhibitory effects of dopamine on the cholinergic interneurones (Lehmann and Langer, 1983). Thus dopamine may increase and acetylcholine decrease competition within a striatal domain.

Although the domain hypothesis is simple-minded in many respects, it has a number of positive points as a heuristic device: It is based on testable assumptions about the anatomy and physiology of the neostriatum, it makes testable predictions about the network dynamics of the neostriatum, and it provides a possible link between the cellular effects of dopamine and the symptoms of dopamine deficiency. In the following sections each of the assumptions and predictions of the domain hypothesis will be examined in the light of recent evidence.

## 2. EVIDENCE RELEVANT TO THE DOMAIN HYPOTHESIS

Electrophysiological studies have shown that striatal neurones with movement-related activity often occur in clusters of cells which are active in association with movements about a single joint (Alexander, 1987). These clusters are physically of similar size to the proposed domains. Activation of the output neurones within these zones by using electrical stimulation, with current intensities in the microampere range, produces discrete movements of individual body parts which are usually restricted to a single joint (Alexander and DeLong, 1985a,b).

The majority of neurones within the striatum are spiny projection neurones, also known as medium-sized, densely spiny neurones (Grofova, 1975; Kitai *et al.*, 1979; Oorschot, 1996, 1997; West *et al.*, 1996). Anatomically, there is sound evidence that each spiny projection neurone gives rise to an extensive arborization of local axons (Preston, Bishop and Kitai, 1980; Wilson and Groves, 1980). The usual role of such local axons is to connect one neurone to neighbouring neurones, in this case each spiny projection neurone to other spiny projection neurones. A number of studies provide indirect evidence for these local connections between spiny projection neurones (Bolam and Izzo, 1988; Wilson and Groves, 1980; Yung *et al.*, 1996). Since the spiny projection neurones all produce the inhibitory neurotransmitter gamma-aminobutyric acid (GABA), this local connectivity seems likely to be mutually inhibitory (Aronin, Chase and DiFiglia, 1986; Kita and Kitai, 1988; Pasik *et al.*, 1988).

Recurrent inhibition, i.e. inhibition of a spiny projection neurone by its own collaterals, has been reported by Park, Lighthall and Kitai (1980). Action potentials evoked in the recorded neurone by a depolarizing current pulse reduced the amplitude of excitatory postsynaptic responses (EPSPs) evoked from stimulation of the substantia nigra. This effect was blocked by GABA antagonists, suggesting that the effect is mediated by GABAergic synapses. Such inhibitory effects were not, however, associated with detectable inhibitory postsynaptic potentials (IPSPs) and did not reduce the amplitude of EPSPs evoked by cortical stimulation. These results suggest that while spiny projection neurones may be inhibited by their own collaterals, the effect of the inhibition is limited to EPSPs arising from the striatal collaterals of corticofugal axons activated by stimulation of the substantia nigra.

If some degree of recurrent inhibition does occur in the striatum, then the spiny projection neurones are likely to exert a similar inhibitory effect on their neighbours. Katayama, Miyazaki and Tsubokawa (1981) showed that antidromic activation of striatal neurones by stimulation in the entopeduncular nucleus produced short latency suppression of firing in other nearby spontaneously firing neurones. The short latency suppression was blocked by GABA antagonists, but was not reduced by removal of the cerebral cortex, suggesting that it might reflect lateral inhibitory interactions between spiny projection neurones. Another possibility is that the suppression is mediated by GABAergic projections from the entopeduncular nucleus to the neostriatum.

In an effort to test the hypothesis that surround inhibition exists in the neostriatum, Jaeger, Kita and Wilson (1994) made simultaneous intracellular recordings from pairs of spiny projection neurones in striatal slices. They found no evidence for inhibitory synaptic interactions. However, the existence of a synaptic contact between the neurones studied was not confirmed by electron microscopy. Since only thirty five pairs of neurones were examined it is important to consider what proportion of the sampled pairs were likely to make synaptic contacts. For comparison, a total of 1163 pairs of pyramidal neurones were studied in cortical slices with only thirty four pairs (i.e. 3%) exhibiting synaptic interactions (Deuchars, West and Thomson, 1994). A similar rate of synaptic interactions has also been reported in hippocampal slices (Deuchars and Thomson, 1996; Sayer, Friedlander and Redman, 1990). Assuming a similar probability (i.e. $p = 0.03$) of detecting synaptic interactions in striatal slices there is a 34% chance of not detecting any interactions in a sample of thirty five pairs. Statistically the chance of detecting no interactions in a sample of

150 pairs of neurones is approximately 1% which suggests that *at least* this many pairs of striatal neurones should be studied before inhibitory interactions can be confidently ruled out.

The difference between the probability of *detecting* a connection and the *actual* probability of a connection may vary according to experimental techniques. Under optimized conditions in cortical slices the probability of detecting a connection can be increased to 21% (Thomson *et al.*, 1996) which is closer to the theoretical connectivity (Braitenberg and Schüz, 1991).

The foregoing argument assumes a uniform probability of synaptic contacts between pairs of striatal neurones. It is possible that synaptic contacts occur preferentially between particular groups of neurones. About half of the striatal neurones project to the globus pallidus, and the other half project to the substantia nigra and entopeduncular nucleus. These two subpopulations are mixed together. There may be differing probabilities of striopallidal to striopallidal, striopallidal to strionigral, strionigral to striopallidal, and strionigral to strionigral synaptic contacts. Non-uniformity may lower the chances of finding pairs of cells that make synaptic contacts.

The effects of antidromic activation of striatal neurones by stimulation of the globus pallidus or substantia nigra were also examined by Jaeger, Kita and Wilson (1994), again with negative results. These studies also used intracellular recording techniques, both *in vivo* and *in vitro*. They are therefore more direct than the studies of Katayama, Miyazaki and Tsubokawa (1981) and the findings are a serious challenge to the domain hypothesis.

## 3. REGULATION OF INHIBITORY INTERACTIONS BY NEUROMODULATORS

It remains possible that inhibitory interactions do not occur under the conditions that normally exist in slices, or in anaesthetised animals (used in the studies referred to above) but might appear in the right conditions. A hint of evidence in support of this was mentioned by Kita (1993) who reported previously unpublished observations that IPSPs could be evoked by antidromic stimulation when synaptic transmission was enhanced by 4-aminopyridine, a substance that increases the effect of synaptic transmission by prolonging action potential duration.

There is some indirect evidence to suggest that inhibitory interactions do occur in the striatum but might be turned on or off by neuromodulators such as dopamine and acetylcholine. When striatal neurones are activated by localized application of excitatory neurotransmitters a surrounding zone of inhibition is also produced. This surrounding zone of inhibition is reduced by dopamine antagonist drugs (Rebec and Curtis, 1988).

A similar phenomenon has been described in the substantia nigra and globus pallidus after dopamine-depleting lesions (Tremblay, Filion and Bedard, 1989). Stimulation in a single striatal area can produce simultaneous excitatory and inhibitory influences on different nigrothalamic neurones (Deniau and Chevalier, 1985). Stimulation in the striatum produces a focus of inhibition in a restricted area of the globus pallidus, with a contrasting surround of excitation at the fringes (Tremblay, Filion and Bedard, 1989). These phenomena may be mediated by lateral inhibition in the neostriatum, and it is of particular interest that the contrasting surround in the globus pallidus is reduced by dopamine depleting lesions (Filion, Tremblay and Bedard, 1988).

There are several possible mechanisms by which dopamine might modulate striatal dynamics. Dopamine has direct effects on potassium and sodium channels in neostriatal neurones and may also act indirectly by reducing acetylcholine release from cholinergic interneurones via an action of dopamine at the D2 receptor subtype (Fujiwara *et al.*, 1987; Hoffman, Talmaciu and Cubeddu, 1986; Stoof *et al.*, 1982; Stoof, Verheijden and Leysen, 1987).

## 4. CHOLINERGIC INTERNEURONES IN STRIATAL DYNAMICS

Cholinergic interneurones receive synaptic input from many different types of axon, involving a range of different neurotransmitters. They receive direct synaptic input from dopaminergic axons (Kubota *et al.*, 1987). They also receive excitatory inputs from thalamostriatal afferents (Wilson, Chang and Kitai, 1990). Corticostriatal afferents form synapses on distal dendrites, but these contacts are infrequent (Dimova *et al.*, 1993) and not always observed (Lapper and Bolam, 1992; Meredith and Wouterlood, 1990).

The cholinergic interneurones give rise to an axon which divides repeatedly to form a dense axonal plexus (Chang, Wilson and Kitai, 1982) the boutons of which form predominantly symmetrical synaptic specializations (Bolam, Ingham and Smith, 1984; Takagi, Somogyi and Smith, 1984). Spiny projection neurones are the major postsynaptic target of cholinergic interneurones (DiFiglia, 1987; Izzo and Bolam, 1988).

Despite being relatively very few in number (Oorschot, 1997) the cholinergic interneurones cause the neostriatum to have the highest concentration of acetylcholine in the brain. They have depolarized resting membrane potentials and low thresholds (Kawaguchi, 1992) and are tonically active *in vivo* (Wilson, Chang and Kitai, 1990). These features suggest a tonic, diffuse modulatory role. The axons of the cholinergic interneurones extend over an area many times larger than the postulated domains. Thus, each cholinergic interneurone probably controls multiple domains (see Figure 7.1A.)

## 5. EFFECTS OF ACETYLCHOLINE ON CELLULAR PROPERTIES OF SPINY PROJECTION NEURONES

There are several biophysical effects of acetylcholine on the spiny projection neurones. Some studies suggest a possible fast nicotinic excitation, and also a muscarinic response that results in a decreased sensitivity to excitatory synaptic activation (Akaike, Sasa and Takaori, 1988; Dodt and Misgeld, 1986). Actions on membrane potassium conductances have been demonstrated, but the findings are difficult to interpret. Dodt and Misgeld (1986) have shown that muscarinic agonists *increase* the membrane resistance of striatal cells, apparently by decreasing a membrane potassium conductance. However, this effect was only seen at high agonist concentrations. In the same paper (Figure 6C, Dodt and Misgeld, 1986) muscarine appears to *decrease* the membrane resistance during hyperpolarization.

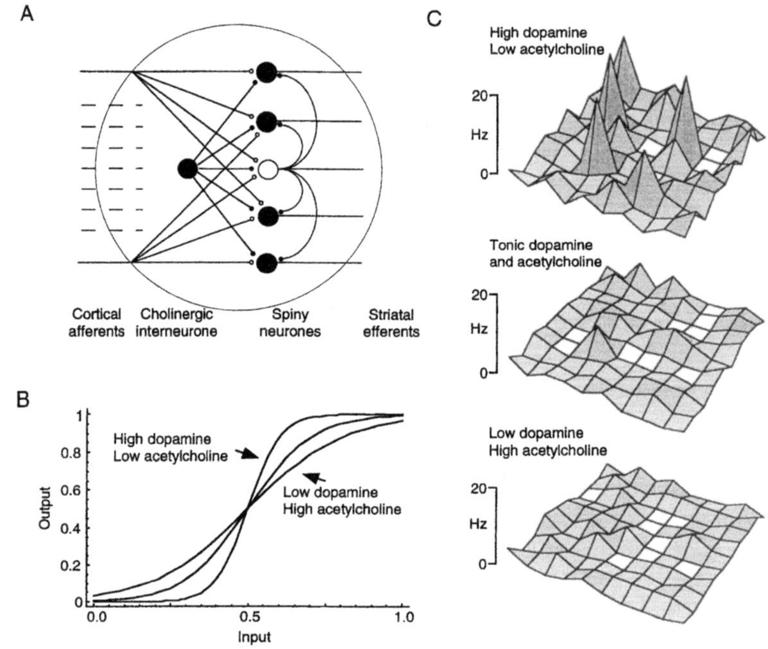


**Figure 7.1.** Outline of the domain concept. **A.** Spiny projection neurones within the range of each other's local axon collaterals have the potential to form synaptic contacts. Those that do form such contacts constitute a domain. Only a single domain is shown, but within the range defined above, multiple domains may form with a variable degree of overlap. One cholinergic interneurone is shown. Each cholinergic interneurone has the capacity to contact spiny projection neurones in multiple domains. **B.** Hypothetical input/output function of a spiny projection neurone depicting the effects of dopamine and acetylcholine. High dopamine/low acetylcholine conditions produce a steeper input/output function. **C.** Computer simulation of the effects of the hypothetical input/output function on the spatial distribution of activity in multiple overlapping domains. A $10 \times 10$ array of neurones is shown. Each is connected to its 20 nearest neighbours, arranged on the surface of a 2-D torus. The rate of action potential of each neurone in response to a uniform synaptic drive is shown. Under high dopamine conditions peaks of high activity form in response to excitatory synaptic input from cortical afferents: a dynamic of competition prevails. Under low dopamine conditions the peaks collapse and a dynamic of coactivation prevails. For details of the computer simulation model see Wickens (1993).

Two types of potassium conductance are revelant here, the M-current and the A-current. The prevailing view has been that acetylcholine probably decreases the M-current in striatal neurones, because that is what it does in other neurones (Adams, Brown and Constanti, 1982). In the hippocampus, for example, the M-current is *decreased* by acetylcholine (McCormick and Prince, 1986). However, it is clear that in different parts of the body, different ion channels may be opened or closed as a result of muscarinic activity (Christie and North, 1988). The actions of acetylcholine in the striatum may be different from its actions in other brain areas.

Voltage-clamp studies of dissociated striatal neurones have provided new information about the actions of the acetylcholine. The A-current, in particular, appears to be sensitive to acetylcholine. Acetylcholine shifts the voltage-dependence of the A-current activation and inactivation curves toward more negative membrane potentials and the peak conductance is increased (Akins, Surmeier and Kitai, 1990). At relatively hyperpolarized resting potentials, acetylcholine alters the A-current so that it suppresses excitatory inputs and further slows discharge rate. At relatively depolarized potentials, acetylcholine increases excitability by removing the A-current through inactivation.

## 6. EFFECTS OF DOPAMINE AND ACETYLCHOLINE ON NEURODYNAMICS

It has previously been suggested that the effects of dopamine and acetylcholine on spiny projection neurones can be represented by variations in a membrane potassium conductance under dopaminergic-cholinergic control. Hypothetically, the effect of increasing acetylcholine levels is to increase potassium conductances, so that the input-output relation of the spiny projection neurones is made flatter (see Figure 7.1B). Conversely, the effect of decreasing acetylcholine levels is to make the input-output relation steeper.

Computer simulation of a network model of a striatal domain showed that varying potassium conductances in this manner could switch the behaviour of the striatal network between two dynamic modes: competition and coactivation (Alexander and Wickens, 1993; Wickens, Alexander and Miller, 1991). Increasing the potassium conductance decreases the steepness of the input-output relation, and decreases the level of competition. This corresponds, in the model, to the effect of increasing acetylcholine and decreasing dopamine levels. Conversely, reducing the potassium conductance increases the steepness of the input-output relation, and increases the level of competition. This corresponds to the effect of decreasing acetylcholine and increasing dopamine levels. The overall effect of increasing dopamine levels is to promote a dynamic mode of competition, and when dopamine is deficient the competitive mode breaks down into one of coactivation (Figure 7.1C).

In principle, the model provides a link between the effects of dopamine deficiency and the symptom of muscular rigidity, known to be brought about by coactivation of mutually antagonistic groups of muscles. This link depends on a number of explicit anatomical assumptions about the connectivity among spiny projection neurones, and the biophysical effects of acetylcholine and dopamine. It is very likely that within the next few years the application of modern methods of anatomical and physiological analysis will force substantial revision of these assumptions. It remains to be seen whether the domain hypothesis and the regulation of competition-coactivation dynamics by neuromodulators will remain as an underlying principle of striatal function. At present, it still seems to offer a possible link between macroscopic phenomena relevant to clinical symptoms and underlying biophysical and anatomical mechanisms.

## REFERENCES

Adams P.R., Brown D.A. and Constanti A. (1982) M-currents and other potassium currents in bullfrog sympathetic neurones. *Journal of Physiology (London)* **330**, 537–572

Akaike A., Sasa M. and Takaori S. (1988) Muscarinic inhibition as a dominant role in cholinergic regulation of transmission in the caudate nucleus. *Journal of Pharmacology and Experimental Therapeutics* **246**, 1129–1136

Akins P.T., Surmeier D.J. and Kitai S.T. (1990) Muscarinic modulation of a transient $K^+$ conductance in rat neostriatal neurons. *Nature (London)* **344**, 240–242

Alexander G.E. (1987) Selective neuronal discharge in monkey putamen reflects intended direction of planned limb movements. *Experimental Brain Research* **67**, 623–634

Alexander G.E. and DeLong M.R. (1985a) Microstimulation of the primate neostriatum: I Physiological properties of striatal microexcitable zones. *Journal of Neurophysiology* **53**, 1401–1416

Alexander G.E. and DeLong M.R. (1985b) Microstimulation of the primate neostriatum: II Somatotopic organization of striatal microexcitable zones and their relation to neuronal response properties. *Journal of Neurophysiology* **53**, 1417–1430

Alexander M.E. and Wickens J.R. (1993) Analysis of striatal dynamics: the existence of two modes of behaviour. *Journal of Theoretical Biology* **163**, 413–438

Aronin N., Chase K. and DiFiglia M. (1986) Glutamic acid decarboxylase and enkephalin immunoreactive axon terminals in the rat neostriatum synapse with striatonigral neurons. *Brain Research* **365**, 151–158

Bolam J.P., Ingham C.A. and Smith A.D. (1984) The section Golgi-impregnation procedure 3. Combination of Golgi-impregnation with enzyme histochemistry to characterize acetylcholinesterase-containing neurons in the rat neostriatum. *Neuroscience* **12**, 687–709

Bolam P. and Izzo P.N. (1988) The postsynaptic targets of substance P-immunoreactive terminals in the rat neostriatum with particular reference to identified spiny striatonigral neurons. *Experimental Brain Research* **70**, 361–377

Braitenberg V. and Schüz A. (1991) *Anatomy of the cortex: Statistics and geometry*. Springer, Berlin

Chang H.T., Wilson C.J. and Kitai S.T. (1982) A Golgi study of rat neostriatal neurons: light microscopic analysis. *Journal of Comparative Neurology* **208**, 107–126

Christie M.J. and North R.A. (1988) Control of ion conductances by muscarinic receptors. *Trends in Pharmacological Science, Supplement* 30–34

Deniau J.M. and Chevalier G. (1985) Disinhibition as a basic process in the expression of striatal functions. II The striatonigral influence on thalamocortical cells of the ventromedial thalamic nucleus. *Brain Research* **334**, 227–233

Deuchars J. and Thomson A.M. (1996) CA1 pyramid-pyramid connections in rat hippocampus in vitro: dual intracellular recordings with biocytin filling. *Neuroscience* **74**, 1009–1018

Deuchars J., West D.C. and Thomson A.M. (1994) Relationship between morphology and physiology of pyramid-pyramid single axon connections in rat neocortex in vitro. *Journal of Physiology (London)* **478**, 423–435

DiFiglia M. (1987) Synaptic organization of cholinergic neurons in the monkey neostriatum. *Journal of Comparative Neurology* **255**, 245–258

Dimova R., Vuillet J., Nieoullon A. and Kerkerian-Le Goff L. (1993) Ultrastructural features of the choline acetyltransferase-containing neurons and relationships with nigral dopaminergic and cortical afferent pathways in the rat striatum. *Neuroscience* **53**, 1059–1071

Dodt H.U. and Misgeld U. (1986) Muscarinic slow excitation and muscarinic inhibition of synaptic transmission in the rat neostriatum. *Journal of Physiology (London)* **380**, 593–608

Filion M., Tremblay L. and Bedard P.J. (1988) Abnormal influences of passive limb movement on the activity of globus pallidus neurons in Parkinsonian monkeys. *Brain Research* **444**, 165–176

Fujiwara H., Kato N., Shuntoh H. and Tanaka C. (1987) D2-Dopamine receptor-mediated inhibition of intracellular $Ca^{++}$ mobilization and release of acetylcholine from guinea-pig neostriatal slices. *British Journal of Pharmacology* **91**, 287–297

Grofova I. (1975) The identification of striatal and pallidal neurons projecting to substantia nigra. An experimental study by means of retrograde axonal transport of horseradish peroxidase. *Brain Research* **91**, 286–291

Hayashi A., Kagamihara Y., Nakajima Y., Narabayashi H., Okuma Y. and Tanaka R. (1988) Disorder in reciprocal innervation upon initiation of voluntary movement in patients with Parkinson's disease. *Experimental Brain Research* **70**, 437–440

Hoffman I.S., Talmaciu R.K. and Cubeddu L.X. (1986) Interactions between endogenous dopamine and dopamine agonists at release modulatory receptors: multiple effects of neuronal uptake inhibitors on transmitter release. *Journal of Pharmacology and Experimental Therapeutics* **238**, 437–446

Hore J. and Villis T. (1980) Arm movement performance during reversible basal ganglia lesions in the monkey. *Experimental Brain Research* **39**, 217–228

Izzo P.N. and Bolam J.P. (1988) Cholinergic synaptic input to different parts of spiny striatonigral neurons in the rat. *Journal of Comparative Neurology* **269**, 219–234

Jaeger D., Kita H., Wilson C.J. (1994) Surround inhibition among projection neurons is weak or nonexistent in the rat neostriatum. *Journal of Neurophysiology* **72**, 2555–2558

Katayama Y., Miyazaki S. and Tsubokawa T. (1981) Electrophysiological evidence favoring intracaudate axon collaterals of GABAergic caudate output neurons in the cat. *Brain Research* **216**, 180–186

Kawaguchi Y. (1992) Large aspiny cells in the matrix of the rat neostriatum in vitro: physiological identification, relation to the compartments and excitatory postsynaptic currents. *Journal of Neurophysiology* **67**, 1669–1682

Kita H. (1993) GABAergic circuits of the striatum. In: *Chemical signalling in the basal ganglia.* (*Progress in Brain Research.* vol 99), edited by G.W. Arbuthnott and P.C. Emson, Amsterdam, Elsevier, pp. 51–72

Kita H. and Kitai S.T. (1988) Glutamate decarboxylase immunoreactive neurons in cat neostriatum: their morphological types and populations. *Brain Research* **447**, 346–352

Kitai S.T., Preston R.J., Bishop G.A. and Koscis J.D. (1979) Striatal projection neurons: morphological and electrophysiological studies. In *The Extrapyramidal System and its Disorders,* edited by L.J. Poirier, New York, Raven Press, pp. 45–51.

Kubota Y., Inagaki S., Shimada S., Kito S., Eckenstein F. and Tohyama M. (1987) Neostriatal cholinergic neurons receive direct synaptic inputs from dopaminergic axons. *Brain Research* **413**, 179–184

Lapper S.R. and Bolam J.P. (1992) Input from the frontal cortex and the parafascicular nucleus to cholinergic interneurons in the dorsal striatum of the rat. *Neuroscience* **51**, 533–45

Lehmann J. and Langer S.Z. (1983) The striatal cholinergic interneuron: synaptic target of dopaminergic terminals? *Neuroscience* **10**, 1105–1120

McCormick D.A. and Prince D.A. (1986) Acetylcholine induces burst firing in thalamic reticular neurons by activating a potassium conductance. *Nature (London)* **319**, 402–405

Meredith G.E. and Wouterlood F.G. (1990) Hippocampal and midline thalamic fibers and terminals in relation to the choline acetyltransferase-immunoreactive neurons in nucleus accumbens of the rat: a light and electron microscopic study. *Journal of Comparative Neurology* **296**, 204–221

Oorschot D.E. (1996) Total number of neurons in the neostriatal, pallidal, subthalamic, and substantia nigral nuclei of the rat basal ganglia: a stereological study using the cavalieri and optical disector methods. *Journal of Comparative Neurology* **366**, 580–599

Oorschot D.E. (1997) Total number of large interneurons within the rat neostriatum: a stereological study using the optical disector and Cavalieri methods. *International Journal of Neuroscience* **89**, 90

Park M.R., Lighthall J.W. and Kitai S.T. (1980) Recurrent inhibition in the rat neostriatum. *Brain Research* **194**, 359–369

Pasik P., Pasik T., Holstein G. andHamori J. (1988) GABAergic elements in the neuronal circuits of the monkey neostriatum: a light and electron microscopic immunocytochemical study. *Journal of Comparative Neurology* **270**, 157–170

Preston R.J., Bishop G.A. and Kitai S.T. (1980) Medium spiny neuron projection from the rat neostriatum: an intracellular horseradish peroxidase study. *Brain Research* **183**, 253–263

Rebec G.V. and Curtis S.D. (1988) Reciprocal zones of excitation and inhibition in the neostriatum. *Synapse* **2**, 633–635

Sayer R.J., Friedlander M.J. and Redman S.J. (1990) The time course and amplitude of EPSPs evoked at synapses between pairs of CA3/CA1 neurones in the hippocampal slice. *Journal of Neuroscience* **10**, 826–836

Stoof J.C., DeBoer T., Sminia P. and Mulder A.H. (1982) Stimulation of D2-dopamine receptors in rat neostriatum inhibits the release of acetylcholine and dopamine but does not affect the release of gamma-aminobutyric acid, glutamate or serotonin. *European Journal of Pharmacology* **84**, 211–214

Stoof J.C., Verheijden P.F.H.M. and Leysen J.E. (1987) Stimulation of D2-receptors in rat nucleus accumbens slices inhibits dopamine and acetylcholine release but not cyclic AMP formation. *Brain Research* **423**, 364–368

Takagi H., Somogyi P. and Smith A.D. (1984) Aspiny neurons and their local axons in the neostriatum of the rat: a correlated light and electron microscopic study of Golgi-impregnated material. *Journal of Neurocytology* **13**, 239–265

Thomson A.M., Est D.C., Hahn J. and Deuchars J. (1996) Single axon IPSPs elicited in pyramidal cells by three classes of interneurones in slices of rat neocortex. *Journal of Physiology (London)* **496**, 81–102

Tremblay L., Filion M. and Bedard B.J. (1989) Responses of pallidal neurons to striatal stimulation in monkeys with MPTP-induced parkinsonism. *Brain Research* **498**, 17–33

West M.J., Ostergaard K., Andreassen O.A. and Finsen B. (1996) Estimation of the number of somatostatin neurons in the striatum: an in situ hybridization study using the optical fractionator method. *Journal of Comparative Neurology* **370**, 11–22

Wickens J.R., Alexander M.E. and Miller R. (1991) Two dynamic modes of striatal function under dopaminergic-cholinergic control: simulation and analysis of a model. *Synapse* **8**, 1–12

Wickens J.R. (1993) *A theory of the striatum.* Oxford, Pergamon Press

Wilson C.J., Chang H.T. and Kitai S.T. (1990) Firing patterns and synaptic potentials of identified giant aspiny interneurons in the rat neostriatum. *Journal of Neuroscience* **10**, 508–19

Wilson C.J. and Groves P.M. (1980) Fine structure and synaptic connection of the common spiny neuron of the rat neostriatum: a study employing intracellular injection of horseradish peroxidase. *Journal of Comparative Neurology* **194**, 599–615

Yung K.K.L., Smith A.D., Levey A.I. and Bolam J.P. (1996) Synaptic connections between spiny neurons of the direct and indirect pathways in the neostriatum of the rat: evidence from dopamine receptor and neuropeptide immunostaining. *European Journal of Neuroscience* **8**, 861–869

# 8 The Domain Hypothesis: A Central Organizing Principle for Understanding Neostriatal Circuitry?

**Dorothy E. Oorschot**

*Department of Anatomy and Structural Biology, School of Medical Sciences, University of Otago, P.O. Box 913, Dunedin, New Zealand.*
*tel: (0064) 3 479-7379; FAX: (0064) 3 479-7254;*
*e-mail: dorothy.oorschot@stonebow.otago.ac.nz*

The domain hypothesis proposes that inhibitory interactions between spiny projection neurones may be a central organizing principle of the neostriatum. The primary aim of this chapter is to consider how the domain hypothesis may help to make sense of the circuitry within the neostriatum. Using the currently known anatomy, biochemistry, pharmacology and physiology of the neostriatal neurones, it is proposed that the major role of the interneurones and the nigral dopaminergic input in this circuit may be as follows: (i) the GABAergic interneurones may assist in the inhibition of the vast majority of spiny projection neurones which are silent at any one point in time as proposed by the domain hypothesis; (ii) the somatostatin / neuropeptide-Y / nitric oxide synthase interneurones may excite the "winning" spiny projection neurones across a number of domains *and* inhibit the non-firing neurons *and* regulate local blood flow at any one point in time; (iii) the nigral dopaminergic input may drive the "winning" spiny projection neurones by increasing competition within a neostriatal domain at any one point in time; and (iv) the role of the cholinergic interneurones may occur *after* a set of spiny projection neurones have fired, each in a different domain. In this context, the role of the cholinergic interneurones may be to 'reset the domains' in readiness for the next movement sequence. In other words, their role may be to 'restore a level playing field' by decreasing the competition that has been generated by the previous firing sequence. The domain hypothesis does not account for all known neostriatal data. Nevertheless, it still seems to offer a conceptual framework upon which most neostriatal circuitry gains coherence.

KEYWORDS: spiny projection neurones, lateral inhibition, neostriatal interneurones, dopamine, competition

## 1. THE DOMAIN HYPOTHESIS

The vast majority of neurones within the neostriatum are the spiny projection cells, also known as the medium-sized, densely spiny neurones (e.g. Grofova, 1975; Kitai *et al.*, 1979; Oorschot, 1996; West *et al.*, 1996; Oorschot, 1997). In the rat these neurones receive monosynaptic inputs from all areas of the cerebral cortex (Webster, 1961; Kemp, 1968; Kitai *et al.*, 1976; Somogyi, Bolam and Smith, 1981; Dube, Smith and Bolam, 1988; MacGeorge and Faull, 1989), from dopamine projections from the midbrain (Andén *et al.* 1964; Kitai *et al.* 1976; Freund, Powell and Smith, 1984), and from some intralaminar nuclei in the thalamus (Nauta, Pritz and Lasek, 1974; Kitai *et al.*, 1976; Wilson, Chang and Kitai, 1983; Dube, Smith and Bolam, 1988). All these inputs converge within the neostriatum and terminate close to one another on the spiny projection neurones (Kitai *et al.*, 1976; Kocsis, Sugimori and Kitai, 1977). The spiny projection neurones also receive input from at least three types of neostriatal interneurones (Phelps, Houser and Vaughn, 1985; Izzo and Bolam, 1988; Vuillet *et al.*, 1989a; Kita, 1993). Since the spiny projection neurones constitute the neostriatal output (e.g. Grofova, 1975), the functional properties of the various synaptic inputs and the response properties of the spiny projection neurones are key determinants of the signal processing operations performed in the neostriatum. The spiny projection neurones therefore represent a key structural and functional unit within the neostriatum.

Anatomically, there is sound evidence that each spiny projection neurone gives rise to an extensive arborization of local axons (e.g. Preston, Bishop and Kitai, 1980; Wilson and Groves, 1980). The usual role of such local axons is to connect one neurone to neighbouring neurones, in this case each spiny projection neurone to other spiny projection neurones. A number of studies provide sound indirect evidence for these local connections between spiny projection neurones (Wilson and Groves, 1980; Bolam and Izzo, 1988; Yung *et al.*, 1996). Measurement of the local axonal spread of spiny projection neurones, together with recent data that around 360 spiny projection neurones and their dendrites are within this axonal spread (Oorschot, 1996), also suggest that local synaptic connectivity is likely. Since the spiny projection neurones all produce the inhibitory neurotransmitter gamma-aminobutyric acid (GABA), this local connectivity seems likely to be mutually inhibitory (Aronin, Chase and DiFiglia, 1986; Kita and Kitai, 1988; Pasik *et al.*, 1988).

The likelihood of local, inhibitory synaptic connectivity has led several authors to propose that interactions between spiny projection neurones may be a central organizing principle in the neostriatum (Park, Lighthall and Kitai, 1980; Groves, 1983; Rolls and Williams, 1986; Wickens, 1993). For example, Wickens (1993) postulated that the basic organizational unit of the neostriatum is a domain of mutual inhibition. Each domain is a population of spiny projection neurones that have mutually inhibitory connections. The prevailing dynamic within a domain would be one of competition, with very few neurones active simultaneously. For the purpose of this Chapter, this postulate has been called the domain hypothesis.

The domain hypothesis may help to make sense of the known anatomy and circuitry within the mammalian basal ganglia, especially with respect to the neostriatum. The aim of this Chapter is to consider how the domain hypothesis may help to make sense of the anatomy of the basal ganglia and the circuitry within the neostriatum. The data reviewed is based primarily on studies in the rat.

## 2. TOTAL NEURONAL NUMBERS WITHIN THE NEOSTRIATUM AND ITS TARGETS

There is a noticeable trend that the neostriatum contains far more output neurones (i.e. medium-sized, spiny projection neurones) than the total number of neurones in all its targets. For example, in the rat there are 29 times more neostriatal output neurones compared with all the pallidal and nigral target neurones (Oorschot, 1996; see Figure 8.1). The resulting anatomical convergence, in both the rat and the human, has been interpreted as a loss of information through the basal ganglia. However, the domain hypothesis challenges this interpretation. In the human, there are thought to be 100 million medium-sized spiny striatal projection neurones. The number per unit volume (i.e. Nv) when corrected for shrinkage is 11,000 per mm$^3$ (Lange *et al.*, 1976). Assuming a radius of contact inhibition for each striatal projection neurone of 250 $\mu$m, Wickens (1993) calculates the number of neurones in a human striatal inhibitory domain to be 720 (i.e., number in a domain = Nv.4$\pi$/3.r$^3$), and the total number of domains to be 139,000. This latter number is remarkably similar to estimates of the number of neurones in the human internal globus pallidus (157,000) and substantia nigra reticulata (160,000: Percheron, Francois and Yelnik, 1987) and is given the interpretation that information is preserved in this pathway (Wickens, 1993). For the rat, the radius of contact inhibition for each striatal projection is likely to be 100 $\mu$m (Penny, Wilson and Kitai, 1988; Kawaguchi, Wilson and Emson, 1989). Coupled with the total number of medium-sized neurones in the striatum of 5.58 million, and their (shrinkagecorrected) Nv of 84,900 per mm$^3$, the number of neurones in a rat domain would be 356 and the total number of domains would be 15,674 (Oorschot, 1996). This number of domains is only 2.4 times the total number of entopeduncular neurones (6,400) and 0.3 of the total number of substantia nigra *pars reticulata* neurones (52,600), thereby implying that information may also be preserved in the rat basal ganglia pathway (Oorschot, 1996).

Thus, a circuit which is somewhat perplexing in terms of information theory at the macroscopic level, becomes less perplexing in the context of the domain hypothesis. Is this also the case when the known neostriatal circuits are considered in the context of the domain hypothesis?

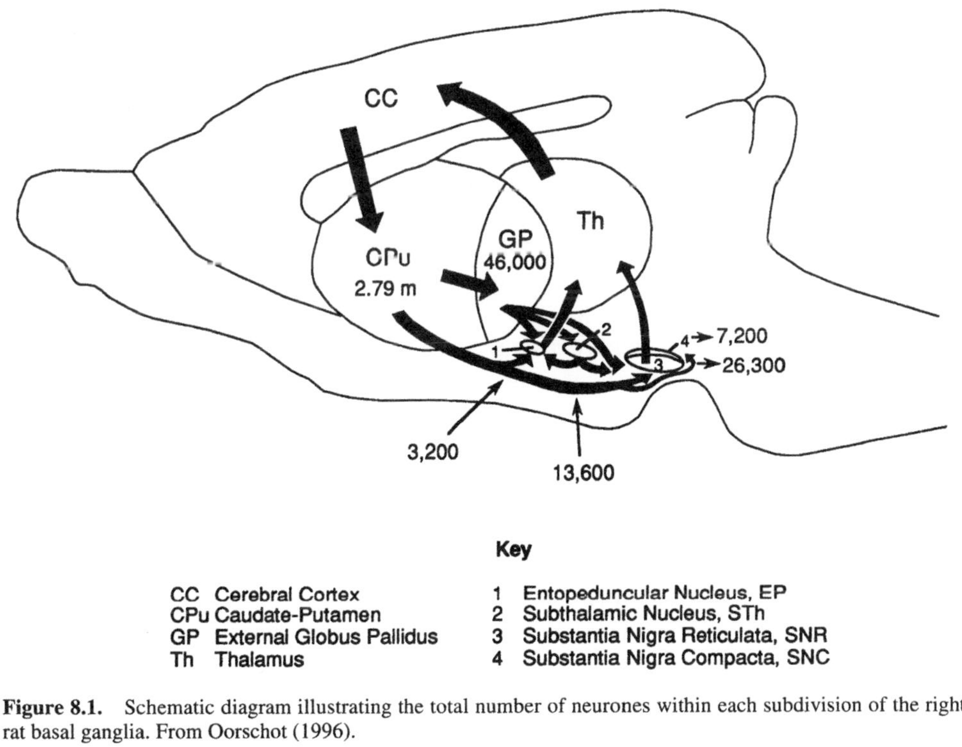


**Figure 8.1.** Schematic diagram illustrating the total number of neurones within each subdivision of the right rat basal ganglia. From Oorschot (1996).

## 3. CIRCUITRY INVOLVING THE NEOSTRIATAL SPINY PROJECTION NEURONES

In addition to their cortical, nigral and thalamic inputs, the spiny projection neurones receive synaptic inputs from at least three types of interneurones within the neostriatum. These are the GABAergic / parvalbumin interneurones (Kita, Kosaka and Heizmann, 1990; Kita, 1993), the somatostatin / neuropeptide Y / nitric oxide synthase interneurones (Vuillet *et al.*, 1989a), and the cholinergic interneurones (Phelps, Houser and Vaughn, 1985; Izzo and Bolam, 1988). There is also a fourth type of rat interneurone, the GABAergic / calretinin neurone (Jacobowitz and Winsky, 1991; Resibois and Rogers, 1992; Bennett and Bolam, 1993), but it is not known if these interneurones synapse with the spiny projection cells (Kawaguchi *et al.*, 1995).

The cortical input to each spiny projection neurone is thought to drive its depolarization. If the domain hypothesis is true, then a few spiny projection neurones within each domain would be preferentially depolarized by this input. This depolarization may ultimately generate a specific movement that is required at a particular joint in the body at a particular time (see also Section 3 of Chapter 7). The depolarized or firing spiny projection neurones would also be inhibiting all their neighbouring spiny projection neurones within the domain. If this is true, what might be the role of the GABAergic / parvalbumin interneurones?

### 3.1. The GABAergic / Parvalbumin Neostriatal Interneurones

The GABAergic neostriatal interneurones which contain the calcium-binding protein parvalbumin receive powerful excitatory input from the cerebral cortex (Kita, 1993; see Figure 8.2). These neurones, in turn, have a dense arborization of local axonal collaterals (Cowan *et al.*, 1990; Kita, Kosaka and Heizmann, 1990) and make synaptic contacts with the soma and dendrites of spiny projection cells (Kita, 1993; Bennett and Bolam, 1994). Based on the data of Kawaguchi (1993), the size of this local axonal field in the rat is approximately the size of 2–3 domains. A rat domain is likely to have a diameter of 200 $\mu$m (Penny, Wilson and Kitai, 1988; Kawaguchi, Wilson and Emson, 1989; Oorschot, 1996). There are also extensive gap junctions amongst parvalbumin-positive GABAergic interneurones, which suggest that these relatively rare neurones might form a continuous network over a much larger scale than one neurone (Kita, Kosaka and Heizmann, 1990). Based on *in vitro* slice recording of these neurones, they exhibit burst firing and have a rapid succession of action potentials of short duration within each burst (Kawaguchi, 1993). Thus, the GABAergic / parvalbumin interneurones seem to be actively-firing neurones that form a network of inhibition that is directed primarily to the spiny projection neurones.

With respect to the domain hypothesis, the primary role of these GABAergic interneurones may be to assist in the inhibition of the vast majority of spiny projection neurones which are silent at any one point in time. Based on their cortical input, there is a possibility that the cortical input serves to drive the spiny projection neurones which need to fire, and, *at the same time*, ensures that the spiny projection neurones which are not required to fire are also inhibited via the GABAergic / parvalbumin interneurones (see Figure 8.2).

The GABAergic / parvalbumin interneurones also make synaptic contacts amongst themselves (Chang and Kita, 1992), which are presumably inhibitory (Kita, 1993). This

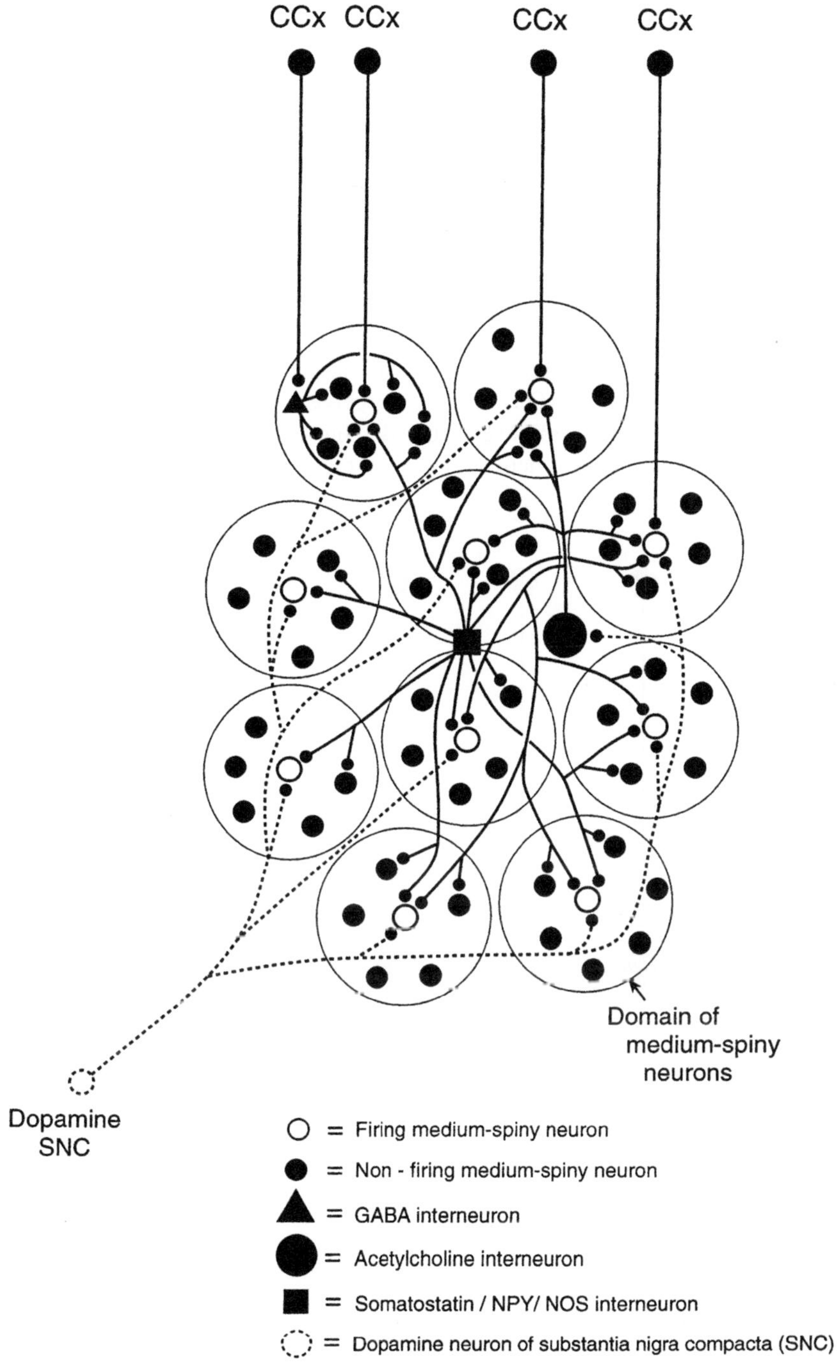


**Figure 8.2.** Schematic diagram illustrating the circuits of the neostriatum in the context of the domain hypothesis. The scheme is based on known circuits and anatomy within the rat neostriatum. Somatostatin interneurones possibly innervate the greatest number of domains, followed by the cholinergic interneurones, and then the GABA interneurones.

could be a way of limiting the duration of this inhibitory input upon the population of spiny projection neurones, thereby allowing a new dynamic to emerge within a domain. This would be necessary for generating the next specific movement required.

Synaptic contacts are also made by the cholinergic interneurones on the GABAergic / parvalbumin interneurones, but seemingly not *vice versa* (Chang and Kita, 1992; Kawaguchi, Aosaki and Kubota, 1997; see Figure 8.3). In addition, the GABAergic / parvalbumin interneurones may receive a nigral dopaminergic input (Kubota *et al.*, 1987a; Kita, 1993, Figure 8.3). These findings are discussed further in Sections 3.3 and 3.4.

### 3.2. The Somatostatin / Neuropeptide Y / Nitric Oxide Synthase Neostriatal Interneurones

Like the GABAergic / parvalbumin interneurones, the somatostatin / neuropeptide Y / nitric oxide synthase (SOM / NPY / NOS) neurones receive direct cortical input (Vuillet *et al.*, 1989b), that might be glutamatergic. Unlike the GABAergic / parvalbumin interneurones, the SOM / NPY / NOS neurones are characterized by a very large axonal

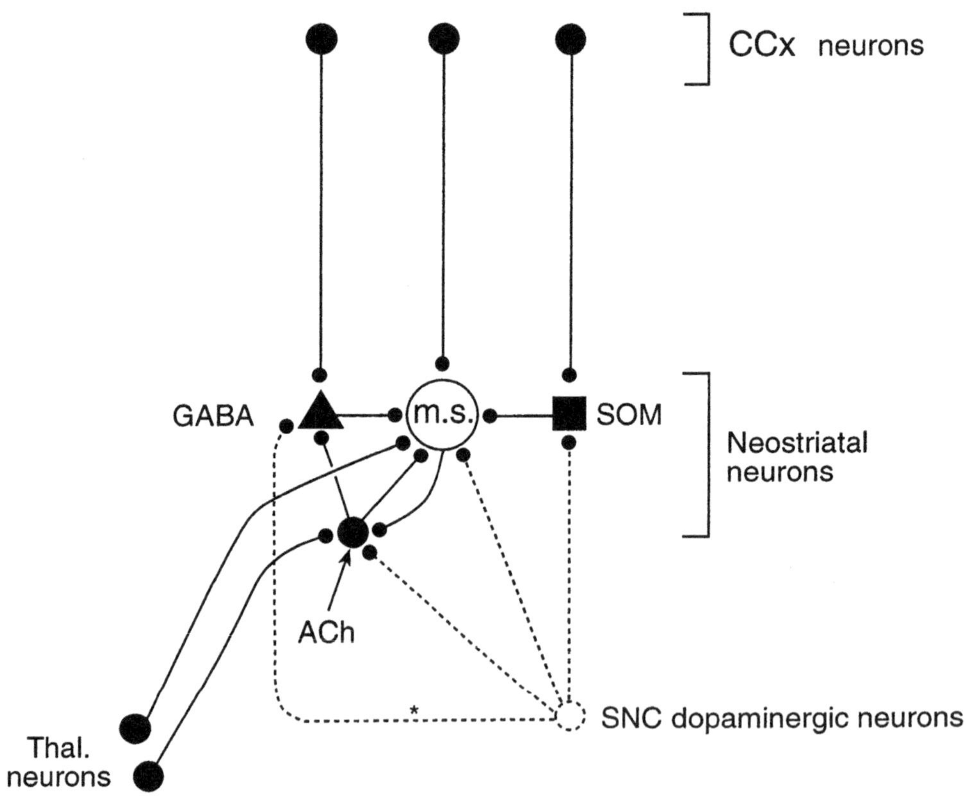


**Figure 8.3.** Schematic diagram illustrating circuits within the rat neostriatum for which there is evidence of synaptic connections. ACh, cholinergic interneurone; CCx, cerebral cortex; GABA, GABAergic interneurone; m.s., medium-spiny projection neurone; SOM, somatostatin / neuropeptide Y / nitric oxide synthase interneurone; SNC, substantia nigra compacta; Thal., Thalamic. *, Note that it is not known if the dopaminergic input is directly to GABAergic / parvalbumin neostriatal interneurones.

field, with the arborization being less dense and relatively unbranched for longer distances (Kawaguchi, 1993; see Figure 8.2). This suggests that each neurone innervates, or crosses, many more than 2–3 domains. Based on the size of their axonal field in the rat (Kawaguchi, 1993), and an estimated depth of this spread of 600 $\mu$m, each neurone may innervate up to 30 domains. The synapses of these neurones make contact with other striatal neurones, including the spiny projection neurones (Vuillet *et al.*, 1989a). When studied electrophysiologically using *in vitro* slices, these neurones are unique among striatal neurones in producing large and persistent low threshold action potentials (Kawaguchi, 1993). They are also unique in that they contain more transmitters and co-transmitters than any other neostriatal neurone.

With respect to the domain hypothesis, what role might these multi-transmitter and persistently-firing neurones play? Due to their large axonal field, and the possibility that SOM has an excitatory influence (Hathway *et al.*, 1998), there is the possibility that these neurones promote the firing of the "winner neurones" in each of a number of different domains at any one point in time (Figure 8.2). Since their axonal territory is large and sparse, with nearby SOM / NPY / NOS neurones very likely to cover overlapping territory, there is the possibility that each SOM neurone is responsible for firing their unique set of "winner neurones" across a number of different domains. Control over which SOM neurones fire, and thus which "winner neurones" fire, could be dependent on the precise set of joint movements needed at any one point in time as directed by the cerebral cortex. The activated neurones in one domain may trigger elbow flexion, and this may occur in conjunction with wrist extension as controlled by activated neurones in another domain, and individual finger flexion as controlled by activated neurones in an array of other domains, as is needed to type this sentence. The co-ordination of this firing across different domains may be the task of the SOM / NPY / NOS neurones.

In addition to facilitating "winners" across domains, the SOM / NPY / NOS neurones may, at the same time, inhibit all the "non-firing" projection neurones in these same domains via the release of nitric oxide (NO). Nitric oxide inhibits glutamatergic NMDA receptors of neostriatal cells *in vitro* (Manzoni *et al.*, 1992), and NMDA receptors are extensively expressed by the spiny projection, and other, striatal neurones. In addition, the majority of striatal projection neurones are enriched in the NO-activating enzyme, guanylate cyclase (Ariano and Matus, 1981; Ariano *et al.*, 1982; Ariano, 1983), which indicates that they are amongst the potential targets for NO produced by the NOS-positive neurones (Kawaguchi *et al.*, 1995).

NO, which is also known as endothelium-derived relaxing factor (Palmer, Ferriga and Moncado, 1987; Snyder and Bredt, 1991; Bredt and Snyder, 1992), is likely also to have a role in the control of local blood flow (Arbuthnott, Kelly and Wright, 1994). Since NO is known to cause vasodilation of blood vessels, by directly acting on guanylate cyclase within the blood vessel's smooth muscle (Bredt and Snyder, 1992), NO may cause local vasodilation around the "winner neurones" at any one point in time. At the same time, the release of NPY from the same SOM neurones may cause vasoconstriction of other blood vessels around the "non-firing" projection neurones at any one point in time. NPY has potent vasoconstrictor effects on pial blood vessels (Edvinsson *et al.*, 1984).

In summary, the known anatomy and chemical identity of these neurones raises the possibility that their role with respect to the spiny projection neurones may be to excite the "winners" in a number of domains *and* inhibit the non-firing neurones *and* regulate local blood flow at any one point in time.

SOM / NPY / NOS interneurones also have a dopaminergic input (Aoki and Pickel, 1988; Kubota *et al.*, 1988; Vuillet *et al.*, 1989b; see Figure 8.3). This circuit is discussed in the next sections.

### 3.3. The Dopaminergic Nigral Input to the Neostriatum

The dopaminergic input from the substantia nigra to the neostriatum (Andén *et al.*, 1964) is one of the best known pathways in the mammalian brain. The nigral input is known to synapse with the neostriatal spiny projection neurones (Freund, Powell and Smith, 1984) and some of the neostriatal interneurones (see Sections 3.2 and 3.4). What might be their role with respect to the domain hypothesis?

As discussed by Wickens and Oorschot in Chapter 7 a possibility is that the role of dopamine may be to increase competition within a striatal domain such that the "winner" spiny projection neurones fire. Or put another way, is the role of dopamine to drive the "winners" with respect to the spiny projection neurones at any one point in time (Figure 8.2)?

With respect to the dopaminergic input to the SOM / NPY / NOS interneurones (see Section 3.2), and the possible dopaminergic input to the GABAergic / parvalbumin interneurones (see Section 3.1), it may be that the role of dopamine is to increase the efficacy of cortical synapses to these neuronal subpopulations, since their primary role may also be to increase competition between spiny projection neurones within neostriatal domains.

### 3.4. The Cholinergic Neostriatal Interneurones

Like the other neostriatal interneurones, the cholinergic neurones constitute less than 1% of the total population of neurones within the neostriatum (Oorschot, 1997). However, the levels of acetylcholine within the neostriatum are the highest in the brain and pharmacological studies since the 1970s have highlighted the profound role that striatal acetylcholine plays in movement control (e.g. Enna *et al.*, 1976).

The axonal field of the rat cholinergic neurones, as shown by Kawaguchi (1993), is of a size similar to that of 4 neostriatal domains. When combined with an estimate of the volume of cholinergic axonal spread in the rat by Bennett and Wilson; (Chapter 6) each cholinergic interneurone may innervate around 20 domains. *In vitro* and *in vivo* electrophysiological studies have shown that these neurones are tonically active (Wilson, Chang and Kitai, 1990; Kawaguchi, 1992, 1993). They also have long afterhyperpolarizations, which prevent the neurone from firing more than once in response to any one synaptic input (Kawaguchi *et al.*, 1995). Another distinctive feature of these cholinergic interneurones is that their input seems to be primarily from the thalamus (i.e. the parafascicular nucleus, or PFN: Herkenham and Pert, 1981; Lapper and Bolam, 1992; Deschenes *et al.*, 1996), rather than the cerebral cortex.

Due to their primary input from the thalamus, there is the possibility that the major role for these neurones occurs *after* a set of spiny projection neurones, each in a different domain, has fired. In this context, their role may be to 'reset the domains' (Figure 8.2) in readiness for the next movement sequence. Or in other words, their role may be to 'restore a level playing field' by decreasing the competition that has been generated by the previous firing sequence. One possible way that this may be achieved by the cholinergic interneurones is through their modulation of one of the potassium conductance channels, the M current, in all the 'non-firing' spiny projection neurones. This would mean that all the 'non-firing' neurones would be bought closer to action potential threshold and thus would be more likely to fire for the next movement sequence. The only evidence for such an effect comes from the computer modelling study of Wickens, Alexander and Miller (1991), in which this precise effect was generated. Contrary evidence comes from Akins, Surmeier and Kitai (1990) in which another potassium conductance channel (the A current) was studied, and the inference drawn that the cholinergic interneurones may have a role in maintaining whatever electrical state a spiny projection neurone is in. The effect of this would be to maintain the competition already present within the domains. Logically, this is problematic however, as it implies that a domain would always be "stuck" in "winner mode for one particular set of spiny projection neurones".

With respect to the long afterhyperpolarizations of the neostriatal cholinergic interneurones, this may allow competition to re-occur within a domain prior to the next depolarization of cholinergic interneurones.

From the foregoing, and from other evidence discussed by Wickens and Oorschot (Chapter 7) we hypothesize that the role of the cholinergic interneurones may be to decrease competition within the domains in readiness for the next movement. Taking this a step further, as the outcome of the competition in domains is relayed through the thalamus and onto the cerebral cortex, the thalamic PFN neurones may also be disinhibited by a lack of inhibitory input from the output neurones of the basal ganglia (i.e. the neurones of the entopeduncular nucleus and the substantia nigra pars reticulata, which are known to synapse in the PFN, Heimer, Zahm and Alheid, 1995). This disinhibition may facilitate the depolarization of these PFN neurones, which may in turn depolarize the striatal ACh neurones in order to "restore baseline" within the domains of spiny projection neurones.

A role for the cholinergic interneurones in decreasing competition, and a role for the dopaminergic input neurones in increasing competition, may also explain why the dopaminergic neurones make synaptic contacts (Kubota *et al.*, 1987b; Chang, 1988; Dimova *et al.*, 1993; see Figures. 8.2 and 8.3) with the cholinergic interneurones. If the role of the dopamine neurones is to increase competition, then it makes sense that they may inhibit the neurones whose actions are to decrease competition. Agonists of the dopamine $D_2$ receptor type, expressed by the cholinergic striatal interneurones, may inhibit acetylcholine release (Stoof *et al.*, 1982; Hoffmann, Talmaciu and Cubeddu, 1986; Fujiwara *et al.*, 1987; Stoof, Verheijden and Leysen, 1987).

If the primary role for the cholinergic interneurones is to decrease competition within neostriatal domains, then it also makes sense that the cholinergic input to the GABAergic / parvalbumin interneurones (see Section 3.1, and Figure 8.3) may modulate the excitability of these interneurones as part of the cholinergic neurone's role in decreasing competition within striatal domains.

Substance P (SP) axonal terminals make synaptic contact with the cholinergic interneurones (Bolam *et al.*, 1983, 1986; see Figure 8.3) and the cholinergic interneurones express SP receptors (Kaneko *et al.*, 1993). Thus, the cholinergic interneurones are likely be innervated by the subpopulation of spiny projection neurones expressing SP. These GABA / SP spiny projection neurones disinhibit the thalamus and thus may be the neurones that amplify the desired movement at any one point in time. By concurrently inhibiting the striatal cholinergic interneurones, they may delay the progression towards decreasing competition, or establishing 'more of a level playing field', within neostriatal domains. This may ensure a spiny projection neurone's dominance for 'now'. Such an effect would be dependent, however, on the relative functional significance of the GABA and SP released, since SP appears to trigger the release of endogenous neostriatal acetylcholine *in vivo* and *in vitro* (Guevara-Guzman, Kendrick and Emson, 1993).

The GABA / enkephalin spiny projection neurones also form synapses, albeit relatively rarely, with the cholinergic interneurones (Martone *et al.*, 1992). The functional consequence of these synaptic interactions may perhaps be similar in principle to those outlined above for the GABA / SP spiny projection neurones.

## 4. MYSTERIES REMAINING AND THE WAY AHEAD

The domain hypothesis and neostriatal circuitry as described above is likely to be an over-simplistic account of biological reality. It also does not account for all known data about the neostriatum. For example, the role of the serotonergic input to the neostriatum (Dahlstrom and Fuxe, 1964; Ungerstedt, 1971) is unknown, as is the link between neostriatal patch and matrix compartments (Pert, Kuhar and Snyder, 1976; Graybiel and Ragsdale, 1978; Herkenham and Pert, 1981; Graybiel, 1983) and neostriatal domains. The exact biophysical consequences of each part of the circuitry also remains to be elucidated. Nonetheless, at present, the domain hypothesis still seems to offer a conceptual framework upon which most neostriatal anatomy, biochemistry, pharmacology and physiology gains coherence.

## REFERENCES


Akins P.T., Surmeier D.J. and Kitai S.T. (1990) Muscarinic modulation of a transient $K^+$ conductance in rat neostriatal neurons. *Nature*, London **344**, 240–242

Anden N.E., Carlsson A., Dahlstrom A., Fuxe K., Hillarp N.A. and Larsson K. (1964) Demonstration and mapping out of nigro-neostriatal dopamine neurons. *Life Sciences* **3**, 523–530

Aoki C. and Pickel V.M. (1988) Neuropeptide Y-containing neurons in the rat striatum: ultrastructure and cellular relations with tyrosine hydroxylase-containing terminals and with astrocytes. *Brain Research* **459**, 205–225

Arbuthnott G.W., Kelly P.A.T. and Wright A.K. (1994) Some consequences of local blockade of nitric-oxide synthase in the rat neostriatum. In *Basal Ganglia IV*, edited by Percheron, G., Plenum Press, pp. 171–178

Ariano M.A. and Matus A.I. (1981) Ultrastructural localization of cyclic GMP and cyclic AMP in rat striatum. *Journal of Cell Biology* **91**, 287–292

Ariano M.A., Lewicki J.A., Brandwein H.J. and Murad F. (1982) Immunohistochemical localization of guanylate cyclase within neurons of rat brain. *Proceedings of the National Academy of Sciences, USA* **79**, 1316–1320

Ariano M.A. (1983) Distribution of components of the guanosine 3',5'-phosphate system in rat caudate-putamen. *Neuroscience* **10**, 707–23

Aronin N., Chase K. and DiFiglia M. (1986) Glutamic acid decarboxylase and enkephalin immunoreactive axon terminals in the rat neostriatum synapse with striatonigral neurons. *Brain Research* **365**, 151–158

Bennett B.D. and Bolam J.P. (1993) Characterization of calretinin-immunoreactive structures in the striatum of the rat. *Brain Research* **609**, 137–148

Bennett B.D. and Bolam J.P. (1994) Synaptic input and output of parvalbumin-immunoreactive neurons in the neostriatum of the rat. *Neuroscience* **62**, 707–719

Bolam J.P. and Izzo P.N. (1988) The postsynaptic targets of substance P-immunoreactive terminals in the rat neostriatum with particular reference to identified spiny striatonigral neurons. *Experimental Brain Research* **70**, 361–377

Bolam J.P., Ingham C.A., Izzo P.N., Levey A.I., Rye D.B., Smith A.D. and Wainer B.H. (1986) Substance P-containing terminals in synaptic contact with cholinergic neurons in the neostriatum and basal forebrain: a double immunocytochemical study in the rat. *Brain Research* **397**, 279–289

Bolam J.P., Somogyi P., Takagi H., Fodor I. and Smith A.D. (1983) Localization of substance P-like immunoreactivity in neurons and nerve terminals in the neostriatum of the rat: a correlated light and electron microscopic study. *Journal of Neurocytology* **12**, 325–344

Bredt D.S. and Snyder S.H. (1992) Nitric oxide, a novel neuronal messenger. *Neuron* **8**, 3–11

Chang H.T. (1988) Dopamine-acetylcholine interaction in the rat striatum: a dual-labeling immunocytochemical study. *Brain Research Bulletin* **21**, 295–304

Chang H.T. and Kita H. (1992) Interneurons in the rat striatum: relationships between parvalbumin neurons and cholinergic neurons. *Brain Research* **574**, 307–311

Cowan R.L., Wilson, C.J., Emson P.C. and Heizmann C.W. (1990) Parvalbumin-containing GABAergic interneurons in the rat neostriatum. *Journal of Comparative Neurology* **302**, 197–205

Dahlstrom A. and Fuxe K. (1964) Evidence for the existence of monoamine-containing neurons in the central nervous system. I. Demonstration of monoamines in the cell bodies of brainstem neurons. *Acta Physiologica Scandinavica Supplement* **232**, 1–55

Deschenes M., Bourassa J., Boan, V.D. and Parent A. (1996) A single-cell study of the axonal projections arising from the posterior intralaminar thalamic nuclei in the rat. *European Journal of Neuroscience* **8**, 329–343

Dimova R., Vuillet J., Nieoullon A. and Kerkerian-Le Goff L. (1993) Ultrastructural features of the choline acteyltransferase-containing neurons and relationships with nigral dopaminergic and cortical afferent pathways in the rat striatum. *Neuroscience* **53**, 1059–1071

Dube L., Smith A.D. and Bolam J.P. (1988) Identification of synaptic terminals of thalamic or cortical origin in contact with distinct medium-size spiny neurons in the rat neostriatum. *Journal of Comparative Neurology* **267**, 455–471

Edvinsson L., Emson P., McCulloch J., Tatemoto K. and Uddman R. (1984) Neuropeptide Y: immunocytochemical localization to and effect upon feline pial arteries and veins in vitro and in situ. *Acta Physiologica Scandinavica* **122**, 155–163

Enna S.J., Bird E.D., Bennett J.P., Bylund D.B., Yamamura H.I., Iversen L.L. and Snyder S.H. (1976) Huntington's chorea: Changes in neurotransmitter receptors in the brain. *New England Journal of Medicine* **294**, 1305–1309

Freund T.F., Powell J.F. and Smith A.D. (1984) Tyrosine hydroxylase-immunoreactive boutons in synaptic contact with identified striatonigral neurons, with particular reference to dendritic spines. *Neuroscience* **13**, 1189–1215

Fujiwara H., Kato N., Shuntoh H. and Tanaka C. (1987) $D_2$-dopamine receptor-mediated inhibition of intracellular $Ca^{2+}$ mobilization and release of acteylcholine from guinea-pig neostriatal slices. *British Journal of Pharmacology* **91**, 287–297

Graybiel A.M. (1983) Compartmental organization of the mammalian striatum. In *Molecular and cellular interactions underlying higher brain functions* (*Progress in Brain Research*, Vol. **58**) edited by J.-P. Changuex, J. Glowinski, M. Imbert and F.E. Bloom Elsevier, Amsterdam, pp. 247–256

Graybiel A.M. and Ragsdale C.W. (1978) Histochemically distinct compartments in the striatum of human, monkey and cat demonstrated by acetylthiocholinesterase staining. *Proceedings of the National Academy of Sciences USA* **75**, 5723–5726

Grofova I. (1975) The identification of striatal and pallidal neurons projecting to substantia nigra. An experimental study by means of retrograde axonal transport of horseradish peroxidase. *Brain Research* **91**, 286–291

Groves P.M. (1983) A theory of the functional organization of the neostriatum and the neostriatal control of voluntary movement. *Brain Research Reviews* **5**, 109–132

Guevara-Guzman R., Kendrick K.M. and Emson P.C. (1993) Effect of substance P on acetylcholine and dopamine release in the rat striatum: a microdialysis study. *Brain Research* **622**, 147–154

Hathway G.J., Emson P.C., Humphrey P.P.A. and Kendrick K.M. (1998) Somatostatin potently stimulates in vivo striatal dopamine and gamma-aminobutyric acid release by a glutamate-dependent action. *Journal of Neurochemistry* **70**, 1740–1749

Heimer L., Zahm D.S. and Alheid G.F. (1995) Basal Ganglia. In *The Rat Nervous System*, edited by G. Paxinos, San Diego, Academic Press, pp. 579–628

Herkenham M. and Pert C.B. (1981) Mosaic distribution of opiate receptors, parafascicular projections and acetylcholinesterase in rat striatum. *Nature*, London **291**, 415–418

Hoffmann I.S., Talmaciu R.K. and Cubeddu L.X. (1986) Interactions between endogenous dopamine and dopamine agonists at release modulatory receptors: multiple effects of neuronal uptake inhibitors on transmitter release. *Journal of Pharmacology and Experimental Therpaeutics* **238**, 437–446

Izzo P.N. and Bolam J.P. (1988) Cholinergic synaptic input to different parts of spiny striatonigral neurons in the rat. *Journal of Comparative Neurology* **269**, 219–234

Jacobowitz D.M. and Winsky L. (1991) Immunocytochemical localization of calretinin in the forebrain of the rat. *Journal of Comparative Neurology* **304**, 198–218

Kaneko T., Shigemoto R., Nakanishi S. and Mizuno N. (1993) Substance P receptor-immunoreactive neurons in the rat neostriatum are segregated into somatostatinergic and cholinergic aspiny neurons. *Brain Research* **631**, 297–303

Kawaguchi Y. (1992) Large aspiny cells in the matrix of the rat neostriatum in vitro: physiological identification, relation to the compartments and excitatory postsynaptic currents. *Journal of Neurophysiology* **67**, 1669–82

Kawaguchi Y. (1993) Physiological, morphological, and histochemical characterization of three classes of interneurons in rat neostriatum. *Journal of Neuroscience* **13**, 4908–4923

Kawaguchi Y., Aosaki T. and Kubota Y. (1997) Cholinergic and GABAergic interneurons in the striatum. *Nihon Shinkei Seishin Yakurigaku Zasshi* **17**, 87–90

Kawaguchi Y., Wilson C.J. and Emson P.C. (1989) Intracellular recording of identified neostriatal patch and matrix spiny cells in a slice preparation preserving cortical inputs. *Journal of Neurophysiology* **62**, 1052–1068

Kawaguchi Y., Wilson C.J., Augood S.J. and Emson P.C. (1995) Striatal interneurones: chemical, physiological and morphological characterization. *Trends in Neurosciences* **18**, 527–535

Kemp J.M. (1968) An electron microscopic study of the terminations of afferent fibres in the caudate nucleus. *Brain Research* **11**, 464–467

Kita H. (1993) GABAergic circuits of the striatum. In *Chemical signalling in the basal ganglia* edited by G.W. Arbuthnott, and P.C. Emson (*Progress in Brain Research* , vol 99), Amsterdam, Elsevier, pp. 51–72

Kita H. and Kitai S.T. (1988) Glutamate decarboxylase immunoreactive neurons in rat neostriatum: their morphological types and populations. *Brain Research* **447**, 346–352

Kita H., Kosaka, T. and Heizmann C.W. (1990) Parvalbumin-immunoreactive neurons in the rat neostriatum: a light and electron microscopic study. *Brain Research* **536**, 1–15

Kitai S.T., Koscis J.D., Preston R.J. and Sugimori M. (1976) Monosynaptic inputs to caudate neurons identified by intracellular injection of horseradish peroxidase. *Brain Research* **109**, 601–606.

Kitai S.T., Preston R.J., Bishop G.A. and Koscis J.D. (1979) Striatal projection neurons: Morphological and electrophysiological studies. In *The Extrapyramidal System and its Disorders*, edited by L.J. Poirier, New York, Raven Press, pp. 45–51

Kocsis J.D., Sugimori, M. and Kitai S.T. (1977) Convergence of excitatory synaptic inputs to caudate spiny neurons. *Brain Research* **124**, 403–413

Kubota Y., Inagaki S., Kito S. and Wu JY. (1987a) Dopaminergic axons directly make synapses with GABAergic neurons in the rat neostriatum. *Brain Research* **406**, 147–56

Kubota Y., Inagaki S., Shimada S., Kito S., Eckenstein F. and Tohyama M. (1987b) Neostriatal cholinergic neurons receive direct synaptic inputs from dopaminergic axons. *Brain Research* **413**, 179–184

Kubota Y., Inagaki S., Kito S., Shimada S., Okayama T., Hatanaka H., Pelletier G., Takagi H. and Tohyama M. (1988) Neuropeptide Y-immunoreactive neurons receive synaptic inputs from dopaminergic axon terminals in the rat neostriatum. *Brain Research* **458**, 389–393

Lange H., Thörner G., Hopf A. and Schröder K.F. (1976) Morphometric studies of the neuropathological changes in choreatic disease. *Journal of Neurological Sciences* **28**, 401–425

Lapper S.R. and Bolam J.P. (1992) Input from the frontal cortex and the parafascicular nucleus to cholinergic interneurons in the dorsal striatum of the rat. *Neuroscience* **51**, 533–545

McGeorge A.J. and Faull R.L.M. (1989) The organization of the projection from the cerebral cortex to the striatum in the rat. *Neuroscience* **29**, 503–537

Manzoni O., Prezeau L., Marin P., Deshager S., Bockaert J. and Fagni L. (1992) Nitric oxide-induced blockade of NMDA receptors. *Neuron* **8**, 653–662

Martone M.E., Armstrong D.M., Young S.J. and Groves P.M. (1992) Ultrastructural examination of enkephalin and substance P input to cholinergic neurons within the rat striatum. *Brain Research* **594**, 253–262

Nauta H.J.W., Pritz M.B. and Lasek R.J. (1974) Afferents to the rat caudoputamen studied with horseradish peroxidase. An evaluation of a retrograde neuroanatomical research method. *Brain Research* **67**, 219–238

Oorschot D.E. (1996) Total number of neurons in the neostriatal, pallidal, subthalamic, and substantia nigral nuclei of the rat basal ganglia: a stereological study using the Cavalieri and optical disector methods. *Journal of Comparative Neurology* **366**, 580–599

Oorschot, D.E. (1997) Total number of large interneurons within the rat neostriatum: a stereological study using the optical disector and Cavalieri methods. *International Journal of Neuroscience* **89**, 90

Palmer R.M.J., Ferriga A.G. and Moncado S. (1987) Nitric oxide release accounts for the biological activity of endothelium-derived relaxing factor. *Nature*, London **327**, 524–526

Park M.R., Lighthall J.W. and Kitai S.T. (1980) Recurrent inhibition in the rat neostriatum. *Brain Research* **194**, 359–369

Pasik P., Pasik T., Holstein G.R. and Hamori J. (1988) GABAergic elements in the neuronal circuits of the monkey neostriatum: a light and electron microscopic immunocytochemical study. *Journal of Comparative Neurology* **270**, 157–170

Penny G.R., Wilson C.J. and Kitai S.T. (1988) Relationship of the axonal and dendritic geometry of spiny projection neurons to the compartmental organization of the neostriatum. *Journal of Comparative Neurology* **269**, 275–289

Percheron G., Francois C. and Yelnik J. (1987) Spatial organization and information processing in the core of the basal ganglia. In *Basal Ganglia II*, edited by M.B. Carpenter and A. Jayaraman pp. 205–226 New York, Plenum Press.

Pert C.B., Kuhar M.J. and Snyder S.H. (1976) Opiate receptor: Autoradiographic localization in rat brain. *Proceedings of the National Academy of Sciences, USA* **73**, 3729–3733

Phelps P.E., Houser C.R. and Vaughn J.E. (1985) Immunocytochemical localization of choline acetyltransferase within the rat neostriatum: a correlated light and electron microscopic study of cholinergic neurons and synapses. *Journal of Comparative Neurology* **238**, 286–307

Preston R.J., Bishop G.A. and Kitai S.T. (1980) Medium spiny neuron projection from the rat striatum: an intracellular horseradish peroxidase stsudy. *Brain Research* **183**, 253–263

Resibois A. and Rogers J.H. (1992) Calretinin in rat brain: an immunohistochemical study. *Neuroscience* **46**, 101–134

Rolls E.T. and Williams G.V. (1986) Sensory and movement-related activity in different regions of the primate striatum. In *Basal ganglia and behavior*, edited by J.S. Schneider and T.I. Lidsky, Stuttgart, Hans Huber, pp. 37–60

Snyder S.H. and Bredt D.S. (1991) Nitric oxide as a neuronal messenger. *Trends in Pharmacological Sciences* **12**, 125–128

Somogyi P. Bolam J.P. and Smith A.D. (1981) Monosynaptic cortical input and local axon collaterals of identified striatonigral neurons. A light and electron microscopic study using the Golgi-peroxidase transport-degeneration procedure. *Journal of Comparative Neurology* **195**, 567–584

Stoof J.C., Verheijden P.F.H.M. and Leysen J.E. (1987) Stimulation of $D_2$-receptors in rat nucleus accumbens slices inhibits dopamine and acetylcholine release but not cyclic AMP formation. *Brain Research* **423**, 364–368

Stoof J.C., DeBoer T., Sminia P. and Mulder A.H. (1982) Stimulation of $D_2$-dopamine receptors in rat neostriatum inhibits the release of acetylcholine and dopamine but does not affect the release of gamma-aminobutyric acid, glutamate or serotonin. *European Journal of Pharmacology* **84**, 211–214

Ungerstedt U. (1971) Stereotaxic mapping of the monoamine pathways in the rat brain. *Acta Physiologica Scandinavica Supplement* **367**, 1–48

Vuillet J., Kerkerian L., Salin P. and Nieoullon A. (1989a) Ultrastructural features of NPY-containing neurons in the rat striatum. *Brain Research* **477**, 241–251

Vuillet J., Kerkerian L., Kachidian P., Bosler O. and Nieoullon A. (1989b) Ultrastructural correlates of functional relationships between nigral dopaminergic or cortical afferent fibers and neuropeptide-Y containing neurons in the rat striatum. *Neuroscience Letters* **100**, 99–104

Vuillet J., Kerkerian L., Salin P. and Nieoullon A. (1989a) Ultrastructural features of NPY-containing neurons in the rat striatum. *Brain Research* **477**, 241–251

Webster K.E. (1961) Cortico-striate interrelations in the albino rat. *Journal of Anatomy* **95**, 532–544

West M.J., Ostergaard K., Andreassen O.A. and Finsen B. (1996) Estimation of the number of somatostatin neurons in the striatum: an in situ hybridization study using the optical fractionator method. *Journal of Comparative Neurology* **370**, 11–22

Wickens J.R. (1993) *A theory of the striatum*, Oxford, Pergamon Press

Wickens J.R., Alexander M.E. and Miller R. (1991) Two dynamic modes of striatal function under dopaminergic-cholinergic control: simulation and analysis of a model. *Synapse* **8**, 1–12

Wilson C.J. and Groves P.M. (1980) Fine structure and synaptic connections of the common spiny neuron of the rat neostriatum: a study employing intracellular injection of horseradish peroxidase. *Journal of Comparative Neurology* **194**, 599–615

Wilson C.J., Chang H.T. and Kitai S.T. (1983) Origins of post synaptic potentials evoked in spiny neostriatal projection neurons by thalamic stimulation in the rat. *Experimental Brain Research* **51**, 217–226

Wilson C.J., Chang H.T. and Kitai S.T. (1990) Firing patterns and synaptic potentials of identified giant aspiny interneurons in the rat neostriatum. *Journal of Neuroscience* **10**, 508–519

Yung K.K.L., Smith A.D., Levey A.I. and Bolam J.P. (1996) Synaptic connections between spiny neurons of the direct and indirect pathways in the neostriatum of the rat: evidence from dopamine receptor and neuropeptide immunostaining. *European Journal of Neuroscience* **8**, 861–869

# 9 Adaptive Classification of Cortical Input to the Striatum by Competitive Learning

**Dietmar Plenz[1] and Stephen T. Kitai**

*Dept. of Anatomy and Neurobiology, University of Tennessee, College of Medicine*
*855 Monroe Avenue, Memphis, TN 38163, USA*
*[1]Corresponding author's address; Unit of Neural Network Physiology LSN/National Institute Mental Health, Building 36 rm 2D-30*
*Bethesda, MD 20892 USA*
*Tel.: +1-301-496-5412 Fax: 1-301-402-8960; e-mail: dplenz@codon.nih.gov*

The anatomy of the corticostriatal system implies that the striatum can classify cortical inputs. Local axon collaterals between inhibitory striatal projection neurones suggest that such classification uses competition, either at the input or at the output of the striatal network. General arguments are put forward that the striatal network, after randomized initialization during ontogeny, has to build an internal representation of the probability distribution in cortical input space before classification can be performed in a robust manner. This internal representation can be achieved by training, using competition, either in the input space or in the output space, combined with local learning rules. Whereas competition in the cortical input space is particularly suited to corticostriatal networks with a high degree of divergence, competition at the output level is favoured in corticostriatal networks with a high degree of convergence. Despite the use of strong lateral inhibition during training, apparent lateral inhibition in the adapted striatal network is weak.

KEYWORDS: Neural Network, competitive learning, lateral inhibition, model

## 1. INTRODUCTION

Early in the history of neuroanatomy, the medium spiny (MS) neurone, the main neuronal class of the mammalian striatum, was characterized as a medium-sized cell body, with densely spiny dendrites, and local intensely-ramifying axon collaterals (Cajal, 1911). The relatively small cell body and intensive local axonal arborization prompted early researchers to consider these neurones to be interneurones (Vogt and Vogt, 1920). Despite some indications from Golgi studies (Leontovich, 1975; DiFiglia, Pasik and Pasik, 1976), it was not until intracellular neuronal tracers became available that it was established that in fact the main axon of MS neurones, besides giving off local collaterals, left the striatum to provide an intense innervation of other nuclei of the basal ganglia (Chang, Wilson and Kitai, 1981; Kawaguchi, Wilson and Emson, 1990). Through this projection, MS neurones inhibit neurones in target nuclei such as the globus pallidus (Park, Falls and Kitai, 1982). The nature of synaptic transmission at the terminals of local MS axon collaterals has not been described so far. Multiple failed attempts to uncover such transmission at the striatal level have left as an open question the functional impact of these local axon collaterals (Jaeger, Kita and Wilson, 1994). Nonetheless, given that no synaptic specialization of collaterals within the striatum has taken place, the prevalent conclusion among most researchers is that local synapses between MS neurones are also inhibitory.

At the beginning of this century, the lack of appropriate tracing methods consigned MS neurones to the status of local circuit neurones. At the end of this century, failure to uncover synaptic transmission between MS neurones through local axon collaterals now regulates MS neurones to the status of relay neurones between the cortex and striatal target nuclei. The controversy about the function of local axon collaterals of MS neurones evolves from the following three statements:

1. Every MS neurone has local axon collaterals.
2. Most striatal models assume interaction between MS neurones through such local axon collaterals.
3. Electrophysiological data, so far, do not ascribe any function to these local axon collaterals.

In fact, the presupposition of recurrent lateral inhibition among MS neurones enables such a large body of network dynamics to be derived that theoreticians are reluctant to abandon any function of local axon collaterals (e.g. Alexander and Wickens, 1993; Plenz, Wickens and Kitai, 1996). In chapter 8, Wickens and Oorschot present a model of striatal function that uses mutual inhibition among striatal MS neurones. Their model starts from a fully differentiated corticostriatal network. Lateral inhibition between medium spiny neurones leads to the coding of mutually exclusive output states as a function of cortical input.

The current chapter will embed the idea of local collateral interaction between MS neurones into the general concept of competitive learning rules. It can only provide a very brief introduction to the field of neural networks. However, several excellent overviews are available (e.g. Hecht-Nielsen, 1990; Hertz, Krogh and Palmer, 1991; Bishop, 1995; Ripley 1996). A formalization of the corticostriatal topography will be developed and used to emphasize several constraints under which neural networks work. The main starting point will be the random initialization of neuronal networks during the early period of brain ontogenesis. Thus, a general approach is taken to show how a network, after random initialization, can evolve to perform the required input-output mapping during normal behaviour. This approach will demonstrate that, before proper classification can take place, a network has to achieve an internal representation of the structure of the input space. In the course of these investigations, it will become obvious that competitive learning rules are suitable to build such an internal network representation. Particularly, the "winner-takes-all" rule, approximated by non-linear lateral inhibition between MS neurones, operates adequately in a corticostriatal topography with high convergence. Within this framework, the striatum is able to extract the correlation structure of high-dimensional cortical states and codes these into mutually exclusive, simple output states. Thus, lateral interactions, either at the input or output level, linked to local learning rules, are necessary in order to attain robust striatal network operation adapted to dynamic input space probability distributions.

## 2. THE LINEAR FORMALIZATION OF THE BASIC CORTICOSTRIATAL NETWORK

Most cortical areas project to the striatum (McGeorge and Faull, 1989; Kemp and Powell, 1970). Autoradiographic and anterograde tracing studies (Künzle, 1975; Selemon and

Goldman-Rakic, 1985; Gerfen, 1989) combined with functional mapping (Parthasarathy, Schall and Graybiel, 1992; Flaherty and Graybiel 1993) have revealed an intricate picture of cortical columns extensively innervating widely-distributed, discrete striatal regions. The majority of these corticostriatal projections originate from small to medium-sized regular spiking pyramidal neurones (Cowan and Wilson, 1994; Plenz and Kitai, 1998). Corticostriatal axons directly synapse on MS neurones (Somogyi, Bolam and Smith, 1981; Frotscher *et al.*, 1981). However, they make only very few contacts *en passant* with each target MS neurone (Kincaid, Zheng and Wilson, 1998).

The above description is the picture of a mature cortex-striatum system, which has experienced considerable learning during development. Despite the heterogeneity encountered in the corticostriatal system on a large scale (cortical layers, striatal compartments), cortical projections to MS neurones are most likely to be governed by randomness on the scale of a few hundred micrometers. On this scale, genetic information does not provide precise connectivity information, and as a simple rule, it will be assumed that *initially every corticostriatal projection neurone contacts every MS neurone.*

Consequently, the activity of a set of corticostriatal pyramidal neurones $c_i$ is described as

$$C = (c_1, c_2, \ldots, c_n),\ i \in [1, \ldots, n] \tag{1}$$

with $c_i$ being either silent ($c_i = 0$) or firing ($c_i > 0$). One particular cortical input constellation e.g. (0,1, 1, 0, ..., 2) will be denoted as $\xi$. Similarly, the activity of a set of striatal medium spiny neurones $s_j$ is defined as

$$S = (s_1, s_2, \ldots, s_m),\ j \in [1, \ldots, m] \tag{2}$$

with $s_j$ being either silent ($s_j = 0$) or firing ($s_j > 0$).

Finally, a corticostriatal pyramidal neurone $c_i$ projects to a striatal medium spiny neurone $s_j$ with a weight factor $w_{ji}$,

$$w_{ji}\colon c_i \to s_j,\ i \in [1,\ldots,n],\ j \in [1,\ldots,m],\ w_{ji} > 0 \tag{3}$$

This weight factor can be seen in physiological terms as, for example, the positive response elicited in the striatal neurone $s_j$ given an instantaneous firing rate in $c_i$. Written in matrix notation, the value of S is determined, to a first linear approximation as

$$S = W\,C^T \tag{4}$$

It is important to see that this relationship can also be expressed as each unit $s_j$ having a corresponding *weighting vector* $W_{sj}$, with

$$W_{sj}\,C^T = \Sigma w_{ji}c_i\ ,\ i \in [1, \ldots, n] \tag{5}$$

If one takes the viewpoint of a single $s_j$ then $W_{sj}$ indicates the contribution of each input $c_i$ to the activity in $s_j$. Thus $W_{sj}$ describes the *receptive field* in the input space C for each $s_j$. This general scheme is the linear formalization that, given a certain cortical input, a medium spiny neurone $s_j$ becomes active.

## 3. THE CLASSIFICATION OPERATOR AND THE WEIGHT MATRIX W

In Figure 9.1, this formalization is illustrated for a small corticostriatal network that will serve for further explication of the various problems in configuring networks for robust classification of inputs. The network consists of two corticostriatal projection neurones $c_1$ and $c_2$ that each project to two striatal neurones $s_1$ and $s_2$. The strength of the synaptic weights is given by the connectivity matrix W. A list of all possible states in C and the resulting activity in S is shown in Figure 9.1B. In the current case, we want the network to give an output at $s_1$ only when $c_1$ is active. In all other cases, the output should be $s_2$. (The trivial situation of no input at all is not considered further.) It is obvious that if at least one $c_i$ is active, then this initially will result in output activity for both $s_1$ and $s_2$. Hence, in order to *classify* the various input patterns from C into the correct output patterns S, a classification operator has to be applied to S. In the present example, the "maximum"-operator (max) will be used, with the result shown in the rightmost column of Figure 9.1B. Simple calculations show that the network will work correctly if $0 < \alpha < \beta < 1$.

However, the question arises why such a classification is necessary at all. Couldn't one design a network "*a priori*" that would do the task without comparison of output stages? The answer relates to a deep functional relation between the classification task and the weight matrix W. In our simple example in Figure 9.1A, certain restrictions had to be applied to the cross weights $\alpha$ and $\beta$ in order for the network to operate correctly. These settings of the matrix W, however, are by no means trivial, and lead directly to the problem of matrix weight modification as a function of input space probability distributions, classification operator, and robustness of network performance.

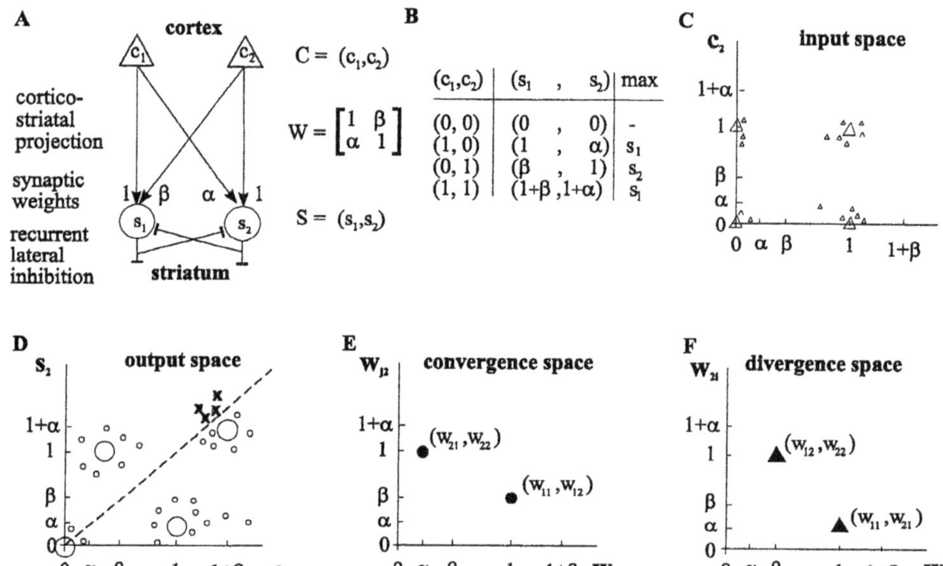


**Figure 9.1.** (**A**) The basic corticostriatal network. Two corticostriatal pyramidal neurones ($c_1$, $c_2$) project to two striatal MS neurones ($s_1$, $s_2$). The pyramidal neurones make synapses on both striatal MS neurones with the synaptic strength specified by the weight matrix W. Both striatal neurones show recurrent lateral inhibition which selects the output neurone with the strongest excitation. (**B**) Table of all possible input constellations ($c_1$, $c_2$; binary states), corresponding activity in ($s_1$, $s_2$) without lateral inhibition, and selection by the maximum operator (max). The network classifies correctly if $0 < \alpha < \beta < 1$. (**C**) Input space C. Large triangles show inputs listed in B, small triangles visualize "noisy" inputs. (**D**) Output space S. Large open circles show output states as listed in B, small open circles visualize outputs under "noisy" input conditions. The broken line indicates the threshold for the correct classification using the max-operator. Crosses indicate wrong classification under noisy conditions. (**E, F**) Convergence and divergence space for the matrix W (see text).

## 4. ROBUST NETWORK PERFORMANCE UNDER CONDITIONS OF NOISY INPUT

The network in Figure 9.1 is able to classify all C into discrete states of S, thereby selecting either MS neurone $s_1$ or $s_2$ (excluding the trivial case of no input). Under noisy conditions, however, the classification strength of this small network is far from being optimal. In biological systems, countless factors will result in a rather fluctuating mapping from C to S. In particular, "noise" in the input space C leads to severe coding problems. During firing *rate* coding (thus on time scales of 10 to 50 ms), pyramidal neurones might fluctuate in their instantaneous firing rates. Alternatively, if single spike timing is important (thus on time scales of 1 to 5 ms), noisy fluctuations in the arrival time of spikes from different corticostriatal neurones will affect the postsynaptic summation in neurones $s_i$. Here, all possible sources of fluctuations in the input space have been summarized and approximated as

$$C\pm = (c_1 \pm \Delta_1, c_2 \pm \Delta_2, \ldots, c_n \pm \Delta_n); 0 < \Delta_i \ll 1 \quad (6)$$

with $\Delta_i$ indicating statistical variance. Now the question arises, how robustly will a network operate under such conditions, i.e. perform reliable classifications?

A convenient way of studying the performance of a network involves geometrical tools. For the network in Figure 9.1A, input constellations in C can be plotted as pairs ($c_1$, $c_2$) in the input space plane (Figure 9.1C). Similarly, activity in S can be visualized in the ($s_1$, $s_2$,) output space (Figure 9.1D). All four possible input constellations in C, and their corresponding output constellations in S are indicated by large open triangles (Figure 9.1C) and large open circles (Figure 9.1D) respectively. Fluctuations in C can now be visualized easily (Figure 9.1C; small open triangles) and the result of such variability can be seen as point-like jitter in $S = [W][C\pm\Delta]^T$ (Figure 9.1D; small open circles). Thus, input constellations of C±Δ map into the neighbourhood of the non-noisy constellations. The broken line drawn in the output space represents the situation where, for all pairs, $s_1 = s_2$, which indicates the threshold for the classification operator max. It is immediately obvious that certain jitters in C will lead to extensive jitter in S and consequently to wrong network classifications (indicated by crossed open circles).

Such false classification can, in principle, be prevented by either reducing $\alpha$ and/or increasing $\beta$. But by how much? Obviously this depends on the size of the input fluctuations. Alternatively, graphically speaking, the more extended the jitter-"clouds" are, the more carefully the individual weights $w_{ij}$ have to be chosen for proper classification. These arguments thus lead to the conclusion that *the probability distribution of the input space determines strongly the values of* $w_{ji} \in W$ *in order to achieve robust network performance.*

## 5. INTERNAL REPRESENTATION OF THE INPUT SPACE PROBABILITY DISTRIBUTION BY LOCAL LEARNING RULES

At this stage it should be clear that the view of the striatum as a neuronal network that classifies cortical inputs involves quite an extended list of non-separable entities: the network architecture (n, m, W), the probability distribution of the input space C, the classification task, and the robustness of operation. The properties of each entity cannot be discussed in isolation from its context. As a consequence, setting up a network, adding some input, and hoping that the network performs in accordance with the targeted classification must be prone to failure. Further elements are lacking and this relates to the fact that each network should contain the history of its performance, and an internal representation of the input space probability distribution. Both goals are generally achieved by the introduction of local learning rules.

The capability of a neural network to build up an internal representation of a particular input space can be visualized using several concepts from computational geometry. So far, the selection of a winner in S given an input was defined as

$$s_{win}(\xi) = \max\,(s_j = w_{j\xi}{}^T, j \in [1, \ldots, m]) \tag{7}$$

It is easy to see that a similar operation can already be applied in the input space C. From the construction of the network,

$$S = W\,C^T \tag{8}$$

if the inverse form of matrix W exists, one can also write

$$[W]^{-1}\,S^T = C \tag{9}$$

A common categorization similar to the maximum rule in S is then applied in the input space C:

$$s_{win}(\xi) = \min\,(\,\|\,W_{sj} - \xi\,\|, j \in [1, \ldots, m]) \tag{10}$$

Geometrically this means that the unit $s_j$ which has inputs matching most closely to its own weight vector $W_{sj}$ is taken as the winner (|| || indicates Euclidean distance). Alternatively, regarding the receptive field, it indicates that the unit whose receptive field matches most closely the stimulus is selected as the winner. In order for the network to build up an internal representation of the input space probability distribution, local learning rules have to be applied. In the current case, the weight vector of a unit $s_{win}$ will be changed according to the following rule:

$$\Delta W_{Swin} = \varepsilon_{(t)}\,(W_{Swin} - \xi\,),\ 0 < \varepsilon_{(t)} < 1 \tag{11}$$

The parameter $\varepsilon_{(t)}$ can be constant, but in order to achieve convergence it is usually considered to be a monotonically-decreasing function over time, thereby ε decreasing the influence of new input constellations on the weight vector for a winner output unit $s_{win}$. The exact value of this function is not critical for our general purposes. In biological terms, $\varepsilon_{(t)}$ can be interpreted either as a learning period during ontogeny or, alternatively, as adaptive learning after a change in input space constellations has taken place. This type of learning, in which a network analyses input data and finds out about some of its properties without being told how well it performs during learning ("unsupervised") was introduced by Kohonen as *vector quantization* (1984; 1997), and is closely related to k-means cluster analysis (Anderberg, 1973).

Formula 11 indicates that the receptive field of $S_{win}$ will change *towards* the input vector of which this unit was a winner. Thus, if Formula 10 and 11 are repetitively used on, for instance, a finite training set of inputs, then the matrix W should eventually store the main structure of the input space probability distribution.

In the following, changes in the weight matrix W will be visualized using two different types of plot. In the *convergence space* (Figure 9.1E), pairs of ($w_{j1}$, $w_{j2}$) will be plotted, which indicate how much input from each $c_i$ converges onto a single $s_j$. In the *divergence space* (Figure 9.1F), pairs of ($w_{1i}$, $w_{2i}$) are plotted indicating how activity from a single $c_i$ diverges to $s_1$ and $s_2$.

## 6. NETWORKS WITH HIGHLY DIVERGENT CORTICOSTRIATAL PROJECTIONS

The adaptive behaviour of more complex neural networks to a particular input space probability distribution can now be visualized. In Figure 9.2, a network with highly divergent corticostriatal projections is shown, that consists of two input units and eight output units. The clustered input space probability distribution is indicated by open triangles (Figure 9.2B). Such an input space could represent (for instance) two cortical cells $c_1$ and $c_2$ which display instantaneous firing rates at different activity modes. How would the neuronal network map this input space onto all eight neurones $s_i$? At the start of the simulation, the weight matrix W was randomly initialized with $w_{ji}$ [0, ..., 40] using a Gamma-distribution (arbitrary units). Black circles indicate the initial weight vectors for each unit $s_j$ (Figure 9.2B). From the mismatch of these weight vectors and the input data cluster, it is obvious that the random initialization reflects only poorly the particular input space probability distribution. The network was then trained for 500 time steps, in which an input vector was randomly chosen from the finite input data set (triangles in Figure 9.2B) and formulae 10 and 11 were applied with $\varepsilon_{(t)}$ = constant = 0.2. After this training session, the receptive fields of units $s_j$ reflect very precisely the main features of the input space probability distribution (Figure 9.2C). This can be judged either from the positioning of output neurones $s_j$ on the various cluster centers of the input space ("codebook vectors": Kohonen, 1997), or, alternatively by the Delaunay-triangulation. This triangulation divides the input space into regions closest to one particular output neurone $s_j$ (neighbourhood) and therefore creates non-overlapping "receptive fields" for each neurone $s_j$. The triangulation taken from the simulation reveals that the "receptive fields" of individual neurones $s_j$ have been rearranged in order to maximize the resolution of the input space probability distribution.

The maximum categorization rule at the output space S ("winner-takes-all", Formula 7) does not lead to such a precise internal representation of the input space. The outcome of such a simulation (same initialization, 500 time steps, formula 7 and 11, $\varepsilon_{(t)}$ = 0.2) is shown in Figure 9.2D. It is obvious that the internal structure of the network does not properly reflect the structure of the input space. Instead, the network has built a dichotomy in which one neurone represents high activity states (Figure 9.2D, $s_4$, open arrow) and all other neurones $s_j$ cluster in a region of low input activity (Figure 9.2D, filled arrow). This simulation indicates that *in a highly divergent corticostriatal projection system, competitive learning in the output space is inferior to competitive learning in the input space.*

This result might provide an explanation for one particular feature of the striatal anatomy. Individual CS-projection neurones can have relatively widespread striatal projection fields, with one projection neurone innervating many MS neurones (high divergence). Between those innervated, but relatively distant MS neurones, direct lateral

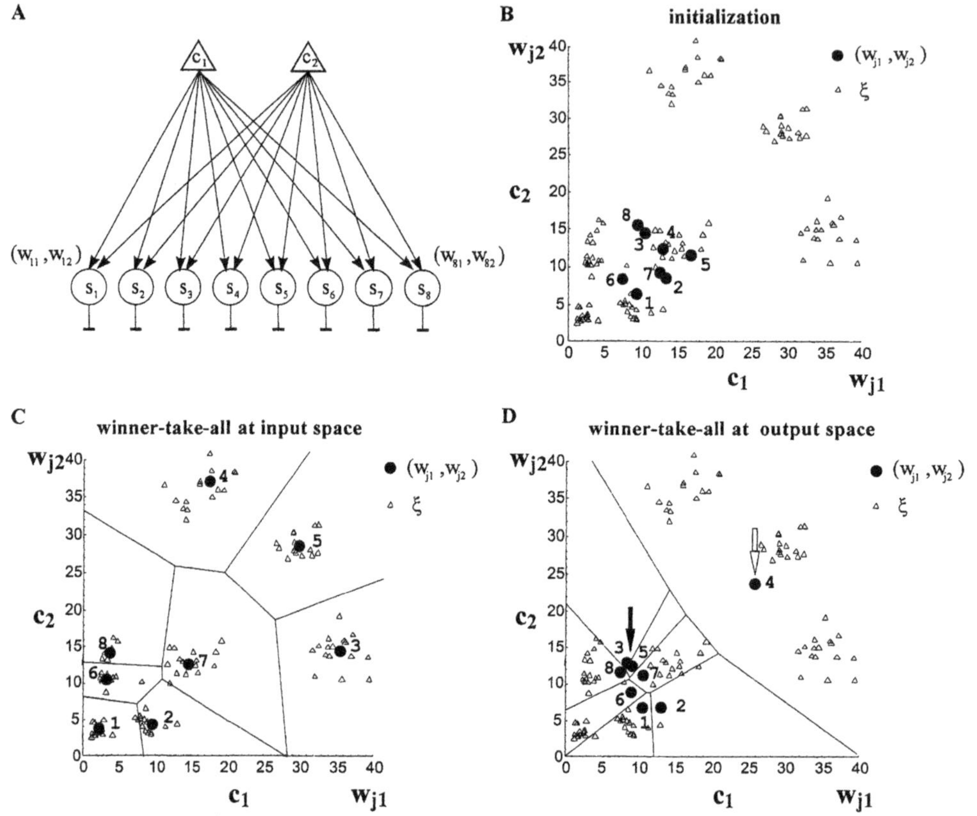


**Figure 9.2.** (**A**) Corticostriatal network with highly divergent corticostriatal projections. Two corticostriatal pyramidal neurones ($c_1$, $c_2$) project to eight striatal MS neurones ($s_1$, ..., $s_8$). (**B**) Distribution of finite input data set (small open triangles). The input distribution shows several clusters simulating (for example) two pyramidal neurones with multiple preferred activity states. Black closed circles indicate the randomized initial distribution of synaptic weights that converge onto each output neurone $s_j$. (**C**) The adaptation of the network to the input space probability distribution after 500 training steps using the finite data set, the "minimum-distance" rule in the input space (formula 10), and local learning rule formula 11 with $e_{(t)} = 0.2$. After the training, the network has built an internal representation of the particular structure of the input space C with individual output neurones $s_j$ having synaptic convergence pairs ($w_{j1}$, $w_{j2}$) positioned at the centre of input space clusters. The graphical net shows the Delaunay-triangulation in which each line indicates the neighbourhood border between individual $s_j$. Note that the neighborhood of each $s_j$ ("receptive field") conveniently partitions the input space probability distribution. (**D**) The same simulation with the maximum operator applied at the output in S (Formula 7, "winner-takes-all"). In this case, the network has stored the input space distribution in a crude dichotomy with one neurone coding for high input activity (open arrow) and all other neurones coding for low input activity modes (filled arrow). Similarly, the Delaunay-triangulation reflects the dichotomy of the network's internal representation, with the neighbourhood of output neurone $s_4$ covering the complete range of high activity modes for both input neurons $c_1$ and $c_2$.

interaction cannot take place because local axonal collaterals of MS neurones on average are mostly confined to their local dendritic field. The absence of lateral interactions is supported by the simulation results which show that under these conditions, a "winner-takes-all" dynamics is not particularly suited to building a precise internal representation of the input space probability distribution. However, this situation changes if one considers networks with highly convergent corticostriatal projections.

## 7. NETWORKS WITH HIGHLY CONVERGENT CORTICOSTRIATAL PROJECTIONS

It has already been demonstrated that, in a highly divergent corticostriatal system, the "winner-takes-all" rule in S is quite inferior to the "minimal-distance" rule in C, and tends to result in a dichotomy with respect to the internal representation of the input space. This particular feature should be advantageous in a network with many corticostriatal neurones converging upon a small set of MS neurones. In Figure 9.3A, such a network is pictured in which eight neurones $c_i$ project to two neurones $s_j$. Recurrent lateral inhibition works between the two units $s_j$ ("winner-takes-all", Formula 7). In the scheme proposed by Wickens and Oorshot, the two output units could (for example) represent two mutually exclusive motor commands such as "stop" and "go". Instead of asking how, under a given input, the network responds with either "go" or "stop", this question is rephrased to how the input space probability distribution–which represents a "go" and a "stop" constellation—is mapped into the network's internal representation and subsequently into mutually exclusive outputs. From the previous paragraphs, it is obvious that this comes down to a modification of the weight matrix as a function of input space, classification operator, and local learning rules.

Similarly as in the case of clustered inputs shown in Figure 9.2A, a clustered input space probability distribution for the network shown in Figure 9.3A was constructed. The set of eight neurones in C was divided into two equal sets of $C_A = (c_1, c_2, c_3, c_4,)$ and $C_B = (c_5, c_6, c_7, c_8)$. Within each set, the probability of one neurone being active was taken from a Bernoulli distribution ($P = 0.5$). The probability that, at the same time, a neurone in the alternative set is active, was taken from a Bernoulli distribution with a lower probability ($P = 0.1$). In this way, a finite data set with 64 input constellations was obtained, in which neurones from one set are more likely to be active together than being active with neurones in the alternative set (Figure 9.3B). Such an input space can be thought of as two different (and noisy) cortical network constellations, which are anticorrelated and are supposed to be identified by the striatal network.

Furthermore, local learning rules were chosen in order for the network to develop an internal representation of the input space and to perform the mutually exclusive classification. After the winner of an input was determined ("winner-takes-all", formula 7), two local learning rules were applied which increased or decreased the individual weights for the currently active neurones $c_i$:

$$\Delta w_{Swin} = \varepsilon_{(t)} \, (w_{max} - w_{Swin}) \tag{12}$$
$$\Delta w_{Sj} = -\varepsilon_{(t)} \, w_{Sj;} \, j \neq win \tag{13}$$

Formula 12 can be viewed as a long-term potentiation (LTP) rule, which increases the individual strength of currently active synaptic connections up to a maximum value $W_{max}$ if these synapses took part in a winning S unit. Similarly, formula 13 can be viewed as a long-term depression (LTD) rule, in which synaptic weights are reduced if they were active on a losing S unit.

In a last step, a graphical representation was chosen to allow for an adequate visualization of the network performance. Whereas in Figure 9.2A, pairs of $(w_{j1}, w_{j2})$ were mapped for each $s_j$, (convergence space), now the divergence space, i.e. the set of $(w_{1i}, w_{2i})$, was mapped for all $c_i$ (Figure 9.3C). The randomized initialization of W was chosen such that

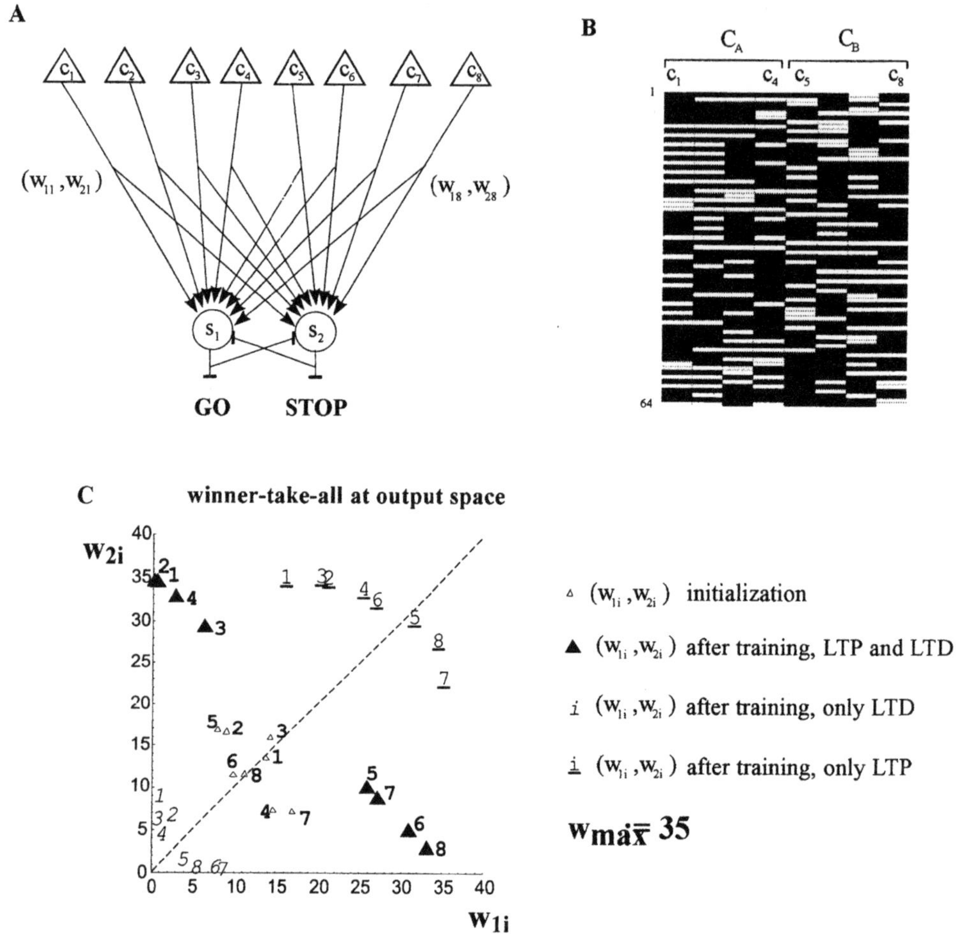


**Figure 9.3.** **(A)** Corticostriatal network with highly convergent corticostriatal projections. Eight corticostriatal pyramidal neurones ($c_1, \ldots, c_8$) project to two striatal MS neurones $s^1$ and $s^2$. **(B)** Density plot of finite input data training set. Each row indicates one input vector. Columns indicate input neurone $c_1$ to $c_8$. White fields indicate active neurones, black fields indicate non-active neurones. Neurones within set $C_A$ or $C_B$ are more likely to be active together than to be active with neurones from the alternative set. **(C)** The adaptation of the neuronal network to the input space correlation shown in B. Open triangles indicate the divergence of individual cortical neurones onto both output unit $s_1$ and $s_2$. Initially, all cortical neurones excite approximately similarly each unit $s_j$. After 500 random presentations of input vectors from the finite data set shown in B, the divergence in the neuronal net has been changed. Cortical input units from set $C_A = (c_1, c_2, c_3, c_4,)$ mainly drive activity in output unit $s_2$, whereas cortical input units from set $C_B = (c_5, c_6, c_7, c_8)$ mainly activate output unit $s_1$. Thus, the striatal network preferentially encodes cortical states in $C_A$ as a "stop" signal, whereas cortical states in $C_B$ are more likely encoded as a "go" signal. The separation of pairs of ($w_{1i}$, $w_{2i}$) into appropriate clusters which reflect the input space correlation structure also works with LTD (italic numbers) or LTP (underlined numbers) alone.

all $w_{ji} \in [0, W_{max}]$ clustered within a central region of the divergence space. The network performance can now be judged easily by the following: If the contribution of (for instance) neurone $c_3$ to $s_1$ decreases, the point ($w_{13}$, $w_{23}$) will move towards the left axis. Alternatively, if the contribution of $c_3$ to $s_2$ increases, the point ($w_{13}$, $w_{23}$) will move away from the bottom axis. Thus, the graphical sketch of the $(w_{1i}, w_{2i})$-plane conveniently summarizes the internal state of the network.

From the construction of the input space, it is clear that individual input units $c_i$ will be more active with one set of neurones than with the alternative set. Consequently, the weight matrix W of the network is expected to change in such a way that more correlated input units ultimately converge onto the same unit $s_j$. This is achieved when input units form two separate clusters in the $(w_{1i}, w_{2i})$-plane perpendicular to the $w_{1i} = w_{2i}$ function (Figure 9.3C, dashed line). The simulation demonstrates that this is indeed the case. After 500 training steps, input units from the set $C_A = (c_1, c_2, c_3, c_4,)$ cluster in the upper left corner (Figure 9.3C). This indicates that these units mainly drive output unit $s_2$. On the other hand, input units from set $C_B = (c_5, c_6, c_7, c_8)$ cluster in the lower right corner, which indicates that these units mainly excite output unit $s_1$. Thus, the network will detect the constellation $C_{A,}$ which results in activity only in unit $s_2$, or constellation $C_B$ which results in activity only in output unit $s_1$. It is important to realize that, in this way, the striatal network is able to *reduce* the eight-dimensional cortical input space to just two dimensions, thereby preserving the correct correlation structure in the input space. Furthermore, the dichotomy in the output reflects the anti-correlation between cortical sets $C_A$ and $C_B$. This reduction in complexity is similar to the reduction achieved by the network shown in Figure 9.2A, in which the complex input space probability distribution was approximated by the receptive field center.

Several important features should be noted. First, the selection of which input set converges onto which particular output unit is governed by randomness, and is sensitive to the initialization of the matrix W. Using learning rules for both LTP and LTP, the network has two asymmetric and stable states $(0, W_{max})$ and $(W_{max}, 0)$ for each input unit $c_i$ (besides the less interesting cases (0,0) and (1,1)). Both states are attractors for both cortical sets, and separate classification is not guaranteed in general. Thus, a robust linkage of one input set to one particular output unit requires additional mechanisms (such as external tutoring by a dopamine pulse as facilitating signals; Wickens *et al.,* 1996). Second, the network adaptation towards the correlation structure in the input space also works with either LTP or LTD alone. This is indicated by underlined numbers (only LTP) and italic numbers (only LTD) shown in Figure 9.3C. However, these are not very stable situations, as continuing training will result in either all $w_{ij} = w_{max}$ (LTP only) or all $w_{ij} = 0$ (LTD only). Third, depending on the exact value of lateral inhibitory connections (including self-inhibition), the classification dynamics of the network can shift from "winner-takes-all" to several winners being simultaneously active (Fukai and Tanaka, 1997). Fourth, although strong mutual inhibition was used during the network adaptation, the *apparent* mutual inhibition among units in S is considerably reduced in the final state. During initial training, a unit $c_i$ excites both units $s_1$ and $s_2$ with similar strength. Under these circumstances, probing, for instance $s_2$ by disrupting local inhibition (e.g. with a $GABA_A$-receptor antagonist) will uncover the underlying excitation from $c_i$. However, in the final state, this very same unit $c_i$ will predominantly excite (for example) $s_1$ and only very weakly unit $s_2$. Removing local inhibition from unit $s_2$ will not result in an increase of firing, due to the reduction of excitatory drive to unit $s_2$. Thus, the *apparent* inhibition (firing inhibition) in the final state is low, despite the fact that inhibitory strength from s1 -> s2 has increased.

## ACKNOWLEDGEMENT

We thank Dr. Jeff Wickens for critical comments on the first version of the manuscript. The simulations were performed in Mathematica (Mathematica 3.0, Wolfram Research) and are available under 'http://nihac.info.nih.gov'.

## REFERENCES

Alexander, M.E. and Wickens, J.R. (1993) Analysis of striatal dynamics: The existence of two modes of behaviour. *Journal of Theoretical Biology* **163**, 413–438.

Anderberg, M.R. (1973) *Cluster Analysis for Applications*. New York: Academic Press, Inc.

Bishop, C.M. (1995) *Neural Networks for Pattern Recognition*. Oxford: Oxford University Press.

Cajal, S.R. (1911) *Histologie du Système Nerveux*. Paris: A.Maloine.

Chang, H.T., Wilson, C.J. and Kitai, S.T. (1981) Single neostriatal efferent axons in the globus pallidus: a light and electron microscopic study. *Science, New York* **213**, 915–918.

Cowan, R.L. and Wilson, C.J. (1994) Spontaneous firing patterns and axonal projections of single cortico-striatal neurons in the rat medial agranular cortex. *Journal of Neurophysiology* **71**, 17–32.

DiFiglia, M., Pasik, P. and Pasik, T. (1976) A Golgi study of neuronal types in the neostriatum of monkeys. *Brain Research* **114**, 245–256.

Flaherty, A.W. and Graybiel, A.M. (1993) Two input systems for body representations in the primate striatal matrix: experimental evidence in the squirrel monkey. *Journal of Neuroscience* **13**, 1120–1137.

Frotscher, M., Rinne, U., Hassler, R. and Wagner, A. (1981) Termination of cortical afferents on identified neurons in the caudate nucleus of the cat. *Experimental Brain Research* **41**, 329–337.

Fukai, T. and Tanaka, S. (1997) A simple neural network exhibiting selective activation of neuronal ensembles. From winner-take-all to winners-share-all. *Neural Computation* **9**, 77–97.

Gerfen, C.R. (1989) The neostriatal mosaic: striatal patch-matrix organization is related to cortical lamination. *Science, New York* **246**, 385–388.

Hecht-Nielsen, R. (1990) *Neurocomputing*. Reading, MA: Addison-Wesley.

Hertz, J. Krogh, A. and Palmer, R. (1991) *Introduction to the theory of neural computation (Santa Fe instituted studies in the sciences of complexity. Lecture notes, Vol.1)*. Redwood City, CA: Addison-Wesley.

Jaeger, D., Kita, H. and Wilson, C.J. (1994) Surround inhibition among projection neurons is weak or non-existent in the rat neostriatum. *Journal of Neurophysiology* **72**, 2555–2558.

Kawaguchi, Y., Wilson, C.J. and Emson, P.C. (1990) Projection subtypes of rat neostriatal matrix cells revealed by intracellular injection of biocytin. *Journal of Neuroscience* **10**, 3421–3438.

Kemp, J.M. and Powell, T.P.S. (1970) The cortico-striate projection in the monkey. *Brain* **93**, 525–546.

Kincaid, A.E., Zheng, T. and Wilson, C.J. (1998) Connectivity and convergence of single corticostriatal axons. *Journal of Neuroscience* **18**, 4722–4731.

Kohonen, T. (1984) *Self-organization and Associative Memory*. Berlin: Springer-Verlag.

Kohonen, T. (1997) *Self-Organizing Maps, 2nd edn*. Berlin: Springer-Verlag.

Künzle, H. (1975) Bilateral projections from precentral motor cortex to the putamen and other parts of the basal ganglia. *Brain Research* **88**, 195–210.

Leontovich, T.A. (1975) Quantitative analysis and classification of subcortical forebrain neurons. In: M. Santini (ed) *Golgi Centennial Symposium: Perspectives in Neurobiology*, New York: Raven Press, pp. 101–122.

McGeorge, A.J. and Faull, R.L.M. (1989) The organization of the projection from the cerebral cortex to the striatum in the rat. *Neuroscience* **29**, 503–537.

Park, M.R., Falls, W.M. and Kitai, S.T. (1982) An intracellular HRP study of the rat globus pallidus. I. response and light microscopic analysis. *Journal of Comparative Neurology* **211**, 284–294.

Parthasarathy, H.B., Schall, J.D. and Graybiel, A.M. (1992) Distributed but convergent ordering of cortico-striatal projections: Analysis of the frontal eye field and the supplementary eye field in the macaque monkey. *Journal of Neuroscience* **12**, 4468–4488.

Plenz, D., Wickens, J.R. and Kitai, S.T. (1996) Basal ganglia control of sequential activity in the cerebral cortex: a model. In *Computational Neuroscience* edited by J.M. Bower San Diego: Academic Press pp. 397–402.

Plenz, D. and Kitai, S.T. (1998) "Up" and "down" states in striatal medium spiny neurons simultaneously recorded with spontaneous activity in fast-spiking interneurons studied in cortex-striatum-substantia nigra organotypic cultures. *Journal of Neuroscience* **18**, 266–283.

Ripley, B.D. (1996) *Pattern recognition and neural networks*. Cambridge: Cambridge University Press.

Selemon, L.D. and Goldman-Rakic, P.S. (1985) Longitudinal topography and interdigitation of corticostriatal projections in the rhesus monkey. *Journal of Neuroscience* **5**, 776–794.

Somogyi, P., Bolam, J.P. and Smith, A.D. (1981) Monosynaptic cortical input and local axon collaterals of identified striatonigral neurons. A light and electron microscopic study using the Golgi-peroxidase transport degeneration procedure. *Journal of Comparative Neurology* **195**, 567–584.

Vogt, C. and Vogt, O. (1920) Zur Lehre der Erkrankungen des striären Systems. *Journal for Psychology and Neurology (Leipzig)* **25**, 627–846.

Wickens, J.R., Begg, A.J. and Arbuthnott, G.W. (1996) Dopamine reverses the depression of rat corticostriatal synapses which normally follows high-frequency stimulation of cortex *in vitro*. *Neuroscience* **70**, 1–5.

# 10 Insights from Gene Regulation into the Functional Role of Dopamine in the Striatum

**Charles R. Gerfen**

*Laboratory of Neurophysiology, National Institute Mental Health*
*Bethesda MD 20892 USA*
*tel: (301) 496–4341; FAX: (301) 402–8960; e-mail: gerfen@helix.nih,gov*

The diverse effects of dopamine on behaviour mediated by basal ganglia function reflects the distinct receptor subtypes through which dopamine activates cellular signal transduction pathways, and the distribution of these receptor subtypes on functionally distinct neurones in the striatum. Neuroanatomical methods have established that the D1 and D2 receptor subtypes are segregated to connectionally distinct striatal neurones, the "direct" and "indirect" projection neurones. Gene regulation studies demonstrate elevated D1-mediated responses in "direct" striatal projection neurones, and decreased D2-mediated responses in "indirect" striatal projection neurones. Thus, dopamine appears to oppositely affect the function of the two major output pathways of the striatum, which regulate the activity through the basal ganglia.

KEYWORDS: dopamine receptor, gene regulation, immediate early genes, Parkinson's Disease

## 1. INTRODUCTION

The functional role of dopamine in the striatum is intriguing. Clearly dopamine has a critical role in maintaining normal behaviour, as evidenced by the profound clinical movement disorders that accompany the degeneration of the nigrostriatal dopamine system in Parkinson's disease (Albin, Young and Penney, 1989). Dopamine has also been implicated in mental disorders such as schizophrenia and attention deficit hyperactivity disorder (ADHD), as well as in the effects of abuse of psychostimulants such as cocaine and amphetamine. However, identifying the specific mechanisms by which dopamine effects such diverse behaviours has been difficult. Physiologic approaches have provided considerable advances in recent years, such that the long held view that dopamine acts to inhibit striatal neurone activity has yielded to more complex concepts, in which it is now appreciated that the effect which dopamine has on neuronal activity is dependent on many factors, involving the concurrent actions of other neurotransmitters. For example, while dopamine may inhibit the response of striatal neurones to glutamate input in some instances, in other instances, such as in combination with NMDA glutamate receptor activation, dopamine may enhance striatal neurone activity (Cepeda, Buchwald and Levine, 1993). An alternative approach to the study of dopamine function in the striatum utilises the effects of dopamine receptor activation on the regulation of genes expressed by striatal neurones. The rationale of this approach and the insights it provides to elucidating the functional role of dopamine in the striatum will be reviewed in this chapter.

The current model of the function of dopamine is based on the concept that dopamine functions to balance the output systems of the striatum, through which the basal ganglia affect behaviour (Albin, Young and Penney, 1989; Bergman, Wichman and DeLong, 1990; Gerfen *et al.*, 1990). This model is grounded in the neuroanatomy of the basal ganglia (Figure 10.1; see Gerfen and Wilson, 1996 for review). The cerebral cortex, together with the intralaminar thalamus, provide the major excitatory input to the striatum, which determines the activity of the GABAergic inhibitory projection neurones of the striatum. These striatal output neurones give rise to two major projection systems, which either directly or indirectly regulate the output of the basal ganglia. The opponent nature of these two pathways is suggested to underlie the involvement of the basal ganglia in behaviour (Albin, Young and Penney, 1989). Cortical activation of the direct striatal output pathway, resulting in disinhibition of the tonic inhibition of basal ganglia output activity has long been suggested to be necessary for the generation of purposive movement and behaviour (Deniau and Chevalier, 1978; Hikosaka and Wurtz, 1983). On the other hand, excessive activity in the indirect pathway, as is suggested to occur in Parkinson's disease, is suggested to be responsible for increased inhibitory activity out of the basal ganglia that underlies Parkinsonian akinesia (Bergman, Wichman and DeLong, 1990). This model is widely criticised for being overly simplistic, particularly as physiologic studies have not yielded a clear substantiation of many of the predictions of this model (Anderson and Turner, 1991; Mink and Thach, 1993). However, this model has had considerable utility in the development of therapeutic approaches to the treatment of Parkinson's disease. For example, neurosurgical and physiologic treatments to affect the activity in the "direct" and "indirect" striatal output pathways have been shown to be highly efficacious in the treatment of Parkinson's disease. While the model may be oversimplified, there is substantial evidence, based on gene regulation studies, of the functional significance of the opponent nature of the striatal output pathway systems and their regulation by dopamine.

The following issues concerning dopamine's function in the striatum will be discussed. *First*, a role for dopamine to function in regulating the output of the basal ganglia is discussed in terms of the neuroanatomical connections within this system. The localization of dopamine receptor subtypes to specific striatal output pathways provides a basis for postulating the opponent nature of dopamine's function in the striatum (Gerfen *et al.*, 1990). *Second*, the functional role of dopamine inferred from studies employing the response of striatal neurones to dopamine receptor activation, as measured by changes in gene expression, will be discussed. Such gene regulation studies provide several types of information. On the one hand they provide information concerning the organization of the striatum, namely the segregation of D1 and D2 dopamine receptor subtypes to connectionally distinct neurones in the striatum, and the consequent effect of dopamine on these neurones. In such studies the acute response of identified striatal output neurones to dopamine receptor agonist and antagonist treatments are measured using the expression of immediate early genes. On the other hand such studies provide a method for examining adaptive responses of striatal neurones that are regulated by dopamine. Such adaptive responses are the subject of a *third* set of issues. Excessive or repeated dopamine activation in the striatum results in differences in adaptive responses of striatal neurones that express D1 dopamine receptors, when these neurones are rendered supersensitive by

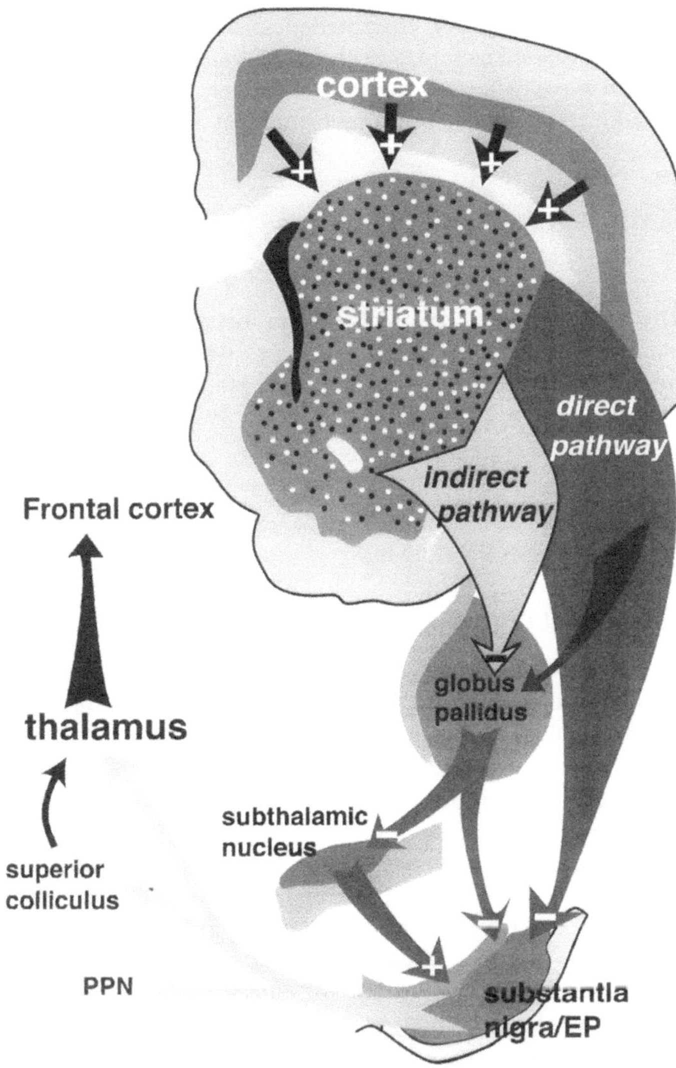


***Direct-Indirect Striatal Output Pathways***

**Figure 10.1.** Summary diagram of the "direct" and "indirect" striatal output pathways. Layer 5 cortical neurones provide excitatory input (+) to the striatum. The direct striatal projection is provided by D1/substance P/dynorphin-containing neurones to the substantia nigra (pars reticulata) and entopeduncular nucleus, and to a lesser degree to the globus pallidus. The indirect striatal projection is provided by D2/enkephalin-containing neurones that project to the globus pallidus. The globus pallidus in turn provides an inhibitory projection to the substantia nigra and to the subthalamic nucleus. The subthalamic nucleus provides an excitatory input to the substantia nigra. Thus, the "direct" and "indirect" pathways provide antagonistic input to the substantia nigra. The GABA neurones in the substantia nigra provides an inhibitory projection to the superior colliculus, pedunculopontine nucleus (not shown) and thalamus. The thalamic nuclei receiving this output project back upon the frontal cortex. The entopeduncular (EP) nucleus is connected in a similar manner to the substantia nigra but is not shown.

dopamine depletion in the striatum, as compared to the normal striatum. These differences display a number of interesting characteristics: Among them is a regional difference between the responses of neurones in the dorsal striatum and those in the nucleus accumbens. Such adaptive responses are discussed in terms of relevance to clinical disorders including Parkinson's disease and psychostimulant drug abuse.

## 2. D1 AND D2 DOPAMINE RECEPTOR LOCALIZATION IN THE STRIATUM.

Dopamine acts primarily through two receptor subtypes in the striatum, the D1 and D2 subtypes. While there are other dopamine receptor subtypes within the striatum, it has been estimated that greater than 95% of the dopamine receptors in the striatum are of either the D1 or D2 subtype (Levey *et al.*, 1993; Hersch *et al.*, 1995). These two receptor subtypes have been reported to be segregated to the two main types of striatal medium spiny projection neurones, the direct and indirect projection neurones (Gerfen, 1992). As this segregation is central to the concept of the opponent nature of dopamine function in the striatum it is worthwhile to review the evidence.

### 2.1. Direct and Indirect Striatal Projection Systems

Critical to the functional significance of the distribution of D1 and D2 receptor subtypes in the striatum is the categorization of connectionally identified striatal projection neurones. As is often stated, greater than 90–95% of the neurones of the striatum are medium spiny projection neurones. These neurones are characterized by their axons, which project out of the striatum, and may be further categorized on the basis of the target of their projection axons. Retrograde tracing studies in the rat, in which fluorescent retrograde tracers such as fluorogold are injected into the main target nuclei of these projections, have shown that approximately 50% of these neurones project to the substantia nigra, whereas the other 50% project to the globus pallidus (Gerfen and Young, 1988). Moreover, those neurones which projected to the substantia nigra were shown to co-label with the mRNA encoding the neuropeptide substance P, whereas those neurones which projected to the globus pallidus were shown to co-label with the mRNA encoding the neuropeptide enkephalin (Gerfen and Young, 1988).

Retrograde tracing studies are subject to methodological limitations. For example there is uptake of tracer by fibres of passage. Also, without exhaustive controls, it is difficult and in some sense impossible, to determine the exact area in which tracer is taken up for retrograde axonal transport. More detailed characterization of striatal medium spiny projection neurones has been provided by a study in which single neurones were intracellularly labelled and their axonal projections traced to their targets (Kawaguchi, Wilson and Emson, 1990). This study has confirmed the general view provided by retrograde tracing studies, namely that medium spiny neurones may be categorized by their axonal projections out of the striatum. This study also provided some significant details of the axonal projections of these neurones. First, it confirmed that there was a subtype of striatal projection neurone whose axon projected out of the striatum and formed an extensive arborization in the globus pallidus but which did not project to either the entopeduncular nucleus or substantia nigra. This type of neurone is classified as a striatopallidal projection neurone. Second, another type of projection neurone was characterized whose axon passed through the globus pallidus and extended to either the entopeduncular nucleus or the substantia nigra. Such neurones are classified as striato-entopeduncular/nigral projection neurones. However, the axon collateral of these neurones that passed through the globus pallidus was determined to have some synapse-bearing branches within the globus pallidus, which were considerably more extensive than had been appreciated before. Thus, it would be inaccurate to describe these neurones as not being striatopallidal projecting neurones. For this reason, a more accurate terminology is adapted, in which neurones are described as being "direct" and "indirect" projection neurones. This terminology is used, since those neurones that provide projections to the entopeduncular nucleus and/or substantia nigra provide "direct" projections to these output nuclei of the basal ganglia, whereas those neurones which provide inputs which terminate in the globus pallidus are connected with these output nuclei only "indirectly" through projections of the globus pallidus itself. This terminology is used to avoid the inaccuracy of the term striatopallidal as excluding the striato-entopeduncular/nigral projecting neurones.

There are number of additional observations that should be made concerning the categorization of striatal projection neurone types. First, there are subtypes of each of the "indirect" and "direct" types. Some of these were described by Kawaguchi, Wilson and Emson (1990). For example, some neurones provide projections to both the entopeduncular nucleus and substantia nigra, and others provide projections to only one of these nuclei. There are differences in the axon collateral of different neurones within the striatum itself. There are also differences in the projections of striatal neurones in the "patch" and "matrix" compartments, which have to do with the axon projections to different parts of the substantia nigra or to different parts of the globus pallidus (Gerfen, 1985,1992). These different types of projections are not to be confused with the categorization of "direct" and "indirect" neurones, in that "patch-matrix" neurones form a subset of "direct" and "indirect" projection neurones (Gerfen and Young, 1988). Second, the finding that "direct" projecting neurones have synapse-bearing axon collaterals within the globus pallidus highlights the limitation of the retrograde transport methods for distinguishing these two classes of striatal projection neurones. One explanation for the failure to label retrogradely the "direct" projecting neurones with some injections into the globus pallidus may be that effective uptake requires an unspecified critical mass of synaptic arbors. This becomes an important issue, as it may also be the case that some "direct" projecting neurones from some regions of the striatum may have more extensive axonal arbors within the globus pallidus, or ventral pallidum. A recent study described the projections of the ventral striatum as being distinct from those of the dorsal striatum, since the substance P projecting neurones were shown to provide inputs to the ventral pallidum. Rather than being distinct, such differences may be more accurately described as being a variant of the "direct-indirect" projection system organization.

While labelling of axons from single neurones in the striatum provides a very elegant way of characterizing striatal projection neurones, this method is limited by sampling size. However, taken together with the retrograde data, there seems to be convincing evidence for the existence of separate "direct" and "indirect" striatal projection systems which arise from roughly equal numbers of neurones. Moreover, these two types of neurones appear to be fairly intermingled, and not segregated from one another in "patch-matrix-type" compartments.

## 2.2. Segregation of D1 and D2 Dopamine Receptors

As described above the striatal projection neurones may be grouped into two major types on the basis of their projection targets, direct and indirect projecting neurones, notwithstanding the variant subtypes. All of these neurones utilize GABA as their transmitter (Kita and Kitai, 1988). However, they express different peptides and dopamine receptor subtypes. Combined retrograde axonal tracing and in situ hybridization histochemistry methods were used to demonstrate that "direct" projecting neurones express the mRNA encoding the D1 dopamine receptor and substance P, whereas "indirect" projecting neurones express the mRNA encoding the D2 dopamine receptor and enkephalin (Gerfen *et al.*, 1990). These original findings have been confirmed in a number of other reports using in situ hybridization histochemistry, which have localized D1 receptors to striatal neurones expressing substance P, and D2 receptors to neurones expressing enkephalin, with minimal evidence of co-expression of D1 and D2 receptors in the same neurones (LeMoine *et al.*, 1990; LeMoine, Normand and Bloch, 1991). Additionally, immunohistochemical localization of both D1 and D2 receptors themselves have confirmed the segregation of these receptor subtypes to separate neurone populations in the striatum (Levey *et al.*, 1993; Hersch *et al.*, 1995). While there was some debate as to the co-expression of D1 and D2 receptors in individual striatal neurones based on single cell mRNA amplification methods (Surmeier *et al.*, 1992), such data do not reconcile with many neuroanatomical and functional studies that have examined this question.

Figure 10.2 depicts the distribution of D1 and D2 dopamine receptor subtypes in "direct" and "indirect" striatal neurones. In this figure adenosine (Ferre *et al.*, 1993) and muscarinic receptors (Bernard, Normand and Bloch, 1994) subtypes are also depicted. In some cases, other receptor subtypes are distributed along the lines of the "direct" and "indirect" neurone subpopulations, as in the case of the adenosine 2A receptor, but in other cases they are not, as in the case of the muscarinic 4 receptor.

# 3. FUNCTIONAL REGULATION OF STRIATAL NEURONES BY D1 AND D2 RECEPTORS

The effects that activation of D1 and D2 receptor subtypes has on striatal neurones may be measured in a variety of ways. These receptors are G-protein coupled receptors. Signal transduction by the D1 receptor subtype is mediated through the trimeric G-protein complex, which includes "G-s" or "G-olf" components. Both of these stimulate adenylate cyclase, and the G-olf component, found in the olfactory tubercle is the variety found in the striatum. The D2 receptor is also linked through the G protein complex which, in this case, includes the Gi components, which inhibits adenylate cyclase (Kebabian and Calne, 1979). One of the effects of such signal transduction is to alter the expression of various genes in these neurones, in part through activation of protein kinase A phosphorylation of CREB, which leads to the induction of immediate early gene transcription factors such as c-fos (Morgan and Curran, 1988; Hyman; 1997; Konradi, Leveque and Hyman, 1996). Thus, the induction of immediate early genes provides a measure of the cellular response to dopamine agonist treatment. This response occurs rapidly, within 15 minutes as measured by the induction of the mRNA encoding immediate early genes, followed by increased levels of protein. Using such methods it has been repeatedly demonstrated that D1 agonist treatments result in the induction of immediate early genes in "direct"

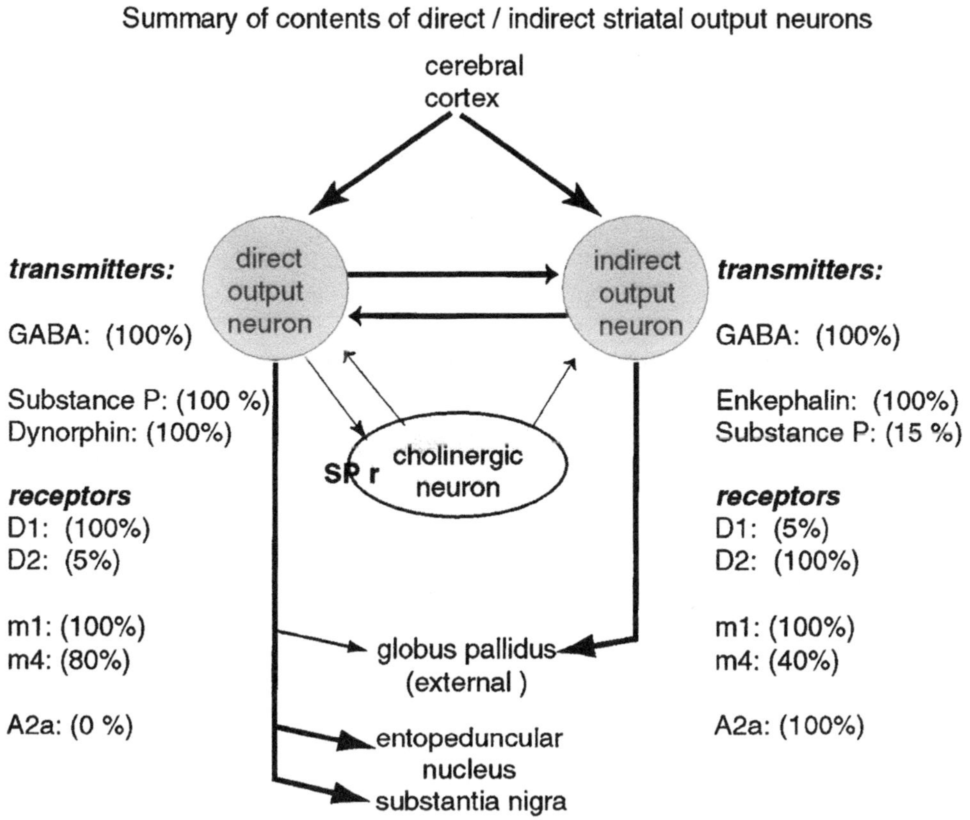


**Figure 10.2.** Diagram depicting the localization of neurotransmitters, peptides, and receptor subtypes in "direct" and "indirect" striatal projection neurones.

projecting striatal neurones, in paradigms utilizing the dopamine-depleted striatum, in which the response to D1 receptor stimulation is supersensitive (Robertson *et al.*, 1989; Robertson, Vincent and Fibiger, 1992; Keefe and Gerfen, 1995; Gerfen, Keefe and Gauda, 1995). Moreover, studies employing other means of excessive stimulation of dopamine receptors, with cocaine or amphetamine, have also demonstrated a selective D1-mediated response. On the other hand, D2 receptor antagonist treatment has been demonstrated to result in the induction of immediate early genes in "indirect" striatal neurones (Dragunow *et al.*, 1990). Similar results are obtained when the level of late expressing genes are measured (Gerfen *et al.*, 1990; Gerfen, McGinty and Young, 1991). Thus, in the dopamine-depleted striatum, D1 agonist treatment selectively elevates the expression of substance P and dynorphin in "direct" projection neurones, and D2 agonist treatment depresses the expression of enkephalin selectively in "indirect" striatal projection neurones. Thus, both the immediate early gene and late gene response is determined by which dopamine receptor subtype the striatal neurone expresses. Moreover, the direction of the change produced by the two receptor subtypes is opposite: D1 activation elevates gene markers, whereas D2 receptor activation suppresses gene markers.

Much of the controversy concerning the segregation of D1 and D2 receptors in separate populations of striatal neurones has been based on physiologic and behavioural studies of the synergistic effects of combined D1 and D2 receptor agonist treatments (Bertorello *et al.*, 1990; Piomelli *et al.*, 1991; Walters *et al.*, 1986; Weick and Walters, 1987; White and Wang, 1986). Such synergistic effects were ascribed to co-expression of D1 and D2 receptors, in stark contrast to the neuroanatomical data. To study this question directly we examined the effects of combined D1 and D2 agonist treatment on immediate early gene expression (Gerfen, Keefe and Gauda, 1995). For this study we measured the expression of the immediate early gene zif268, which has a constitutive level of expression in all striatal neurones, and we identified the type of striatal neurone with a double in situ hybridization labelling technique (Figure 10.3). In the normal striatum, zif268 mRNA is expressed by both "direct" and "indirect" striatal neurones. Following dopamine depletion, zif268 mRNA levels are elevated in D2/indirect striatal neurones and suppressed in D1/direct striatal neurones. This is consistent with the altered pattern of gene expression of peptide mRNA in these neurones (Gerfen *et al.*, 1990; Gerfen, McGinty and Young, 1991). D1 agonist treatment alone results in the elevation of zif268 selectively in D1/direct projecting neurones. Combined D1 and D2 agonist treatment results in a further elevation of zif268 in D1/direct projecting neurones, and a concurrent suppression of zif268 mRNA in D2/indirect striatal neurones. Thus, in this case in which both D1 and D2 agonists are applied in combination, the cellular response of the two neurones is again dependent on the dopamine receptor subtype they express: D1/direct neurones display an increased response, and D2/indirect neurones display a decreased response. While the mechanism responsible for the increased response in D1 neurones is still unclear (Jaeger, Kita and Wilson, 1994), what is clear is that the "direct" and "indirect" neurone populations show opposite cellular responses to concurrent stimulation of D1 and D2 receptors.

Based on neuroanatomical localization data, and on functional studies of the cellular responses to selective D1 and D2 agonist treatment, there appears to be strong evidence that dopamine selectively and oppositely regulates the function of "direct" and "indirect" striatal projection neurones. There are a number of limitations inherent in the use of gene regulation. The measure of cellular response used in these studies does not address the question of the physiologic activity of these neurones. To date there has been no study which has demonstrated unequivocally that increased expression of immediate early genes in response to dopamine D1 agonist stimulation is related to increased physiologic activity. There is some indirect evidence that this is the case. For example, D2 agonist treatments, which result in decreased immediate early gene responses in indirect striatal projection neurones, have been shown to result in c-fos induction in the globus pallidus, which is consistent with decreased activity in the striatopallidal projection pathway (Marshall, Cole and LaHoste, 1992). On the other hand, there are a number of conditions in which there appears to be little to no immediate early gene induction when there are measurable physiologic responses. For example, cocaine treatment results in little c-fos induction in the nucleus accumbens, but there are clear physiologic effects on the activity of neurones in this area. Thus, in the absence of direct measures of the relationship between immediate early gene induction and physiologic activity, this issue remains unresolved.

Another issue is that the most compelling evidence for the opposite cellular responses in direct and indirect striatal neurones to D1 and D2 receptor activation is provided in studies utilizing the dopamine-depleted striatum. This model system is used for a number

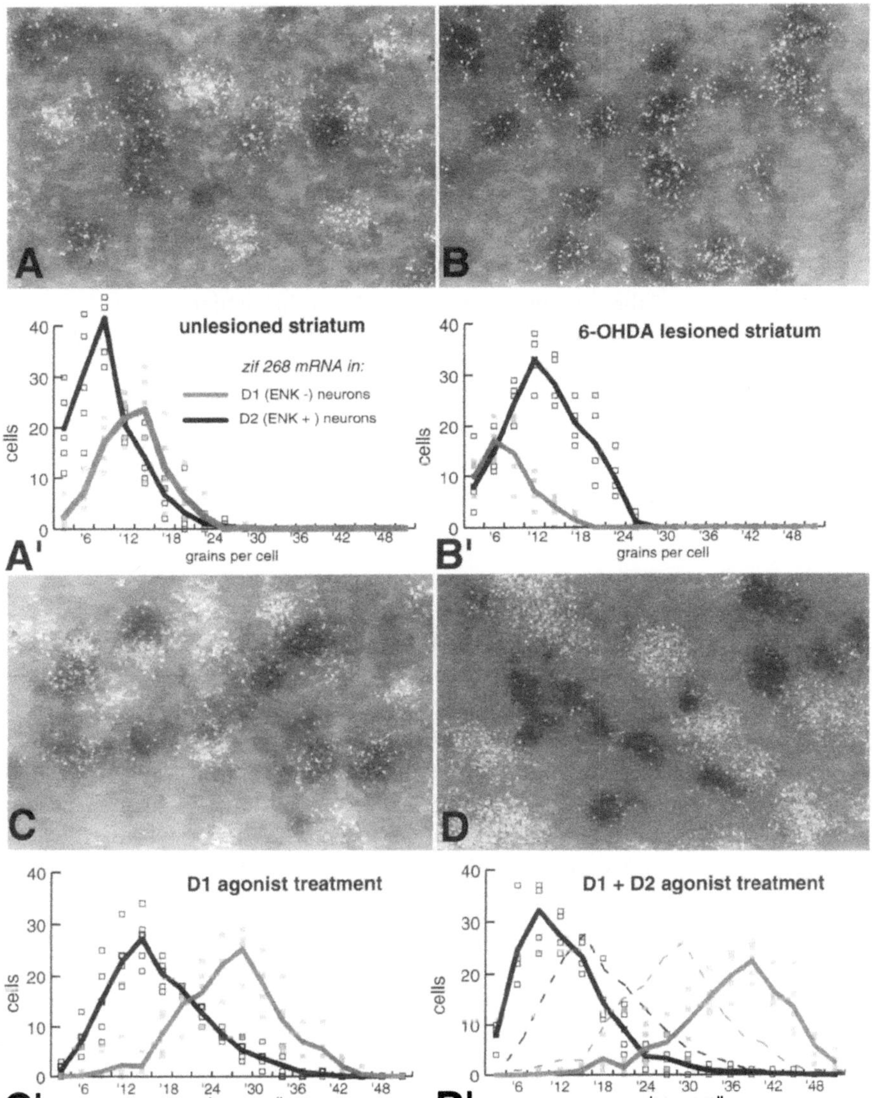


**Figure 10.3.** Photomicrographs and frequency distribution of the amount of zif268 mRNA labeling in the striatum. Photomicrographs A–D show zif268 mRNA labelling (35S-generated silver grains: white) in section labeled for enkephalin mRNA (dark cells labeled with alkaline phosphatase reaction product). Frequency distribution graphs (A′–D′) provide data on the average amount of zif268 mRNA label per cell for enkephalin-positive cells, which are putative D2-containing neurones (ENK+, individual cases marked as white squares, average marked as black line), and enkehpalin-negative, putative D1-containing cells (ENK–, individual cases marked as grey squares, average marked in gray line).

of reasons. First, it provides a model system in which the effects of D1 and D2 receptor activation may be isolated with selective agonists, in the absence of endogenous dopamine (which acts on both receptor subtypes). Thus, this model has been useful for determining that synergistic responses to combined D1 and D2 agonist treatments involve interneuronal interactions, rather than the consequence of intracellular interactions (Gerfen, Keefe and Gauda, 1995). Second, the D1 response is robust and thus provides a reproducible measurable response that is present in most direct projecting striatal neurones. While the neuronal expression of D1 and D2 dopamine receptor subtypes in direct and indirect striatal neurones occurs in the dopamine-depleted striatum (Gerfen *et al.*, 1990), certain aspects of the response are distinct from that which occurs in the normal striatum. These differences do not detract from the conclusions concerning the segregation of the D1 and D2 receptors in different striatal neurones, nor from those about the opposite effects stimulation of these receptors has on the responses of these neurones as measured by alterations in gene expression. The differences in response that occur in the dopamine depleted striatum are in fact instructive in revealing the adaptive responses of neurones to dopamine receptor stimulation, particularly when compared to the normally innervated striatum.

## 4. ADAPTIVE RESPONSES OF STRIATAL NEURONES TO DOPAMINE RECEPTOR STIMULATION

One of the most striking observations in studies utilizing D1 agonist treatments in animals with unilateral lesions of the nigrostriatal dopamine system is the robust immediate early gene response in the lesioned striatum, and the rather meagre response that is observed in the normal striatum, even with relatively high doses of a D1 agonist. The normal striatum appears to be relatively resistant to displaying alterations in gene regulation, except to pharmacological treatments that result in excessive dopamine receptor stimulation. An example of such pharmacological treatment is provided by cocaine, which blocks the dopamine transporter and reuptake of the transmitter, leading to excess dopamine receptor stimulation. This results in the induction of immediate early genes in the striatum (Cenci *et al.*, 1992; Graybiel, Moratalla and Robertson, 1990; Steiner and Gerfen, 1993; Young, Porrino and Iadarola, 1990). We have used this paradigm to study the adaptive responses of striatal neurones to dopamine receptor overstimulation (Steiner and Gerfen, 1993, 1994). Cocaine (20 mg/kg) treatment results in the induction of mRNAs encoding c-fos and zif268 immediate early genes in the dorsal striatum (Figure 10.4). Cocaine induction of striatal c-fos has been shown to be a consequence of activation of D1 dopamine receptors (Cenci *et al.*, 1992; Graybiel, Moratalla and Robertson, 1990; Steiner and Gerfen, 1994). This response diminishes over the course of successive days of repeated cocaine treatment, such that by 4 days the induction of c-fos by cocaine (20 mg/kg) is negligible. We made the observation that the regional pattern of cocaine-induced c-fos mRNA is complementary to the distribution of the level of expression of mRNA encoding dynorphin. This peptide is expressed by D1/direct projecting striatal neurones throughout the striatum, but the normal level of expression is higher, on a "per cell" basis, in the ventral striatum including the nucleus accumbens. During the repeated treatment with cocaine over several days the level of dynorphin expression is increased in the area of the striatum in which cocaine induces c-fos mRNA. Other peptides, such as substance P and enkephalin, which are also display dopamine receptor-mediated regulation, do not show the same regional and temporal relationship to cocaine treatments. This led to the suggestion that the adaptive response of dorsal striatal neurones to excessive D1-receptor stimulation caused by cocaine is to increase their expression of dynorphin, which functions to blunt the action of cocaine. To test this, the kappa opiate receptor agonist spiradoline was injected prior to an initial injection of cocaine, to attempt to mimic the effect of increased dynorphin expression. This treatment resulted in the blockade of the initial cocaine-induced cfos mRNA in the dorsal striatum. In a subsequent study a similar result was obtained by infusing spiradoline directly into the striatum to ensure that the effect in blocking cocaine-induction of cfos mRNA was occurring within the striatum (Steiner and Gerfen, 1994). Thus, these studies demonstrated a causal relationship between the altered expression of dynorphin and the response of striatal neurones to excessive stimulation of D1 receptors by cocaine. It is important to emphasise that the diminished response we are referring to is that indicated by the induction of immediate early genes. The effect on physiologic activity has not been addressed.

The adaptive response that is seen to repeated cocaine treatment, in which dorsal striatal neurones display a diminished response, is in contrast to what occurs in the dopamine-depleted striatum. Repeated D1 agonist treatment (SKF38393, 10 mg/kg/day daily for 8 days) results in increased expression of dynorphin, to levels as high as 300% above normal, with little or no decrease in the induction of c-fos mRNA to D1 agonist treatment (Steiner and Gerfen, 1994). The absence of a compensatory decreased response in the dopamine-depleted striatum might suggest that in the normal striatum dynorphin acts presynaptically on dopamine afferents. Alternatively it may be that the action which dynorphin exerts on striatal neurones is not strong enough to overcome the D1 supersensitive response. At this point we simply do not know enough about the mechanisms involved. Another interesting observation in the response pattern in the dopamine depleted striatum to repeated D1 agonist treatment is that, while the immediate early gene induction to D1 agonists does not desensitize in the dorsal striatum, it does in the ventral striatum including the nucleus accumbens (Steiner and Gerfen, 1994). These different experimental paradigms for manipulating the striatal dopamine system reveal regional differences in the acute response to excessive D1 receptor stimulation (in the cocaine paradigm) and in the adaptive response (or lack of it) to repeated excessive stimulation (in both the cocaine and supersensitive D1 agonist paradigm).

## 5. SUMMARY

Understanding the functional role of dopamine in the striatum remains a perplexing exercise. Using the response of striatal neurones to dopamine receptor agonist treatments as measured by changes in gene regulation provides some insight. However, there are inherent limitations in this approach. Most notably is the difficulty of reconciling changes in gene regulation responses to the effects of dopamine agonists on the physiologic activity of striatal neurones. Nonetheless, gene regulation studies have been instructive.

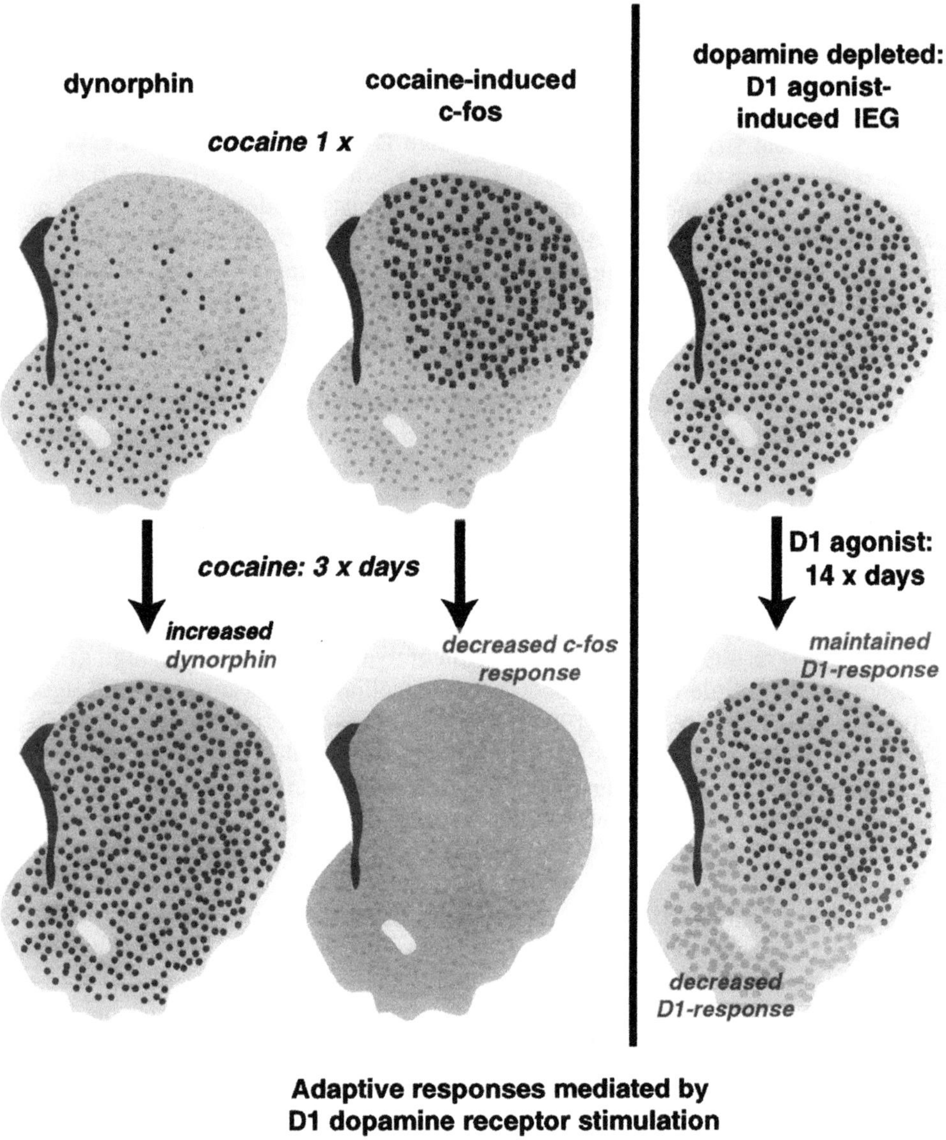


**Figure 10.4.** Diagram of the regional response within the striatum to the indirect dopamine agonist cocaine, demonstrating the functional role of dynorphin in modulating this response. The basal level of dynorphin expression shows a higher level in ventral and medial striatal regions. A single injection of cocaine induces the immediate early gene c-fos by a D1 mediated mechanism in the dorsal lateral striatal region, complementary to the area showing high levels of dynorphin. Repeated treatment with cocaine (single daily injections of 30 mg/kg for 3 days) results in an increase in dynorphin levels in the dorsal striatal region, which has low basal expression, and a marked reduction of c-fos induction in this area, in which c-fos had previously been induced. These data suggest that dynorphin blunts the response of neurones to D1 receptor stimulation. Further studies have shown that this effect of dynorphin is mediated through kappa opiate receptors. (from Steiner and Gerfen, 1993). On the right is a diagram illustrating the maintenance of D1-dopamine immediate early gene (IEG) response to D1 agonist treatment following 14 daily injections of the D1 agonist SKF38393 (5 mg/kg) in the dorsal striatum but not in the ventral striatum (from Steiner and Gerfen, 1996).

For example, the segregation of D1 and D2 receptor subtypes to connectionally distinct populations of striatal neurones has been verified by the functional response of these neurones. Moreover, the adaptive responses of striatal neurones to excessive D1 receptor stimulation provides a strategy for understanding the role of dopamine in animal models that are relevant to clinical disorders such as Parkinson's disease and psychostimulant drug abuse.

## REFERENCES

Albin R.L., Young A.B. and Penney J.B. (1989): The functional anatomy of basal ganglia disorders. *Trends in Neuroscience* **12**, 366–375.

Anderson M.E. and Turner R.S. (1991) Activity of neurons in cerebellar-receiving and pallidal-receiving areas of the thalamus of the behaving monkey. *Journal of Neurophysiology* **66**, 879–893

Bergman H., Wichmann T. and DeLong M.R. (1990): Reversal of experimental parkinsonism by lesions of the subthalamic nucleus. *Science, New York* **249**, 1436–1438.

Bernard V., Normand E. and Bloch B. (1992) Phenotypical characterization of the rat striatal neurons expressing muscarinic receptor genes. *Journal of Neuroscience* **12**, 3591–600.

Bertorello A.M., Hopfield J.F., Aperia A. and Greengard P. (1990) Inhibition by dopamine of $(Na^+K^+)$ATPase activity in neostriatal neurons through $D_1$ and $D_2$ dopamine receptor synergism. *Nature, London* **347**, 386–388.

Cenci M.A., Campbell K., Wictorin K. and Björklund A. (1992) Striatal *c-fos* induction by cocaine or apomorphine occurs preferentially in output neurons projecting to the substantia nigra in the rat. *European Journal of Neuroscience* **4**, 376–380.

Cepeda C., Buchwald N.A. and Levine M.S. (1993) Neuromodulatory actions of dopamine in the neostriatum are dependent upon the excitatory amino acid receptor subtypes activated. *Proceedings of the National Academy of Sciences, USA* **90**, 9576–9580.

Deniau J.M. and Chevalier G. (1985) Disinhibition as a basic process in the expression of striatal functions. II. The striato-nigral influence on thalamocortical cells of the ventromedial thalamic nucleus. *Brain Research* **334**, 227–233.

Dragunow M., Robertson G.S., Faull R.L.M., Robertson H.A. and Jansen K. (1990) $D_2$ dopamine receptor antagonists induce Fos and related proteins in rat striatal neurons. *Neuroscience* **37**, 287–294.

Ferre S., O'Connor W.T., Fuxe K. and Ungerstedt U. (1993) The striopallidal neuron: a main locus for adenosine-dopamine interactions in the brain. *Journal of Neuroscience* **13**, 5402–5406.

Gerfen C.R. (1985) The neostriatal mosaic. I. Compartmental organization of projections from the striatum to the substantia nigra in the rat. *Journal of Comparative Neurology,* **236**, 454–476.

Gerfen C.R. (1989) The neostriatal mosaic: striatal patch-matrix organization is related to cortical lamination. *Science, New York,* **246**, 385–388.

Gerfen C.R. (1991) Substance P (neurokinin–1) receptor mRNA is selectively expressed in cholinergic neurons in the striatum and basal forebrain. *Brain Research* **556**, 165–170.

Gerfen C.R. (1992) The neostriatal mosaic: multiple levels of compartmental organization. *Trends in Neuroscience* **15**, 133–139.

Gerfen C.R., Engber T.M., Mahan L.C., Susel Z., Chase T.N., Monsma F.J. Jr. and Sibley D.R. (1990) D1 and D2 dopamine receptor-regulated gene expression of striatonigral and striatopallidal neurons. *Science, New York,* **250**, 1429–1432.

Gerfen C.R., Keefe K.A. and Gauda E.B. (1995) D1 and D2 dopamine receptor function in the striatum: Co-activation of D1- and D2-dopamine receptors on separate populations of neurons results in potentiated immediate early gene response in D1-containing neurons. *Journal of Neuroscience* **15**, 8167–8176.

Gerfen C.R., McGinty J.F. and Young W.S., III (1991) Dopamine differentially regulates dynorphin, substance P, and enkephalin expression in striatal neurons: In situ hybridization histochemical analysis. *Journal of Neuroscience* **11**, 1016–1031.

Gerfen C.R. and Young W.S. (1988) Distribution of striatonigral and striatopallidal peptidergic neurons in both patch and matrix compartments: an *in situ* hybridization histochemistry and fluorescent retrograde tracing study. *Brain Research* **460**, 161–167.

Graybiel A.M., Moratalla R. and Robertson H.A. (1990) Amphetamine and cocaine induce drug-specific activation of the c-fos gene in striosome-matrix compartments and limbic subdivisions of the striatum. *Proceedings of the National Academy of Sciences, USA* **87**, 6912–6916.

Hersch S.M., Ciliax B.J., Gutekunst C.-Y., Rees H.D., Heilman C.J., Uung K.K.L., Bolam J.P., Ince E., Yi H. and Levey A.I. (1995) Electron microscopic analysis of D1 and D2 dopamine receptor proteins in the dorsal striatum and their synaptic relationships with motor corticostriatal afferents. *Journal of Neuroscience* **15**, 5222–5237.

Hikosaka O, Wurtz RH (1983) Visual and oculomotor functions of monkey substantia nigra pars reticulata. I. Relation of visual and auditory responses to saccades. *Journal of Neurophysiology* **49**, 1230–1253.

Hyman S.E. (1996 ) Addiction to cocaine and amphetamine. *Neuron* **16**, 901–4

Kawaguchi Y., Wilson C.J. and Emson P.C. (1990) Projection subtypes of rat neostriatal matrix cells revealed by intracellular injection of biocytin. *Journal of Neuroscience* **10**, 3421–3438.

Kebabian J.W. and Calne D.B. (1979) Multiple receptors for dopamine. *Nature (London)* **277**, 93–96.

Keefe K. and Gerfen C.R (1995) Synergistic response to combined D1- and D2-dopamine receptor stimulation in striatum: Immediate early gene response to intrastriatal drug adminstration. *Neuroscience* **66**, 903–913.

Kita H. and Kitai S.T. (1988) Glutamate decarboxylase immunoreactive neurons in rat neostriatum: their morphological types and populations. *Brain Research* **447**, 346–352.

Konradi C., Leveque J.C. and Hyman S.E. (1996) Amphetamine and dopamine-induced immediate early gene expression in striatal neurons depends on postsynaptic NMDA receptors and calcium. *Journal of Neuroscience* **16**, 4231–4239

Le Moine C., Normand E. and Bloch B. (1991) Phenotypical characterization of the rat striatal neurons expressing the D1 dopamine receptor gene. *Proceedings of the National Academy of Sciences, USA* **88**, 4205–4209.

Le Moine C., Normand E., Guitteny A.F., Fouque B., Teoule R. and Bloch B. (1990) Dopamine receptor gene expression by enkephalin neurons in rat forebrain. *Proceedings of the National Academy of Sciences, USA* **87**, 230–234.

Levey A.I., Hersch S.M., Rye D.B., Sunahara R.K., Niznik H.B., Kitt C.A., Price D.L., Maggio R., Brann M.R., Ciliax, B.J. *et al.* (1993) Localization of D1 and D2 dopamine receptors in brain with subtype-specific antibodies. *Proceedings of the National Academy of Sciences, USA* **90**, 8861–8865.

Jaeger D., Kita H. and Wilson C.J. (1994) Surround inhibition among projection neurons is weak or nonexistent in the rat neostriatum. *Journal of Neurophysiology* **72**, 2555–2558.

Mink J.W. and Thach W.T. (1991) Basal ganglia motor control. I. Nonexclusive relation of pallidal discharge to five movement modes. *Journal of Neurophysiology* **65**, 273–300.

Marshall J.F., Cole B.N. and LaHoste G.J. (1993) Dopamine D2 receptor control of pallidal fos expression: comparisons between intact and 6-hydroxydopamine-treated hemispheres. *Brain Research* **632**, 308–313

Morgan J.I. and Curran T. (1989) Stimulus-transcription coupling in neurons: role of cellular immediate-early genes. *Trends in Neuroscience* **12**, 459–462.

Piomelli D., Pilon C., Giros B., Sokoloff P., Martres M.-P. and Schwartz J.-C. (1991) Dopamine activation of the arachidonic acid cascade as a basis for D1/D2 receptor synergism. *Nature (London)* **353**, 164–167.

Robertson G.S., Herrera D.G., Dragunow M. and Robertson H.A. (1989) L-dopa activates c-fos in the striatum ipsilateral to a 6-hydroxydopamine lesion of the substantia nigra. *European Journal of Pharmacology* **159**, 99–100.

Robertson G.S., Vincent S.R. and Fibiger H.C. (1992) $D_1$ and $D_2$ dopamine receptors differentially regulate c-fos expression in striatonigral and striatopallidal neurons. *Neuroscience* **49**, 285–296.

Steiner H. and Gerfen C.R. (1993) Cocaine-induced c-*fos* messenger RNA is inversely related to dynorphin expression in striatum. *Journal of Neuroscience* **13**, 5066–5081.

Steiner H. and Gerfen C.R. (1994) kappa opioid receptor inhibition of D1 dopamine receptor mediated induction of immediate early genes in striatum *Journal of Comparative Neurology* **353**, 200–212.

Steiner H. and Gerfen C.R. (1996) Dynorphin regulates D1 dopamine receptor-mediated responses in the striatum: relative contributions of pre- and postsynaptic mechanisms in dorsal and ventral striatum demonstrated by altered immediate-early gene induction. *Journal of Comparative Neurology* **376**, 530–541

Surmeier D.J., Eberwine J., Wilson C.J., Cao Y., Stefani A. and Kitai S.T. (1992) Dopamine receptor subtypes colocalize in rat striatonigral neurons. *Proceedings of the National Academy of Sciences, USA* **89**, 10178–10182.

Walters J.R., Bergstrom D.A., Carlson J.H., Chase T.N. and Braun A.R. (1987) D1 dopamine receptor activation required for postsynaptic expressin of D2 agonist effects. *Science, New York* **236**, 719–722.

Weick B.G. and Walters J.R. (1987) Effects of D1 and D2 dopamine receptor stimulation on the activity of substantia nigra pars reticulata neurons in 6-hydroxydopamine lesioned rats: D1/D2 coactivation induces potentiated responses. *Brain Research* **405**, 234–246.

White F.J. and Wang R.Y. (1986) Electrophysiologic evidence for the existence of both D–1 and D–2 dopamine receptors in the nucleus accumbens. *Journal of Neuroscience* **6**, 274–280.

Wilson C.J. and Groves P.M. (1980) Fine structure and synaptic connections of the common spiny neuron of the rat neostriatum: a study employing intracellular inject of horseradish peroxidase. *Journal of Comparative Neurolology,* **194**, 599–615.

Young S.T., Porrino L.J. and Iadarola M.J. (1991) Cocaine induces striatal c-Fos immunoreactive proteins via dopaminergic D1 receptors. *Proceedings of the National Academy of Sciences, USA* **88**, 1291–1295.

Young W.S. III, Bonner T.I. and Brann M.R. (1986) Mesencephalic dopaminergic neurons regulate the expression of neuropeptide mRNAs in the rat forebrain. *Proceedings of the National Academy of Sciences, USA,* **83**, 9827–9831.

# 11 Dopaminergic Regulation of Striatal Physiology


## D. James Surmeier

*Department of Physiology, Northwestern Medical School, Searle-5-447,*
*320 E-Superior St., Chicago Il 60611, USA*
*tel: (312)503-4904; FAX:(312) 503-7345; e-mail: j-surmeier@nwu.edu*



Significant progress has been made recently in understanding the neuromodulatory effects of dopamine in striatal neurones. This progress has come about by taking a fundamentally more reductionistic approach than previous electrophysiological studies. Studies combining single cell RT-PCR profiling of mRNA and voltage-clamp analysis of dopamine-induced changes in ionic conductances have proven particularly valuable. Although significant questions remain unanswered, at this point in time these studies have provided significant insights into the effects of dopamine on cholinergic interneurones and medium spiny neurones that project to the substantia nigra. Three topics will be discussed: The first is how dopamine regulates cholinergic interneurones. The second is how dopamine influences medium spiny neurones that express D1 dopamine receptors. The third is how dopamine and acetylcholine interact in regulating the activity of medium spiny neurones. These findings are discussed within the context of current debates about dopamine's action within the striatum, and an attempt is made to reconcile these discrepant views. A central conclusion of these studies is that dopamine should not be thought of as an excitatory or inhibitory neurotransmitter in the classical sense.



KEYWORDS: Dopamine, GABA, single cell RT-PCR, basal ganglia, voltage clamp, neuromodulation, phosphorylation


## 1. INTRODUCTION

Dopamine has long been known to be an important neurotransmitter in the striatum. It has been implicated in the pathophysiology of Parkinson's disease (PD) (Hornykiewcz, 1973), dystonia (Nygaard, Marsden and Duvoisin, 1988), Tourette's syndrome (Singer and Walkup, 1991), schizophrenia (Nemeroff and Bissette, 1988; Seeman *et al.*, 1997) and drug addiction (Nestler and Aghajanian, 1997). In spite of its prominence, the physiological impact of dopamine has remained something of a mystery, particularly within the striatum. Claims have been made that dopamine is excitatory or inhibitory or neither.

There are several potential reasons for the discrepancies that have arisen about dopamine. One is that there are at least five dopamine receptors (D1-D5) (Sibley, 1995). These receptors can be grouped into D1- (D1, D5) and D2-classes (D2, D3, D4) on the basis of structural homologies and G-protein coupling. In addition, the D2 receptor comes is short- and long-splice variants. Expression studies in heterologous systems have suggested that receptors within each class may differ in functionally significant ways, raising the possibility at least that these differences could account for varied functional linkages *in situ*.

Another possible origin of discrepant findings is that the striatum is richly interconnected, making it difficult to exclude the possibility that changes in activity or gene expression seen in any one population of neurones are attributable to dopamine indirectly acting upon them in the usual experimental paradigms.

A third possible explanation for many of the discrepancies is based upon how G-protein coupled receptors (like the dopamine receptors) exert their effects on cellular excitability and function (Hille, 1994). Unlike classical neurotransmitters, dopamine does not directly gate an ion channel. That is, dopamine does not produce excitatory postsynaptic potentials or inhibitory synaptic potentials in the conventional sense. G-protein coupled receptors act through G-protein and second messenger cascades to alter the properties of voltage-dependent and ligand-gated conductances. In other words, they alter the response to other stimuli, be they alterations in membrane voltage that accompany synaptic input, or release of a classical transmitter. This feature of dopaminergic signalling significantly complicates the analysis of what dopamine is normally doing, because it cannot be studied in isolation and will depend upon the ongoing behaviour of the cell. To make matters worse, other neuromodulators may also affect the consequences of DA receptor stimulation. There is considerable evidence, for example, for interaction between signalling cascades triggered by D1 receptors and those of m1 muscarinic receptors or NMDA receptors (Greengard and Browning, 1988; Halpain, Girault and Greengard, 1990) (see below).

## 2. WHAT IS THIS CHAPTER ABOUT?

In spite of these problems, significant progress has been made recently in understanding the neuromodulatory effects of dopamine in striatal neurones. This progress has come by taking a fundamentally more reductionistic approach than previous electrophysiological studies. Our work has had three broad goals. One is to eliminate indirect or circuitry mediated effects by studying isolated neurones. The second is to use voltage-clamp techniques to generate quantitative estimates of aggregate channel behaviour in the presence and absence of dopaminergic agonists. The third is to generate molecular descriptions of receptors, channels and other phenotyping proteins involved in dopaminergic signalling in identified striatal cell types. These molecular studies have not only provided insight into the mechanisms mediating dopaminergic effects on neurones but also have allowed information, gathered with the reduced preparations, to be placed in a systems context.

As dealt with in greater detail elsewhere in this book, the neostriatum contains projection neurones and interneurones (Graybiel, 1990). The projection neurones consitute roughly 95% of all neurones in the rodent brain, and are medium-sized with spiny dendrites. Based upon expression of releasable peptides (substance P and enkephalin), there are at least three subtypes of medium spiny neurone (Surmeier, Song and Yan, 1996). In addition, there are at least three types of interneurone that have been described (Kawaguchi, 1993). One of these is a very large, aspiny neurone that releases acetylcholine. Such cholinergic interneurones constitutes 1–3% of all striatal neurones (chapter 6, this volume) but have extremely rich intrastriatal connections (Bolam and Bennett, 1995). Here I focus on the dopaminergic regulation of cholinergic interneurones and the medium spiny neurones. Although significant questions remain unanswered, at this point in time we have enough insight into these cell types to sketch the outlines of how dopamine regulates subsets of these neurones, and how they might interact. In particular, three topics will be discussed. The first will be how dopamine regulates cholinergic interneurones. The second will be how dopamine influences medium spiny neurones. The third will be how dopamine and acetylcholine interact in regulating the activity of medium spiny neurones.

## 3. HOW DOES DOPAMINE MODULATE CHOLINERGIC INTERNEURONES?

Early clinical observations in PD patients made it clear that cholinergic interneurones were important determinants of striatal function (Lehmann and Langer, 1983). These observations showed that the symptoms of PD could be ameliorated by anti-muscarinic drugs, suggesting that dopamine and acetylcholine were somehow out of balance in the PD striatum. Until recently, the cellular mechanisms mediating this interaction have been undefined. One of the goals of this chapter is partially to fill this gap.

How does dopamine regulate the activity of cholinergic interneurones? The first step toward answering this question is to determine what dopamine receptor subtypes are expressed by cholinergic interneurones. Previous work had suggested that cholinergic interneurones express D2 receptors, but no (or very few) D1 receptors (LeMoine, Tison and Bloch, 1990). Single cell RT-PCR analysis of cholinergic interneurones confirmed the presence of D2 receptor mRNA (both short and long isoforms), but revealed that virtually all cholinergic interneurones co-express high levels of D5 receptor mRNA (Figure 11.1AB) (Yan, Song and Surmeier, 1997). The functional studies described below show that both D2 and D5 receptors have potent physiological consequences.

Activation of D2 receptors produces no noticeable change in the properties of $Na^+$ or depolarization-activated $K^+$ currents in cholinergic interneurones. However, they do greatly reduce voltage-dependent $Ca^{2+}$ currents (Yan, Song and Surmeier, 1997) (Figure 11.1C,D). This reduction is mediated by a membrane delimited G-protein pathway that selectively targets N-type $Ca^{2+}$ channels. The qualitative characteristics of this modulation are similar to those of the m2 muscarinic autoreceptor (Yan and Surmeier, 1996). Like the m2 receptor, D2 receptors diminish ACh release (Bertorelli *et al.*, 1992; Stoof *et al.*, 1992), presumably by inhibiting $Ca^{2+}$ influx. However, there are differences in the D2 signalling pathway that may prove to be functionally important. Unlike the m2 receptor modulation of N-type $Ca^{2+}$ channels, the D2 modulation is not relieved by depolarization or (by inference) by repetitive spiking. Another difference is that the D2 modulation is not reversed by activation of protein kinase C (PKC). That is, PKC can effectively block the autoreceptor functions of the m2 muscarinic receptor without affecting the ability of D2 receptors to inhibit $Ca^{2+}$ channels and acetylcholine release (by inference). The full functional consequences of this difference remain to be worked out because the receptors linked to PKC have yet to be identified. Our unpublished work has shown that NK1 receptors do not appear to be coupled efficiently to phospholipase C in cholinergic interneurones, in spite of the fact they do in other cell types.

Unlike the situation in medium spiny neurones, activation of D5 (or D1-class receptors) does not detectably alter the properties of $Ca^{2+}$ or $Na^+$ channels. Their possible

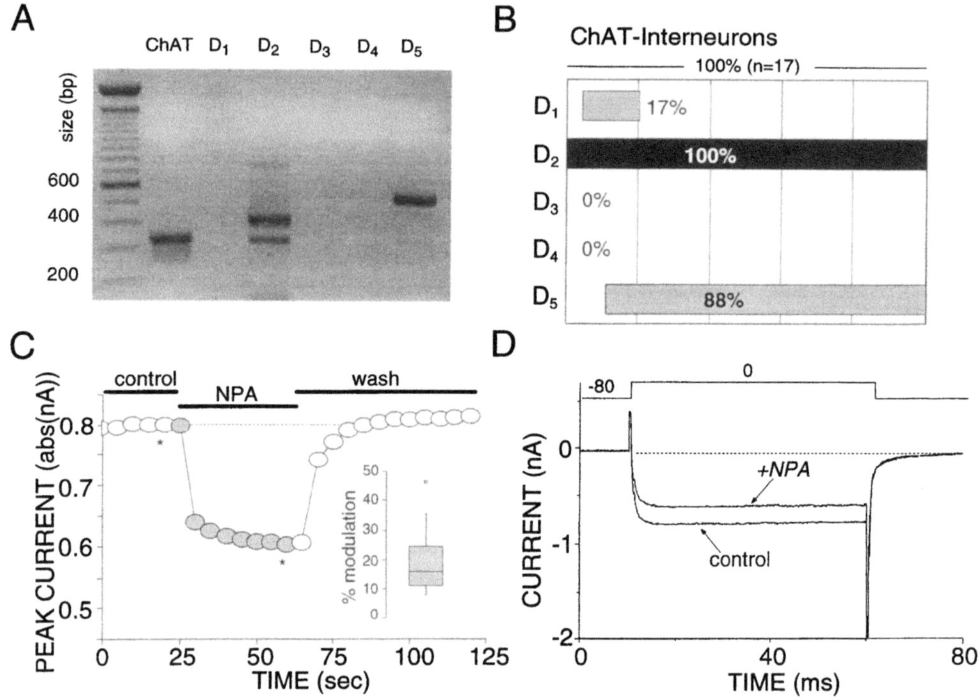


**Figure 11.1.** D2-class agonists produce a reversible decrease in $Ca^{2+}$ currents in cholinergic interneurones. **A.** PCR profile of a single ChAT-positive neurone having detectable levels of D2 and D5 mRNAs. **B.** Bar plot showing the coordinated expression of dopamine receptor subtypes D1 -D5 in ChAT-positive neurones detected by the multiplex PCR method (n = 17). **C.** Plot of peak current evoked by a voltage step to 0 mV as a function of time and agonist application. D2-class agonist NPA (10 $\mu$M) rapidly and reversibly reduced peak currents. Inset is a box plot, showing the reduction in peak current by D2-class agonists in a sample of 51 large neurones. **D**. Current traces from the data used to construct **C**, the points in the record used are denoted by asterisks.

linkage to depolarization activated $K^+$ channels, which are dominated in the somatodendritic membrane by the Kv4.2/4.1 varieties of potassium channels (Song *et al.*, 1997), remains to be adequately tested. Nevertheless, D5 receptors are strongly coupled to another channel type, namely $GABA_A$ receptors (Yan and Surmeier, 1997), that is an important regulator of interneuronal activity (Figure 11.2). Acting through cAMP and protein kinase A (PKA), D5 receptor stimulation results in the enhancement of GABA-evoked currents. Although PKA is necessary for the enhancement, it does not appear to be sufficient. Blocking the activity of protein phosphatase 1 (PP1) also blocks the effects of receptor stimulation, suggesting that dephosphorylation of $GABA_A$ receptors (or closely associated proteins) is responsible for the alteration in evoked current. PKA-mediated regulation of PP1 activity has been described previously in muscle (Hubbard and Cohen, 1993) and in medium spiny neurons (Surmeier *et al.*, 1995), but this is the first example of $GABA_A$ receptors being targeted by this type of signalling cascade.

Another interesting feature of this modulation is that it targets a specific subset of $GABA_A$ receptors. The application of extracellular low-micromolar concentrations of $Zn^{2+}$ reduces GABA-evoked currents, presumably by blocking receptors that lack a $\gamma$-subunit

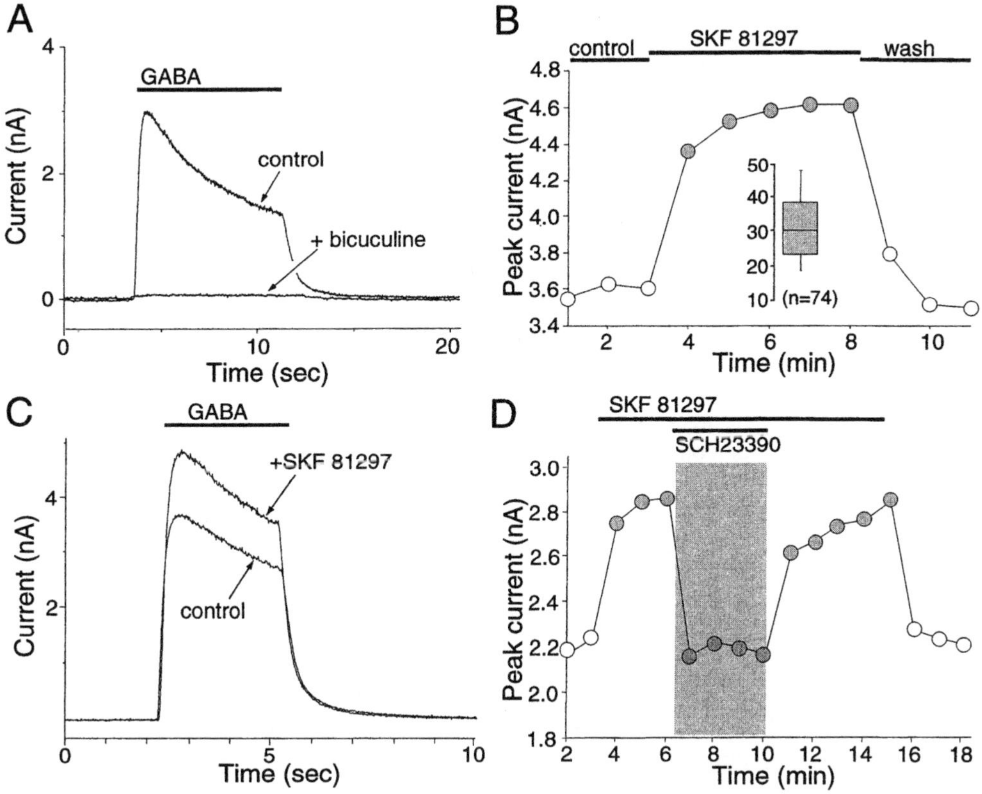


**Figure 11.2.** Activation of D5 dopamine receptors reversibly enhanced $GABA_a$ currents. **A.** Currents elicited by the application of GABA (100 mM ) and the co-application of GABA and the $GABA_a$ receptor antagonist bicuculline (30 mM). **B.** Plot of peak current evoked by GABA (100 $\mu$M) as a function of time and agonist application. The D1-class agonist SKF 81297 (10 $\mu$M) reversibly enhanced $GABA_a$ currents. The inset is a box plot summary of the percent enhancement of peak $GABA_a$ currents produced by D1-class agonists in a sample of 74 cholinergic interneurones. **C.** Current traces taken from the records used to construct B. **D.** Plot of peak $GABA_a$ current as a function of time and ligand application. The D1-class antagonist SCH 23390 (10 $\mu$M)

(Draguhn *et al.*, 1990). When this subset of channels is blocked, D5 receptor stimulation fails to exert any effect on evoked currents, suggesting that such channels are the obligate targets of the modulation. Whether the D5 receptors are broadly distributed, or are limited to a subset of synaptic or extra-synaptic sites, remains to be determined. However, the high levels of vesicular $Zn^{2+}$ located in the terminals of corticostriatal afferent fibres (Vincent and Semba, 1989; Mengual *et al.*, 1995) suggests that they may be regulated by both dopaminergic and cortical activity.

In any event, this result suggests that D1 agonists acting directly on cholinergic interneurones should slow ambient discharge and reduce ACh release. Although this is at odds with the effects of systemic D1 agonist administration on ACh release, it is consistent with new data showing that intrastriatal D1 agonist application reduces ACh release (Abercrombie and DeBoer, 1997). Moreover, this result means that the actions of dopamine acting through D2 and D5 receptors are similar, in that both should diminish cholinergic tone. The result also provides a partial explanation of why the loss of the dopaminergic innervation of the striatum should result in elevated ACh tone.

There is another phenomenon that may be explained by these signalling pathways. The work of Kimura, Graybiel and colleagues has suggested that cholinergic interneurones act as coordinators of striatal activity in associative learning paradigms (Aosaki *et al.*, 1994; Graybiel *et al.*, 1994). As the animal learns the association between conditioned and unconditioned stimuli, the activity pattern of cholinergic neurones becomes bound to the conditioned stimulus. This plasticity is dependent upon dopaminergic input. Schultz's group has shown that the activity of dopaminergic neurones undergoes similar changes (Schultz, 1994; Mirenowicz and Schultz, 1996). Our results suggest a mechanism by which this plasticity might occur. As cholinergic neurones become bound to the conditioned stimulus, their activity changes in a characteristic way: It is transiently suppressed, as if an inhibitory input to this neurone were potentiated. Since a substantial subset of neostriatal GABAergic neurones are linked to sensory events (Alexander, DeLong and Strick, 1986), it is our conjecture that activation of D5 dopamine receptors potentiates this input. Repeated pairing of D5 receptor activity and GABAergic input may lead to lasting potentiation of this GABAergic input. A similar phenomenon has already been described in cerebellar Purkinje neurons (Kano and Konnerth, 1992).

## 4. DOPAMINERGIC SIGNALLING IN MEDIUM SPINY NEURONES

The principal neurones of the neostriatum are medium spiny neurones. There is a number of ways in which this group has been subdivided (e.g., by projection site, by releaseable peptides, or by dopamine receptor). Several of these indices are correlated with one another. For example, the releaseable peptides, substance P and enkephalin, are strongly correlated with projection site (Gerfen, 1992 and chapter 10). Using single cell RT-PCR techniques (Surmeier, Song and Yan, 1996), we have identified three groups of neurone, based upon SP and ENK expression: neurones expressing SP alone, neurones expressing ENK alone and neurones co-expressing SP and ENK. The first two groups are roughly equal in size (ca. 40%) whereas the latter group is smaller (ca. 20%).

Dopamine receptor expression is strongly correlated with the expression of releaseable peptide mRNAs in our single cell RT-PCR studies. SP neurones express high levels of D1 mRNA, and ENK neurones express high levels of D2 mRNA (both short and long isoforms) (Figure 11.3A-D). Neurones that co-localize these peptides, also co-localize D1 and D2 receptors (Figure 11.3E,F). D3, D4 and D5 mRNAs are less robustly expressed by medium spiny neurones, but they still represent a significant receptor complement. For example, D3 mRNA is found at relatively high levels in a subset of SP containing neurones (ca., 50%). D4 receptor mRNA is less common (ca. 25%) and present at lower levels. In contrast, D5 receptor mRNA is rarely seen in SP containing neurones. In neurones expressing only ENK mRNA, D5 mRNA is seen at relatively high levels in a subset (ca. 25%) whereas D3 and D4 mRNA are present at lower levels in a largely overlaping subset (20–40%). This pattern of mRNA expression suggests that approximately 70% of all medium spiny neurons colocalize D1- and D2-class receptors with less than half of the colocalization being attributable specifically to D1 and D2 receptors.

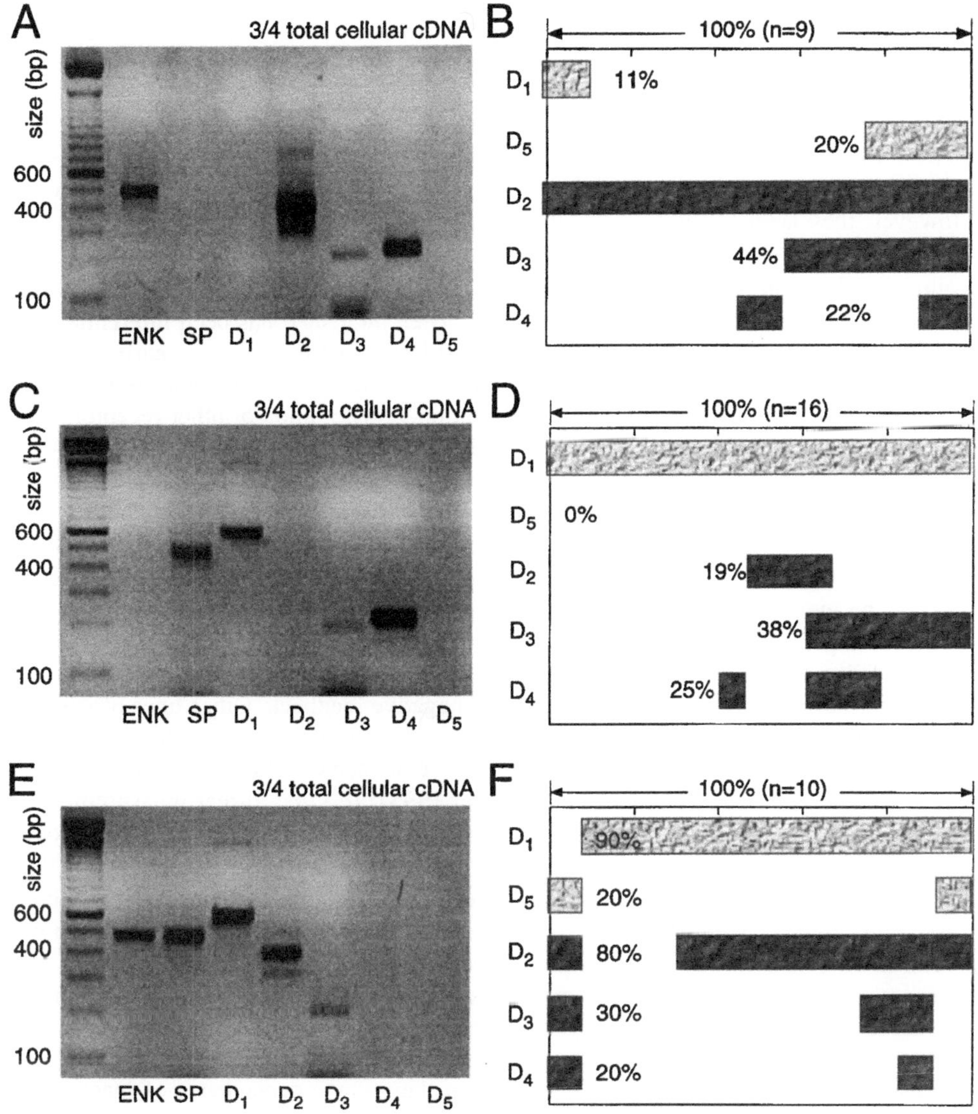


**Figure 11.3.** Medium spiny neurones co-express D1 and D2-class receptor mRNAs. **A**. Photograph of an ethidium bromide stained agarose gel in which dopamine receptor and peptide mRNA amplicons from a single ENK+/SP– medium spiny neurone have been separated by electrophoresis. A multiplex procedure employing 3/4 of the total cellular cDNA was used for detection of DA receptor mRNAs. **B**. Summary of the coordinated DA receptor mRNA expression in 9 neurones expressing enkephalin but not SP. Co-expression for any set of mRNAs can be deduced by the extent to which shaded bars in their lanes overlap. **C**. Photograph of a gel containing amplicons from a single SP+/ENK– medium spiny neurone in which a multiplex procedure employing 3/4 of the total cellular cDNA was used for detection of DA receptor mRNAs. Note that in these amplifications, the D1 primer set yielding the longer amplicon was used. **D**. Summary of co-expression detected with the multiplex procedure in 16 ENK–/SP+ neurones. **E**. Photograph of a gel containing amplicons from a single ENK+/SP+ medium spiny neurone in which a multiplex procedure employing 3/4 of the total cellular cDNA was used for detection of DA receptor mRNAs. **F**. Summary of co-expression detected with the multiplex procedure in 10 ENK+/SP+ neurones.

This level of receptor colocalization is largely consistent with physiological studies examining the responsiveness of medium spiny neurones to class-specific dopaminergic agonists (Uchimura, Higashi and Nishi, 1986; Akaike *et al.*, 1987; Ohno, Sasa and Takaori, 1987; Calabresi *et al.*, 1992; Cepeda, Buchwald and Levine, 1993). Biochemical studies have also argued for receptor colocalization (Stoof and Kebabian, 1984; Bertorello, *et al.* 1990). These studies and those examining the distribution of receptor mRNAs are consistent with the concept of multiple parallel striatal efferent pathways. However, they are not consistent with models in which D1 and D2 receptors (and by inference D1- and D2-classes) are strictly segregated (Gerfen, 1992; chapter 10, this volume). Although appealing in its simplicity, there has never been much compelling data to argue for this model. Most of what has been published has been inferential. For example, alterations in mRNA levels for SP and ENK, or for immediate early genes, in response to dopamine depletion or exogenously applied dopaminergic agonists are related in undefined ways to dopamine receptor activation, co-activation of other receptors (e.g. of glutamatergic receptors) and neuronal activity. The assumption has repeatedly been made that changes in gene expression (regardless of the gene in question) are reflections of increased spike activity. This is an unproven proposition.

One of the most frequently repeated inferences drawn from this body of literature is that D1 receptor stimulation excites SP-containing medium spiny neurones (Gerfen, 1992). The evidence for this view stems largely from the decreased expression of SP mRNA and peptide after dopamine depletion. It is assumed this reflects decreased neuronal activity. But of course, one could argue just as convincingly that this is a consequence of elevated activity and a reflection of negative feedback. Other lines of study *in vivo*, using more direct measurements of neuronal activity, have been less consistent, some seeing evidence of excitation, some seeing evidence of inhibition (Bernardi *et al.*, 1978; Engber *et al.*, 1990; Pierce and Rebec, 1995). *In vitro*, direct measurement of changes in electrical activity following application of D1 agonists has consistently found supression in evoked discharge (Uchimura, Higashi and Nishi, 1986; Akaike *et al.*, 1987; Calabresi *et al.*, 1987). Our voltage-clamp studies have attempted to resolve this controversy by examining the effects of D1-class agonists on each conductance system known to contribute to the discharge of medium spiny neurons.

These studies have lead us to a somewhat startling conclusion: D1-receptor activation can be both excitatory and inhibitory. How? D1 receptor activation results in changes in voltage-dependent $Na^+$, $Ca^{2+}$ and $K^+$ conductances simultaneously. $Na^+$ currents are decreased, as a consequence of a small shift in the voltage-dependence of inactivation and a reduction in available channels (Surmeier *et al.*, 1992; Cepeda *et al.*, 1995; Schiffmann, Lledo and Vincent, 1995) (Figure 11.4). This modulation appears to be mediated by direct PKA phosphorylation of type II $Na^+$ channels. Inwardly rectifying $K^+$ currents are increased through a PKA-dependent mechanism (Pacheco-Cano *et al.*, 1996; unpublished observations). These two effects are responsible for the suppression in evoked discharge seen from negative membrane potentials (ca.–85 mV). Thus, at hyperpolarized membrane potentials, D1 receptor activation results in an inhibiton (Figure 11.5A). However, if the cell is more depolarized, the inward rectifier currents begin to shut off, and $Ca^{2+}$ conductances play more of a role in dictating activity. D1 receptor activation increases L-type $Ca^{2+}$ currents (Surmeier *et al.*, 1995), resulting in a slow inward current at depolarized potentials, which elevates discharge rate. In addition, $Ca^{2+}$ currents that appear to be linked to $Ca^{2+}$-dependent $K^+$ channels (N/P/Q type) are

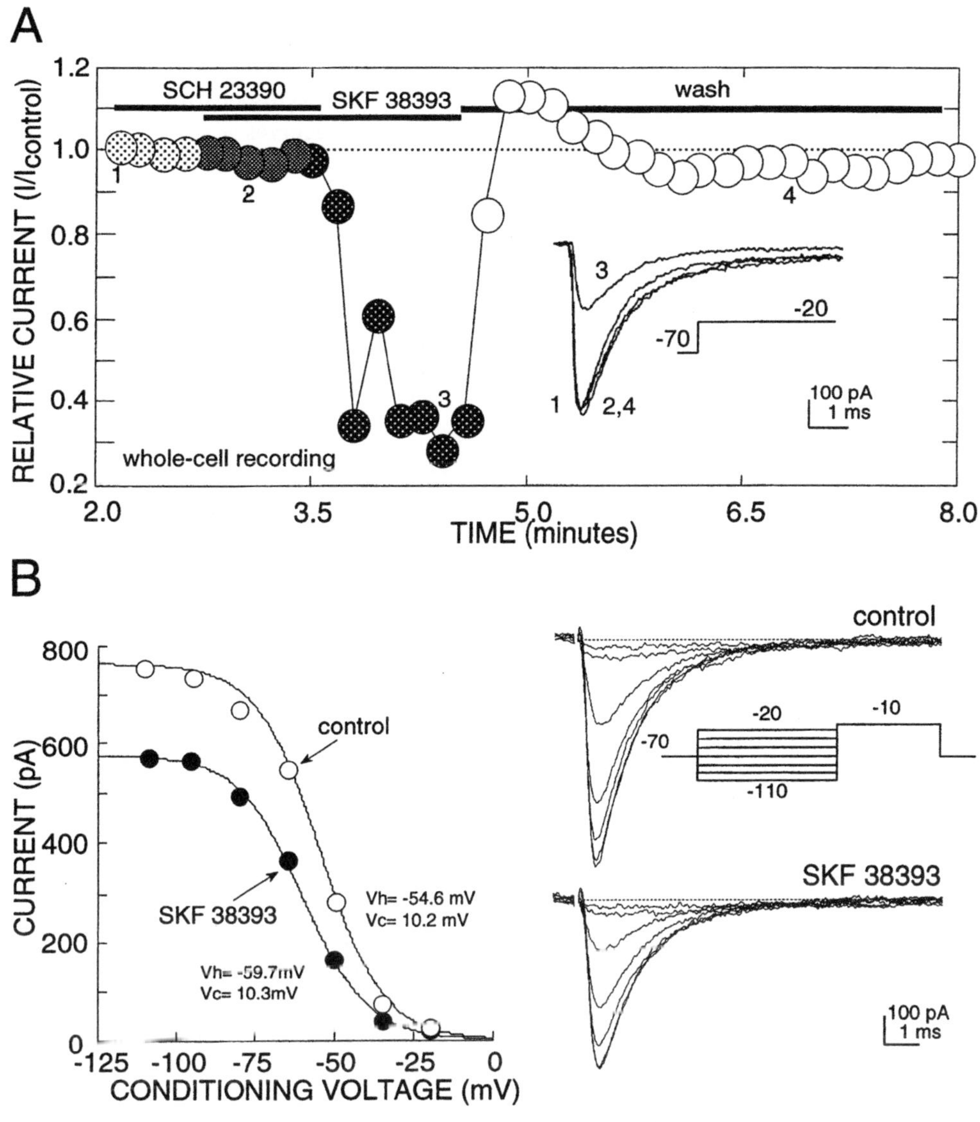


**Figure 11.4.** D1 receptor activation decreases $Na^+$ currents in medium spiny neurones. **A.** Plot of peak $Na^+$ current evoked by a slowly repeated voltage step to 0 mV in a retrogradely identified striatonigral medium spiny neurone. The D1 receptor antagonist SCH 23390 (5 $\mu$M) had no effect on currents but blocked the effects of the D1 agonist SKF 38393 (5 $\mu$M), which was revealed after washing off the antagonist. Inset are representative current traces. **B.** The D1 agonist SKF 38393 (5 $\mu$M) produced a small shift in the voltage dependence of steady state inactivation and reduced peak current. Representative currents evoked in the steady-state inactivation protocol in control and agonist conditions are shown at the right.

diminished (Surmeier *et al.*, 1995), resulting in smaller after-hyperpolarizing potentials. Thus, at depolarized potentials, D1 receptor activation results in an excitation (Figure 11.5B). These observations effectively reconcile the two discrepant points of view on the actions of D1 receptors (Hernandez-Lopez *et al.*, 1997). Since the global potential shifts in membrane potential that occur in medium spiny neurones are controlled by cortex, they also underline the importance of glutamatergic input in understanding the actions of dopamine.

What are some of the implications of this type of excitatory/inhibitory modulation? This work suggests that quiescent medium spiny neurones will be less responsive to cortical input when D1 receptor tone is high. However, if a neurone receives a strong, maintained excitatory input, the response to this input will be augmented, particularly in the latter phases of the response. Similarly, cells that have been active will be more responsive. What this effectively should do is to increase the signal-to-noise ratio of striatal outflow, if we assume that strong, coherent inputs contain the principal message arising from cortex.

## 5. INTERACTIONS BETWEEN CHOLINERGIC AND DOPAMINERGIC SIGNALLING IN MEDIUM SPINY NEURONES

Based upon what we know about cellular physiology at this point, dopaminergic and cholinergic signalling pathways appear to be largely antagonistic. This is, of course, in agreement with clinical observations. Nearly all medium spiny neurones express the m1

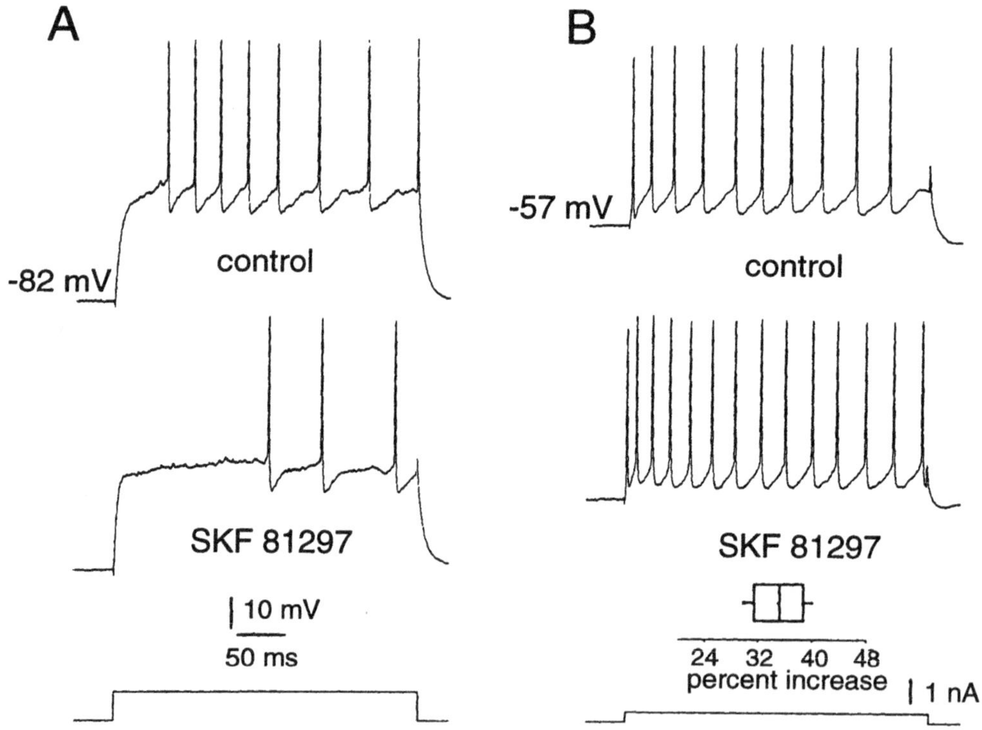


**Figure 11.5.** D1 receptor activation inhibits evoked activity from negative membrane potentials but enhances it from positive membrane potentials. **A.** Current-clamp recordings from medium spiny neurones in striatal slices using conventional sharp microelectrodes. The application of the D1 agonist SKF 81297 (10 $\mu$M) decreases the spike discharge evoked by current injection from a resting potential of –82 mV. **B.** At more depolarized membrane potentials (–57 mV), SKF 81297 increases the discharge evoked by current injection. Inset is a box plot summary of the percent increase in evoked spikes at depolarized membrane potentials produced in the presence of SKF 81297.

muscarinic receptor (our unpublished observations; Bernard, Normand and Bloch, 1992). This receptor is positively coupled to phospholipase C$\beta$ isoforms that elevate intracellular inositol trisphosphate (IP3) levels (Akins, Surmeier and Kitai, 1990a). IP3 promotes the release of $Ca^{2+}$ from intracellular stores. This signalling pathway appears to underlie the reduction in L-type $Ca^{2+}$ currents produced by muscarinic agonists (Howe and Surmeier, 1995). Preliminary data suggests that this same pathway, presumably acting through calcineurin, results in a reduction in inwardly rectifying $K^+$ currents (unpublished observations). Both of these effects will antagonize the modulations brought about by D1 receptor stimulation (see above). The story may not be quite this simple however. Muscarinic receptors also influence $Na^+$ (unpublished observations) and $K^+$ currents (Akins, Surmeier and Kitai, 1990b) in ways that do not appear to be simply antagonistic to D1 receptor actions. In addition, m4 receptors are expressed by a substantial complement of medium spiny neurons (Bernard, Normand and Bloch, 1992; and our unpublished observations) and these receptors reduce N- and P/Q-type $Ca^{2+}$ currents (Howe and Surmeier, 1995), as do D1 receptors. So, while many of the actions of acetylcholine and dopamine are antagonistic, at least on cells expressing D1 receptors, the interaction between the two transmitter systems may be more complicated than previously thought.

## ABBREVIATIONS

ACh: Acetylcholine
cAMP: Cyclic adenosine monophosphate
DA: Dopamine
D1, D2, D3, D4, D5: Dopamine receptor subtypes
ENK: Enkephalin
GABA: Gamma-aminobutyric acid
IP3: Inositol triphosphate
m2: muscarinic receptor subtype
NK1: Neurokinin–1
NMDA: N-Methyl-d-aspartic acid
PD: Parkinson's Disease
PKA: Phosphokinase-A
PKC: Phosphokinase-C
PP1: Protein phosphatase–1
RT-PCR: Reverse transcriptase polymerase chain reaction
SP: Substance P

## REFERENCES


Abercrombie E.D. and DeBoer P. (1997) Substantia nigra D1 receptors and stimulation of striatal cholinergic interneurons by dopamine: A proposed circuit mechanism. *Journal of Neuroscience* **17**, 8498–505.

Akaike A., Ohno Y., Sasa M. and Takaori S. (1987) Excitatory and inhibitory effects of dopamine on neuronal activity of the caudate neurons in vitro. *Brain Research* **418**, 262–272.

Akins P.T., Surmeier D.J. and Kitai S.T. (1990a) The M1 muscarinic acetylcholine receptor in cultured rat neostriatum regulates phosphoinositide hydrolysis. *Journal of Neurochemistry* **54**, 266–273.

Akins P.T., Surmeier D.J. and Kitai S.T. (1990b) Muscarinic modulation of the transient potassium current in rat neostriatal neurons. *Nature (London)* **344**, 240–242.

Alexander G.E., DeLong M.R. and Strick P.L. (1986) Parallel organization of functionally segregated circuits linking basal ganglia and cortex. *Annual Review of Neuroscience* **9**, 357–381.

Aosaki T., Tsubokawa H., Ishida A., Watanabe K., Graybiel A.M. and Kimura M. (1994) Responses of tonically active neurons in the primate's striatum undergo systematic changes during behavioral sensorimotor conditioning. *Journal of Neuroscience* **14**, 3969–84.

Bernard V., Normand E. and Bloch B. (1992) Phenotypical characterization of the rat striatal neurons expressing muscarinic receptor genes. *Journal of Neuroscience* **12**, 3591–3600.

Bernardi G., Marciani M.G., Morocutti C., Pavone F. and Stanzione P. (1978) The action of dopamine on rat caudate neurones intracellularly recorded. *Neuroscience Letters* **8**, 235–240.

Bertorelli R., Zambelli M., Di Chiara G. and Consolo S. (1992) Dopamine depletion preferentially impairs D1- over D2-receptor regulation of striatal in vivo acetylcholine release. *Journal of Neurochemistry* **59**, 353–7.

Bertorello A.M., Hopfield J.F., Aperia A. and Greengard P. (1990) Inhibition by dopamine of (Na-K) ATPase activity in neostriatal neurons through D1 and D2 dopamine receptor synergism. *Nature (London)* **347**, 386–388.

Bolam J.P. and Bennett B.D. (1995) Microcircuitry of the neostriatum. In: M.A. Ariano and D.J. Surmeier (eds) *Molecular and Cellular Mechanisms of the Neostriatum* Austin, Texas, R.G. Landes, pp. 1–20.

Calabresi P., Maj R., Mercuri N.B. and Bernardi G. (1992) Coactivation of D1 and D2 dopamine receptors is required for long-term synaptic depression in the striatum. *Neuroscience Letters* **136**, 1–5.

Calabresi P., Mercuri N., Stanzione P., Stefani A. and Bernardi G. (1987) Intracellular studies on the dopamine-induced firing inhibition of neostriatal neurons in vitro: Evidence for D1 receptor involvement. *Neuroscience* **20**, 757–771.

Cepeda C., Buchwald N.A. and Levine M.S. (1993) Neuromodulatory actions of dopamine in the neostriatum are dependent upon the excitatory amino acid receptor subtypes activated. *Proceedings of the National Academy of Science, U.S.A.* **90**, 9576–9580.

Cepeda C., Chandler S.H., Shumate L.W. and Levine M.S. (1995) Persistent Na+ conductance in medium-sized neostriatal neurons: characterization using infrared videomicroscopy and whole cell patch-clamp recordings. *Journal of Neurophysiology* **74**, 1343–8.

Draguhn A., Verdorn T.A., Ewert M., Seeburg P.H. and Sakmann B. (1990) Functional and molecular distinction between recombinant rat GABAA receptor subtypes by $Zn^{2+}$. *Neuron* **5**, 781–8.

Engber T.M., Susel Z., Kuo S. and Chase T.N. (1990) Chronic Levodopa treatment alters basal and dopamine agonist-stimulated cerebral glucose utilization. *Journal of Neuroscience* **10**, 3889–3895.

Gerfen C.R. (1992) The neostriatal mosaic: multiple levels of compartmental organization in the basal ganglia. *Annual Review of Neuroscience* **15,** 285–320.

Graybiel A.M. (1990) Neurotransmitters and neuromodulators in the basal ganglia. *Trends in Neuroscience* **13**, 244–254.

Graybiel A.M., Aosaki T., Flaherty A.W. and Kimura M. (1994) The basal ganglia and adaptive motor control. *Science, New York* **265**, 1826–1831.

Greengard P. and Browning M.D. (1988) Studies of the physiological role of specific neuronal phosphoproteins. *Advances in Second Messenger and Phosphoprotein Research* **21**, 133–46.

Halpain S., Girault J.A. and Greengard P. (1990) Activation of NMDA receptors induces dephosphorylation of DARPP–32 in rat striatal slices. *Nature (London)* **343**, 369–72.

Hernandez-Lopez S., Bargas J., Surmeier D.J., Reyes A. and Galarraga E. (1997) D1 receptor activation enhances evoked discharge in neostriatal medium spiny neurons by modulating an L-type Ca2+ conductance. *Journal of Neuroscience* **17**, 3334–42.

Hille B. (1994) Modulation of ion-channel function by G-protein-coupled receptors. *Trends in Neuroscience* **17**, 531–536.

Hornykiewcz O. (1973) Dopamine in the basal ganglia. Its role and therapeutic implications. *British Medical Bulletin* **29**, 172–178.

Howe A.R. and Surmeier D.J. (1995) Muscarinic receptors modulate N-, P- and L-type $Ca^{2+}$ currents in rat striatal neurons through parallel pathways. *Journal of Neuroscience* **15**, 458–469.

Hubbard M.J. and Cohen P. (1993) On target with a new mechanism for the regulation of protein phosphorylation. *Trends in Biochemical Science* **18**, 172–176.

Kano M. and Konnerth A. (1992) Potentiation of GABA-mediated currents by cAMP-dependent protein kinase. *NeuroReport* **3**, 563–6.

Kawaguchi Y. (1993) Physiological, morphological and histochemical characterization of three classes of interneurons in rat neostriatum. *Journal of Neuroscience* **13**, 4908–4923.

Lehmann J. and Langer S.Z. (1983) The striatal cholinergic interneuron: Synaptic target of dopaminergic terminals? *Neuroscience* **10**, 1105–1120.

LeMoine C., Tison F. and Bloch B. (1990) D2 dopamine receptor gene expression by cholinergic neurons in the rat striatum. *Neuroscience Letters* **117**, 248–252.

Mengual E., Casanovas-Aguilar C., Perez-Clausell J. and Gimenez-Amaya J.M. (1995) Heterogeneous and compartmental distribution of zinc in the striatum and globus pallidus of the rat. *Neuroscience* **66**, 523–37.

Mirenowicz J. and Schultz W. (1996) Preferential activation of midbrain dopamine neurons by appetitive rather than aversive stimuli. *Nature (London)* **379**, 449–51.

Nemeroff C.B. and Bissette G. (1988) Neuropeptides, dopamine and schizophrenia. *Annals of the New York Academy of Sciences* **537**, 273–291.

Nestler E.J. and Aghajanian G.K. (1997) Molecular and cellular basis of addiction. *Science, New York* **278**, 58–63.

Nygaard T.G., Marsden C.D. and Duvoisin R.C. (1988) Dopa-responsive dystonia. *Advances in Neurology* **50**, 377–84.

Ohno Y., Sasa M. and Takaori S. (1987) Coexistence of inhibitory dopamine D–1 and excitatory D–2 receptors on the same caudate nucleus neurons. *Life Science* **40**, 1937–1945.

Pacheco-Cano M.T., Bargas J., Hernandez-Lopez S., Tapia D. and Galarraga E. (1996) Inhibitory action of dopamine involves a subthreshold Cs+-sensitive conductance in neostriatal neurons. *Experimental Brain Research* **110**, 205–211.

Pierce R.C. and Rebec G.V. (1995) Iontophoresis in the neostriatum of awake, unrestrained rats: differential effects of dopamine, glutamate and ascorbate on motor- and nonmotor-related neurons. *Neuroscience* **67**, 313–324.

Schiffmann S.N., Lledo P.M. and Vincent J.D. (1995) Dopamine Dl receptor modulates the voltage-gated sodium current in rat striatal neurones through a protein kinase A. *Journal of Physiology, London* **483**, 95–107.

Schultz W. (1994) Behavior related activity of primate dopamine neurons. *Revue Neurologique, Paris* **150**, 634–9.

Seeman P., Guan H.C., Nobrega J., Jiwa D., Markstein R., Balk J.H., Picetti R., Borrelli E. and Van Tol H.H. (1997) Dopamine D2-like sites in schizophrenia, but not in Alzheimer's, Huntington's, or control brains, for [3H]benzquinoline. *Synapse* **25**, 137–46.

Sibley D.R. (1995) Molecular biology of dopamine receptors. In: M.A. Ariano and D.J. Surmeier (eds.) *Molecular and cellular mechanisms of neostriatal function*. Austin, Texas, R.G. Landes, pp 255–272.

Singer H.S. and Walkup J.T. (1991) Tourette syndrome and other tic disorders. Diagnosis, pathophysiology, and treatment. *Medicine* **70**, 15–32.

Song W.-J, Tkatch T., Baranauskas G. and Surmeier D.J. (1997) Somatodendritic depolarization-activated potassium currents in rat neostriatal cholinergic interneurons are predominantly of the A-type and attributable to co-expression of Kv4.2 and Kv4.1 subunits. *Journal of Neuroscience* **18**, 3124–3137.

Stoof J.C., Drukarch B., de Boer P., Westerink B.H. and Groenewegen H.J. (1992) Regulation of the activity of striatal cholinergic neurons by dopamine. *Neuroscience* **47**, 755–70.

Stoof J.C. and Kebabian J.W. (1984) Two dopamine receptors: Biochemistry, physiology and pharmacology. *Life Science* **35**, 2281–2296.

Surmeier D.J., Bargas J., Hemmings Jr. H.C., Nairn A.C. and Greengard P. (1995) Modulation of calcium currents by a D1 dopaminergic protein kinase/phosphatase cascade in rat neostriatal neurons. *Neuron* **14**, 385–397.

Surmeier D.J., Eberwine J., Wilson C.J., Stefani A. and Kitai S.T. (1992) Dopamine receptor subtypes colocalize in rat striatonigral neurons. *Proceedings of the National Academy of Science, U.S.A.* **89**, 10178–10182.

Surmeier D.J., Song W.-J. and Yan. Z. (1996) Coordinated expression of dopamine receptors in neostriatal medium spiny neurons. *Journal of Neuroscience* **16**, 6579–6591.

Uchimura N., Higashi H. and Nishi S. (1986) Hyperpolarizing and depolarizing actions of dopamine via D–1 and D–2 receptors on nucleus accumbens neurons. *Brain Research* **375**, 368–372.

Vincent S.R. and Semba K. (1989) A heavy metal marker of the developing striatal mosaic. *Brain Research (Developmental Brain Research)* **45**, 155–9.

Yan Z., Song W.-J. and Surmeier D.J. (1997) D2 dopamine receptor reduce N-type $Ca^{2+}$ currents in rat neostriatal cholinergic interneurons though a membrane-delimited, protein kinase C-insensitive pathway. *Journal of Neurophysiology* **78**, 1003–1015.

Yan Z. and Surmeier D.J. (1996) Muscarinic (m2/m4) receptors reduce N- and P-type Ca2+ currents in rat neostriatal cholinergic interneurons through a fast, membrane-delimited, G-protein pathway. *Journal of Neuroscience* **16**, 2592–2604.

Yan Z. and Surmeier D.J. (1997) D5 Dopamine receptors Enhance $Zn^{2+}$-sensitive $GABA_a$ currents in striatal cholinergic interneurons through a PKA/PP1 cascade. *Neuron* **19**, 1115–1126

# 12 Striatal Contention Scheduling and the Split Circuit Scheme of Basal Ganglia-Thalamocortical Circuitry: From Anatomy to Behaviour

**Daphna Joel and Ina Weiner***

*Department of Psychology, Tel Aviv University, Ramat-Aviv, Tel Aviv 69978 Israel*
*tel: (00972) 3-640-8993; FAX: (00972) 3-640-7391; e-mail: weiner@freud.tau.ac.il*
**Corresponding author*

While models of basal ganglia-thalamocortical circuitry have become central to research and theoretical approaches aimed at understanding normal and abnormal brain functions, the significance of this circuitry for information processing in the forebrain and the specific contribution of the striatum and the frontal cortex to such processing, have remained obscure. Furthermore, it has remained unclear how the different basal ganglia-thalamocortical circuits interact to produce integrated and coherent behaviour which includes motor, cognitive and limbic elements. The present model focuses on the contribution of the motor, associative and limbic basal ganglia-thalamocortical circuits to information processing in the forebrain. It describes (1) the striatal contribution to behaviour, (2) the interaction between corresponding striatal and frontal regions within each of the basal ganglia-thalamocortical circuits in the production of circuit-specific behaviour, and (3) the interaction between the different circuits in the production of coordinated behaviour which includes motor, cognitive and limbic components. For this purpose, the model brings together (1) Norman and Shallice's (1986) model of the control of action; (2) models of information processing in the striatum; and (3) an open-interconnected scheme of basal ganglia-thalamocortical circuitry (Joel and Weiner 1994, 1997). The major theme of the present model is that the basal ganglia-thalamocortical split circuits provide the brain machinery for the selection and execution of goal-directed routine behaviour, with routine behaviour being a product of striato-frontal interactions within each circuit and the coordinated interaction between the circuits.

KEYWORDS: Basal ganglia; basal ganglia-thalamocortical circuitry; Split circuits; Open-interconnected; Striatum; Frontal cortex; Information processing; Behaviour; Routine Behaviour; Contention scheduling; Supervisory Attentional System.

## 1. INTRODUCTION

The demonstration in the 1960s and 70s (Divac 1968; Divac, Rosvold and Scwarcbart, 1967; Rosvold 1972) that lesions to restricted striatal areas result in the same kind of behavioural deficits as lesions of the anatomically-associated frontocortical areas, has been seminal to the promulgation of the idea that the frontal cortex and the striatum operate in tandem to produce behavioural output. This idea has received a powerful impetus from the notion, pioneered by DeLong and Georgopoulos (1981) that anatomically and functionally associated regions of the striatum and the frontal cortex are linked within several basal ganglia-thalamocortical circuits.

To date, the most comprehensive "circuit" scheme has been presented by Alexander, DeLong and Strick (1986). According to this scheme, anatomically and functionally distinct frontocortical, basal ganglia, and thalamic areas are connected in several parallel segregated circuits, leading from a particular frontocortical area via specific regions of the striatum, the internal segment of the globus pallidus (GPi) (or the rat analogue, the entopeduncular nucleus), the substantia nigra *pars reticulata* (SNR), thalamic nuclei, and back to the cortical area from which each circuit originates. Based on the specific regions of the frontal cortex that contribute to the individual circuits, Alexander and colleagues (Alexander, Crutcher and DeLong, 1990; Alexander, DeLong and Strick, 1986) described five basal ganglia-thalamocortical circuits: two motor ("motor" and "oculomotor"), two associative ("dorsolateral prefrontal" and "lateral orbitofrontal"), and one limbic ("anterior cingulate"). Within each circuit, striatal output reaches the basal ganglia output nuclei (SNR and GPi) via a "direct" pathway, and via an "indirect pathway" which traverses the external segment of the globus pallidus (GPe) and the subthalamic nucleus (STN) (Albin, Young and Penney, 1989; Alexander, Crutcher and DeLong, 1990; Alexander and Crutcher, 1990; DeLong, Crutcher and Georgopoulos, 1985; Penney and Young, 1986).

While models of basal ganglia-thalamocortical circuitry have become central to research and theoretical approaches aimed at understanding normal and abnormal brain functions (e.g. Albin, Young and Penney, 1989; DeLong and Georgopoulos, 1981; Goldman-Rakic and Selemon, 1990; Gray *et al.*, 1991; Graybiel, 1990; Groenewegen *et al.*, 1990; Parent, 1990; Penney and Young, 1983,1986; Robbins and Brown, 1990; Swerdlow and Koob, 1987), the significance of the "loops" for information processing in the forebrain, and the specific contribution of the striatum and the frontal cortex to such processing, have remained obscure. Furthermore, it has remained unclear how the circuits interact to produce integrated and coherent behaviour which includes motor, cognitive and limbic elements. Indeed, the principle of parallel segregated organization poses a serious problem for the notion of between-circuits interaction.

It is our intention to offer a tentative model of the control of behaviour by the basal ganglia-thalamocortical circuits, with an emphasis on striato-frontal interaction. Of the gallery of functions attributed to the frontal cortex and striatum, one figures most prominently, namely, the sequential organization (or programming) of behaviour (e.g. Berridge, 1989; Berridge and Fentress, 1987; van der Bos and Cools, 1989; Cools, 1980; Cools *et al.*, 1984; Fuster, 1990; Gray *et al.*, 1991; Graybiel, 1990; Heimer, Switzer and Van Hosen, 1982; Iversen, 1984; Jaspers *et al.*, 1984; Marsden, 1982; Mogenson, Jones and Yim, 1980; Penney and Young, 1983; Phillips, Pfauss and Blaha, 1991; Robbins and Brown, 1990; Rolls and Williams, 1987). It is also accepted that the frontal cortex and the striatum have distinct contributions to this function, with the frontal cortex having a central role in flexible behaviour, planning and decision making (Case, 1992; Fuster, 1990; Goldman-Rakic, 1987a,b; Karnath and Wallesch, 1992; Kolb and Whishaw, 1990; Levine, Leven and Prueitt, 1992; Luria, 1973; Milner, 1963; Pribram, 1973; Shallice, 1982; Shallice and Burgess, 1991; Stuss, 1992), and the striatum subserving routine or automatic aspects of behaviour (e.g. Berridge, 1989; Berridge and Fentress, 1987; Berridge and Whishaw, 1992; Cools, 1980; Cools *et al.*, 1984; Gabrieli, 1995; Graybiel *et al.*, 1994; Hikosaka, 1994; Iversen, 1984; Jaspers *et al.*, 1984; Levine, Leven and Prueitt, 1992; Marsden, 1982;

Marsden and Obeso, 1994; Miller and Wickens, 1991; Norman and Shallice, 1986; Robbins and Brown, 1990; Rolls and Williams, 1987).

Taking as its starting point the premise that the striatum and the frontal cortex subserve automatic versus non-automatic aspects of behaviour, respectively, the present model describes (1) the striatal contribution to behaviour, (2) the interaction between corresponding striatal and frontal regions within each of the basal ganglia-thalamocortical circuits in the production of behaviour, and (3) the interaction between the different circuits in the production of co-ordinated behaviour which includes motor, cognitive and limbic components. For this purpose, the model brings together (1) Norman and Shallice's (1986) model of the control of action; (2) models of information processing in the striatum; and (3) an open-interconnected scheme of basal ganglia-thalamocortical circuitry (Joel and Weiner, 1994, 1997)[1]. The model is based on data from rats and primates.

## 2. CONTROL OF BEHAVIOUR: NORMAN AND SHALLICE'S MODEL

Norman and Shallice's (1986) model posits two levels of behavioural control. Contention scheduling is a mechanism responsible for the automatic selection of schemas, i.e., motor and cognitive actions. A schema is selected once its activation level exceeds a threshold. The activation level of a schema is influenced by the satisfaction of trigger conditions, thus allowing for precise environmental control of performance. Activated schemas compete with one another, and the competition is effected through lateral activation and inhibition. When automatic response selection is not possible, such as in situations which are ill-learned, dangerous or difficult, contain novel sequences of action or involve planning and decision-making, a Supervisory Attentional System (SAS) is called into action. This system can influence the activation level of schemas, thus biasing their selection by the contention scheduling mechanism (Norman and Shallice, 1986; Shallice, 1982). In line with the widely held views of the roles of the frontal cortex and striatum in behaviour, Norman and Shallice (1986; Shallice, 1982; Shallice and Burgess, 1991) attributed contention scheduling to the striatum, and the supervisory process to the frontal cortex.

## 3. STRIATAL CONTRIBUTION TO BEHAVIOUR

### 3.1. The Striatum and Contention Scheduling

The striatum is thought to select motor routines according to the current environmental, behavioural and motivational state, and this selection is molded by reward-driven association learning (Arbib and Dominey, 1995; Graybiel and Kimura, 1995; Houk and Wise, 1995; Kimura, 1987, 1995; Pennartz, Groenewegen and Lopes da Silva, 1994; Plenz and Aertsen, 1994; Schultz *et al.*, 1995a,b; Wickens and Kötter, 1995). Recent models of information processing in the striatum point to several characteristics of the neural

[1] It should be noted that our model does not incorporate information processing within the output nuclei of the basal ganglia (SNR and GPi) and the thalamus. This is not to imply that these structures do not contribute to information processing within the basal ganglia-thalamocortical circuits.

organization of the striatum and its cortical and dopaminergic inputs which may subserve such functions (Groves *et al.*, 1995; Houk, 1995; Houk and Wise, 1995; Houk, Adams and Barto, 1995; Miller and Wickens, 1991; Pennartz, Groenewegen and Lopes da Silva, 1994; Plenz and Aertsen, 1994; Wickens and Kötter, 1995). First, the striatum is innervated by virtually all cortical areas, and thus receives perceptual, cognitive and limbic information. Second, the orderly divergence of cortical projections into several striatal zones results in different combinations of cortical inputs at different striatal zones (Flaherty and Graybiel, 1993, 1994; Gerfen and Wilson, 1996; Graybiel *et al.*, 1994; Graybiel and Kimura, 1995; Pennartz, Groenewegen and Lopes da Silva, 1994; Wickens and Kötter, 1995). Third, due to the sparse cortical innervation of each striatal neurone and the electrophysiological properties of striatal neurones, a temporally and/or spatially synchronized input from a relatively large subset of the cortical afferents is required for the activation of a striatal neurone (Gerfen and Wilson, 1996; Houk, 1995; Pennartz, Groenewegen and Lopes da Silva, 1994; Wilson, 1995; see also chapter 6 of this volume).

These characteristics have led to the suggestion that striatal neurones are well suited for the detection of specific cortical contexts, i.e., specific activity patterns in different frontal, posterior and limbic cortical regions, which represent a particular behavioural, perceptual, or motivational state (Houk, 1995; Houk and Wise, 1995; Houk, Adams and Barto, 1995; Pennartz, Groenewegen and Lopes da Silva, 1994; Plenz and Aertsen, 1994; Schultz *et al.*, 1995a). In addition, these neurones and their circuits may allow the formation of associations between different sets of cortical inputs, in particular, between sensory (somatosensory, visual and auditory) information and motor output (Flaherty and Graybiel, 1994; Kimura, 1987; Pennartz, Groenewegen and Lopes da Silva, 1994; Schultz *et al.*, 1995a; Wickens and Kötter, 1995).

Associative learning in the striatum is considered to take place via long term changes in corticostriatal synaptic efficacy, guided by a reinforcement signal provided by the dopaminergic input to the striatum (Groves *et al.*, 1995; Graybiel *et al.*, 1994; Houk, 1995; Houk, Adams and Barto, 1995; Kimura, 1995; Miller and Wickens, 1991; Pennartz, 1995; Pennartz, Groenewegen and Lopes da Silva, 1994; Schultz *et al.*, 1995a,b; Wickens, 1990; Wickens and Kötter, 1995). More specifically, learning occurs when, in a specific cortical context, a set of striatal neurones is activated and the resulting behaviour leads to favourable outcomes to the organism, which are signalled to the striatum by increased dopaminergic input (see Chapter 1 by Hyland; Chapter 4 by Wickens). As a result, the activated corticostriatal synapses onto these activated striatal neurones are strengthened, so that in the future the same set of neurones is more likely to be activated, that is, the same behaviour is more likely to be selected and executed, in the same or a similar context. Such a reward-based learning mechanism ensures that the striatum selects the "most appropriate" behaviour, i.e., behaviour which according to past experience is expected to maximize reward in a given environmental situation (Wickens and Kötter, 1995). The dopaminergic input may also enable the execution of well-learned behaviours (Graybiel and Kimura, 1995; Kimura, 1995; Koob, 1996; Sabol *et al.*, 1985; but see Schultz *et al.*, 1995b).

The extensive inhibitory axon collaterals of the striatal projection neurones—which can suppress the firing of activated striatal neurones (Wilson, 1995)—or feed-forward inhibition by striatal interneurones (Kita, 1996), were suggested to subserve competition between activated striatal projection neurones, thus restricting the number of activated neurones in a given situation to a small subset (Groves, 1983; Miller and Wickens, 1991;

Pennartz, Groenewegen and Lopes da Silva, 1994; Penney and Young, 1983; Swerdlow and Koob, 1987; Wickens and Kötter, 1995; Wilson, 1995). This facilitates selection of the most activated neurones, which in well-learned situations will lead to the selection of the most appropriate response, while in novel situations it will limit learning to a small set of striatal neurones (Barto, 1995; Groves *et al.,* 1995; Houk, Adams and Barto, 1995; Pennartz, Groenewegen and Lopes da Silva, 1994; Wickens and Kötter, 1995).

While facilitation of appropriate responses is attributed to striatal neurones of the direct pathway, striatal neurones of the indirect pathway are thought to suppress unwanted responses (e.g. Alexander and Crutcher, 1990; Hikosaka, 1994; Jackson and Houghton, 1995; Kimura, 1995; Marsden and Obeso, 1994; Matsumura *et al.*, 1992; Penney and Young, 1986; Smith *et al.*, 1994). Houk and Wise (1995) have suggested that the striatum may "learn" to select not only the most appropriate movement to be executed, but also the movements that should be suppressed. More specifically, in a given cortical context, some striatal neurones of both the direct and the indirect pathways are activated. Activated neurones of the direct pathway facilitate some movements, while activated neurones of the indirect pathway suppress some other movements. If the resultant behaviour leads to favourable outcomes, active corticostriatal synapses onto both types of neurones will be strengthened, so that in the future, in this or a similar context, the same set of neurones is more likely to be activated, simultaneously facilitating wanted movements and suppressing unwanted movements.

Activity of a subset of striatal neurones, including neurones of the direct and indirect pathways, leads to a specific pattern of inhibition and excitation of a corresponding subset of neurones in GPi/SNR. This results in disinhibition of a corresponding subset of thalamic neurones. The latter will in turn activate frontocortical neurones whose activity patterns represent actions (in motor and premotor cortical regions) and motor programs (in the prefrontal cortex [PFC]) (Arbib and Dominey, 1995; Houk, 1995; Houk and Wise, 1995; Levine, Leven and Prueitt, 1992; Miller and Wickens, 1991; Schultz *et al.*, 1995a). Thalamic neurones synapse either on frontocortical neurones which were part of the cortical context which had activated the striatal neurones, or on different frontocortical neurones. In the former case, the activity of striatal neurones contributes to sustained activity in a group of frontocortical neurones which were part of the original cortical context. In the latter, the activity of striatal neurones activates a group of frontocortical cells which were not part of the original cortical context, and in this manner, provides a mechanism whereby one cortical context (that which originally activated the striatal cells) leads to the activation of another cortical context. Such sequential activation of subgroups of frontal neurones may provide a mechanism for the sequencing of behaviour by the striatum.

*Electrophysiological Evidence:*

It is widely documented that striatal neurones fire in relation to the initiation and execution of well-learned movements, both externally- and self-generated (e.g. Aldridge, Jaeger and Gilman, 1991; Brotchie, Iansek and Horne, 1991a,b; DeLong, Crutcher and Georgopoulos, 1985; Kimura, 1987; Lidsky, Manetto and Schneider, 1985; Marsden and Obeso, 1994; Mink and Thach, 1991a,b,c; Nambu, Yoshida and Jinnai, 1990; Rolls and Williams, 1987; Schultz and Romo, 1988; Trouche *et al.*, 1994). Consistent with the suggestion that striatal neurones "learn" to produce a motor output in response to specific sensory inputs, is the finding that some striatal neurones respond to external stimuli only when they elicit movement (Kimura, 1987; Schultz and Romo, 1988; Schultz *et al.*,

1995a), and may thus code stimulus-response associations (Kimura, 1987). Interestingly, some striatal neurones were reported to fire in relation to the suppression of a movement in response to a sensory stimulus (Rolls and Johnstone, 1992; Schultz and Romo, 1988), in accord with the suggestion that striatal neurones of the direct and indirect pathways perform a similar function but neurones of the direct pathway "learn" to associate a given cortical context with the activation of a specific movement, while neurones of the indirect pathway "learn" to associate a given cortical context with the suppression of a specific movement. Recently, the context-dependency of the activity of striatal cells has become particularly apparent, as it has been shown that their activity is not simply related to certain responses or stimuli, but depends on different aspects of the behavioural situation (Houk, 1995; Kimura, 1995; Lidsky, Manetto and Schneider, 1985; Rolls and Johnstone, 1992; Rolls and Williams, 1987; Schultz *et al.*, 1995a).

### 3.2. Striatal Contention Scheduling and the Frontal Cortex

The above scheme of striatal functioning fits many of the characteristics of Norman and Shallice's (1986) contention scheduling mechanism. A schema corresponds to a set of striatal neurones, its trigger conditions are provided by the specific cortical context which activates this set of striatal neurones, and competition between activated striatal neurones (i.e., between activated schemas), takes place via their inhibitory axon collaterals and/or via feed-forward inhibition by striatal interneurones. The biasing effect of SAS on contention scheduling is implemented via the direct projections from the frontal cortex to the striatum.

However, striatal contention scheduling as conceived here departs from Norman and Shallice's model in several respects. First, the degree of activation of a given schema by a specific cortical context can be modified by experience, and this learning is guided by reinforcement. The incorporation of a learning mechanism into striatal contention scheduling has two important consequences: (a) Once the association between a particular cortical context and a particular behavioural output is learned, part of the cortical context may be sufficient to activate, that is, select this behaviour. It should be noted that since, in this case, the current cortical context is only partly similar to the original cortical context, the behaviour selected is expected to be modified according to the current cortical context. (b) As a result of reward-driven learning, the striatum selects the most appropriate behaviour in a given context, that is, the behaviour which according to past experience is expected to maximize reward.

The second major departure concerns the relations between the selection and execution of a schema. In Norman and Shallice's (1986) model the selection of a schema by contention scheduling entails its execution. The descending pathways from the striatum to the pallidum and nigra, and their subsequent projections to motor centres in the brainstem (such as the superior colliculus and pedunculopontine nucleus) (Graybiel and Kimura, 1995; Hikosaka, 1994; Mogenson *et al.*, 1993; Pennartz, Groenewegen and Lopes da Silva, 1994; Skinner and Garcia-Rill, 1993), may provide a route whereby the selection of a schema entails its execution. However, these projections are limited to relatively simple motor output (e.g. Hikosaka, 1994; Kalivas, Churchill and Klitenick, 1993; Mogenson, Jones and Yim, 1980; Mogenson *et al.*, 1993; Pennartz, Groenewegen and Lopes da Silva, 1994; Skinner and Garcia-Rill, 1993; Swerdlow *et al.*, 1993). The major striatal projections to the frontal lobes do not have direct influence on behavioural output but rather bias the selection of activity patterns of frontocortical neurones (Arbib and Dominey, 1995; Houk, 1995; Houk and Wise, 1995; Levine, Leven and Prueitt, 1992; Miller and Wickens, 1991; Schultz *et al.*, 1995a) towards the execution of the selected behaviour.

Although the striatum provides the frontal cortex with information about the most appropriate behaviour in the current behavioural, perceptual, cognitive and motivational context, this information does not necessarily translate into behavioural output. This is because the frontal cortex receives in addition to striatal information, information about the current context from other cortical regions. Therefore, the frontal cortex is continuously biased by the two inputs. Whether the actual behavioural output is the one selected by the striatum, depends on two factors: (1) the strength of the striatal biasing effect on the frontal cortex, which is a function of the degree of activation of striatal neurones, and (2) the degree of correspondence between the striatal and cortical biasing effects on the activity pattern in the frontal cortex.

In well-learned/highly familiar situations, striatal neurones which encode the most appropriate behaviour, as well as striatal neurones which encode incompatible behaviours, are expected to be strongly activated, thus facilitating via the direct pathway the selection of the most appropriate behaviour by the frontal cortex, and inhibiting via the indirect pathway the selection of inappropriate behaviours. This leads to a strong striatal biasing effect on the activity pattern of frontal neurones towards the execution of the routine behaviour for these situations. In addition, in familiar situations the inputs to the frontal cortex from other cortical regions are expected to bias the activity pattern in the frontal cortex in the same direction as the striatum. Thus, in well-learned/familiar situations there is a strong striatal biasing effect and a high degree of correspondence between the striatal and cortical biasing effects. The result is an effortless production of routine behaviour.

A mirror process occurs in novel or ill-learned situations. In such situations striatal neurones are expected to be weakly activated, and in addition, their (weaker) biasing effect on the activity of frontal neurones is unlikely to coincide with the biasing effects of other inputs to the frontal cortex. Under such circumstances the selection of behavioural output cannot be achieved automatically, but requires a supervisory process which will yield alternative ways of action.

Intermediate situations are those in which the cortical context is partly familiar, either because some components are missing or new components appear. In the former type of situations, striatal neurones will be activated, albeit less strongly than they would be in the fully familiar context, and their biasing effect will be similar to the cortical biasing effect. These biasing effects may be sufficient to automatically activate the routine behaviour for the fully familiar cortical context, most probably with some modifications suited to the current (partial) cortical context (see above). Striatal neurones may be strongly activated also in the latter type of situations (new components in a familiar context), but their biasing effect is less likely to correspond to the cortical biasing effect. Two outcomes can ensue when striatal and cortical biasing effects do not correspond. (1) The striatal biasing effect is sufficiently strong to lead to the execution of the routine behaviour. For example, what is usually referred to as a "capture error", i.e., performing the routine behaviour instead of a behaviour one intended to perform, is one such outcome. In this case, the added component in a familiar cortical context is the intention to behave differently, but the resulting change in the cortical biasing effect is not sufficient to counteract the strong biasing effect exerted by the striatum; as a result the routine behaviour is executed. (2) The cortical biasing effect may be sufficiently strong to counteract the striatal biasing effect. In this case the routine behaviour will not be performed, and the supervisory process will intervene.

The preceding analysis has implications for specifying when the supervisory process becomes involved in the production of behavioural output. Thus, the supervisory process ensues when routine behaviour is not possible (i.e., when the selection of behaviour by the striatum does not lead to its execution), yielding non-routine behaviour. Importantly, in the present scheme the frontal cortex is responsible for the execution of both routine and non-routine behaviour. As detailed above, the former is achieved via the interaction of the frontal cortex with the striatum. The latter involves interaction of the frontal cortex with other brain regions, with possible candidates being posterior association regions and high-order limbic cortices. Furthermore, since frontal output is continuously influenced by both striatal and cortical biasing effects, routine and non-routine behaviour are viewed here as lying on a continuum, rather than being two independent modes of operation.

### 3.3. Summary: The Interaction Between the Striatum and the Frontal Cortex

The striatum, which operates as a contention scheduling mechanism, selects the most appropriate behaviour in the current internal and external environment. As a result of reward-driven learning, the selected behaviour is the one that maximizes reward under these conditions, and in this sense is the "most appropriate". The output of this selection process is channelled to the frontal cortex where it acts to bias the selection of the actual behavioural output of the organism. Under well-learned/routine situations, the biasing effect of the striatum is strong and leads to the automatic selection and execution of this behaviour. In contrast, when a strong biasing effect of the striatum is in conflict with a strong cortical biasing effect, or when the striatal biasing effect is relatively weak, an automatic selection of behaviour is not possible, and if an action is made, it is the result of the supervisory process, subserved by the interaction of the frontal cortex with other brain regions. Importantly, the activity of the supervisory mechanism is itself a part of the current cortical context, and as such is channelled to the striatum, which selects the most appropriate way to carry out this action in the current situation, and biases the frontal cortex towards this way of action.

Continuous dynamic interplay, then, is a major feature of the striato-frontal interaction as conceived here. The striatum is continuously fed with information about the current perceptual, cognitive and motivational context from different cortical regions and with information about the current behavioural context from the frontal cortex. On the basis of this information the striatum selects the most appropriate behaviour and transmits it to the frontal cortex. The same cortical information is channelled to the frontal cortex together with information from the striatum about the most appropriate behaviour in this context. In this manner, there is a constant updating of the striatum with information from the frontal cortex and of the frontal cortex with information from the striatum. However, the output of neither is determined solely by this information, as both regions receive in addition information from other brain regions. Finally, the actual behaviour, routine and non-routine, is determined mainly by the frontal output: When routine behaviour is executed, frontal output is largely determined by its interaction with the striatum, while in other situations frontal output largely reflects its interactions with other brain regions.

## 4. STRIATO-FRONTAL INTERACTION AND BASAL GANGLIA-THALAMOCORTICAL CIRCUITS

Up to this point we have treated the striatum as a single entity. The projections from motor, associative, and limbic cortical areas, however, impose a functional organization upon the striatum, so that the striatum can be divided into motor, associative and limbic territories, with high degree of convergence of cortical information within each territory, but a high degree of segregation between the territories (DeLong and Georgopoulos, 1981; Haber, Lynd-Balta and Spooren, 1994; Parent and Hazrati, 1995a). Furthermore, these anatomo-functional striatal subterritories are components of three anatomically and functionally corresponding motor, associative and limbic basal ganglia-thalamocortical circuits (DeLong and Georgopoulos, 1981; Haber, Lynd-Balta and Spooren, 1994; Joel and Weiner, 1994; Parent and Hazrati, 1995a).

It has been argued that the similar anatomical and physiological organization of the different basal ganglia-thalamocortical circuits suggests that they carry out a similar function but on different types of information, namely, motor, cognitive, or limbic (e.g. Alexander, DeLong and Strick, 1986; Alexander, Crutcher and DeLong, 1990; Gabrieli, 1995; Gray *et al.,* 1991; Jackson and Houghton, 1995; Strick, Dum, and Picard, 1995). Consistent with this notion, we propose that the striato-frontal interaction described above takes place within each circuit. More specifically, each circuit-engaged striatal subdivision subserves contention scheduling of the circuit-specific behaviour (motor, cognitive or limbic), and biases activity patterns in the corresponding frontal subregion towards the execution of the selected behaviour. In addition, we propose that the integration of circuit-specific behavioural elements (motor, cognitive or limbic) into coherent behavioural output is subserved by the connections between the circuits. To date, the split circuit scheme of basal ganglia-thalamocortical organization is the only scheme in which interaction is inherent in the neural architecture of the basal ganglia-thalamocortical circuits. In the following sections, we will describe the anatomical organization of the motor, associative and limbic split circuits, and their joint operation in the production of routine and non-routine behaviour.

### 4.1. The Split Circuit Scheme: Anatomy

A schematic diagram of a split circuit is presented in Figure 12.1. A split circuit contains one frontocortico-striatal projection and two striato-frontocortical pathways. One striato-frontocortical pathway re-enters the frontocortical area which is a source of cortical input to this striatal subregion and thus forms a closed circuit, which subserves segregated processing. The second striato-frontocortical pathway terminates in a frontocortical area which innervates a different striatal subregion and thus forms an open pathway, which subserves integrated processing. Thus, connectivity is inherent in the split circuits, in the sense that the same set of connections (striato-frontocortical) subserves either segregated or integrated processing, depending on the specific target of these connections (Joel and Weiner, 1994).

The open interconnected principle governs also the organization of the indirect pathways (i.e., the connections between the striatum, pallidum, nigra and STN; Joel and Weiner, 1997). Thus, there are two types of indirect pathways (see Figure 12.2). An indirect pathway which terminates in the same subregion as the direct pathway forms a closed

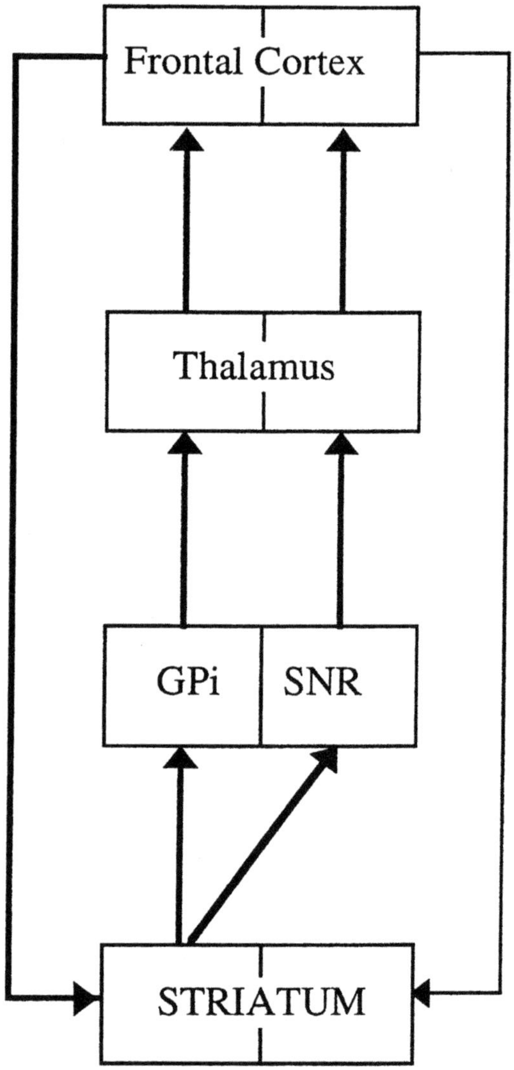


**Figure 12.1.** A schematic diagram of a split circuit containing a closed circuit, formed by the striato-frontocortical pathway that re-enters the frontocortical area that is the source of the frontocortico-striatal projections, and an open striato-frontocortical pathway that terminates in a different frontocortical area.

indirect pathway, and thus contributes to segregated processing. An indirect pathway which terminates in a subregion different from the direct pathway forms an open indirect pathway, and thus contributes to integrated processing. Thus, also the internal connections of the basal ganglia, which involve the STN, provide a neural substrate for the transfer of information within as well as between basal ganglia-thalamocortical circuits.

The organization of the connections of the striatum with the dopaminergic system also fits well the open interconnected principle. Thus, the basic design of these connections is that of a "loop" comprising the projections from a striatal subregion to a subregion of the

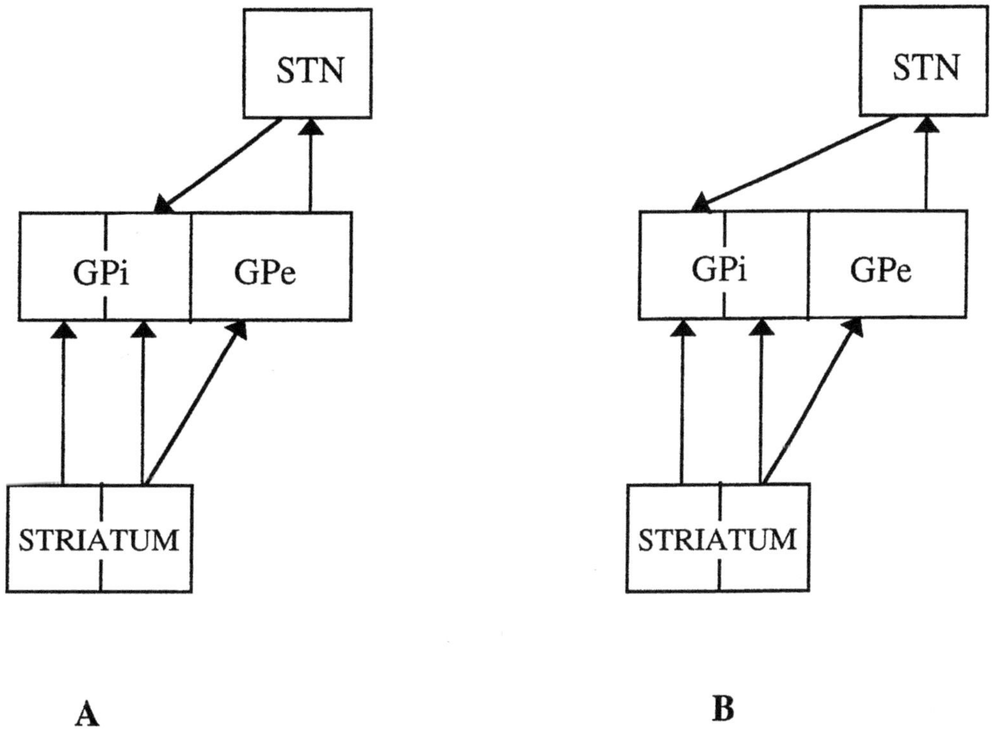


**Figure 12.2.** A schematic diagram of a direct and an indirect pathway. Both pathways originate in the same striatal subregion, but the former leads directly to its target in the output nuclei, whereas the latter traverses parts of GPe and STN before reaching the output nuclei. **2A**. A closed indirect pathway terminates in the same subregion in the output nuclei as the direct pathway arising from the same striatal subregion. **2B.** An open indirect pathway terminates in a subregion in the output nuclei different from the direct pathway arising from the same striatal subregion.

dopaminergic system and from this subregion to a striatal subregion. A loop which terminates in the striatal subregion from which it originates forms a "closed loop", and thus contributes to segregated processing. A loop which terminates in a striatal subregion different from that of its origin, forms an "open loop", and thus contributes to integrated processing (Joel and Weiner, submitted; for related views see, Groenewegen, Berendse and Wouterlood, 1994; Haber *et al.*, 1990; Haber, Lynd-Balta and Spooren, 1994; Jimenez-Castellanos and Graybiel, 1987, 1989; Lynd-Balta and Haber, 1994a,b; Mogenson, Jones and Yim, 1980; Nauta *et al.*, 1978; Parent and Hazrati, 1995a; Swanson and Mogenson, 1981).

On the basis of available anatomical data, and using a tripartite subdivision of the striatum, pallidum and STN into motor, associative and limbic subregions, we tentatively identified a motor, an associative, and a limbic split circuit, each containing a closed circuit and an open pathway (Figure 12.3). **The associative split circuit:** the associative striatum is innervated by the associative PFC (including the frontal eye field [FEF] and dorsolateral PFC), and in addition receives sensory information from higher order somatosensory, auditory and visual cortical regions, as well as from association cortices in the parietal and temporal lobes. One striato-frontocortical pathway, traversing SNR, reenters the associative PFC, thus forming a closed associative circuit. The other striato-frontocortical pathway, traversing the associative GPi, terminates in the premotor cortex (PMC), which projects to the motor striatum, thus forming an open associative pathway. **The motor split circuit:** the motor striatum is innervated by motor and premotor cortical regions, and in addition receives somatosensory, auditory and visual information from primary and secondary cortical regions. One striato-frontocortical pathway, traversing the motor GPi, re-enters the primary motor cortex and the supplementary motor area (SMA), thus forming a closed motor circuit. The other striato-frontocortical pathway, traversing the SNR, terminates in the associative PFC, thus forming an open motor pathway. Since only the striatonigral portion of this pathway belongs exclusively to the motor circuit, whereas the nigro-thalamo-cortical portion is also part of the associative split circuit we termed the striatonigral portion an "open motor route". **The limbic split circuit:** the limbic striatum is innervated by the limbic PFC (including the orbitofrontal cortex and anterior cingulate area) as well as by limbic regions (including the amygdala, the hippocampus and related cortical regions) and association cortices. One striato-frontocortical pathway, traversing the ventral pallidum (VP), re-enters the limbic PFC, thus forming a closed limbic circuit. The other striato-frontocortical pathway, traversing the SNR terminates in the associative PFC, thus forming an open limbic route. There may be an additional open limbic pathway, via the rostromedial GPi, which terminates in the motor/premotor cortices.

There is a closed indirect pathway within each of the closed circuits (i.e., motor, associative, and limbic) as well as within the open associative pathway. In addition, there is an open indirect pathway which connects the associative striatum to the motor GPi, via the associative GPe and motor STN, and there may be an additional open indirect pathway connecting the associative striatum to the VP, via the associative GPe and limbic STN.

The motor, associative and limbic striatum are the source of three closed loops with the dopaminergic system. In addition, the limbic striatum is also the source of two open loops, one terminating in the associative striatum and the other terminating in the motor striatum.

Figure 12.3 presents a summary diagram of the structural organization of the motor, associative, and limbic split circuits. As can be seen, this structural organization enables four modes of between-circuit interaction: via open pathway, via open route, via open indirect pathway, and via open loop. (A summary of the terminology we adopt is defined in the legend to Figure 12.3.) It is noteworthy that the three split circuits differ in their modes of connectivity. Thus, the motor split circuit is connected to the associative split circuit at the level of SNR (via the open motor route). The associative split circuit is connected to the motor split circuit at the level of the cortex and pallidum (via the open associative pathway and the open indirect pathway, respectively). The associative split circuit may be connected to the limbic split circuit at the level of the pallidum (via an open indirect pathway). The limbic split circuit is connected to the associative split circuit at the level of SNR and the striatum (via the open limbic route and an open loop, respectively), and to the motor split circuit at the level of the striatum (via an open loop) and maybe also at the level of the cortex (via an open limbic pathway). These differences in the levels of connectivity between the split circuits are likely to have functional significance (see below).

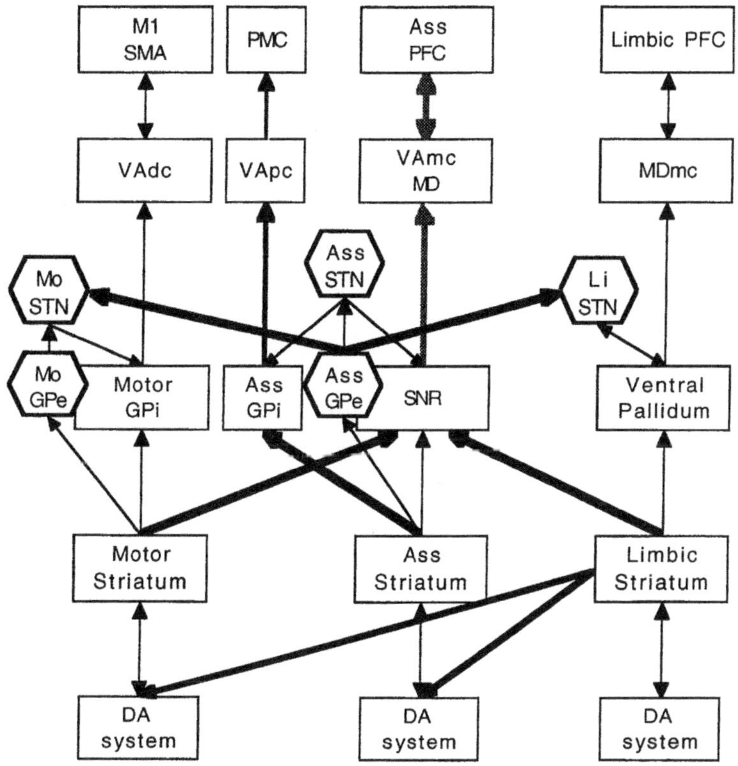


**Figure 12.3.** A summary diagram of the structural organization of the motor, associative, and limbic split circuits. Each split circuit contains a closed circuit and an open pathway or an open route. Included within each of the closed circuits as well as within the open associative pathway is a direct and a closed indirect pathway. In addition, the associative split circuit contains an open indirect pathway which connects it with the motor split circuit, and possibly an open indirect pathway which connects it with the limbic split circuit. Each split circuit has a closed loop with the DA system, and in addition there are two open loops connecting the limbic split circuit with the motor and the associative split circuits. Pathways connecting between circuits are demarcated in thick lines.

The external segment of the globus pallidus and the subthalamic nucleus appear within hexagons for clarity of presentation. Abbreviations used in the figure: Ass: associative; DA: dopamine; M1: primary motor cortex; MD: mediodorsal thalamic nucleus; MDmc: MD, magnocellular subdivision; Mo: motor; Li: limbic; VAdc: ventral anterior thalamic nucleus, densicellular subdivision; VAmc: ventral anterior thalamic nucleus, magnocellular subdivision; VApc: ventral anterior thalamic nucleus, parvicellular subdivision.

**Terminology:**

*Closed circuit*: A striato-frontocortical pathway that reenters the frontocortical area which is the source of cortical input to this striatal subregion.

*Open pathway*: A striato-frontocortical pathway that terminates in a frontocortical area which innervates a different striatal subregion.

*Open route*: The striatonigral portion of an open pathway.

*Closed indirect pathway*: An indirect pathway (striatum-GPe-STN-GPi/SNR) which connects functionally corresponding subregions of the basal ganglia, that is, which terminates in the same subregion of the basal ganglia output nuclei as the direct pathway

*Open indirect pathway*: An indirect pathway (striatum-GPe-STN-GPi/SNR) which connects functionally non-corresponding subregions of the basal ganglia, that is, which terminates in a different subregion of the basal ganglia output nuclei than the direct pathway.

*Closed loop*: A loop (striatum-DA system-striatum) which terminates in the striatal subregion from which it originates.

*Open loop*: A loop (striatum-DA system-striatum) which terminates in a different striatal subregion than that from which it originates.

### 4.2. The Split Circuit Scheme: Function

We suggest that each of the three (motor, associative and limbic) split circuits subserves the selection and execution of behavioural output in a given cortical context. The limbic split circuit receives information about the external and internal environment, and selects goals. The associative split circuit receives highly processed sensory information as well as information about the goals of the organism, and selects and executes motor programs to achieve these goals. The motor split circuit receives sensory and motor information, including information on specific motor steps of the current motor program, and selects and executes the simple motor acts needed to achieve these motor steps. The connections within each split circuit subserve the transfer of information about the most appropriate, circuit-specific behaviour (goal, motor program, or motor act) from the striatum to the frontal cortex, and about the intended behaviour from the frontal cortex to the striatum. Via the connections between the split circuits, the different aspects of behaviour, which are subserved by the different circuits, are integrated to produce coherent behavioural output.

#### *4.2.1. The motor and associative split circuits*

A major function of the striatum is the selection and initiation of the elements of a motor program, as well as their serial ordering into a co-ordinated sequential pattern (e.g. Aldridge, Jaeger and Gilman, 1991; Brotchie, Iansek and Horne, 1991a,b; DeLong, Crutcher and Georgopoulos, 1985; Hikosaka, 1991; Kimura, 1987; Lidsky, Manetto and Schneider, 1985; Marsden, 1982; Marsden and Obeso, 1994; Mink and Thach, 1991a,b,c; Nambu, Yoshida and Jinnai, 1990; Robbins and Brown, 1990; Rolls and Johnstone, 1992; Rolls and Williams, 1987; Trouche *et al.*, 1994). In recent years, it has become widely accepted that motor programs include also cognitive components, since, at certain points along the motor sequence, memory must be consulted, attention must be switched to a certain location or stimulus ('programmed attention' in Gray *et al.*'s [1991] terminology), etc. The sequencing of the motor and cognitive components of motor programs are usually ascribed to the motor and associative regions of the basal ganglia, respectively (Ballard, Hayhoe and Pelz, 1995; Berridge and Whishaw, 1992; Gabrieli, 1995; Gray *et al.*, 1991; Graybiel *et al.*, 1994; Hikosaka, 1994; Jackson and Houghton, 1995; Miller and Wickens, 1991). However, the way in which co-ordinated execution of both components is achieved has not been detailed.

##### *4.2.1.1. The motor split circuit*

The motor striatum receives information about relatively simple sensory stimuli from primary and sensory cortical regions, and about current and intended movements from motor and premotor cortical regions. Its output is directed to the primary motor cortex and SMA; both regions project directly to the spinal cord, and are involved in the production of movement. Thus, the motor striatum may serve to select the most appropriate movement in the context of current environmental stimuli, current movement and intended movement. Following the selection by the motor striatum, wanted movements are facilitated via the direct pathway, and unwanted movements are suppressed via the indirect pathway. Once the association between a particular cortical context and a particular movement is learned, part of the cortical context may be sufficient to activate, that is, select the movement. Thus, the movement may be triggered under several circumstances: (1) When the particular environmental stimuli occur. This is in line with the striatal role in the formation of habits, i.e., stimulus-response associations (Flaherty and Graybiel, 1994; Kimura, 1987; Schultz *et al.*, 1995a; Wickens and Kötter, 1995); (2) When a movement occurs. This is in line with the striatal role in the sequencing of movements, since performance of one movement triggers the following movement; (3) When the intention to act occurs (see next section).

*4.2.1.2. The associative split circuit*

The associative striatum is innervated by the associative PFC (i.e., the dorsolateral PFC and FEF), which is involved in planning, attention and working memory (Arbib and Dominey, 1995; Gabrieli, 1995; Goldman-Rakic, 1995; Houk, 1995; Jackson and Houghton, 1995; Levine, Leven and Prueitt, 1992; Schultz *et al.*, 1995a), and by cortical regions involved in high levels of unimodal and multimodal sensory processing. Its output is directed, via the closed associative circuit, to the associative PFC, and via the open associative pathway, to the PMC, which projects to the motor striatum. Thus, the architecture of the associative split circuit is ideally suited for sequencing both the cognitive and motor components of a motor program, via the closed associative circuit and the open associative pathway, respectively.

The involvement of the closed associative circuit, namely, the striato-nigro-thalamo-cortical projections to the associative PFC, in cognitive processes is well documented. Thus, the projections to the dorsolateral PFC and FEF contribute to working memory and to the selection of plans (Alexander, Crutcher and DeLong, 1990; Arbib and Dominey, 1995; Gabrieli, 1995; Goldman-Rakic, 1995; Houk, 1995; Levine, Leven and Prueitt, 1992; Schultz *et al.*, 1995a), as well as to attention (Alexander, Crutcher and DeLong, 1990; Arbib and Dominey, 1995; Jackson and Houghton, 1995), respectively. In addition, SNR projections to the superior colliculus have been suggested to contribute to the direction of saccades (Alexander, Crutcher and DeLong, 1990; Arbib and Dominey, 1995; Graybiel and Kimura, 1995; Hikosaka, 1991). As for the involvement of the associative split circuit in motor processes, we have previously suggested that the open associative pathway, namely, the striato-pallido-thalamo-cortical projections to the PMC, directs the execution of the motor steps of motor programs by the motor circuit (Joel and Weiner, 1994) (see next section).

*4.2.1.3. The co-ordinated execution of motor programs*

The associative striatum is involved in the selection of the motor program which (based upon past experience) is the most suitable for the current situation, and in the co-ordinated execution of the motor and cognitive elements of this program. Once the association between a particular environment, i.e., cortical context, and a particular schema, i.e., motor program, is learned, part of the cortical context may be sufficient to activate, that is, select the motor program. Thus, the motor program may be triggered under the following circumstances: (1) By particular environmental stimuli, including verbal information; (2) By particular thoughts, represented in the activity of posterior association cortices; and (3) By the intention to act, represented in the activity of the associative PFC.

The selection of a well-learned motor program by the associative striatum, i.e., activation of a subset of associative striatal neurones, leads, via the closed associative circuit, to the sequential activation of the cognitive elements in the associative PFC, and via the open associative pathway, to the sequential activation of the corresponding motor steps in the PMC. The closed indirect pathway of the closed associative circuit assists in the execution of the cognitive elements by suppressing inappropriate cognitive elements of the executed program as well as elements of competing motor programs. The closed indirect pathway of the open associative pathway aids in selecting the appropriate motor step of the program by suppressing inappropriate motor steps of the selected program as well as motor steps of competing motor programs. Information about the subsequent motor step, encoded in the PMC, is transferred to the motor striatum which also receives information from motor and sensory cortical areas about the current perceptual and motor state of the organism. The motor striatum selects the best set of specific movements to achieve this motor step in the current situation (see a similar suggestion in Levine, Leven and Prueitt, 1992). In this way the motor program encoded in the associative striatum does not need to encode specific movements but rather encodes specific motor steps, while the motor details for achieving each motor step are selected by the motor striatum according to the current situation and past experience (see a similar suggestion in Mogenson, Jones and Yim, 1980).

The open indirect pathway, which connects the associative striatum to the motor split circuit, assists in the ending of each motor step of the current motor program, so that the next step can be initiated, thus enabling the smooth transition between different steps in the motor program. In this context it should be noted that the motor STN receives, in addition to projections from the associative striatum (via the associative GPe), projections from the primary motor cortex. The former provides the motor STN with information about the intended motor step, while the latter provides information as to the actual movement that has just been made. Such combined information seems essential for the decision when to terminate the execution of a motor step, so that the next motor step can be initiated.

Via the open motor route, which connects the motor striatum to SNR, the motor striatum may fine tune the activity of SNR neurones according to the current status of the motor step executed. In this way temporal deviations in the execution of the motor steps will not result in desynchronization between the cognitive and motor components of the motor program.

Via the open limbic route, which connects the limbic striatum to SNR, the limbic striatum may bias nigral output according to the current goal of the organism, and also may override the execution of the current motor program under certain conditions, such as occurrence of a novel or surprising event (see also below).

In sum, it is suggested that the interconnected architecture of the motor and associative split circuits subserves the co-ordinated execution of motor programs. The closed associative circuit and the closed motor circuit subserve the execution of the cognitive and motor elements of the program, respectively, while their open components subserve the co-ordination between the two circuits; the open associative pathway transfers information from the associative striatum to the motor circuit about the next motor step, the open indirect pathway enables smooth transition between steps, and the open motor route transfers information from the motor striatum to the associative circuit about the status of the execution of the current motor step, so as to keep the execution of the motor and cognitive steps synchronized.

#### *4.2.1.4. Electrophysiological evidence*

Single cell recordings in the striatum have revealed that response patterns of neurones in the motor parts of the striatum and pallidum are more related to performance of single movements, while response patterns of neurones in the associative parts of the striatum and pallidum are more related to performance of chains of movements (e.g. Aldridge,

Jaeger and Gilman, 1991; Alexander, DeLong and Strick, 1986; Alexander, Crutcher and DeLong, 1990; Brotchie, Iansek and Horne, 1991a,b; DeLong, Crutcher and Georgopoulos, 1985; Kimura, 1987; Marsden and Obeso, 1994; Mink and Thach, 1991a, b, c; Nambu, Yoshida and Jinnai, 1990; Trouche *et al.,* 1994). In addition, the activity of neurones in the associative striatum has been found to be strongly related to switching of behavioural states in a well-learned task, while in the motor striatum the activity was more related to movements and to stimuli that trigger movement (Aldridge Jaeger, and Gilman, 1991; Kimura, 1987). These findings are consistent with the suggestion that the closed motor circuit is involved in the execution of movements, while the open associative pathway (originating from the associative striatum and traversing the associative GPi) is involved in sequencing motor steps. Recordings of associative striatal neurones revealed the following types: (i) Neurones which respond in relation to future events, that is, increase their discharge rate before predictable task-related signals in a well-learned task (Aldridge, Jaeger and Gilman, 1991; Schultz *et al.*, 1995a); (ii) neurones with activity related to instructional cues, that is, cues that inform the animal what to do when an additional (trigger) stimulus appears (Schultz and Romo, 1988; Schultz *et al.*, 1995a); (iii) neurones that fire in relation to the preparation of movements, including eye and head movements (Aldridge, Jaeger and Gilman, 1991; Rolls and Williams, 1987; Schultz and Romo, 1988); as well as (iv) neurones that change (increase or decrease) their firing rate before memory guided saccades (Hikosaka, 1991). Neurones with such response characteristics may be involved in functions attributed to the associative split circuit. This includes: recruiting (via the closed associative circuit) the dorsolateral PFC for the retention in working memory of this information; directing attention (via striatonigral projections and subsequent nigral projections to FEF and superior colliculus) to the location at which the next stimulus is expected to appear or the next movement should be made; or preparing (via the open associative pathway to the PMC) the motor response that should be released once the trigger stimulus appears. Finally, some associative striatal neurones fire before the spontaneous, i.e., internally generated, performance of a well-learned act (Schultz and Romo, 1988), in line with the suggestion that after an action has been well-learned, it can be triggered by the intention to act.

### *4.2.2. The limbic split circuit*

Until now we have presented a model in which the motor and associative striatum subserve contention scheduling. The relevant 'environment' for the associative striatum is highly processed sensory, motor and cognitive information and the "schemata" it selects are motor programs, including motor and cognitive components. The relevant "environment" for the motor striatum is lower-order sensory and motor information, including information on specific motor steps from the PMC, and the "schemata" it selects being simple motor acts. In both cases, as a result of reward-driven learning, the striatum selects the most appropriate schema in a given environment.

The major cortical input to the limbic striatum arises from the limbic PFC, which contributes to emotional and social behaviours and provides the organism with motivation, that is, propels it to action in the purpose of achieving specific goals (de Bruin, 1990; Eslinger and Damasio, 1985; Frith, 1992; Grafman, 1994; Hecaen, 1964; Heilman, 1994; Kolb and Whishaw, 1990; Kunishio and Haber, 1994; Luria, 1973; Reep, 1984; Rolls, 1985). The limbic PFC provides the limbic striatum with information about current goals, and about the social and emotional significance of the current environment. In addition, the limbic striatum receives information about the emotional and motivational significance of stimuli, as well as about basic emotions (such as fear) from the amygdala (Cador *et al.*, 1991; Everitt *et al.*, 1991; Kunishio and Haber, 1994; Pennartz, Groenewegen and Lopes da Silva, 1994; Pribram, 1992; Rolls, 1985); and about the expected outcomes of situations and behaviours, as well as about discrepancies between expected and actual outcomes, from the hippocampus (Gray *et al.*, 1991; Gray and Rawlins, 1986; Pribram, 1992). The output of the limbic striatum is directed to the limbic PFC.

In light of the above, we suggest that the limbic striatum subserves the contention scheduling of goals, i.e., the selection and sequencing of the most appropriate goal in a given internal and external environment. Complementarily, the limbic PFC subserves a SAS-like mechanism, that is, it is called into action in novel, ill-learned, dangerous or non-routine situations and subserves the deliberate process of deciding what to do.

As was suggested for the motor and associative striatum, the dopaminergic input to the limbic striatum is assumed to act both as an activating agent in well-learned situations, and as a "teaching" signal when learning occurs. In the latter condition, active excitatory synapses onto activated limbic striatal neurones are expected to be strengthened whenever the attainment of a goal (encoded in the activity of striatal cells) selected in a specific cortical context (representing a specific external and internal environment) had led to rewarding outcomes for the organism. In this way the limbic striatum "learns" to select the most appropriate goal in a given situation, i.e., the goal whose attainment will maximize reward. Once the association between a particular cortical context and a particular goal is learned, part of the cortical context may be sufficient to activate, that is, select, that goal. Thus, a goal may be triggered in the following ways: (1) By different types of environmental information, including verbal and social information; (2) By internal stimuli, such as drives and emotions; (3) By particular thoughts, encoded in posterior association cortices; and (4) By intentions to achieve the goal, encoded in the limbic PFC.

The proposition that goals are selected by a contention scheduling mechanism has several implications: (1) Goals are selected according to their activation level which is determined by external and internal information (provided by the inputs to the limbic striatum); (2) As a result of a reward-driven learning mechanism, the most appropriate goal is selected, that is, the goal that according to past experience is expected to maximize reward in the present situation; (3) In routine, i.e., well-learned, situations, the selection of goals is automatic and effortless; (4) In novel, ill-learned or dangerous situations, in which automatic selection of goals is not possible, a SAS-like mechanism, residing in the limbic PFC, selects a goal. These characteristics are in line with current views of goal-directed behaviour, according to which goals are activated by environmental and internal factors, and selected in a way that maximizes expected value[2]. In addition, it is accepted that when activity is well organized and routine, action moves from goal to goal fairly smoothly, without requiring a deliberate "choice" or "decision" to change goals. The effortful process of selecting a goal is required under unusual internal or external stimulation (for review, see Pervin, 1983, 1996).

[2] Goal selection is commonly assumed to be a rational, conscious process, although it is admitted that goals do not have to be conscious (Pervin, 1983, 1996). Contention scheduling is a mechanism whereby goals which maximize reward are selected without the involvement of a rational, conscious process.

The proposal that the contention scheduling of goals is subserved by the limbic striatum is consistent with the long-known role of this region and its dopaminergic input in goal-directed behaviour and motivation (Berridge, 1996; Cador, Robbins and Everitt, 1989; Cador *et al.*, 1991; Cole and Robbins, 1989; Everitt *et al.*, 1991; Gray *et al.*, 1991; Houk, Adams and Barto, 1995; Kalivas, Churchill and Klitenick, 1993; Koob, 1996; Koob *et al.*, 1993; Mogenson, Jones and Yim, 1980; Nauta *et al.*, 1978; Pennartz, Groenewegen and Lopes da Silva, 1994; Phillips, Pfauss and Blaha, 1991; Rolls and Williams, 1987; Salamone, 1994; Schacter *et al.*, 1989; Schultz *et al.*, 1992, 1993, 1995a,b; Swerdlow *et al.*, 1993). The mesolimbic dopaminergic system has been implicated mainly in the "energizing" aspects of motivation (e.g., Berridge, 1996; Cador *et al.*, 1991; Cole and Robbins, 1989; Gray *et al.*, 1991; Koob, 1996; Koob *et al.*, 1993; Phillips, Pfauss and Blaha, 1991; Salamone, 1994; Scheel-Kruger and Willner, 1991), in both positively and negatively motivated behaviours (e.g., Berridge, 1996; Koob, 1996; Mirenowicz and Schultz, 1996; Pennartz, Groenewegen and Lopes da Silva, 1994; Salamone, 1994; Scheel-Kruger and Willner, 1991; Schultz *et al.*, 1993, 1995b; Zacharko and Anisman, 1991). The limbic striatum is considered to contribute to the translation of limbic information to action, that is, in the "directional" aspects of motivation (e.g., Cador, Robbins and Everitt, 1989; Cador *et al.*, 1991; Everitt *et al.*, 1991; Kalivas, Churchill and Klitenick, 1993; Lavoie and Mizumori, 1992; Mogenson, Jones and Yim, 1980; Nauta *et al.*, 1978; Pennartz, Groenewegen and Lopes da Silva, 1994; Schacter *et al.*, 1989; Scheel-Kruger and Willner, 1991; Schultz *et al.*, 1992, 1995a). In this context, the selection of a goal can be viewed as a crucial link between motivation and behaviour, and the suggestion that the limbic striatum subserves contention scheduling of goals is therefore in line with leading views of its function. Furthermore, although earlier writings have emphasized the descending projections of the limbic striatum and VP in the translation of motivation to action (Kalivas, Churchill and Klitenick, 1993; Mogenson, Jones and Yim, 1980; Mogenson *et al.*, 1993; Skinner and Garcia-Rill, 1993; Swerdlow *et al.*, 1993), recent accounts suggest that the ascending projections via the thalamus to the frontal cortex are also involved (Cador *et al.*, 1991; Kalivas, Churchill and Klitenick, 1993; Pennartz, Groenewegen and Lopes da Silva, 1994), in line with the present focus on striato-frontal interaction.

#### *4.2.2.1. Electrophysiological evidence*

Electrophysiological data are in line with the suggestion that the activity of limbic striatal neurones encodes goals but not the motor programs needed to attain these goals. Thus, the activity of these neurones is related to stimuli that predict reward, and to behaviours that produce reward, but not to the motor action required to attain reward (Henriksen and Giacchino, 1993; Pennartz, Groenewegen and Lopes da Silva, 1994; Schultz *et al.*, 1992, 1993). While such reward expectancy-related activity can be seen as reflecting involvement in a mechanism for computing a reinforcement signal (e.g., Barto, 1995; Houk, Adams and Barto, 1995), it may alternatively reflect the current goal of the animal, namely, achieving this reinforcement. This is in line with the findings that the relation between cell firing and behaviours that produce reward disappears following extinction (Henriksen and Giacchino, 1993), and that some cells in the limbic striatum respond to stimuli which predict the initiation of a task (Rolls and Williams, 1987), and may thus encode the goal of performing the task. Limbic striatal cells were also found to respond to novel stimuli and to stimuli with emotional or motivational significance, and some of these neurones show such responses only in a particular environmental context (Rolls and Johnstone, 1992; Rolls and

Williams, 1987). Since novel and emotional stimuli propel an animal into action, that is, activate goals, the activity of these cells may be related to such activated goals. Interestingly, in rats exploring a novel environment, there are cells which fire in relation to the exploration of specific stimuli, and the introduction of novel objects or a favourite food frequently results in dramatic decrease in their activity (Henriksen and Giacchino, 1993). This change (decrease) in activity may reflect a change in goals following the introduction of the novel stimulus or food.

*4.2.2.2. How do goals, selected by the limbic split circuit, translate into motor programs, selected and executed by the associative and motor split circuits?*

For both animals and people, the daily flow of behaviour is patterned, organized and apparently directed toward end points or goals (e.g., James, 1892; Pervin, 1983, 1996; Tolman, 1932). Goals, which are desired end points, should be distinguished from motor programs, which represent routes towards attaining a goal. In the present scheme, the limbic split circuit selects goals, without specifying the specific motor program by means of which these goals are to be achieved. The latter is suggested to be the function of the associative split circuit. However, via its connections with the associative and motor split circuits, the limbic split circuit directs and "energizes" the selection and execution of motor programs towards achieving the selected goals.

The selection of a goal in the limbic striatum biases, via the closed limbic circuit, the selection of goals in the limbic PFC, and contributes to sustained activity of limbic PFC neurons, which maintain active goals and intentions in the absence of the external and internal stimuli which arouse them.

Via the open limbic route, the limbic striatum may bias nigral output according to the current goal of the organism. In this way it can affect the transfer of information in the striato-nigro-thalamo-cortical pathway to the associative PFC, which is involved in the selection of motor programs and the execution of the cognitive elements of motor programs, as well as the nigral output to the superior colliculus. The latter effect may contribute to the automatic reallocation of attention when the goal is changed, such as when a novel or surprising stimulus appears (see Fabre-Thorpe and Montaron's [1994] suggestion that the limbic striatum is involved in maintaining the balance between attending the current task and attending other stimuli).

The open limbic pathway connecting the limbic striatum to motor cortical regions (if such a pathway exists, see above) may serve to bias directly the selection of motor acts in these cortical regions according to the present goal of the organism. This pathway may be particularly involved in the production of species-specific behaviours (Joel and Weiner, 1994), together with limbic striatal projections to brainstem motor regions (e.g., Kalivas, Churchill and Klitenick, 1993; Mogenson, Jones and Yim, 1980; Mogenson *et al.*, 1993; Skinner and Garcia-Rill, 1993; Swerdlow *et al.*, 1993).

Via the open and closed loops connecting the limbic striatum to the dopaminergic system, the limbic striatum can modulate its own dopaminergic input as well as the dopaminergic input to the motor and associative striatum. We have recently suggested that via each of the loops, closed or open, the striatum exerts a direct inhibitory effect on DA cells as well as an indirect disinhibitory effect, i.e., facilitation of burst firing in DA cells (Joel and Weiner, submitted). Thus, the activation of a set of limbic striatal neurones encoding a specific goal is expected to directly inhibit dopaminergic neurones, and this inhibition can counteract the excitatory input to the dopaminergic cells when this goal is attained. Such inhibition may serve to prevent the firing of dopaminergic neurones to predictable rewards (Schultz *et al.*, 1993, 1995b; Wickens and Kötter, 1995), and in this way restrict striatal learning in well-learned situations. Once a behaviour is well learned, the indirect disinhibitory effect provides a mechanism for the maintenance of striatal dopamine level required for the execution of such behaviour. Therefore, via the closed loop the limbic split circuit can regulate its own "energizing" DA effect, and via the open loops it can direct learning and regulate the "energizing" DA effect in the motor and associative striatum according to the current goal.

In addition to the different open pathways connecting the limbic split circuit with the associative and motor split circuits, information about goals can be channelled from the limbic split circuit to the associative split circuit via corticocortical connections between the limbic PFC and the associative PFC. The projections from the limbic PFC to the associative PFC may directly bias the selection of motor programs by the associative PFC according to the current goals, encoded in the limbic PFC. As both PFC regions subserve a SAS-like mechanism, this link may be particularly important for the effortful and deliberate process of deciding what to do and how to do it. This is in contrast with the transfer of information via the different open pathways which may contribute to an automatic, effortless process by which goals can affect different levels of behaviour, from the selection of motor programs to the execution of simple motor acts.

While the pathways described above connect the limbic split circuit to the associative and motor split circuits, there may be an open indirect pathway connecting the associative split circuit to the limbic split circuit, passing from the associative striatum, via the associative GPe and limbic STN, to the VP. Via this pathway, activity in the associative striatum may lead to disinhibition of limbic STN neurones, and thus to increased activity of VP neurones, and in this manner to reduce information flow in the closed limbic circuit. This attenuating effect may be a relatively general one, since subthalamopallidal projections are distributed through relatively large regions of the pallidum (Haber, Lynd-Balta and Spooren, 1994; Hazrati and Parent, 1992; Parent and Hazrati, 1995b). The attenuation of information transfer in the closed limbic circuit may function to prevent strong activation of the limbic PFC in situations in which there is a strong activity in the associative striatum, i.e., during the execution of a routine motor program. In such a case, the activity of limbic striatal neurones encoding a goal serves to provide a directing influence on the functioning of the associative and motor split circuits via the open limbic route and pathway, respectively, as well as an energizing effect via the open loops, without recruiting the limbic PFC. In this way, behaviour is directed by current goals in an automatic and effortless way. The limbic PFC will be recruited once activity in limbic striatal cells increases and/or activity in associative striatal cells decreases. Both changes are expected to occur following a change in cortical context, particularly when the new cortical context is not a familiar one.

## 5. SUMMARY

The present model focuses on the contribution of the basal ganglia-thalamocortical split circuits to information processing in the forebrain. The major theme of this model is that these circuits provide the brain machinery for the selection and execution of goal-directed routine behaviour. More specifically, routine behaviour is a product of striato-frontal interactions within each circuit and the co-ordinated interaction between circuits.

There is a dynamic striato-frontal interplay within each circuit, with a constant updating of the striatum with information from the frontal cortex, and of the frontal cortex with information from the striatum. The striatum operates as a reward-driven contention scheduling mechanism and continuously biases the frontal cortex towards the selection and execution of the most appropriate behaviour in the current cortical context. The limbic striatum biases the limbic PFC towards the selection of the most appropriate goal, the associative striatum biases the associative PFC and the PMC towards the selection and execution of the most appropriate motor program and motor step, respectively, and the motor striatum biases the primary motor cortex and SMA towards the selection and execution of the most appropriate motor act. When the striatal biasing effect is sufficiently strong and is not counteracted by other inputs to the frontal cortex, striatum-selected behaviour is executed by the frontal cortex, i.e., the routine behaviour for these circumstances is performed. When the striatal biasing effect is relatively weak or counteracted by other inputs to the frontal cortex, non-routine behaviour, which is a product of interactions of the frontal cortex with other brain regions, is executed.

The connections between the three split circuits serve to co-ordinate their action so that the production of complex goal-directed routine behaviour is possible. The limbic split circuit directs, via its connections with the associative split circuit, the selection of motor programs which can attain the goal selected by the limbic split circuit; the associative split circuit directs, via its connections with the motor split circuit, the selection of motor acts which can achieve the motor steps of the selected motor program; the motor split circuit updates the associative split circuit with regard to the current state of the execution of the motor step so that the execution of the motor and cognitive elements of the motor program remains synchronized.

Thus, during the routine performance of goal-directed behaviour the activity of the three split circuits is highly synchronized, so that the motor acts executed by the motor split circuit are the most appropriate in the current environment for the current motor program; the latter is the most appropriate motor program in the current environment for attaining the current goal, which is the most appropriate in the current internal and external environment. Moreover, the connections between the circuits enable the execution of routine behaviour with only limited involvement of the associative and limbic PFC. Thus, the associative split circuit directs the execution of the motor steps via the open associative pathway without the involvement of the associative PFC. The limbic split circuit directs the selection of the motor program via the open limbic route and the open loops without the involvement of the limbic PFC. In addition, it is possible that the associative split circuit reduces information flow from the limbic striatum to the limbic PFC when the routine motor program is executed, further reducing the involvement of the limbic PFC.

## ACKNOWLEDGEMENTS

The authors are indebted to the generous support of the Joseph Buchmann Doctoral Fellowship Fund to D.J.

## REFERENCES

Albin, R.L., Young, A.B. and Penney, J.B. (1989) The functional anatomy of basal ganglia disorders. *Trends in Neuroscience,* **12**, 366–375.

Aldridge, J.W., Jaeger, D. and Gilman, S. (1991) A comparison of single unit activity in primate caudate nucleus and putamen in a sensory cued motor task. In *The Basal Ganglia III*, edited by G. Bernardi, M.B. Carpenter, G. Di Chiara, M. Morelli, and P. Stanzione, New York: Plenum Press, pp. 303–310.

Alexander, G.E. and Crutcher, M.D. (1990) Functional architecture of basal ganglia circuits: neural substrates of parallel processing. *Trends in Neuroscience,* **13**, 266–271.

Alexander, G.E., Crutcher, M.D. and Delong, M.R. (1990) Basal ganglia-thalamocortical circuits: parallel substrates for motor, oculomotor, "prefrontal" and "limbic" functions. In *The Prefrontal cortex: its structure, function and pathology (Progress in Brain Research vol. 85),* edited by H.B.M. Uylings, C.G. van Eden, J.P.C. de Bruin, M.A. Corner and M.G.P. Feenstra Elsevier, Amsterdam, pp. 119–146.

Alexander, G.E., Delong, M.R. and Strick, P.L. (1986) Parallel organization of functionally segregated circuits linking basal ganglia and cortex. *Annual Reviews of Neuroscience,* **9**, 357–381.

Arbib, M.A. and Dominey, P.F. (1995) Modeling the role of basal ganglia in timing and sequencing saccadic eye movements. In *Models of information processing in the basal ganglia*, edited by J.C. Houk, J.L. Davis, and D.G. Beiser J.C. Houk, J.L. Davis, and D.G. Beiser (eds), Cambridge: MIT Press, pp. 149–162.

Ballard, D.H., Hayhoe, M.M. and Pelz, J. (1995) Memory limits in sensorimotor tasks. In: J.C. Houk, J.L. Davis, and D.G. Beiser (eds) *Models of information processing in the basal ganglia*, Cambridge: MIT Press, pp. 295–313.

Barto, A.G. (1995) Adaptive critic and the basal ganglia. In *Models of information processing in the basal ganglia*, edited by J.C. Houk, J.L. Davis, and D.G. Beiser, Cambridge: MIT Press, pp. 215–232.

Berridge, K.C. (1989) Substantia nigra 6-OHDA lesions mimic striatopallidal disruption of syntactic analysis of sequence control. *Psychobiology,* **17**, 377–385.

Berridge, K.C. (1996) Food reward: brain substrates of wanting and liking. *Neuroscience and Biobehavioral Reviews,* **20**, 1–25.

Berridge, K.C. and Fentress, J.C. (1987) Disruption of natural grooming chains after striatopallidal lesions. *Psychobiology,* **15**, 336–342.

Berridge, K.C. and Whishaw, I.Q. (1992) Cortex, striatum and cerebellum—control of serial order in a grooming sequence. *Experimental Brain Research,* **90**, 275–290.

Brotchie, P., Iansek, R. and Horne, M.K. (1991a) Motor function of the monkey globus pallidus: neuronal discharge and parameters of movement. *Brain,* **114**, 1667–1683.

Brotchie, P., Iansek, R. and Horne, M.K. (1991b) Motor function of the monkey globus pallidus: cognitive aspects of movement and phasic neuronal activity. *Brain,* **114**, 1685–1702.

Cador, M., Robbins, T.W. and Everitt, B.J. (1989) Involvement of the amygdala in stimulus-reward associations—interaction with the ventral striatum. *Neuroscience,* **30**, 77–86.

Cador, M., Robbins, T.W., Everitt, B.J., Simon, H., LeMoal, M. and Stinus, L. (1991) Limbic-striatal interactions in reward-related processes: modulation by the dopaminergic system. In: *The mesolimbic dopamine system: from motivation to action*, edited by P. Willner and J. Scheel-Kruger, Chichester: John Wiley pp. 225–250.

Case, R. (1992) The role of the frontal lobes in the regulation of cognitive development. *Brain and Cognition,* **20**, 51–73.

Cole, B.J. and Robbins, T.W. (1989) Effects of 6-hydroxydopamine lesions of the nucleus accumbens septi on performance of a 5-choice serial reaction time task in rats: implications for theories of selective attention and arousal. *Behavioral Brain Research* **33**, 165–179.

Cools, A.R. (1980) Role of the neostriatal dopaminergic activity in sequencing and selecting behavioral strategies: facilitation of processes involved in selecting the best strategy in a stressful situation. *Behavioral Brain Research,* **1**, 361–378.

Cools, A.R., van den Bercken, J.H.L., Horstink, M.W.I., van Spandonck, K.P.M. and Berger, H.J.C. (1984) Cognitive and motor shifting aptitude disorder in Parkinson's disease. *Journal of Neurology, Neurosurgery and Psychiatry,* **47**, 443–453.

de Bruin, J.P.C. (1990) Social behaviour and the prefrontal cortex. In*The Prefrontal cortex: its structure, function and pathology (Progress in Brain Research vol. 85)* edited by H.B.M. Uylings, C.G.van Eden, J.P.C. de Bruin, M.A.Corner and M.G.P. Feenstra, Elsevier, Amsterdam, pp. 485–497.

DeLong, M.R., Crutcher, M.D. and Georgopoulos, A.P. (1985) Primate globus pallidus and subthalamic nucleus: functional organization. *Journal of Neurophysiology,* **53**, 530–543.

DeLong, M.R. and Georgopoulos, A.P. (1981) Motor functions of the basal ganglia. In *Handbook of Physiology, Vol. II*, edited by J.M. Brookhart, V.B. Mountcastle, and V.B. Brooks, Bethseda: American Physiological Society, pp. 1017–1061.

Divac, I. (1968) Effects of prefrontal and caudate lesions on delayed response in rats. *Acta Biologiae Experimentalis,* **28**, 149–167.

Divac, I., Rosvold, H.E. and Scwarcbart, M.K. (1967) Behavioural effects of selective ablation of the caudate nucleus. *Journal of Comparative and Physiological Psychology,* **63**, 184–190.

Eslinger, P.J. and Damasio, A.R. (1985) Severe distrubance of higher cognition after bilateral frontal lobe ablation: patient EVR. *Neurology,* **35**, 1731–1741.

Everitt, B.J., Morris, K.A., O'Brien, A. and Robbins, T.W. (1991) The basolateral amygdala-ventral striatal system and conditioned place preference: Further evidence of limbic-striatal interactions underlying reward-related processes. *Neuroscience,* **42**, 1–18.

Fabre-Thorpe, M. and Montaron, M.-F. (1994) From preparation to action: involvement of the ventral striatum. In *The basal ganglia IV: New ideas and data on structure and function* edited by G. Percheron, J.S. McKenzie, and J. Feger, New York: Penum Press, pp. 297–303.

Flaherty, A.W. and Graybiel, A.M. (1993) Two input systems for body representations in the primate striatal matrix: experimental evidence in the squirrel monkey. *Journal of Neuroscience* **13**, 1120–1137.

Flaherty, A.W. and Graybiel, A.M. (1994) Input-output organization of the sensorimotor striatum in the squirrel monkey. *Journal of Neuroscience,* **14**, 599–610.

Frith, C.D. (1992). *The cognitive neuropsychology of schizophrenia.* Hillsdale: Lawrence Erlbaum Associates Ltd.

Fuster, J.M. (1990) Behavioral electrophysiology of the prefrontal cortex of the primate. In *The Prefrontal cortex: its structure, function and pathology (Progress in Brain Research vol. 85)* edited by H.B.M. Uylings, C.G.van Eden, J.P.C. de Bruin, M.A. Corner and M.G.P. Feenstra, Amsterdam Elsevier pp. 313–324.

Gabrieli, J. (1995) Contribution of the basal ganglia to skill learning and working memory in humans. In *Models of information processing in the basal ganglia,* edited by J.C. Houk, J.L. Davis, and D.G. Beiser, Cambridge: MIT Press, pp. 277–294.

Gerfen, C.R. and Wilson, C.J. (1996) The basal ganglia. In *Handbook of chemical neuroanatomy, Vol. 12,* edited by L.W. Swanson, A. Bjorklumd, and T. Hokfelt, Elsevier Science B.V, pp. 371–468.

Goldman-Rakic, P.S. (1987a) Circuitry of the prefrontal cortex and the regulation of behavior by representational memory. In *Handbook of Physiology, Vol. V,* edited by F. Planum and V.B. Mountcastle, Bethesda: American Physiological Society, pp. 373–417.

Goldman-Rakic, P.S. (1987b) Motor control function of the prefrontal cortex. In *Motor areas of the cerebral cortex,* (Ciba Foundation Symposium, No. 132), Chichster, Wiley Interscience, pp. 187–200.

Goldman-Rakic, P.S. (1995) Cellular basis of working memory. *Neuron,* **14**, 477–485.

Goldman-Rakic, P.S. and Selemon, L.D. (1990) New frontiers in basal ganglia research. *Trends in Neuroscience* **13**, 241–244.

Grafman, J. (1994) Neuropsychology of the prefrontal cortex. In *Neuropsychology,* edited by W.D. Zaidel, Academic Press, New York, pp. 159–181.

Gray, J.A., Feldon, J., Rawlins, J.N.P., Hemsley, D.R. and Smith, A.D. (1991) The neuropsychology of schizophrenia. *Behavioral and Brain Science* **14**, 1–84.

Gray, J.A. and Rawlins, J.N.P. (1986) Comparator and buffer memory: an attempt to integrate two models of hippocampal function. In *The Hippocampus, Vol. 4* edited by R.L. Isaacson and K.H. Pribram, New York: Plenum Press, pp. 159–201.

Graybiel, A.M. (1990) Neurotransmitters and neuromodulators in the basal ganglia. *Trends in Neuroscience,* **13**, 244–254.

Graybiel, A.M., Aosaki, T., Flaherty, A.W. and Kimura, M. (1994) The basal ganglia and adaptive motor control. *Science, New York* **265**, 1826–1831.

Graybiel, A.M. and Kimura, M. (1995) Adaptive neural networks in the basal ganglia. In *Models of information processing in the basal ganglia.* edited by J.C. Houk, J.L. Davis, and D.G. Beiser, Cambridge: MIT Press, pp. 103–116.

Groenewegen, H.J., Berendse, H.W., Wolters, J.G. and Lohman, A.H.M. (1990) The anatomical relationship of the prefrontal cortex with the striatopallidal system, the thalamus and the amygdala: evidence for a parallel organization. In *The Prefrontal cortex: its structure, function and pathology (Progress in Brain Research vol. 85)* edited by H.B.M. Uylings, C.G. van Eden, J.P.C. de Bruin, M.A. Corner and M.G.P. Feenstra, Amsterdam, Elsevier, pp. 95–118.

Groenewegen, H.J., Berendse, H.W. and Wouterlood, F.G. (1994) Organization of the projections from the ventral striato-pallidal system to ventral mesencephalic dopaminergic neurons in the rat. In *The basal ganglia IV. New ideas and data on structure and function* edited by G. Percheron, J.S. McKenzie, and J. Feger, New York: Plenum Press, pp. 81–93.

Groves, P.M. (1983) A theory of the functional organization of the neostriatum and the neostriatal control of voluntary movement. *Brain Research. Reviews* **5**, 109–132.

Groves, P.M., Garcia-Munoz, M., Linder, J.C., Manley, M.S., Martone, M.E. and Young, S.J. (1995) Elements of the intrinsic organization and information processing in the neostriatum. In *Models of information processing in the basal ganglia* edited by J.C. Houk, J.L. Davis, and D.G. Beiser, Cambridge: MIT Press, pp. 51–96.

Haber, S.N., Lynd, E., Klein, C. and Groenewegen, H.J. (1990) Topographic organization of the ventral striatal efferent projections in the rhesus monkey: an anterograde tracing study. *Journal of Comparative. Neurology* **293**, 282–298.

Haber, S.N., Lynd-Balta, E. and Spooren, W.P.J.M. (1994) Integrative aspects of basal ganglia circuitry. In *The basal ganglia IV: New ideas and data on structure and function* edited by G. Percheron, J.S. McKenzie, and J. Feger, New York: Penum Press, pp. 71–80.

Hazrati, L.N. and Parent, A. (1992) Convergance of subthalamic and striatal efferents at pallidal level in primates: an anterograde-labeling study with biocytin and PHA-L. *Brain Research* **569**, 336–340.

Hecaen, H. (1964) Mental symptoms associated with tumors of the frontal lobe. In *The frontal granular cortex and behavior* edited by J.M. Warren and K. Akert, New York: McGraw Hill, pp. 335–352.

Heilman, K.M. (1994) Emotion and the brain—a distributed modular network mediating emotional experience. In *Neuropsychology*, edited by W.D. Zaidel, New York. Academic Press, pp. 139–158.

Heimer, L., Switzer, R.D. and Van Hosen, G.L. (1982) Ventral striatum and ventral pallidum. Components of the motor system? *Trends in Neuroscience* **5**, 83–87.

Henriksen, S.J. and Giacchino, J. (1993) Functional characteristics of nucleus accumbens neurons: evidence obtained from *in vivo* electrophysiological recordings. In *Limbic motor circuits and neuropsychiatry* edited by P.W. Kalivas and C.D. Barnes, Boca Raton: CRC Press, pp. 101–124.

Hikosaka, O. (1991) Basal ganglia- possible role in motor coordination and learning. *Current Opinion in Neurobiology,* **1**, 638–643.

Hikosaka, O. (1994) Role of basal ganglia in control of innate movements, learned behavior and cognition. In *The basal ganglia IV: New ideas and data on structure and function.* edited by G. Percheron, J.S. McKenzie, and J. Feger, New York: Penum Press, pp. 589–596.

Houk, J.C. (1995) Information processing in modular circuits linking basal ganglia and cerebral cortex. In *Models of information processing in the basal ganglia*, edited by J.C. Houk, J.L. Davis, and D.G. Beiser, Cambridge: MIT Press, pp. 3–9.

Houk, J.C., Adams, J.L. and Barto, A.G. (1995) A model of how the basal ganglia generate and use reward signals that predict reinforcement. In *Models of information processing in the basal ganglia* edited by J.C. Houk, J.L. Davis, and D.G. Beiser, Cambridge: MIT Press, pp. 249–270.

Houk, J.C. and Wise, S.P. (1995) Distributed modular architectures linking basal ganglia, cerebellum, and cerebral cortex: their role in planning and controlling action. *Cerebral Cortex,* **2**, 95–110.

Iversen, S.D. (1984) Behavioral effects of manipulation of basal-ganglia neurotransmitters. In *Functions of the Basal Ganglia, (Ciba Foundation Symposium, no. 107)*, edited by D. Evered and M. O'Connor, London: Pitman, pp. 183–200.

Jackson, S. and Houghton, G. (1995) Sensorimotor selection and the basal ganglia: a neural network model. In *Models of information processing in the basal ganglia* edited by J.C. Houk, J.L. Davis, and D.G. Beiser, Cambridge: MIT Press, pp. 337–367.

James, W. (1892/1983) *The principles of psychology.* Cambridge: Harvard University Press.

Jaspers, R., Schwarz, M., Sontag, K.H. and Cools, A.R. (1984) Caudate nucleus and programming behavior in cats: role of dopamine in switching motor patterns. *Behavioral Brain Research* **14**, 17–28.

Jimenez-Castellanos, J. and Graybiel, A.M. (1987) Subdivisions of the dopamine-containing A8-A9-A10 complex identified by their differential mesostriatal innervation of striosomes and extrastriosomal matrix. *Neuroscience,* **23**, 223–243.

Jimenez-Castellanos, J. and Graybiel, A.M. (1989) Evidence that histochemically distinct zones of the primate substantia nigra pars compacta are related to patterned distributions of nigrostriatal projection neurons and striatonigral fibers. *Experimental Brain Research* **74**, 227–238.

Joel, D. and Weiner, I. (1994) The organization of the basal ganglia-thalamocortical circuits: open interconnected rather than closed segregated. *Neuroscience,* **63**, 363–379.

Joel, D. and Weiner, I. (1997) The connections of the primate subthalamic nucleus: indirect pathways and the open-interconnected scheme of basal ganglia-thalamocortical circuitry. *Brain Research Reviews* **23**, 62–78.

Joel, D. and Weiner, I. (submitted) The connections of the dopaminergic system with the striatum in rats and primates: an analysis with respect to the functional and compartmental organization of the striatum.

Kalivas, P.W., Churchill, L. and Klitenick, M.A. (1993) The circuitry mediating the translation of motivational stimuli into adaptive motor responses. In *Limbic motor circuits and neuropsychiatry*, edited by P.W. Kalivas and C.D. Barnes, Boca Raton: CRC Press, pp. 237–287.

Karnath, H.O. and Wallesch, C.W. (1992) Inflexibility of mental planning—a charateristic disorder with prefrontal lobe lesions? *Neuropsychologia,* **30**, 1011–1016.

Kimura, M. (1987) The putamen neuron: activity and the association of a sensory stimulus with movement in the monkey. In *The basal ganglia II: Structure and function-current concepts* edite by M.B. Carpenter and A. Jayaraman, New York: Plenum Press, pp. 337–347.

Kimura, M. (1995) Role of basal ganglia in behavioral learning. *NeuroscienceResearch* **22**, 353–358.

Kita, H. (1996) Two pathways between the cortex and the basal ganglia output nuclei and the globus pallidus. In *The basal ganglia V*, edited by C. Ohye, M. Kimura, and J.S. McKenzie, New York: Plenum Press pp. 72–94.

Kolb, B. and Whishaw, I.Q. (1990) *Fundamentals of human neuropsychology*. New York: W.H. Freeman and Company.

Koob, G.F. (1996) Hedonic valence, dopamine and motivation. *Molecular Psychiatry*. **1**, 186–189.

Koob, G.F., Robledo, P., Markou, A. and Caine, S.B. (1993) The mesocorticolimbic circuit in drug dependence and reward—a role for the extended amygdala? In *Limbic motor circuits and neuropsychiatry*, edited by P.W. Kalivas and C.D. Barnes, Boca Raton: CRC Press, pp. 289–309.

Kunishio, K. and Haber, S.N. (1994) Primate cingulostriatal projection: limbic striatal versus sensorimotor striatal input. *Journal of Comparative Neurology*., **350**, 337–356.

Lavoie, A.M. and Mizumori, S.J.Y. (1992) Spatial, movement- and reward-sensitive discharge by medial nucleus accumbens. *Society for Neuroscience Abstracts* **19**, 707.

Levine, D.S., Leven, S.J. and Prueitt, P.S. (1992) Integration, disintegration and the frontal lobes. In *Motivation, emotion, and goal direction in neural networks*, edited by D.S. Levine and S.J. Leven, Hillsdale: Lawrence Erlbaum Associates, Publishers, pp. 301–335.

Lidsky, T.I., Manetto, C. and Schneider, J.S. (1985) A consideration of sensory factors involved in motor functions of the basal ganglia. *Brain Research. Reviews*. **9**, 133–146.

Luria, A.R. (1973) The frontal lobes and the regulation of behavior. In *Psychophysiology of the frontal lobes*, edited by K.H. Pribram and A.R. Luria, New York: Academic Press, pp. 3–26.

Lynd-Balta, E. and Haber, S.N. (1994a) The organization of midbrain projections to the striatum in the primate: sensorimotor-related striatum versus ventral striatum. *Neuroscience*, **59**, 625–640.

Lynd-Balta, E. and Haber, S.N. (1994b) Primate striatonigral projections: a comparison of the sensorimotor-related striatum and the ventral striatum. *Journal of Comparative Neurology*, **345**, 562–578.

Marsden, C.D. (1982) The mysterious motor function of the basal ganglia. *Neurology*, **32**, 514–539.

Marsden, C.D. and Obeso, J.A. (1994) The functions of the basal ganglia and the paradox of stereotaxic surgery in Parkinson's disease. *Brain*, **117**, 877–897.

Matsumura, M., Kojima, J., Gardiner, T.W. and Hikosaka, O. (1992) Visual and oculomotor functions of monkey subthalamic nucleus. *Journal of Neurophysiology* **67**, 1615–1632.

Miller, R. and Wickens, J.E. (1991) Corticostriatal cell assemblies in selective attention and in representation of predictable and cotrollable events. *Concepts in Neuroscience*, **2**, 65–95.

Milner, B. (1963) Effects of different brain lesions on card sorting, the role of the frontal lobes. *Archives of Neurology*, **9**, 90–100.

Mink, J.W. and Thach, W.T. (1991a) Basal ganglia motor control. I. Nonexclusive relation of pallidal discharge to five movement modes. *Journal of Neurophysiology*, **65**, 273–300.

Mink, J.W. and Thach, W.T. (1991b) Basal ganglia motor control. II. Late pallidal timing relative to movement onset and inconsistent pallidal coding of movement parameters. *Journal of Neurophysiology*, **65**, 301–329.

Mink, J.W. and Thach, W.T. (1991c) Basal ganglia motor control. III. Pallidal ablation: normal reaction time, muscle cocontraction, and slow movement. *Journal of Neurophysiology*, **65**, 330–351.

Mirenowicz, J. and Schultz, W. (1996) Preferential activation of midbrain dopamine neurons by appetitive rather than aversive stimuli. *Nature, London*, **379**, 449–451.

Mogenson, G.J., Brudzynski, S.M., Wu, M., Yang, C.R. and Yim, C.C.Y. (1993) From motivation to action: a review of dopaminergic regulation of limbic -> nucleus accumbens -> ventral pallidum -> peduncolopontine nucleus circuitries involved in limbic-motor integration. In *Limbic motor circuits and neuropsychiatry*, edited by P.W. Kalivas and C.D. Barnes, Boca Raton: CRC Press, pp. 193–236.

Mogenson, G.J., Jones, D.L. and Yim, C.Y. (1980) From motivation to action: functional interface between the limbic system and the motor system. *Progress Neurobiology*, **14**, 69–97.

Nambu, A., Yoshida, S. and Jinnai, K. (1990) Discharge pattern of pallidal neurons with input from various cortical areas during movement in the monkey. *Brain Research*, **519**, 183–191.

Nauta, H.J.W., Smith, G.P., Faull, R.L.M. and Domesick, V.B. (1978) Efferent connections and nigral afferents of the nucleus accumbens septi in the rat. *Neuroscience*, **3**, 385–401.

Norman, D.A. and Shallice, T. (1986) Attention to action: willed and automatic control of behavior. In *Consciousness and self-regulation: advances in research, Vol. IV.*, edited by R.J. Davidson, G.E. Schwartz, and D. Shapiro, New York: Plenum Press, pp. 1–18.

Parent, A. (1990) Extrinsic connections of the basal ganglia. *Trends in Neuroscience*, **13**, 254–258.

Parent, A. and Hazrati, L.N. (1995a) Functional anatomy of the basal ganglia. I. the cortico-basal ganglia-thalamo-cortical loop. *Brain Research Reviews* **20**, 91–127.

Parent, A. and Hazrati, L.N. (1995b) Functional anatomy of the basal ganglia. II. The place of subthalamic nucleus and external pallidum in basal ganglia circuitry. *Brain Research Reviews*., **20**, 128–154.

Pennartz, C.M.A. (1995) The ascending neuromodulatory systems in learning by reinforcement: comparing computational conjectures with experimental findings. *Brain Research Reviews,* **21**, 219–245.

Pennartz, C.M.A., Groenewegen, H.J. and Lopes da Silva, F.H. (1994) The nucleus accumbens as a complex of functionally distinct neuronal ensembles: an integration of behavioural, electrophysiological and anatomical data. *Progress in Neurobiology* **42**, 719–761.

Penney, J.B. and Young, A.B. (1983) Speculations on the functional anatomy of basal ganglia disorders. *Annual. Reviews of Neuroscience* **6**, 73–97.

Penney, J.B. and Young, A.B. (1986) Striatal inhomogeneities and basal ganglia function. *Movement Disorders,* **1**, 3–15.

Pervin, L.A. (1983) The stasis and flow of behavior: towards a theory of goals. In *Pesonality: current theory and research,* edited by M.M. Page, Lincoln: University of Nebraska Press, pp. 1–53.

Pervin, L.A. (1996). *The science of personality.* New York: John Wiley and Sons, Inc.

Phillips, A.G., Pfauss, J.G. and Blaha, C.D. (1991) Dopamine and motivated behavior: insights provided by in vivo analyses. In*The Mesolimbic Dopamine System: From Motivation to Action,* edited by P. Willner and J. Scheel-Kruger, Chichester: Wiley J. and Sons, pp. 199–224.

Plenz, D. and Aertsen, A. (1994) The basal ganglia: "minimal coherence detection" in cortical activity distribution. In *The basal ganglia IV: New ideas and data on structure and function.,* edited by G. Percheron, J.S. McKenzie, and J. Feger, New York: Penum Press, pp. 579–588.

Pribram, K. (1992) Familiarity and novelty: the contributions of the limbic forebrain to valuation and the processing of relevance. In *Motivation, emotion, and goal direction in neural networks,* edited by D.S. Levine and S.J. Leven, Hilsdale, New Jersey: Lawrence Erlbaum Associates, pp. 337–365.

Pribram, K.H. (1973) The primate frontal cortex—executive of the brain. In *Psychophysiology of the frontal lobes,* edited by K.H. Pribram and A.R. Luria, New York: Academic Press, pp. 293–314.

Reep, R. (1984) Relationship between prefrontal and limbic cortex: a comparative anatomical review. *Brain Behavior and Evolution* **25**, 5–80.

Robbins, T.W. and Brown, V.J. (1990) The role of the striatum in the mental chronometry of action: a theoretical review. *Reviews of Neuroscience* **2**, 181–213.

Rolls, E.T. (1985) Connections, functions and dysfunctions of limbic structures, the pre-frontal cortex and hypothalamus. In *The scientific basis of clinical neurology,* edited by M. Seash and C. Kennard, London: Curchill Livingston, pp. 201–213.

Rolls, E.T. and Johnstone, S. (1992) Neurophysiological analysis of striatal function. In *Neuropsychological Disorders with Subcortical Lesions,* edited by C. Wallesch and G. Vallar, Oxford: University Press, pp. 61–97.

Rolls, E.T. and Williams, G.V. (1987) Sensory and movement-related neuronal activity in different regions of the primate striatum. In *Basal Ganglia and Behavior: Sensory aspects and motor functioning* edited by J.S. Schneider and T.T. Lidsky, Bern: Hans Huber, pp. 37–59.

Rosvold, H.E. (1972) The frontal lobe system: corticosubcortical interrelationships. *Acta Neurobiologiae Experimentalis,* **32**, 439–460.

Sabol, K.E., Neil, D.B., Wages, S.A., Church, W. and Justice, J.B. (1985) Dopamine depletion in a striatal sub-region disrupts performance of a skilled motor task in the rat. *Brain Research* **335**, 33–43.

Salamone, J.D. (1994) The involvement of nucleus accumbens dopamine in appetitive and aversive motivation. *Behavioral Brain Research* **61**, 117–133.

Schacter, G.B., Yang, C.R., Innis, N K. and Mogenson, G.J. (1989) The role of the hippocampal-nucleus accumbens pathway in radial-arm maze performance. *Brain Research,* **494**, 339–349.

Scheel-Kruger, J. and Willner, P. (1991) The mesolimbic system: principles of operation. In *The Mesolimbic Dopamine System: From Motivation to Action,* edited by P. Willner and J. Scheel-Kruger, Chichester: Wiley J. and Sons, pp. 559–597.

Schultz, W., Apiccella, P., Romo, R. and Scarnati, E. (1995a) Context-dependent activity in primate striatum reflecting past and future behavioral events. In *Models of information processing in the basal ganglia,* edited by J.C. Houk, J.L. Davis, and D.G. Beiser, Cambridge: MIT Press, pp. 11–27.

Schultz, W., Apiccela, P., Scarnati, E. and Ljungberg, T. (1992) Neuronal activity in monkey ventral striatum related to the expectation of reward. *Journal of Neuroscience* **12**, 4595–4610.

Schultz, W., Apicella, P., Ljungberg, T., Romo, R. and Scarnati, E. (1993) Reward-related activity in the monkey striatum and substantia nigra. In *Chemical Signalling in the Basal Ganglia, (Progress in Brain Reseach, Vol. 99),* edited by G.W. Arbuthnott and P.C. Emson, Elsevier, Amsterdam, pp. 227–235.

Schultz, W. and Romo, R. (1988) Neuronal activity in the monkey striatum during the initiation of movements. *Experimental Brain Research* **71**, 431–436.

Schultz, W., Romo, R., Ljungberg, T., Mirenowicz, J., Hellerman, J.R. and Dickinson, A. (1995b) Reward-related signals carried by dopaminergic neurons. In *Models of information processing in the basal ganglia,* edited by J.C. Houk, J.L. Davis, and D.G. Beiser, Cambridge: MIT Press, pp. 233–248.

Shallice, T. (1982) Specific impairments of planning. *Philosophical Transactions of the Royal Society* (London), Section B. **298**, 199–209.

Shallice, T. and Bugress, P.W. (1991) Deficits in strategy application following frontal lobe damage in man. *Brain,* **114**, 727–741.

Skinner, R.D. and Garcia-Rill, E. (1993) Mesolimbic interactions with mesopontine modulation of locomotion. In *Limbic motor circuits and neuropsychiatry*, edited by P.W. Kalivas and C.D. Barnes, Boca Raton: CRC Press, pp. 155–191.

Smith, C.G., Beninger, R.J., Mallet, P.E., Jhamandas, K. and Boegman, R.J. (1994) Basal forebrain injections of the benzodiazepine partial inverse agonist FG 7142 enhance memory of rats in the double y-maze. *Brain Research,* **666**, 61–67.

Strick, P.L., Dum, R.P. and Picard, N. (1995) Macro-organization of the circuits connecting the basal ganglia with the cortical motor areas. In *Models of information processing in the basal ganglia,* edited by J.C. Houk, J.L. Davis, and D.G. Beiser, Cambridge: MIT Press, pp. 117–130.

Stuss, D.T. (1992) Biological and Psychological Development of Executive Functions. *Brain and Cognition,* **20**, 8–23.

Swanson, L.W. and Mogenson, G.J. (1981) Neural mechanisms for the functional coupling of autonomic, endocrine and somatomotor responses in adaptive behavior. *Brain Research Reviews.,* **3**, 1–34.

Swerdlow, N.R., Braff, D.L., Caine, S.B. and Geyer, M.A. (1993) Limbic cortic-striato-pallido-pontine substrates of sensorimotor gating in animal models and psychiatric disorders. In *Limbic motor circuits and neuropsychiatry*, edited by P.W. Kalivas and C.D. Barnes, Boca Raton: CRC Press, pp. 311–328.

Swerdlow, N.R. and Koob, G.F. (1987) Dopamine, schizophrenia, mania and depression: toward a unified hypothesis of cortico-striato-pallido-thalamic function. *Behavioral Brain Research* **10**, 215–217.

Tolman, E.C. (1932). Purposive behaviour in animals and men. New York: Century.

Trouche, E., Viallet, F., Apicella, P., Alamy, M., Pons, J.-C. and Legallet, E. (1994) Pallidal and nigral hypokinesia: an experimental analysis in the monkey. In *The basal ganglia IV: New ideas and data on structure and function*, edited by G. Percheron, J.S. McKenzie, and J. Feger, New York: Penum Press, pp. 337–348.

van der Bos, R. and Cools, A.R. (1989) The involvement of the nucleus accumbens in the ability of rats to switch to cue-directed behaviours. *Life Sciences,* **44**, 1697–1704.

Wickens, J. (1990) Striatal dopamine in motor activation and reward-mediated learning: Steps towards a unifying model. *Journal of Neural Transmission (General Section.),* **80**, 9–31.

Wickens, J. and Kotter, R. (1995) Cellular models of reinforcement. In *Models of information processing in the basal ganglia,* edited by J.C. Houk, J.L. Davis, and D.G. Beiser, Cambridge: MIT Press, pp. 187–214.

Wilson, C.J. (1995) The contribution of cortical neurons to the firing pattern of striatal spiny neurons. In *Models of information processing in the basal ganglia*, edited by J.C. Houk, J.L. Davis, and D.G. Beiser, Cambridge: MIT Press, pp. 29–50.

Zacharko, R.M. and Anisman, H. (1991) Stressor-provoked alterations of intracranial self-stimulation in the mesocorticolimbic system: an animal model of depression. In *The Mesolimbic Dopamine System: From Motivation to Action*, edited by P. Willner and J. Scheel-Kruger, Chichester: Wiley J. and Sons pp. 411–442.

# 13 Motor and Non-Motor Roles of the Cortico-Basal Ganglia Circuitry


**Emilio Salinas, Ioan Opris, Antonio Zainos, Adrián Hernández and Ranulfo Romo***

*Instituto de Fisiología Celular Universidad Nacional Autónoma de México, 04510 México, D.F., México*
** Corresponding author's address; Ranulfo Romo, Instituto de Fisiología Celular Universidad Nacional Autónoma de México, Apartado Postal 70-253, 04510 México, D.F. México*
*Tel: (525) 622-5586; FAX: (525) 622-5607; e-mail: rromo@ifcsun1.ifisiol.unam.mx*



Numerous areas of the cerebral cortex provide input to the basal ganglia, whose activity is largely relayed back to the cortex, via the thalamus. The function of this feedback circuitry has remained elusive. Numerous anatomical, lesion and electrophysiological results point to a role in motor control. However, few neurophysiological studies have been able to reveal unique features in the responses of neurones in the basal ganglia; in most cases similar activity has also been found in cortex. The evidence in support of this idea is reviewed, and is further illustrated in the case of neuronal activity associated with the preparation for a voluntary movement. Similarities between neural responses in cortex and basal ganglia have also been observed in paradigms that stress cognitive *versus* strictly motor processes. Several examples are discussed, including experiments in which neuronal activity correlates with the categorization of somatosensory stimuli. One interpretation of these results is that the cortico-basal ganglia circuit is part of a large, multifunctional network, that operates in a parallel fashion, and participates in motor and non-motor processes. Alternatively, the basal ganglia might be involved in unique but subtle aspects of higher-order functions, which can be fully revealed only by novel experimental paradigms.



KEYWORDS: basal ganglia, cerebral cortex, motor control, primates, preparatory activity, sensory categorization.


## 1. INTRODUCTION

A standard assumption about the basal ganglia is that they are implicated in the regulation of motor activity. A large literature has been devoted to their cellular properties: Neuronal types, neurotransmitters and single-cell anatomy and neurophysiology have all been extensively studied (for a recent review, see Mink, 1996). However, the functional role of these subcortical structures has been difficult to pinpoint. The anatomy of the circuitry in which the basal ganglia are embedded is well known. In essence, information originating in the neocortex passes through the basal ganglia, and returns, via the thalamus, to specific areas in the frontal lobe (Alexander, DeLong and Strick, 1986;

Alexander and Crutcher, 1990a). The neocortical input areas include various domains that subserve sensory, motor, associative and limbic functions (Parent and Hazrati, 1995a,b). How information is modified, before and after its transit through the basal ganglia, is mostly unknown. Why this circuit utilizes two inhibitory synapses in sequence is not known either. On the basis of lesion and neurophysiological studies (DeLong and Georgopoulos, 1981; DeLong, 1990), it has been suggested that the basal ganglia are crucially involved in the control, and especially in the planning of goal-driven movements, which is consistent with the massive input they receive from frontal cortex (Parent and Hazrati, 1995a,b). The wide diversity of the sources that send information to them also suggests integrative functions.

However, these are extremely broad possibilities. Consider, for example, the following observations made by Alexander and Crutcher (1990b): To reach an object, the brain has to translate the spatial information specified by a goal into a well-defined set of muscle activation patterns that will carry the limb to the target. This process requires a sequence of computations regarding (a) target location, (b) selection of limb trajectory, (c) limb kinematics and dynamics to acquire the target, and (d) the patterns of muscle activation that satisfy the dynamics. Additionally, the neural substrate of motor behaviour employs not only a serially interconnected chain of highly specialized motor structures, but also a parallel organization that processes information in a distributed fashion. Thus, individual motor structures may participate simultaneously at several levels of a motor task, whose execution may simultaneously depend on multiple cortical, basal ganglia, thalamic and cerebellar modules (Alexander and Crutcher, 1990a; Hoover and Strick, 1993; Houk and Wise, 1995; Middleton and Strick, 1994). Based on the current literature, the precise contribution of the basal ganglia to the processing chain between target determination and muscle activation is difficult to disentangle. On the other hand, recent studies have shown that they may also participate in higher aspects of motor performance, such as the cognitive processes leading to the intention to make a movement (Schultz and Romo, 1992). Thus, an interesting question that may serve as a guideline for future research is: to what extent are the basal ganglia involved in motor *versus* non-motor functions? By non-motor we mean those cognitive processes complementary to strictly motor actions, such as sensory decision-making, motivational driving, attentional and volitional modulation of other neural structures, etc. This chapter addresses the above question by contrasting electrophysiological activity recorded in the basal ganglia and in various neocortical and thalamic areas during motor and non-motor processes. We review the connectivity and possible functions of several elements of the cortico-basal ganglia circuit, and underline the differences and similarities between activity recorded in the various components of the circuit. Regarding motor paradigms, we focus the discussion on the role of preparatory activity; regarding non-motor paradigms, we mainly discuss the results of experiments in which monkeys learned to classify somatosensory stimuli.

## 2. FUNCTIONAL OVERVIEW OF THE CORTICO-BASAL GANGLIA CIRCUITRY

As mentioned above, the basal ganglia are in the middle of a circuit in which, grossly described, information circulates from the cortex into the basal ganglia, then to the thalamus, and then back to the cortex. For short we will refer to this feedback circuit as the cortico-basal ganglia loop. The principal components of the basal ganglia are: the putamen and caudate nucleus (CN)—collectively known as the neostriatum (NS)—the globus pallidus (GP), the substantia nigra *pars compacta* (SNpc) and *pars reticulata* (SNpr) and the subthalamic nucleus (STN). According to current anatomical evidence, virtually the whole cerebral cortex sends convergent and divergent projections to NS, which functions as a gate through which a large fraction of cortical inputs has to enter to access the rest of the basal ganglia (Alexander, DeLong and Strick, 1986; Graybiel and Kimura, 1995; Parent and Hazrati, 1995a). Many of these cortical inputs originate in the frontal lobe and are organized topographically (McGuire, Bates and Goldman-Rakic, 1991; Strick, Dum and Picard, 1995).

A distinction has been drawn between two parallel channels connecting the cortex and the basal ganglia, these are the 'direct' and 'indirect' pathways (DeLong, 1990; but see Parent and Hazrati, 1995a). The direct pathway conveys activity from NS to the cortex via GP (internal segment) and thalamus. The indirect pathway also goes from cortex to NS to GP (external segment), but then it passes through STN and back to GP (internal segment) before reaching the thalamus. In both cases the returning thalamocortical connections seem to reach precisely the regions of frontal cortex that contribute as inputs to NS (Strick, Dum and Picard, 1995). There are at least five separate cortico-basal ganglia loops, which have been identified according to the part of frontal cortex that serves as a target: the skeletomotor, oculomotor, dorsolateral prefrontal, lateral orbitofrontal and anterior cingulate circuits (Alexander, DeLong and Strick, 1986; Hoover and Strick, 1993). These loops seem to be, to a good extent, functionally segregated (Alexander, DeLong and Strick, 1986; but see Chapter 12 by Joel and Weiner). It should also be kept in mind that numerous interconnections amongst the various elements of the basal ganglia have been described (Parent and Hazrati, 1995a,b) .

The excitatory connections from cortex to NS use glutamate as a neurotransmitter, whereas those from NS to GP use GABA, exerting an inhibitory effect on pallidal neurones that are themselves inhibitory. These two inhibitory synapses in series preceding the thalamus in effect act as an excitation: when NS is activated the GP neurones disinhibit the thalamocortical projections and thus presumably facilitate the initiation of movement; suppression of unwanted movements probably ensues through inhibition of the thalamocortical projections, via the indirect pathway (Wichmann, Bergman and DeLong, 1994; Bergman, Wichmann and DeLong, 1990; Graybiel, 1995), because the projections from STN to the internal segment of GP are excitatory (Nakanishi, Kita and Kitai, 1991). Thus, excitatory and inhibitory influences converge on GP, and their balance might be important (Bergman, Wichmann and DeLong, 1990; DeLong, 1990). Neostriatal activity may be modulated in several ways. Probably the most important neuromodulator is dopamine, which is provided by a nigrostriatal projection from SNpc.

The best known pathways in the cortico-basal ganglia loop, both from the anatomical and neurophysiological points of view, are the skeletomotor and oculomotor circuits (Alexander and Crutcher, 1990b,c; Hikosaka and Wurtz, 1983a,b,c; Hikosaka, Sakamoto and Usui, 1989a,b; Hikosaka, Sakamoto and Miyashita, 1993; Romo and Schultz, 1987; Schultz and Romo, 1988). Throughout the skeletomotor loop, neurones have been found with responses that correlate with the parameters of arm motion, in paradigms in which animals performed self-initiated movements or movements triggered by sensory cues of different modalities (Alexander and Crutcher, 1990a; Schultz and Romo, 1992; Romo and Schultz, 1992). Some traditionally non-motor aspects of this pathway have also been investigated (Merchant *et al.*, 1997; Romo *et al.*, 1995; Schultz and Romo, 1988, 1990, 1992; Schultz *et al.*, 1992; Schultz, Apicella and Ljungberg, 1993; Schultz, Dayan and Montague, 1997), but both anatomical (Strick, Dum and Picard, 1995) and lesion studies (DeLong and Georgopoulos, 1981; DeLong, 1990) are in general consistent with the involvement of the basal ganglia in movement control (but see also Wise, 1996). The dopamine cells in SNpc stand out from the general case because they turned out to be related to reward expectation and learning, rather than to movement initiation (Ljungberg, Apicella and Schultz, 1992; Mirenowicz and Schultz, 1994; Romo and Schultz, 1990; Schultz and Romo, 1990; Schultz, Apicella and Ljungberg, 1993; Hyland, chapter 1, this volume). These neurones apparently encode the error or difference between the predicted reward and the actual reward associated with an action, very much along the lines of algorithms used in learning and adaptive-control theories (Mirenowicz and Schultz, 1994, 1996; Schultz, Dayan and Montague, 1997). In contrast, the putamen and CN have indeed been found to react during movement initiation in a context-dependent way (Romo and Schultz, 1992; Schultz and Romo, 1992).

### 2.1. Electrophysiological Activity in the Skeletomotor Loop

Here we briefly review neurophysiological results showing how the various structures in the skeletomotor loop participate in motor behaviour. In cortex, the main characters in this respect are the supplementary motor area (SMA, medial segment of Brodmann's area 6) and the premotor (PM, dorsal and ventrolateral segments of area 6) and primary motor (M1, area 4) cortices, which exhibit activity related to the preparation, initiation and execution of movements. The basal ganglia are very heavily interconnected with them, particularly with SMA (Künzle, 1975, 1977, 1978; Selemon and Goldman-Rakic, 1985; McGuire, Bates and Goldman-Rakic, 1991; Strick, Dum and Picard, 1995).

It seems likely that prefrontal and anterior cingulate cortices, which are involved in the cognitive aspects of action (Fuster, 1973; Niki and Watanabe, 1976; Shima *et al.*, 1991), initiate the programs for the planning and preparation for movement found in SMA and PM (Kurata and Wise, 1988; Okano and Tanji, 1987; Romo and Schultz, 1987, 1992). In the three motor-related cortical areas, movement- and set-related activity reflecting the amplitude, direction and kinematics of limb motion has been found (Alexander and Crutcher, 1990a; Ashe and Georgopoulos, 1994; Georgopoulos, 1995; Kurata, 1989, 1993). Neuronal activity in SMA and M1 may appear similar, especially in simple motor tasks, but SMA neurones seem to encode more complex aspects of motor behaviour (Tanji and Mushiake, 1996). Tanji and Shima (1994) reported evidence indicating that SMA neurones fire selectively in relation to the particular order in which a sequence of movements is carried out; the temporal sequencing of multiple movements is in fact a typical example of the kind of motor task in which SMA is strongly engaged (Tanji and Mushiake, 1996). Using brain imaging techniques, Roland *et al.* (1980) found intense activity in SMA during mental rehearsal of a sequence of finger movements. There is also evidence that SMA is involved in bimanual coordination (Tanji, Okano and Sato, 1987), and monkeys with lesions in SMA cannot solve a task in which a piece of food needs to be pushed out of a hole with one hand and caught with the other, so the food does not fall down (Brinkman, 1984). PM and M1 also respond similarly in a number of paradigms (Caminiti *et al.*, 1991; Shen and Alexander, 1997a,b), but PM seems responsible for more abstract forms of motor control.

Monkeys with premotor lesions cannot reach for a piece of food when the arm needs to go around a barrier (Moll and Kuypers, 1977). Interestingly, a large proportion of PM neurones—but not M1 neurones—are modulated by the direction of gaze, and seem to encode arm-movement direction in gaze-centered co-ordinates (Mushiake, Tanatsugu and Tanji, 1997). Additionally, a smaller percentage of M1 neurones show preparatory activity before impending movements, compared to PM (Tanji and Evarts, 1976; Weinrich and Wise, 1982), and their roles in the control of posture may also be different (Crammond and Kalaska, 1995). It is well known that the discharge of many M1 cortical cells is strongly influenced by the spatial attributes of limb trajectory, such as direction and force (Georgopoulos, 1995; Georgopoulos *et al.*, 1992), and by the geometry and mechanics of the arm (Scott and Kalaska, 1997). Therefore, these three areas probably deal with motor performance at different levels of abstraction, such that M1 is more responsible for the moment to moment control of motor output (Kalaska and Crammond, 1992). Nevertheless, even this process might require, for example, complex visuospatial transformations (Zhang *et al.*, 1997; Shen and Alexander, 1997a).

In many cases, neuronal responses in NS replicate cortical activity. For example, Graziano, Yap and Gross (1994) found bimodal neurones in PM with somatosensory and visual receptive fields that are always in register with each other, and very similar responses were found in the putamen (Graziano and Gross, 1993). Several laboratories have reported activity in both the putamen and CN related to preparation for movement, programming target acquisition and movement initiation and execution (Alexander and Crutcher, 1990a; Crutcher and Alexander, 1990; Kimura, 1990; Gardiner and Nelson, 1992; Romo, Scarnati and Schultz, 1992; Schultz and Romo, 1992; Rolls, 1994). Using the positron emission tomography (PET) imaging technique, cerebral activation in subjects speaking their native language was compared to cerebral activation when the same subjects spoke a second, foreign language that was learned later than their mother tongue (Klein *et al.*, 1994). The only site that was activated differentially when the subjects spoke the second language was the left putamen. This result is consistent with traditional proposals for striatal function, such as fine control or selection of motor sequences. Kermadi and Joseph (1995) described a large variety of responses in CN in a paradigm in which monkeys had to perform remembered sequences of movements. The specific sequence in a given trial was indicated by the order in which LED lights on three targets were flashed. Some neurones were selective for the onset or offset of visual targets, others were related to reward or reward expectation, others were specific for movement sequences, and many responses were modulated by specific targets, but only when these occupied certain positions in the sequence. These results are certainly reminiscent of the activity found in SMA, as described above.

Inactivation of GP leads to a lack of stability in posture (Inase, Buford and Anderson, 1996), but whether pallidal neurones directly encode movement variables or not was the source of some controversy (Mink and Thach, 1991). However, Turner and Anderson (1997) showed, in a task involving reaching movements in two dimensions, that the skeletomotor portions of both external and internals segments of GP indeed reflect movement kinematics. Other studies are consistent with these results. Jaeger, Gilman and Aldridge (1993, 1995) showed that pallidal neurones have some characteristics in common with cortical and striatal neurones, such as activity related to movement preparation or execution. Kimura *et al.* (1996) demonstrated that, during the execution of learned sequences of movements, information is transferred from neurones in the putamen (that respond phasically) to neurones in GP. The observations of Mushiake and Strick (1995) are most interesting in this respect. They found pallidal responses selective for specific, internally-generated movement sequences, as in the SMA (Tanji and Shima, 1994) and CN (Kermadi and Joseph, 1995) studies. However, in addition, they found neurones that were phase-specific: one of these cells would always signal, for example, the third movement of whatever sequence. In this study, about the same numbers of neurones selective for the first, second and third movements were found. This means that an ensemble of such neurones may encode the spatiotemporal development of a sequential movement that is not guided by sensory cues. Contrary to most movement-related studies of the basal ganglia, it seems that phase-specific neurones with these characteristics have not been found elsewhere.

Studies in cats (Cheruel, Dormont and Farin, 1996) and monkeys (Wichmann, Bergman and DeLong, 1994) have revealed neuronal responses in STN that correlate with the preparation and initiation of movement. However, microstimulation of the STN core failed to elicit movements (Wichmann, Bergman and DeLong, 1994). This is consistent with the finding that the 'indirect' projection from STN to GP is excitatory (Nakanishi, Kita and Kitai, 1991), and hence produces a net inhibition of the thalamic neurones targeted by GP. In this way the thalamus could be finely controlled by the balance of excitatory and inhibitory influences, respectively, from the direct and indirect pathways (Bergman, Wichmann and DeLong, 1990; DeLong, 1990).

Neuronal activity from the anterior ventrolateral nucleus of the thalamus (VLa), which is the link between the GP and the SMA, has been recorded in monkeys executing ballistic movements (Forlano *et al.*, 1993; Buford, Inase and Anderson, 1996). The responses found confirmed the participation of this structure in the control of movements, but it appeared that these responses did not encode movement parameters directly. Their role is probably to modulate the activity of downstream connections. There is some evidence indicating that the nucleus ventralis posterior lateralis pars oralis (VPLo) of the thalamus is also involved in movement execution (Aizawa and Tanji, 1994). Considering that the thalamus is a major link between the basal ganglia and the cortex, it is surprising that so few studies have investigated its neurophysiological properties; its responses should provide some insight into how the basal ganglia modify the information provided by the cortex.

With regard to the oculomotor loop, Hikosaka and collaborators have contributed a series of important studies. Using a delayed visuomotor task, they recorded electrophysiological activity from CN, SNpr and STN (Hikosaka and Wurtz, 1983a,b,c; Hikosaka, Sakamoto and Usui, 1989a,b; Matsumura *et al.*, 1992). They found a large variety of neuronal responses, some modulated by saccade and/or visual (sensory) activity, some that signalled the initiation or termination of oculomotor behaviour, and others with memory-contingent activity related to visual stimuli or to saccadic events. Some of these responses were task- or context-dependent. Bruce and Goldberg (1985; Bruce *et al.*, 1985) found similar types of activity in the oculomotor cortical fields of the frontal lobe. It is well known that the frontal eye fields set in motion a neural signal that is transferred to the superior colliculus via CN and SNpr (Hikosaka and Wurtz, 1983c; Hikosaka, Sakamoto and Usui, 1989a,b). This (direct) signal serves to control voluntary eye movements. In fact, the frontal eye fields may participate in the decision process by which one particular location or object is selected as the target for the next saccade (Schall *et al.*, 1995; Thompson *et al.*, 1996). Electrical microstimulation of this area can elicit saccades with as little as 10 μA of injected current (Bruce *et al.*, 1985). Apart from this route, another projection from the SNpr returns to the frontal eye fields via the thalamus. This pathway has the classical feedback organization found in the cortico-basal ganglia circuitry. Matsumura *et al.* (1992) found that STN neurones in the indirect pathway are also related to visual and oculomotor functions. They act through excitatory projections to SNpr, which tonically inhibits the saccadic burst cells in the superior colliculus.

In summary, a number of relatively well segregated subcircuits form part of the cortico-basal ganglia loop. Almost all structures involved exhibit activity related in one way or another to the parameters of motor behaviour. In general, the same kinds of neuronal responses can be found in the basal ganglia and in the cortex, but it is unfortunate that information flow through the various elements of the circuit has not been studied in the same delay task. The apparent lack of functional specificity might be due to the intrinsically parallel nature of the neural machinery that controls movements. However another possibility is that, to reveal true specificity, more complicated paradigms involving particular cognitive processes are required, as has been elegantly demonstrated in the case of the dopamine neurones in SNpc.

## 3. PREPARATION FOR MOVEMENT

As discussed above, the electrophysiological activity circulating between the neocortex, basal ganglia and thalamus appears to be involved in the programming, initiation and execution of intended movements. We now explore one aspect of this activity which may play an important role in delayed-reaction tasks. The common denominator that has emerged in all delayed-movement paradigms is preparatory activity before the animal initiates a behavioural motor reaction. Typically, preparatory activity builds up after an instruction signal that allows the monkey to plan a movement, it rises until the onset of movement, and afterwards it decreases—sometimes very abruptly (see Figure 13.1). Evarts, Shinoda and Wise (1984) interpreted this activity as a 'state of readiness' that allows an animal to elaborate plans for impending actions. This form of activity was predicted from the human studies made by Kornhuber and Deecke (1965), and indeed has been observed in SMA (Alexander and Crutcher, 1990a,b; Kurata and Wise, 1988; Okano and Tanji, 1987; Romo and Schultz, 1987, 1992), PM (Kurata and Wise, 1988; Okano and Tanji, 1987; Romo and Schultz, 1987) and M1 (Alexander and Crutcher, 1990a,b; Tanji and Evarts, 1976), as well as in NS (Alexander and Crutcher, 1990a,b; Romo, Scarnati and Schultz, 1992; Schultz and Romo, 1988, 1992) and GP (Jaeger, Gilman and Aldridge, 1993, 1995). Preparatory activity may last up to several tens of seconds, which can be seen by progressively delaying the onset of movement (Schultz and Romo, 1992). In contrast, activity related to movement initiation is phasic, it starts before electromyographic (EMG) activity and lasts until movement onset. Activity related to movement execution usually starts at EMG onset and is sustained throughout the movement.

The study by Deiber *et al.* (1996) gives a global picture of the central processes involved in motor preparation. Using PET imaging, they looked at the brains of human subjects performing simple finger movements in a reaction time paradigm. The timing of events was designed to emphasize the motor preparation component. They measured an increase in regional cerebral blood flow during movement preparation in the following contralateral regions: (a) frontal cortex (M1, dorsal PM, cingulate, and SMA), (b) parietal cortex (anterior and posterior regions), (c) basal ganglia, (d) thalamus, and (e) ipsilateral

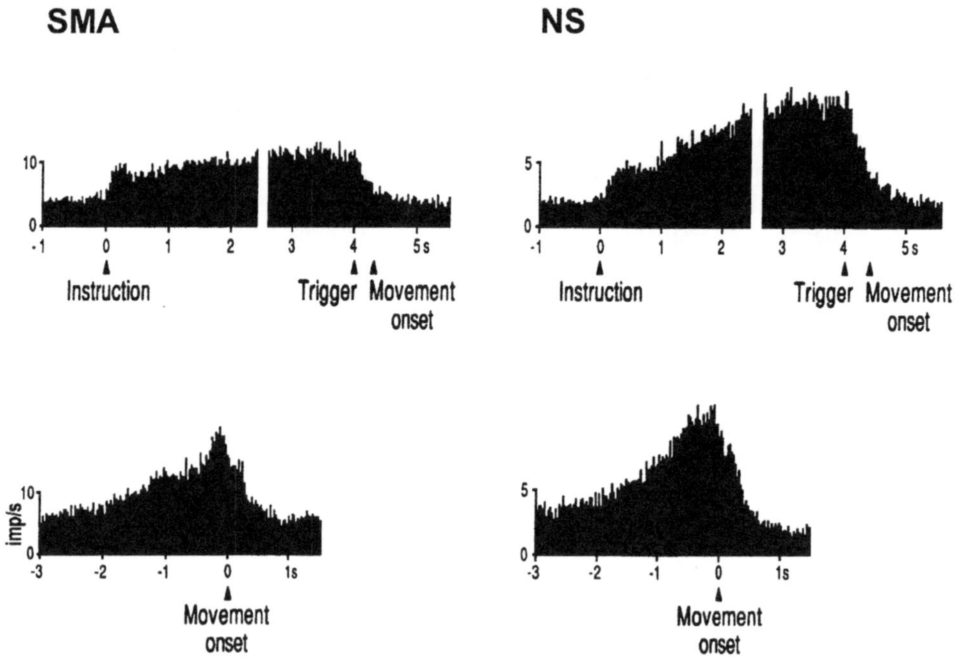


**Figure 13.1.** Preparatory activity in the supplementary motor area (SMA, left column) and in the neostriatum (NS, right column). Each of the four histograms represents the averaged response of a neuronal population; the individual histograms (averaged over trials) of all neurones with statistically significant increases in activity were added bin by bin, and the resulting sum was divided by the number of neurones. Populations include between 43 and 92 neurones, with 10–25 trials per neurone. In all cases, the x axis indicates time and the y axis indicates instantaneous firing rate, in impulses/s. The two histograms on the top row correspond to responses during "go" trials of a visually-instructed delay task. At the time of the instruction the monkeys got ready to make a movement, but they executed it only after a trigger stimulus. Preparatory activity in this condition starts shortly after the instruction and builds up until the trigger appears; afterward it rapidly decreases. The plots are split because the time between instruction and trigger signals was variable; the left sides are aligned with instruction onset and the right sides with trigger onset. Neurones that responded transiently to the instruction, trigger or during movement were excluded. The two histograms on the bottom row correspond to responses during self-initiated movements. In this condition, there were no external cues whatsoever. The monkey periodically—at his own pace—reached toward a small box to check whether a small piece of food was there. The food box was covered, so its contents could not be seen. The histograms show an increase in spiking activity starting about two seconds before the onset of movement. The timecourse of activity is similar in the two areas for both internally-generated and externally-triggered movements. (Data reprinted from Romo and Schultz [*Experimental Brain Research,* 1992 **91**, 396–407, Figure 7,B and C] and Schultz and Romo [*Experimental Brain Research,* 1992, **91**, 363–384, Figure 16, B and C] with permission, copyright: Springer-Verlag)

cerebellum. Except for the parietal areas, all these structures were activated similarly when the subjects were given in advance partial or full information specifying the impending movement; for example, knowing either the direction of motion, or which finger to move, produced activation similar to that when knowing both finger and direction. However, the reaction times were shorter when the full information was available. They also found that SMA was the main area preferentially involved in self-initiated (*versus* cued) movements.

In neurophysiological studies, the structures of the skeletomotor loop with the highest percentages of neurones responding to the preparation of visually-guided limb movements were the SMA and dorsal PM cortices, with more than 50%; these were followed by M1 (37%), putamen (33%) and prefrontal cortex (16%) (Boussaoud and Wise, 1993; Alexander and Crutcher, 1990a,b,c). Jaeger, Gilman and Aldridge (1993) reported similar numbers: 30% in the putamen, 31% in CN, and 27% in GP (internal). The external segment of GP appeared to have fewer neurones with preparatory responses (Jaeger, Gilman and Aldridge, 1993). Schultz and Romo (1992) found a smaller proportion of neurones displaying preparatory activity in the putamen (14%) and CN (10%), but their population pool was larger and their task was somewhat different from the one used by Alexander and Crutcher (1990b,c). The average onsets and offsets of preparatory activity in SMA and M1 occurred significantly earlier than in the putamen. However, preparatory neurones in these three areas were active simultaneously throughout most of the postinstruction interval, consistent with the concept of parallel, distributed processing (Alexander and Crutcher, 1990a,b,c). Similar differences in timing were also seen in the responses related to movement execution (Crutcher and Alexander, 1990), suggesting that SMA neurones are engaged first, and M1 and putamen neurones are recruited later. In these studies, preparatory activity related to arm-movement direction was dissociated from activity related to goal or target direction. Both preparatory neurones selective for movement direction and preparatory neurones selective for target direction were found in all three areas investigated (Alexander and Crutcher, 1990c). Similar results were obtained in M1 and PM when the paradigm was extended to two-dimensional movements (Shen and Alexander, 1997a,b).

Rather than correlating neuronal activity with the parameters of arm motion, Romo and Schultz (1992) investigated how preparatory activity is generated. They compared externally-induced movements triggered by sensory cues and internally-induced or self-initiated movements (see also Kimura *et al.*, 1992). Schultz and Romo (1992) recorded from SMA, PM and NS (CN and putamen) using delayed go/nogo paradigms. Transient and sustained preparatory activity was found in both conditions. In all cases there were roughly twice as many preparatory neurones in SMA as in NS. Transient responses had similar latencies and durations in the two structures, but sustained activity started and peaked earlier in SMA; this was the case for both externally-instructed and self-initiated movements. Nevertheless, due to their long durations, sustained responses in the two structures overlapped substantially. Figure 13.1 illustrates these similarities

Concerning the oculomotor loop, we will just mention that activity in CN related to the preparation for saccades has some similarity with skeletomotor preparatory activity, in the sense that it starts after the target cue, then gradually builds up during the delay period and ceases abruptly at saccade onset (Hikosaka, Sakamoto and Usui, 1989a). It appears that CN actively contributes to the initiation of saccadic eye movements (Hikosaka, Sakamoto and Usui, 1989a,b; Hikosaka, Sakamoto and Miyashita, 1993).

### 3.1. Reverberating Activity along the Cortico-basal Ganglia Circuitry

Preparatory activity may originate in frontal cortical areas with which SMA has heavy reciprocal connections, such as prefrontal, premotor and anterior cingulate cortices. Since preparatory responses show up earlier in SMA than in the striatum, this may imply a sequential development of premovement activity from SMA to striatum (Romo and Schultz, 1992). However, the large temporal overlap of activity in both structures preceding externally- and internally-triggered movements suggests that preparatory activity may develop through feedback interactions in neuronal loops linking several cortical and subcortical structures (Alexander and Crutcher, 1990a; Romo and Schultz, 1992). As discussed above, preparatory activity has been found at practically all levels of the cortico-basal ganglia loop. Dopamine neurones in SNpc stand apart, because they exert an enabling effect on their target areas, such as NS and prefrontal cortex (Romo and Schultz, 1990; Schultz and Romo, 1990). However, in general, it has not been possible to test hypotheses concerning the functional role of preparatory activity.

Instruction-induced premovement activity in SMA and NS often begins with a few impulses, increases slowly and culminates with movement onset (Romo and Schultz, 1992). This is even more evident during self-initiated movements. Figure 13.1 illustrates this development in both situations. The first few impulses in these activity bursts are likely to occur within one component of the loop, or may enter through some of the many inputs to the cortex or basal ganglia. They remain as small fluctuations in spontaneous activity, dying out without further consequences, unless they can synaptically activate the next component of the loop, and keep propagating. This self-boosting effect might enable a slow and steady increase in premovement activity that might built up through successive reverberations between cortex and striatum. In the macaque monkey, the time necessary for one full reverberation from cortex to NS to GP to thalamus and back to cortex could be as short as 35 ms. Preparatory discharges in SMA and striatum begin about 1.3 seconds before self-initiated movements, which would allow about 30 reverberations before the onset of muscle activity and about 40 before the onset of movement. Thus, the feedback architecture of the cortico-basal ganglia loops may be suitable for the slow build up and amplification of neuronal activity preceding voluntary movements.

## 4. NONMOTOR FUNCTIONS

Since the basal ganglia receive inputs from a wide variety of neocortical domains, it seems reasonable to assume that they process information associated with higher functions accompanying motor behaviour. These may include sensory decision-making, learning, memory, attention, motivation, emotion and volitional behaviour (Evarts and Wise, 1984; Rolls, 1994; Nishino *et al.*, 1991; Schultz *et al.*, 1995a,b; Saint-Cyr, Taylor and Nicholson, 1995; Saper, 1996). The experiments on reward-based learning performed by Schultz and colleagues are a prime example of non-motor neuronal behaviour (see Chapter 1 by Hyland). Several other studies have also reported neurones in the ventral striatum that respond to stimuli associated with reinforcement or novelty; some of these neurones reflect the expectation or occurrence of a reward (Apicella *et al.*, 1991, 1992; Rolls, 1994; Schultz *et al.*, 1992; Schultz, Apicella and Ljungberg, 1993). Notice that the character of these reward-related responses is different from that of the dopaminergic neurones in SNpc, which signal the difference between expected and actual reward (Mirenowicz and Schultz, 1994, 1996; Schultz, Dayan and Montague, 1997). We will not elaborate any further on the function of SNpc, but will discuss other cases illustrating that the basal ganglia may be involved in neuronal computations that precede movement or relate to movement only indirectly.

Attention is one form of dynamic neuronal modulation that has been studied in a variety of modalities. However, the participation of the basal ganglia in attentive processes has not been thoroughly explored. Of particular importance is the work of

Kermadi and Boussaoud (1995), who designed a task in which the same visual stimuli, which were presented twice in each trial, could have two different meanings. On its first presentation, the stimulus appeared at a spatial location to which the monkey had to attend; after a variable delay period, the stimulus appeared again and its colour indicated the direction of an arm movement. This paradigm could thus dissociate responses related to visual stimuli, movement or attentional shifts. About 40% of the neurones in NS and 20% of those in PM responded to a stimulus only when it signalled that a spatial location had to be attended, and not when the same stimulus indicated movement direction. Hence these responses seemed to be related to shifts in the focus of attention—and sometimes they were also modulated by spatial location. These neurones were more abundant in NS than in PM, but the finding suggests that these two related structures are engaged in attentional shifts. Attentional effects have been described in other areas controlling eye movements (Bruce and Goldberg, 1985; Bon and Lucchetti, 1997). In an imaging study using PET, it was shown that prefrontal and cingulate cortices are preferentially activated during the learning of new sequences of finger movements, but not during the rehearsal of prelearned sequences (Jueptner *et al.*, 1997). Another structure that was strongly activated during the learning stage was CN. During rehearsal CN was silent, whereas the putamen became active. Activation of the putamen did not come as a surprise, given that it had been previously implicated in the planning or control of sequential movements. On the other hand, one of the possible interpretations of the CN result considered by the authors is that the increase in CN activity was related to reinforcement learning or reward outcome, consistent with the neurophysiological findings of Schultz *et al.* (1992). In fact, Graybiel (1995; Graybiel *et al.*, 1994; Graybiel and Kimura, 1995) has emphasized that the basal ganglia may participate in diverse memory-based learning processes. For example, tonically active striatal neurones can become selective for conditioning stimuli during Pavlovian, classical conditioning (Aosaki *et al.*, 1994; Graybiel *et al.*, 1994). These neurones respond by making a pause in their firing shortly after a conditioned stimulus is presented. They do not respond to the reward itself; they can retain their responsiveness after weeks of intermission; they can lose their selectivity or shift it from one type of stimulus to another; and they can acquire their selectivity within tens of minutes of training. These NS neurones thus quickly learn to signal the reward value of a sensory stimulus. Given the large number of studies in which activity in the basal ganglia is duplicated in the cortex, an interesting question is whether the responses related to conditioned stimuli are confined to NS or can also be found, say, in prefrontal cortex.

The studies discussed above may explain—at the cellular level—the psychophysical results from combined behavioural and lesion experiments (Gaffan, 1996). Monkeys were trained in a visual discrimination task in which correct choices were indicated by a white line on the screen—the secondary reinforcer. A reward was delivered only after four correct trials in a row. To solve the task, the animals had to associate the visual discriminative images with the secondary reinforcer, not with the primary food reward. A combination of lesions was performed such that the only significant output pathway left from the visual association (inferotemporal) cortex was to NS , the amygdala and connections to the hippocampus and prefrontal cortex having been eliminated. Interestingly, inferotemporal cortex also seems to be a major target of SNpr via the thalamus (Middleton and Strick, 1996). The remaining cortico-striatal pathway was capable of mediating visual learning in the task in an almost normal way; this was also true when the secondary reinforcer was switched to an auditory cue (Gaffan, 1996). Thus, there is substantial neurophysiological and psychophysical evidence implicating the basal ganglia in learning and associative processes.

### 4.1. Sensory Decision Making

The different delay tasks designed to study motor functions have used different sensory stimuli as preparatory cues or trigger signals. It has consistently been observed that some neurones in the frontal motor areas (SMA, PM, M1) and NS (putamen and CN) show responses that are time-locked to the sensory stimuli that trigger the behavioural motor reaction. This is a fascinating observation, considering that this kind of response may reflect a link between sensory information and the transformation or redirection of this information to guide motor behaviour. That the basal ganglia may be involved in sensory processing is also supported by anatomical studies demonstrating the existence of cortical input to the putamen from primary somatosensory cortex (Künzle, 1977; Flaherty and Graybiel, 1991). In view of this evidence, the cortico-basal ganglia circuitry could be closely associated with the higher-order sensory information processing needed to perform voluntary movements. Although there has been much progress in understanding how primary sensory cortices process information, and how the motor areas of the brain construct a neural representation of movements, much less is known about the link between sensory perception and motor action. One way of addressing this problem is to use psychophysical paradigms in which these two aspects are strongly dissociated, for example by introducing a decision-making step between the sensory and motor stages of the task.

With these considerations in mind, Romo and collaborators designed a categorization task in which a monkey classified the speed of a moving tactile stimulus as either low or high, and the animal indicated its choice by pressing one of two switches in front of it (Romo *et al.*, 1996). The stimuli were delivered by a computer-controlled tactile stimulator on the tip of one finger of the left, restrained hand. Five of ten possible speeds were designated as low (12, 14, 16, 18 and 20 mm/s) and the rest as high (22, 24, 26, 28 and 30 mm/s). Movements toward the push-buttons were performed with the right hand-arm, contralateral to the stimulus. Low speeds corresponded to a medial button and high speeds to a lateral one. In this task, the stimulus does not directly indicate which button to press; a function of the stimulus, its category, needs to be computed to produce an adequate motor reaction. Extracellular recordings during performance of the task revealed the same repertoire of responses in SMA (Romo *et al.*, 1993, 1997), M1 (Salinas and Romo, 1997) and also in the putamen (Romo *et al.*, 1995; Merchant *et al.*, 1997). Some neurones discharged in relation to the arm movements, as expected; some others increased their activity during stimulation and/or during the reaction time, but this increase was not a function of stimulus speed: these neurones seemed to respond to the existence of a stimulus, ignoring its physical characteristics. However, in all of these areas a fraction of the neurones, between 1/4 and 1/7, responded selectively to the speed categories. For example, some neurones increased their firing rates significantly for all speeds belonging to the 'low' category, but not for speeds belonging to the 'high' category. Another population of neurones did exactly the opposite. The similarity between these differential or category-tuned responses found in cortex and in the putamen is illustrated in Figure 13.2.

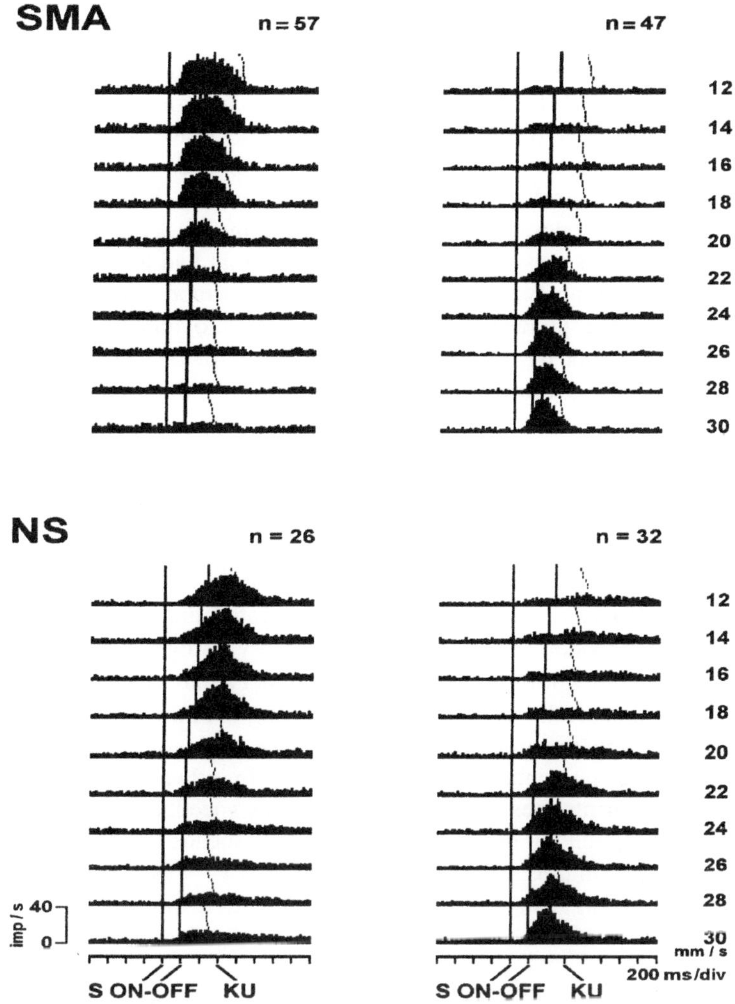


**Figure 13.2.** Activity associated with the categorization of tactile stimuli in SMA (top row) and NS (bottom row). Each of the four panels consists of ten histograms representing the average responses of neuronal populations; the numbers of neurones (n) are indicated above each panel. During the task, monkeys had to categorize the speed of a moving tactile stimulus as either low (12–20 mm/s) or high (22–30 mm/s). They indicated their choice by pressing one of two switches; the medial one corresponded to low speeds and the lateral one to high speeds. Each individual histogram corresponds to a different speed, indicated on the right. In all cases, the y axis indicates instantaneous firing rate, in impulses/s, and the x axis indicates time. All responses are aligned with the onset of probe movement, ie. stimulus onset (S-ON), indicated by the continuous, vertical lines. Stimulus offset (S-OFF) and movement onset (KU) are also indicated in all histograms. Stimulus offset changes with speed because the distance traversed by the stimulating probe was kept constant. Movement onset was the moment at which the monkey released an immovable key and started to project his arm toward one of the switches. The responses shown are selective for the speed categories. Populations on the left column increase their activity when the stimuli belong to the "low" category, whereas those on the right column increase their activity when the stimuli belong to the "high" category. The majority of these responses were task-dependent: they vanished when the stimuli were delivered passively—no arm movements were made—or when the same arm movements were triggered by visual cues. These differential responses, most likely associated with the output of the categorization process, were very similar in SMA and NS. (Data from NS are unpublished results from Romo *et al.*; data from SMA is reprinted, with permission, from Romo *et al.* (*Cerebral Cortex*, 7, 317–326, 1997).

Three lines of evidence indicated that the differential responses were neither purely sensory nor exclusively related to the parameters of arm motion: (1) None of the differential neurones ever recorded in these areas responded during passive stimulation. In this condition the right hand of the animal was restrained. The same tactile stimuli used for categorization were delivered, but the monkey made no movements and there was no reward. This result demonstrated that the differential responses were not simply stimulus driven. (2) In another control situation the animal performed the same arm movements but was guided by visual cues. In this case no tactile stimuli were delivered. In each trial one of the target switches was illuminated, and the animal had to press it after the light had turned off. More than two thirds of the category-tuned neurones either did not respond in this task, or responded identically for the two movements. Hence their differential activation could not be interpreted as preference for one of the two arm movements. (3) Modeling and computer-simulation results indicated that the monkey's psychophysical performance in the task could be accurately predicted from the differential responses. In other words, an observer measuring the responses of a population of category-tuned neurones in each trial could estimate the speed category—using decoding techniques—with the same accuracy as the monkey; the percent of correct categorizations as a function of speed of this hypothetical observer matched very precisely the percent of correct categorizations exhibited by the monkeys during the recording sessions (Salinas and Romo, 1997). These results suggested that the category-tuned neurones act as a link between the output of the sensory categorization process, which encodes the selected category, and the neural machinery that generates a motor action, which physically indicates the decision. It is interesting to compare these responses with those in the primary sensory area. In the same task, neurones of primary somatosensory cortex contralateral to the stimulated hand responded in either of two ways: some neurones increased their firing rates linearly with increasing speed, and others increased their activity during stimulation but did so identically for all speeds (Romo *et al.*, 1996). This representation of the stimulus contrasts sharply with what was observed in the motor-related areas, which suggests that the category-tuned responses are indeed highly processed signals that are derived from the sensory stimulus, but that are not strictly sensory.

These findings provide another example in which the putamen is activated during non-motor processes. However, again there is a large overlap between its activity and the activity that is observed in the motor-related areas of the cortex.

## 5. CONCLUSION

The current neurophysiological literature definitely suggests that the basal ganglia are involved in diverse aspects of motor control, particularly during complex sequences of movements. On the other hand, at least some of its components also play a role in non-motor aspects of behaviour, such as stimulus-stimulus and stimulus-reward association, attentional shifting and sensory-to-motor transfer of information. Therefore, one asks: is there anything unique about the basal ganglia? It appears that, whenever a direct comparison has been possible, most responses recorded in the basal ganglia had similar counterparts in one or another region of frontal cortex. Few exceptions to this general rule can be counted. The reverberating cortico-basal ganglia loops thus appear to be part of a large, parallel network that may subserve a multitude of functions, not necessarily motor, including sensory, associative and limbic processes. However, an alternative possibility is that the basal ganglia have not been properly probed by the existing behavioural paradigms; new tasks emphasizing other cognitive processes may be needed to uncover a general functional principle underlying the feedback organization of the cortico-basal ganglia circuits.

## ACKNOWLEDGEMENTS

The research of R. Romo was supported in part by an International Research Scholars Award from the Howard Hughes Medical Institute and grants from DGAPA-UNAM, CONACyT and Fundación Miguel Alemán, A.C.

## REFERENCES

Aizawa, H. and Tanji, J. (1994) Corticocortical and thalamocortical responses of neurons in the monkey primary motor cortex and their relation to a trained motor task. *Journal of Neurophysiology*, **71**, 550–560.

Alexander, G.E. and Crutcher, M.D. (1990a) Functional architecture of basal ganglia circuits: Neural substrates of parallel processing. *Trends in Neurosciences*, **13**, 266–271.

Alexander, G.E. and Crutcher, M.D. (1990b) Preparation for movement: Neural representations of intended direction in three motor areas of the monkey. *Journal of Neurophysiology*, **64**, 133–150.

Alexander, G.E. and Crutcher, M.D. (1990c) Neural representations of the target (goal) of visually guided arm movements in three motor areas of the monkey. *Journal of Neurophysiology*, **64**, 164–178.

Alexander, G.E., Delong, M.R. and Strick, P.L. (1986) Parallel organization of functionally segregated circuits linking basal ganglia and cortex. *Annual Review of Neuroscience*, **9**, 357–381.

Aosaki, T., Tsubokawa, H., Ishida, A., Watanabe, K., Graybiel, A. M. and Kimura, M. (1994) Responses of tonically active neurons in the primate striatum undergo systematic changes during behavioral sensorimotor conditioning. *Journal of Neuroscience*, **14**, 3969–3984.

Apicella, P., Ljungberg, T., Scarnati, E. and Schultz, W. (1991) Responses to reward in monkey dorsal and ventral striatum. *Experimental Brain Research*, **85**, 491–500.

Apicella, P., Ljungberg, T., Scarnati, E. and Schultz, W. (1992) Neuronal activity in the monkey striatum related to the expectation of predictable environmental events. *Journal of Neurophysiology*, **68**, 945–960.

Ashe, J. and Georgopoulos, A.P. (1994) Movement parameters and neural activity in motor cortex and area 5. *Cerebral Cortex*, **6**, 590–600.

Bergman, H., Wichmann, T. and DeLong, M.R. (1990) Reversal of experimental parkinsonism by lesions of the subthalamic nucleus. *Science, New York*, **249**, 1436–1438.

Bon, L. and Lucchetti, C. (1997) Attention related neurons in the supplementary eye field of the macaque monkey. *Experimental Brain Research*, **113**, 180–185.

Boussaoud, D. and Wise, S.P. (1993) Primate frontal cortex: effects of stimulus and movement. *Experimental Brain Research*, **95**, 28–40.

Brinkman, C. (1984) Supplementary motor area of the monkeyís cerebral cortex: Short- and long-term deficits after unilateral ablation and the effects of subsequent callosal section. *Journal of Neuroscience*, **4**, 918–929.

Bruce, C.J. and Goldberg, M.E. (1985) Primate frontal eye fields. I. Single neurons discharging before saccades. *Journal of Neurophysiology*, **53**, 603–635.

Bruce, C. J., Goldberg, M.E., Bushnell, M.C. and Stanton, G.B. (1985) Primate frontal eye fields. II. Physiological and anatomical correlates of electrically evoked eye movements. *Journal of Neurophysiology*, **54**, 714–734.

Buford, J.A., Inase, J.A. and Anderson, M.E. (1996) Contrasting locations of pallidal-receiving neurons and microexcitable zones in primate thalamus. *Journal of Neurophysiology*, **75**, 1105–1116.

Caminiti, R., Johnson, PB., Galli, C., Ferraina, S. and Bournod, Y. (1991) Making arm movements within different parts of space: The premotor and motor cortical representations of a coordinate system for reaching to visual targets. *Journal of Neuroscience*, **11**, 1182–1197.

Cheruel, F., Dormont, J.F. and Farin, D. (1996) Activity of neurons of the subthalamic nucleus in relation to motor performance in the cat. *Experimental Brain Research*, **108**, 206–220.

Crammond, D.J. and Kalaska, J.F. (1995) Differential relation of discharge in primary motor cortex and premotor cortex to movements versus actively maintained postures during a reaching task. *Experimental Brain Research*, **108**, 45–61.

Crutcher, M.D. and Alexander, G.E. (1990) Movement-related neuronal activity selectively coding either direction or muscle pattern in three motor areas of the monkey. *Journal of Neurophysiology*, **64**, 151–163.

Deiber, M.P., Ibanez, V., Sadato, N. and Hallett, M. (1996) Cerebral structures participating in motor preparation in humans: a positron emission tomography study. *Journal of Neurophysiology*, **75**, 233–47.

DeLong, M.R. (1990) Primate models of movement disorders of basal ganglia origin. *Trends in Neuroscience*, **13**, 281–285.

DeLong, M.R. and Georgopoulos, A.P. (1981) Motor functions of the basal ganglia. In *Handbook of Physiology. Cerebellum and basal ganglia. vol. 2, sect. 1*, edited by V.B. Brooks Bethesda, MD: American Physiological Society, pp. 1017–1061.

Evarts, E.V. and Wise, S.P. (1984) Basal ganglia outputs and motor control. In *Functions of the Basal Ganglia. Ciba Foundation Symposium, no. 107*, edited by D. Evered and M. O'Connor, London, Pitman, pp. 83–102.

Evarts, E.V., Shinoda, Y. and Wise, SP (1984) *Neurophysiological Approaches to Higher Brain Function*. John Wiley & Sons, Inc.

Flaherty, A.W. and Graybiel, A.M. (1991) Corticostriatal transformations in the primate somatosensory system. Projections from physiologically mapped body-part representations. *Journal of Neurophysiology*, **66**, 1249–1263.

Forlano, L.M., Horne, M.K. Butler, E.G. and Finkelstein, D. (1993) Neural activity in the monkey anterior ventrolateral thalamus during trained, ballistic movements. *Journal of Neurophysiology*, **70**, 2276–2288.

Fuster, J.M. (1973) Unit activity of prefrontal cortex during delayed-response performance: neuronal correlates of transient memory. *Journal of Neurophysiology*, **36**, 61–78.

Gaffan, D. (1996) Memory, action and the corpus striatum: current developments in the memory-habit distinction. *Seminars in the Neurosciences*, **8**, 33–38.

Gardiner, T.W. and Nelson, R.J. (1992) Striatal neuronal activity during the initiation of hand movements made in response to visual and vibratory cues. *Experimental Brain Research*, **92**, 15–26.

Georgopoulos A.P. (1995) Current issues in directional motor control. *Trends in Neuroscience*, **18**, 506–510.

Georgopoulos A.P., Ashe J., Smyrnis N., Taira M (1992) The motor cortex and the coding of force. *Science, New York*, **256**, 1692–1695.

Graybiel, A.M. (1995) Building action repertoires: memory and learning functions of the basal ganglia. *Current Opinion in Neurobiology*, **5**, 733–741.

Graybiel, A.M., Aosaki, T., Flaherty, A.W. and Kimura, M. (1994) The basal ganglia and adaptive motor control. *Science, New York*, **265**, 1826–1831.

Graybiel, A.M. and Kimura, M. (1995) Adaptive neural networks in the basal ganglia. In *Models of Information Processing in the Basal Ganglia*, edited by J.C. Houk, J.L. Davis and D.G. Beiser, Cambridge, MA: MIT Press, pp. 103–116.

Graziano, M.S. and Gross, C.G. (1993) A bimodal map of space: somatosensory receptive fields in the macaque putamen with corresponding visual receptive fields. *Experimental Brain Research*, **97**, 96–109.

Graziano, M.S., Yap, G.S. and Gross, C.G. (1994) Coding of visual space by premotor neurons. *Science, New York*, **266**, 1054–1057.

Hikosaka, O. and Wurtz, R.H. (1983a) Visual and Oculomotor functions of monkey substantia nigra pars reticulata. I. Relation of Visual and auditory responses to saccades. *Journal of Neurophysiology*, **49**, 1230–1253.

Hikosaka, O. and Wurtz, R.H. (1983b) Visual and Oculomotor functions of monkey substantia nigra pars reticulata. III. Memory-contingent visual and saccade responses. *Journal of Neurophysiology*, **49**, 1268–1284.

Hikosaka, O. and Wurtz, R.H. (1983c) Visual and Oculomotor functions of monkey substantia nigra pars reticulata. IV. Relation of substantia nigra to superior colliculus. *Journal of Neurophysiology*, **49**, 1254–1267.

Hikosaka, O., Sakamoto, M. and Usui, S. (1989a) Functional properties of monkey caudate neurons. I. Activities related to saccadic eye movements. *Journal of Neurophysiology*, **61**, 780–798.

Hikosaka, O., Sakamoto, M. and Usui, S. (1989b) Functional properties of monkey caudate neurons. III. Activities related to expectation of target and reward. *Journal of Neurophysiology*, **61**, 814–832.

Hikosaka, O., Sakamoto, M. and Miyashita, N. (1993) Effects of caudate nucleus stimulation on substantia nigra cell activity. *Experimental Brain Research*, **95**, 457–472.

Hoover, J.E. and Strick, P.L.(1993) Multiple output channels in the basal ganglia. *Science, New York*, **259**, 819–821.

Houk J.C. and Wise, S.P. (1995) Distributed modular architectures linking basal ganglia, cerebellum and cerebral cortex: their role in planning and controlling action. *Cerebral Cortex*, **2**, 95–110.

Inase, M., Buford, J.A., and Anderson, M.E. (1996) Changes in the control of arm position, movement and thalamic discharge during local inactivation in the globus pallidus of the monkey. *Journal of Neurophysiology*, **75**, 1087–1104.

Jaeger, D., Gilman, S. and Aldridge, J.W. (1993) Primate basal ganglia activity in a precued reaching task: preparation for movement. *Experimental Brain Research*, **95**, 51–64.

Jaeger, D., Gilman, S. and Aldridge, J.W. (1995) Neuronal activity in the striatum and pallidum of primates related to the execution of externally cued reaching movements. *Brain Research*, **694**, 111–127.

Jueptner, M., Frith, C.D., Brooks, D.J., Frackowiak, R.S. and Passingham, R.E. (1997b) Anatomy of motor learning. II. Subcortical structures and learning by trial and error. *Journal of Neurophysiology*, **77**, 1325–1337.

Kalaska, J.F. and Crammond, D.J. (1992) Cerebral cortical mechanisms of reaching movements. *Science, New York*, **255**, 1517–1522.

Kermadi, I. and Boussaoud, D. (1995) Role of the primate striatum in attention and sensorimotor processes: comparison with premotor cortex. *NeuroReport*, **6**, 1177–1181.

Kermadi, I. and Joseph, J.P. (1995) Activity of caudate nucleus of monkey during spatial sequencing. *Journal of Neurophysiology*, **74**, 911–933.

Kimura, M. (1990) Behaviorally contingent property of movement-related activity of the primate putamen. *Journal of Neurophysiology*, **63**, 1277–1296

Kimura, M., Aosaki, T., Hu, Y., Ishida, A. and Watanabe, K. (1992) Activity of primate putamen neurons is selective to the mode of voluntary movement: visually guided, self-initiated or memory-guided. *Experimental Brain Research*, **89**, 473–477.

Kimura, M., Kato, M., Shimazaki, H., Watanabe, K. and Matsumoto, N. (1996) Neural information transferred from putamen to the globus pallidus during learned movement in the monkey. *Journal of Neurophysiology*, **76**, 3771–3786.

Klein, D., Zatorre., R. J., Milner, E., Meyer, E. and Evans, A.C. (1994) Left putaminal activation when speaking a second language: evidence from PET. *NeuroReport*, **5**, 2295–2297.

Kornhuber, H.H. and Deecke, L. (1965) Hirnpotentialänderungen bei Wilkürbewegungen und passiven Bewegungen des Menschen: Bereitschaftspotential und reafferente Potentiale. *Pflügers Archives*, **284**, 1–17.

Künzle, H. (1975) Bilateral projections from precentral motor cortex to the putamen and other parts of the basal ganglia. An autoradiographic study in Macaca fascicularis. *Brain Research*, **88**, 195–209.

Künzle, H. (1977) Projections from the primary somatosensory cortex to the putamen and other parts of the basal ganglia. *Experimental Brain Research*, **30**, 481–492.

Künzle, H. (1978) An autoradiographic analysis of the efferent connections from premotor and adjacent prefrontal regions (area 6 and 9) in Macaca fascicularis. *Brain Behavior and Evolution*, **15**, 185–234.

Kurata, K. (1989) Distribution of neurons with set- and movement-related activity before hand and foot movements in the premotor cortex of rhesus monkeys. *Experimental Brain Research*, **77**, 245–256.

Kurata, K. (1993) Premotor cortex of monkeys: set- and movement-related activity reflecting amplitude and direction of wrist movements. *Journal of Neurophysiology*, **69**, 187–200.

Kurata, K. and Wise, S.P. (1988) Premotor cortex of rhesus monkeys: set-related activity during two conditional motor tasks. *Experimental Brain Research*, **88**, 283–291.

Ljungberg, T., Apicella, P. and Schultz, W. (1992) Responses of monkey dopamine neurons during learning of behavioral reactions. *Journal of Neurophysiology*, **67**, 145–163.

Matsumura, M., Kojima J., Gardiner, T.W. and Hikosaka, O. (1992) Visual and oculomotor functions of monkey subthalamic nucleus. *Journal of Neurophysiology*, **67**, 1615–1632.

McGuire, P.K., Bates J.F. and Goldman-Rakic, P.S. (1991) Interhemispheric integration: II. Symmetry and convergence of the corticostriatal projections of the left and the right principal sulcus (PS) and the left and the right supplementary motor area (SMA) of the rhesus monkey. *Cerebral Cortex*, **1**, 408–417.

Merchant, H., Zainos, A., Hernåndez, A., Salinas, E. and Romo, R. (1997) Functional properties of primate putamen neurons during the categorization of tactile stimuli. *Journal of Neurophysiology*, **77**, 1132–1154.

Middleton, F.A. and Strick, P.L. (1994) Anatomical evidence for cerebellar and basal ganglia involvement in higher cognitive function. *Science, New York*, **266**, 458–461.

Middleton, F.A. and Strick, P.L. (1996) The temporal lobe is a target of output from the basal ganglia. *Proceedings of the National Academy of Sciences, U.S.A.*, **93**, 8683–8687.

Mink, J.W. (1996) The basal ganglia: focused selection and inhibition of competing motor programs. *Progress in Neurobiology*, **50**, 381–425.

Mink, J.W. and Thach, W.T. (1991) Basal ganglia motor control. I. Nonexclusive relation of pallidal discharge to five movement modes. *Journal of Neurophysiology*, **65**, 273–300.

Mirenowicz, and Schultz, W. (1994) Importance of unpredictability for reward responses in primate dopamine neurons. *Journal of Neurophysiology* **72**, 1024–1027.

Mirenowicz, and Schultz, W. (1996) Preferential activation of midbrain dopamine neurons by appetitive rather than aversive stimuli. *Nature, London,* **379**, 449–451.

Moll, L. and Kuypers, H.G.J.M. (1977) Premotor cortical ablations in monkeys: contralateral changes in visually guided reaching behavior. *Science, New York,* **198**, 317–319.

Mushiake, H. and Strick, P.L. (1995) Pallidal neuron activity during sequential arm movements. *Journal of Neurophysiology,* **74**, 2754–2758.

Mushiake H., Tanatsugu Y. and Tanji J. (1997) Neuronal activity in the ventral part of premotor cortex during target-reach movement is modulated by direction of gaze. *Journal of Neurophysiology,* **78**, 1132–1154.

Nakanishi, H., Kita, H. and Kitai, S.T. (1991) Intracellular study of the rat entopeduncular nucleus neurons in an in vitro slice preparation: response to subthalamic stimulation. *Brain Research,* **549**, 285–291.

Niki, H. and Watanabe, M. (1976) Cingulate unit activity during timing behavior in the monkey. *Brain Research,* **117**, 213–224.

Nishino, H., Hattori, S., Muramoto, K. and Ono, T. (1991) Basal ganglia neural activity during operant feeding behavior in the monkey: relation to sensory integration and motor execution. *Brain Research Bulletin,* **27**, 463–468.

Okano, K. and Tanji, J. (1987) Neuronal activities in the primate motor fields of the agranular frontal cortex preceding visually triggered and self-paced movements. *Experimental Brain Research,* **66**, 155–166.

Parent, A. and Hazrati, L.N. (1995a) Functional anatomy of the basal ganglia. I. The cortico-basal ganglia-thalamo-cortical loop. *Brain Research Reviews,* **20**, 91–127.

Parent, A. and Hazrati, L.N. (1995b) Functional anatomy of the basal ganglia. II. The place of subthalamic nucleus and external pallidum in basal ganglia circuitry. *Brain Research Reviews,* **20**, 128–154.

Roland, P.E., Larsen, B., Lassen, N.A., Skinh¯j, E. (1980) Supplementary motor area and other cortical areas in organization of voluntary movements in man. *Journal of Neurophysiology,* **43**, 118–136.

Rolls, E.T. (1994) Neurophysiology, and cognitive functions of the striatum. *Revue de Neurologie (Paris),* **150**, 648–60.

Romo, R., Merchant, H., Ruiz, S., Crespo, P. and Zainos, A. (1995) Neuronal activity of primate putamen during categorical perception of somaesthetic stimuli. *NeuroReport,* **6**, 1013–1017.

Romo, R., Merchant, H., Zainos, A., Hernández, A. (1997) Categorical perception of somesthetic stimuli: psychophysical measurements correlated with neuronal events in primate medial premotor cortex. *Cerebral Cortex,* **7**, 317–326.

Romo, R., Merchant, H., Zainos, A., Hernández, A. (1996) Categorization of somaesthetic stimuli: sensorimotor performance and neuronal activity in primary somatic sensory cortex of awake monkeys. *NeuroReport,* **7**, 1273–1279.

Romo, R., Ruiz, S., Crespo, P., Zainos, A. and Merchant, H. (1993) Representation of tactile signals in primate supplementary motor area. *Journal of Neurophysiology,* **70**, 2690–2694.

Romo, R., Scarnati, E., and Schultz, W. (1992) Role of primate basal ganglia and frontal cortex in the internal generation of movements. II. Movement-related activity in the anterior striatum. *Experimental Brain Research,* **91**, 385–395.

Romo, R. and Schultz, W. (1987) Neuronal activity preceding self-initiated or externally timed arm movements in area 6 of monkey cortex. *Experimental Brain Research,* **67**, 656–662.

Romo, R. and Schultz, W. (1990) Dopamine neurons of the monkey midbrain: contingencies of responses to active touch during self-initiated arm movements. *Journal of Neurophysiology,* **63**, 592–606.

Romo, R. and Schultz, W. (1992) Role of primate basal ganglia and frontal cortex in the internal generation of movements. III. Neuronal activity in the supplementary motor area. *Experimental Brain Research,* **91**, 396–407.

Saint-Cyr, J. A., Taylor, A.E. and Nicholson, K. (1995) Behavior and the basal ganglia. *Advances in Neurology,* **65**, 1–28.

Salinas, E. and Romo. R. (1998) Conversion of sensory signals into motor commands in primary motor cortex. *Journal of Neuroscience,* **18**, 499–511.

Saper, C.B. (1996) Role of the cerebral cortex and striatum in emotional motor response. In: *The Emotional motor system. (Progress in Brain Research, vol. 107),* edited by C. Holstege, R. Bamdler and C.B. Saper, Amsterdam, Elsevier, pp. 537–550.

Schall, J.D., Hanes, D.P., Thompson, K.G. and King, D.J. (1995) Saccade target selection in frontal eye field of macaque. I. Visual and premovement activation. *Journal of Neuroscience,* **15**, 6905–6918.

Schultz, W., Apicella, P., Scarnati, E. and Ljungberg, T. (1992) Neuronal activity in monkey ventral striatum related to the expectation of reward. *Journal of Neuroscience,* **12**, 4595–4610.

Schultz W., Apicella, P. and Ljungberg, T. (1993) Response of monkey dopamine neurons to reward and conditioned stimuli during successive steps of learning a delayed response task. *Journal of Neuroscience,* **13**, 900–913.

Schultz, W., Apicella, P., Romo, R. and Scarnati, E. (1995a) Context-dependent activity in primate striatum reflecting past and future behavioral events. In *Models of information processing in the basal ganglia*, edited by J.C. Houk, J.L Davis and D.G. Beiser, Cambridge, MA: MIT Press, pp. 11–27.

Schultz, W., Dayan, P. and Montague, P.R. (1997) A neural substrate of prediction and reward. *Science, New York*, **275**, 1593–1599.

Schultz, W. and Romo, R. (1988) Neuronal activity in the monkey striatum during the initiation of movements. *Experimental Brain Research*, **71**, 431–436.

Schultz, W. and Romo, R. (1990) Dopamine neurons of the monkey midbrain: contingencies of responses to stimuli eliciting immediate behavioral reactions. *Journal of Neurophysiology*, **63**, 607–624.

Schultz, W. and Romo, R. (1992) Role of primate basal ganglia and frontal cortex in the internal generation of movements. I. Preparatory activity in the anterior striatum. *Experimental Brain Research*, **91**, 363–384.

Schultz, W., Romo, R., Ljungberg, T., Mirenowicz, J., Hollerman, J.R. and Dickinson, A. (1995b) Reward-related signals carried by dopamine neurons. In: *Models of information processing in the basal ganglia*, edited by J.C. Houk, J.L. Davis and D.G. Beiser, Cambridge, MA: MIT Press, pp. 233–248.

Scott, S.H. and Kalaska, J.F. (1997) Reaching movements with similar hand paths but different arm orientations. *Journal of Neurophysiology*, **77**, 826–852.

Selemon L.D. and Goldman-Rakic, P.S. (1985) Longitudinal topography and interdigitation of corticostriatal projections in the rhesus monkey. *Journal of Neuroscience*, **5**, 776–794.

Shen, L. and Alexander, G.E. (1997a) Neural correlates of a spatial sensory-to-motor transformation in primary motor cortex. *Journal of Neurophysiology*, **77**, 1171–1194.

Shen, L. and Alexander, G.E. (1997b) Preferential representation of instructed target location versus limb trajectory in dorsal premotor area. *Journal of Neurophysiology*, **77**, 1195–1212.

Shima, K., Aya, K., Mushiake, H., Inase, H., Aizawa, H. and Tanji, J. (1991) Two movement-related foci in the primate cingulate cortex observed in signal-triggered and self-paced forelimb movements. *Journal of Neurophysiology*, **65**, 188–202.

Strick, P.L., Dum, R.P. and Picard, N. (1995) Macro-organization of the circuits connecting the basal ganglia with the cortical motor areas. In *Models of Information Processing in the Basal Ganglia*, edited by J.C. Houk, J.L. Davis and D.G. Beiser, Cambridge, MA: MIT Press, pp. 117–130.

Tanji, J. and Evarts, E.V. (1976) Anticipatory activity of motor cortex neurons in relation to direction of an intended movement. *Journal of Neurophysiology*, **39**, 1062–1068.

Tanji, J. and Mushiake, H. (1996) Comparison of neuronal activity in the supplementary motor area and primary motor cortex. *Cognitive Brain Research*, **3**, 143–150.

Tanji, J. and Shima, K. (1994) Role for supplementary motor area cells in planning several movements ahead. *Nature, London*, **371**, 413–416.

Tanji, J, Okano, K. and Sato, K.C. (1987) Relation of neurons in the nonprimary motor cortex to bilateral hand movement. *Nature, London*, **327**, 618–620.

Thompson, K.G., Hanes, D.P., Bichot, N.P. and Schall, J.D. (1996) Perceptual and motor processing stages identified in the activity of macaque frontal eye field neurons during visual search. *Journal of Neurophysiology*, **76**, 4040–4055.

Turner, R.S. and Anderson, M.E. (1997) Pallidal discharge related to the kinematics of reaching movements in two dimensions. *Journal of Neurophysiology*, **77**, 1051–1074.

Weinrich, M. and Wise, S.P. (1982) The premotor cortex of the monkey. *Journal of Neuroscience*, **2**, 1329–1345.

Whichmann, T., Bergman, H., DeLong, M.R. (1994) The primate subthalamic nucleus. I. Functional properties. *Journal of Neurophysiology*, **72**, 494–506.

Wise, S. (1996) The role of the basal ganglia in procedural memory. *Seminars in the Neurosciences*, **8**, 39–46.

Zhang, J., Riehle, A., Requin, J. and Kornblum, S. (1997) Dynamics of single neuron activity in monkey primary motor cortex related to sensorimotor transformation. *Journal of Neuroscience*, **17**, 2227–2246.

# 14 Discussion Section

After receiving chapters from contributors to this book the editors sent out letters, with questions on matters arising from each chapter, or seeking clarification on unresolved issue. Copies of various chapters of other authors were also despatched along with these letters, to encourage the discussion. The following chapter contains an edited version of the questions posed, and the responses received. In a few places the editors felt it was relevant for them also to participate in the discussion. However, it is made clear below when the editors are writing as protagonists, and when they are acting as editors. For clarity, the discussion section is subdivided into broad subjects area

## 1. PSYCHOLOGY OF INCENTIVE LEARNING

**Editors:** The chapter by Beninger and Olmstead discussed some rather fundamental issues about the relation between reinforcing effects of stimuli, the stimuli themselves, and the motivational states they may be related to. It is quite important to have the case presented that stimuli in themselves do not have intrinsic reinforcing properties, but only acquire such properties when they occur in such a way as to predict favourable changes in internal motivational states. In the chapter by Miller, this case was accepted as far as appetitive reinforcement goes. However it is uncertain whether the argument applies generally to other sorts of reinforcer. There may be some stimuli which, in themselves, are inherently aversive, so that one does not need to have the mediation of a separate motivational state. Beninger and Olmstead were asked to comment on this issue.

**Beninger/Olmstead:** In our chapter, we outlined the work of Dickinson and colleagues, which proposes that the influence of primary motivational states on goal directed actions is indirect. That is, the incentive value of a stimulus, as it relates to a particular motivational state, must be acquired through experience with the stimulus in the relevant state. This account necessitates the existence of neuronal representations of the action, the outcome, and the motivational significance of the latter. Incentive learning mediates the association that is formed between the neuronal representation of an outcome and its incentive value. It is critical to note that the representation of the incentive value of the outcome must be modifiable for incentive learning to occur.

Our chapter dealt almost entirely with appetitive reinforcement and we are asked to comment on the influence of aversive stimuli on operant responding. The fact that operant responding for a sucrose solution that was previously paired with an injection of lithium chloride is reduced only if animals are re-exposed to the sucrose solution prior to testing suggests that incentive learning mediates the food-nausea (stimulus-aversion) association. On the other hand, it seems intuitive that some stimuli (i.e. shock) are inherently aversive such that the incentive value of these stimuli does not need to be learned through experience with the stimuli in the relevant motivational state. The difficulty with this interpretation is that, unlike the motivational significance of food, shock is aversive in all states. With reference to Dickinson's model, the neuronal representation of the stimulus is not modifiable. Theoretically, if the aversiveness of such stimuli could be alleviated, would incentive learning mediate the re-assignment of incentive value to the stimuli? For example, if animals were trained to lever press to avoid shock and were then presented with the shock under the influence of analgesics, would lever pressing be reduced when animals were not re-exposed to the shock under the influence of the drug?

The question then arises, "Under what conditions can motivational states (either appetitive or aversive) directly influence goal-directed behaviours?" Two lines of research are relevant to this issue. First, it is important to consider that the effectiveness of incentive learning on instrumental performance varies with the level of training (Dickinson *et al.*, 1995). In general, extended training reduces the ability of re-exposure to a goal object in a shifted motivational state to alter response rates during an extinction test. To account for this over-training effect, Dickinson *et al.* (1995) propose that instrumental performance may be governed by two independent processes. The action-outcome mechanism controls intentional, goal-directed behaviour, through knowledge about the relationship between the instrumental action and its consequences. In contrast, the stimulus-response mechanism controls behaviour through a strengthened association between environmental stimuli and the operant response. Extended training shifts control of the instrumental response from the action-outcome to the stimulus-response system. Because incentive learning mediates the former, but not the latter, response levels of well-trained animals are directly sensitive to shifts in motivational state.

Second, although they emphasize a role for incentive learning, Dickinson and colleagues acknowledge that Pavlovian conditioning processes may also contribute to the motivational control of instrumental performance (Dickinson and Balleine, 1994; Dickinson *et al.,* 1995). Through their association with the goal object, contextual or discriminative stimuli that are present during operant training may acquire the capacity to support instrumental performance during an extinction test. The incentive properties of these conditioned stimuli, as measured in a Pavlovian instrumental transfer test, appear to be sensitive to shifts in motivational states, independent of incentive learning processes (Dickinson and Balleine, 1990; Dickinson and Dawson, 1987). Thus, during an extinction test when animals were thirsty, instrumental responses associated with a food reward were increased in the presence of a cue previously paired with the presentation of a sucrose solution, even though the animals had no experience with the sucrose solution in a thirst state. The effect was not observed when animals were tested hungry. These results suggest that, unlike the association between an operant response and its outcome, associations between contextual or discriminative stimuli and the outcome are not mediated through an incentive learning process. More importantly, the Pavlovian processes that contribute to instrumental responding are directly sensitive to shifts in motivational states. Because motivational states directly influence responding in the Pavlovian instrumental transfer test, but not in tests of incentive learning, the associations that are formed between contextual stimuli and the outcome cannot be represented within the same memory system as that which encodes the instrumental contingency.

On the topic of Pavlovian associations, we should make reference to the work of Davidson (1993; 1998). He objects to a motivational or incentive account of goal directed actions (and we believe to the theoretical construct of motivation in general). In contrast to the incentive learning account, he suggests that the effects of motivational states on feeding behaviour may be explained by the same principles that describe how animals solve conventional Pavlovian discrimination problems. This conditioned modulatory account proposes that motivational states influence behaviour by modulating the strength of the association that is formed between the conditioned stimulus (CS) and the unconditioned stimulus (US). More specifically, interoceptive cues such as hunger serve as Pavlovian occasion-setters in instrumental tasks. Hunger would promote feeding behaviour (and responding for food) by making it easier for a food CS to excite the neural representation of a postingestive US.

## 2. THE ROLE OF DOPAMINE IN LEARNING AND SYNAPTIC PLASTICITY

**Editors:** The chapter by Beninger and Olmstead reviews data on the role of dopamine at a behavioural level in reward-mediated incentive learning. The chapter by Rebec reviews single unit evidence on the effect of amphetamine on the behavioural correlates of single neurone firing in the striatum. The chapter by Wickens deals with the role of dopamine in mediating synaptic strengthening in excitatory cortico-striatal synapses. We seek for unity amongst these three accounts. Are the sort of synaptic changes dependent on dopamine described by Wickens and co-workers able to account for the effects of amphetamine on behavioural correlates? What is the relationship between the correlations described by Beninger and the role of dopamine in the electrophysiological experiments described by Rebec.

**Beninger/Olmstead:** For an animal to learn to press a lever for food (or other rewards), and for the animal to continue to press the lever for food, the lever and lever-related stimuli must be able to control responding (i.e., incentive learning must occur). We know that dopamine plays a role in establishing and maintaining this ability of the lever and lever-related stimuli to control responding. The model of Wickens and co-workers provides many details of the molecular interactions at striatal synapses that occur when dopamine produces incentive learning. Amphetamine is known to increase synaptic concentrations of dopamine and probably preserves the dopamine signal associated with the presentation of reward. The observation that amphetamine leads to an increase in responding for a number of rewards could be stated in terms of incentive learning; thus, under the influence of amphetamine, lever pressing is enhanced because amphetamine enhances the ability of the lever and lever-related stimuli to control responding. From the point of view of Wickens' model, the effects of amphetamine are to augment the processes subserving this learning. Wickens' model would appear to provide a good account of the effects of amphetamine.

The work reviewed by Rebec fits well with ours in that the general message is that dopamine is a modulator of afferent input, adjusting the gain by which other neurotransmitters, especially glutamate, influence striatal output activity. The finding that it is the response of the motor units to glutamate that is modified by dopamine suggests that incentive learning directly affects motor activity; this is consistent with the definition of incentive learning as a change in the ability of reward-related stimuli to elicit approach and other responses in the future.

**Rebec:** There is no question that the ability of amphetamine to activate striatal units involves some type of dopamine-glutamate interaction. In fact, without glutamate input (i.e., in rats with cortical ablations) the excitatory effects of amphetamine are significantly attenuated (Tschanz *et al.,* 1994). Interestingly, cortical ablations do nothing to the relatively small population of striatal units inhibited by amphetamine; we interpret this to mean that some striatal units may not receive much glutamate input, and that the amphetamine-induced inhibition of these cells mainly reflects a dopaminergic action (Haracz *et al.*, 1998). Also noteworthy is evidence that, when applied iontophoretically to intact, awake animals, amphetamine almost always inhibits striatal units (Kiyatkin and Rebec, 1997). This again suggests that in striatum the excitatory effects of systemic amphetamine require glutamatergic — most likely corticostriatal— activation.

Despite the apparent importance of glutamate in the amphetamine-induced activation of striatal units, dopamine still plays a major role in this effect, in that it is blocked by dopamine antagonists. In the light of our iontophoretic evidence from behaving animals, we feel that striatal dopamine functions mainly as a modulator, enhancing the strength of the glutamate signal relative to the level of background activity. Dopamine, for example, may weakly inhibit basal firing rate, and even attenuate the absolute magnitude of the glutamate response, but relative to basal activity, the glutamate signal is larger during dopamine iontophoresis (i.e., an increase in the glutamate signal-to-noise ratio). Wickens *et al.* (1996) also suggest a dopamine-mediated enhancement of corticostriatal glutamatergic transmission. Although these authors are using a very different preparation (striatal slices) to collect very different data (intracellular records), our results are supportive. In fact, in voltammetric studies of behaving animals, we have shown a brief (~6 s) increase in dopamine signal during free-choice entry into a novel environment (Rebec *et al.*, 1997a). Although this novelty-related signal is highly localized (it appears in accumbal shell, but not in its core or in the dorsal striatum), the data support the view that rewarding stimuli are linked to a transient burst of dopaminergic activity. We feel that this brief dopamine response helps to strengthen concomitant activity in glutamate afferents to facilitate behavioral output.

This model also seems to agree with Beninger's proposed mechanism of striatal dopamine involvement in incentive or reward-related learning. Beninger further proposes a critical role for D1 receptors in the control of behaviour by rewarding stimuli. We have recently tested the effects of SCH-23390, a D1 antagonist, on basal activity of striatal neurones, as well as their response to dopamine and glutamate iontophoresis (Rebec and Kiyatkin, in press). Whereas this drug elevated basal activity and completely blocked the dopamine response, it markedly increased the magnitude of glutamate-induced excitations. SCH-23390 also enhanced movement-related increases in striatal activity. In contrast, eticlopride, a D2 antagonist, had little or no effect on both basal firing rate and dopamine iontophoresis, but reduced the magnitude of the glutamate response. Taken together, these data suggest that dopamine acts via D1 receptors to exert a tonic inhibition or restraining action on basal firing rate. This action of dopamine increases the signal-to-noise ratio of the glutamate response, even though the absolute magnitude of the glutamate excitation may be attenuated (see Kiyatkin and Rebec, 1996).

The critical issue is not whether dopamine excites or inhibits striatal cells, but how dopamine influences the response of these units to other, including glutamatergic, inputs relative to the level of background activity. Noteworthy in this regard is evidence that only low doses of dopamine (i.e., low iontophoretic dopamine ejection currents) have a facilitating effect on the glutamate signal (Kiyatkin and Rebec, 1996). Relatively high doses of dopamine either have no effect or an opposing effect. In fact, amphetamine iontophoresis, which applies a high concentration of the drug near dopamine terminals and presumably causes massive dopamine release, mimics the effect of a high dose of dopamine even at low amphetamine ejection currents (Kiyatkin and Rebec, 1997). The level of dopamine release under physiological conditions is presumably well below that induced by amphetamine. Such endogenous release, acting at D1 receptors, may be essential for the relative increase in the glutamate signal necessary for reward-related learning.

## 3. CONTROL OF DOPAMINE CELL FIRING

**Editors:** Joel and Weiner make the suggestion that inhibitory links from the striatum to midbrain dopaminergic neurones could be influential in control of dopamine cell firing. In particular, the suppression of firing when a rewarding stimulus comes to be familiar, might be achieved by this inhibitory link. A chapter written by one of us (RM) refers to the processing which might occur on the input side of the dopamine neurones. This focusses on the amygdala, and makes a suggestion that a process akin to mathematical differentiation for motivationally significant stimuli might occur therein, so that the dopamine neurones are controlled by favourable changes rather than by steady and predictable rewarding stimuli. This is a somewhat different view from the one inferred by Joel and Weiner. Joel and Weiner were invited to comment.

**Joel/Weiner:** Miller suggests that the selective firing of dopaminergic neurones to unpredictable rewards is largely determined by the amygdala, which provides the dopaminergic neurones with information about motivationally favourable changes in the environment, rather than about predictable favorable events. This suggestion seems to obliviate the need for an inhibitory process which prevents the firing of dopaminergic neurones to predictable rewards, and therefore to be inconsistent with our view of the limbic striatum as providing such an inhibitory input (see Wickens and Kötter [1995] for a similar suggestion). Two points seem to be relevant in this respect: (1) There is electrophysiological evidence suggesting that such an inhibitory process exists, i.e., the responses of dopaminergic neurones to the omission of predictable rewards (Schultz *et al.*, 1993, 1995b). (2) There are other excitatory inputs to the dopaminergic neurones which may signal primary and higher order rewards, such as those arising from the hypothalamus and prefrontal cortex. Such inputs necessitate the existence of a counteracting inhibitory input to the dopaminergic neurones. Indeed, Miller also states that the selective firing of dopaminergic neurones to unpredictable primary and conditioned reinforcers implies that "processes must exist which not only give the secondary stimulus control of the dopamine neurones, but also inhibit the responsiveness of these neurones to the "primary" rewarding stimuli, when these become predictable".

**Editors:** Can anything be concluded at this stage about the exact mechanisms or information processing algorithms used to control dopamine cell firing in behaving animals?

**Hyland:** There are insufficient data to resolve this at present. However, several papers in this volume have addressed particular aspects of this problem. To discuss these I find it most useful to return to the specific responses of dopamine cells that have been described, and deal with them sequentially.

(1) *Dopamine cells in the monkey respond specifically to rewarding stimuli.* The specificity for rewarding stimuli reflects a high degree of sensory processing in the information reaching the dopamine cells. For instance, in experiments where monkeys reach into a hidden space, the touch of a food morsel held on a wire activates the cell, but the touch of the empty wire does not (Romo and Schultz, 1990). Such recognition is presumably based upon tactile discrimination of texture, shape etc, and comparison with memories of food objects, and so involves information flow along the lines of the cartoon in Figure. 14.1A. It is often assumed that cortical processing is implicated in such detailed processing of sensory attributes. Use of this information to guide appropriate behaviour, and the addition of "emotional value" to the percept may require involvement of prefrontal cortex and the amygdala (as discussed by Miller, in chapter 5). Other sources of input to the dopamine cells also exist which could be involved, such as the pedunculopontine nucleus and hypothalamus. Whatever route is employed for processing and transmission of this information, it is extremely rapid, with median latencies of 65 ms (and minimal latencies of 20 ms) reported (Romo and Schultz, 1990). This can be compared to latency of neural responses to simple tactile stimuli in primary and secondary somatosensory cortex (around 10 and 30 ms, respectively (Burton and Sinclair, 1990: Jiang *et al.*, 1991)). The latency to delivery of fluid rewards to the mouth seems longer in the Schultz series of studies; perhaps processing of taste takes longer.

Interestingly, reward-driven responses are also seen in the striatum (Apicella *et al.*, 1991). I think it is generally assumed that this activity does not drive the dopamine cell excitations (which it theoretically could, via phasic suppression of a tonic inhibitory input from the *pars reticulata*). This assumption is supported to some extent by the finding that, as a population, striatal reward responses seem to have longer latencies than do dopamine cells to equivalent stimuli. However, attempting to dissect out possible directions of information flow on the basis of response latencies is fraught with difficulties, owing to the broad range of latencies seen, and the difficulty is compounded when comparing across studies using different methods. For instance, reward-related responses occur in prefrontal cortex (e.g. Inoue *et al.*, 1985) and amygdala (e.g. Nakano *et al.*, 1987), but different methods of reward delivery were used in these studies, making comparison with the dopamine cells difficult. Of course, a pure linear approach to data flow is usually too simplistic; structures projecting to the dopamine cells receive dopaminergic input, which at least for the amygdala has been shown to modify reward-driven responses (Nakano *et al.*, 1987).

Finally, the Editors raises the issue of what it is about a stimulus that confers reward. For appetitive rewards delivered to a hungry animal, it might be assumed that ultimately it would be the actual or implied caloric value of the reward that would be the important variable. However, even in naive animals this might not be the case: unexpected intravenous infusions of glucose do not appear to activate dopamine cells (Trulson *et al.*, 1983; Strecker *et al.*, 1983). It would be interesting to know if such infusions show evidence of being behaviourally rewarding. For example, can they support self-administration behaviour?

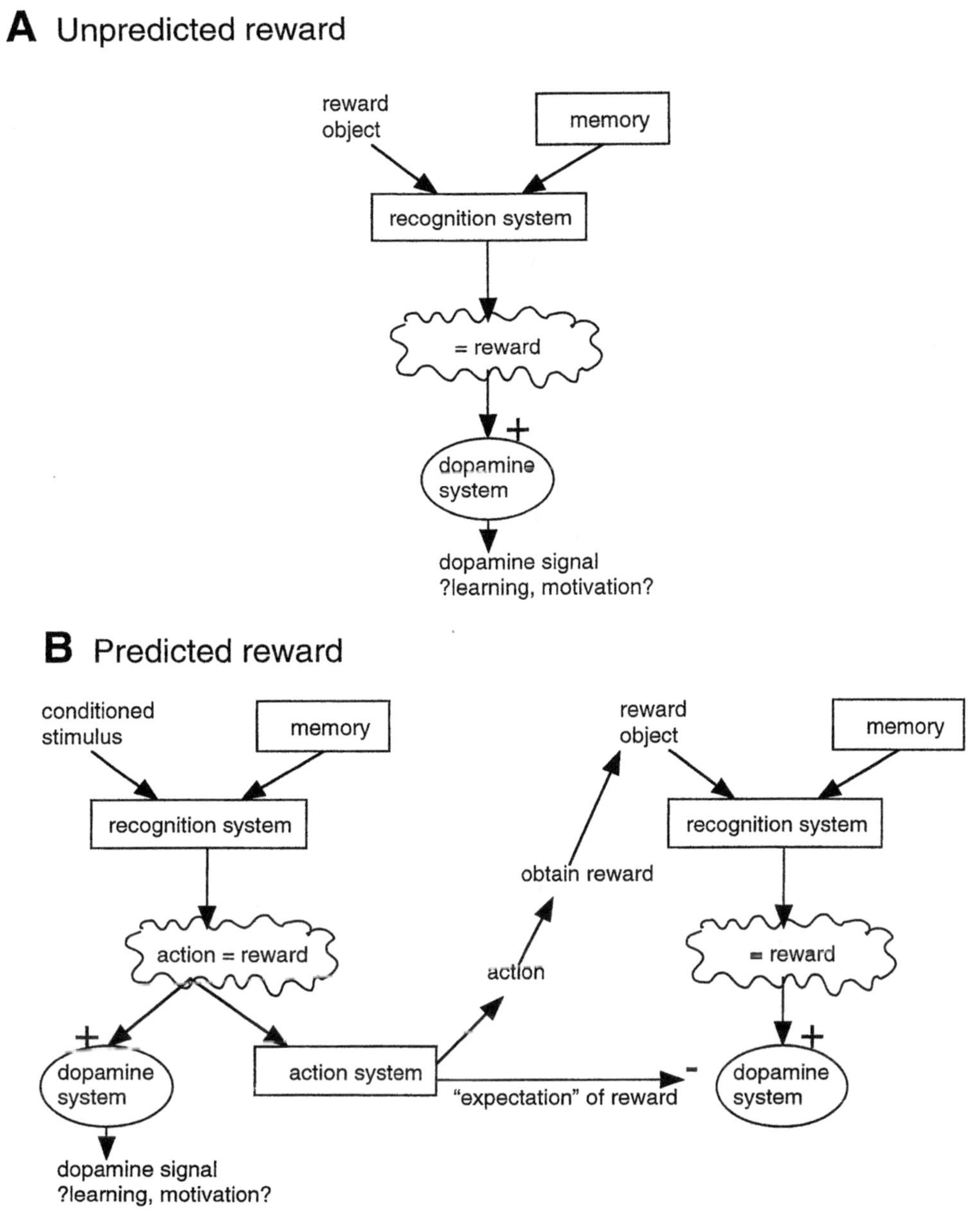


**Figure 14.1.** Diagram summarising possible information flows involved in driving activity in dopamine cells. **(A)** Responses to unpredicted rewards are selective, and so require prior recognition of sensory stimuli as having certain attributes. **(B)** When reward attainment is predicted, dopamine cells respond to the cue, but not to the reward itself. The diagram shows the situation in which reward attainment is "predicted" on the basis of being acquired by an action an animal performs in response to a sensory cue, having learnt that such a response will produce a reward. According to Joel and Wiener (1999, this volume) the striatum would be the relevant structure indicated by the "action system" in this diagram. Inhibitory responses following lack of delivery of expected rewards are directly accounted for in this model. Alternative models, in which the expectation of reward simply removes the excitatory drive to the dopamine cells in response to primary reward, would require activation of a separate inhibitory input to account for such suppressions.

(2) *Dopamine cells that respond to rewards do not respond when the reward is predictable*

The simplest way to account for this observation might be to assume that "predictability" is factored into processing by "upstream" structures such as the PFC and amygdala. In this case predicted rewards simply do not generate an excitatory drive to the dopamine cells (Miller, Chapter 5). However, there are always many ways in which observed patterns of neural activity could be generated. Joel and Wiener (Chapter 12) propose an alternative explanation, which arises from their consideration of striatal function. They suggest that all rewards send an excitatory input to dopamine cells, but that, in the case of rewards achieved by action of the animal, this is countered by a simultaneous inhibition from the striatum. This scenario is summarised in Figure 14.1B. A subset of striatal neurones are indeed activated with expectation of reward (Schultz *et al.,* 1992), which is at least consistent with this hypothesis. A critical test would be to demonstrate that these particular neurones project to the dopamine cells. This proposal does have the advantage of incorporating an explanation for the next observation to be considered, that is:

(3) *When expected reward is omitted, dopamine cell activity is inhibited to below baseline levels*

This finding implies an active inhibitory input to dopamine cells. In the model put forward by Joel and Wiener, the absence of expected reward would leave the already proposed inhibitory input acting alone on the dopamine cells, and therefore produce suppression. However it is not possible at present to rule out the possibility that upstream centres such as prefrontal cortex and amygdala process information regarding predictability, and then organise a suppression of dopamine cell activity. This could be achieved by reducing a tonic excitatory drive to the dopamine cells below normal baseline levels, or by activating a specific inhibitory pathway.

The idea that dopamine cells integrate input has the appeal of leaving them with some processing function, rather than as simply dopamine-containing conduits of signals generated elsewhere. It is interesting to note that, although not couched in the same terms, the processes implied for dopamine cells by Joel and Wiener appear identical to the "reward-prediction error signal" proposed in the temporal-difference (TD) model of reward related learning (Schultz *et al.*, 1997; Montague *et al.*, 1996). In both cases the dopamine cell integrates two inputs, one relating to the presence of a reward, and one which can negate this if a reward is "expected". Beyond this similarity, there are some interesting differences. Joel and Wiener consider an inhibitory input derived from striatal circuits involved in *generating an action* by the animal. The TD model makes reference to inputs from cortex concerned simply with detecting the presence of the an arbitrary signal in a classical conditioning (signal = reward) paradigm. In the Joel/Weiner model, appropriate timing of the inhibitory input is achieved naturally, since reward attainment will depend on the action of the animal, which the striatum may be directly involved in. In the work on the temporal-difference algorithm, no action is modelled, and a hypothetical temporally-extended representation of the signal is required to achieve the required timing. Whatever the underlying mechanisms, it remains possible that the striatum could be the structure involved in both kinds of behavioural paradigm. This would require that the striatum be involved in processing of learnt signal-reward associations that do not require action by the animals, such as classically conditioned associative learning. Is this known?

*(4) Dopamine cells respond to signals predicting reward*

This observation would be most easily accounted for by assuming that learning of an association between a signal and the reward occurs in upstream structures so that the signal acquires rewarding properties (see Miller, Chapter 5). As summarised in Figure 14.1B these structures would then pass on an excitatory signal to the dopamine cells. The sites at which the association is learnt could be the same structures that respond to the primary reward event, but this is not necessarily the case. An interesting question to ask is, what would happen during learning of the association? The most obvious suggestion arising from the informal "black-box" approach of Figure 14.1B is that the response to the primary reward would gradually decline in amplitude, as a response to the signal simultaneously grew. No particular shift in latency would be expected, *a priori*. This sets up an obvious contrast with the prediction of the temporal difference model, in which a progressive backwards shift in activation timing is predicted by the mechanism used in the model. It should be possible to experimentally verify which of these two patterns are in fact seen during learning, and thereby test the biological validity of the specific mechanistic details of the temporal difference algorithm.

## 4. THE "STATE OF READINESS"

**Editors:** In the chapter by Joel and Weiner, one of the main messages is that different parts of the striatum are involved at different stages of the cascade of events leading to purposeful behaviour. In particular the dorsal parts of the striatum are involved in learning motor acts and immediate objectives, while the ventral striatum is involved in selection of particular motives or goals. The last point is also discussed a little in the chapter by Miller. The question arises whether a similar learning rule for the striatum can apply at all levels of the cascade of events in purposeful behaviour. In particular, in the acquisition stage for learning of motor acts or approach of an immediate objective, the time which elapses between emission of behaviour and it becoming clear that it has motivationally favourable consequences may be quite short. However, an animal may have to wait much longer to be informed whether a particular goal or motive which has been selected is appropriate to a particular circumstance. The difference may be merely a quantitative one as far as psychological description goes, but may require qualitatively different processes at the neuronal level. Is this as a problem? Comments were invited from two contributors.

**Beninger/Olmstead:** We do not think that there is a problem for a striatal learning rule with the multiple split circuit model of striatal function, elaborated by Joel and Wiener. In your question, you stated that some time may elapse between the selection of a goal or motive and information about its appropriateness. However, because the circuits are parallel, and because dopamine acts similarly on all of them to produce learning, whenever a reward is encountered, the entire constellation of neuronal representations in each of the associative, motor and limbic circuits will be strengthened. The limbic striatum will switch from goal to goal under the influence of changing internal and external stimuli. This will influence the stimuli in the environment that are attended to, and the motor patterns that are activated. When a combination of goal, appropriate stimuli and appropriate motor acts is realized which leads to procurement of reward, the dopamine system will be activated and the incentive learning mechanism will operate at every level. From this point of view, there is no differential time delay for the different parallel circuits. Thus, "a similar learning rule for the striatum can apply at all levels of the cascade of events in purposeful behaviour".

**Joel/Weiner:** A prolonged state of readiness in the limbic striatum is one solution to the problem of delayed reinforcement of goal selection. However, this assumption may not be necessary in view of the suggested role of the dopaminergic input to the striatum (including the limbic striatum) in providing a reinforcement signal. As suggested by Barto and colleagues (Barto, 1995; Houk *et al.,* 1995) this immediate reinforcement signal is determined not only by current primary reinforcements, but also by all predictors of future reinforcement, such as conditioned reinforcements. This is best exemplified in the discharge of dopamine neurones, which "ratchets backward in time in a sequence of familiar events, so as to respond to earlier and earlier predictors of reinforcement" (Barto, 1995). The same mechanism may underlie the signaling of delayed reinforcement associated with goal selection/acquisition. Thus, subgoals may be functionally similar to conditioned reinforcement in that both are predictors of future reinforcement and can maintain behaviour. In this respect it may be relevant that humans' behaviour is governed by subgoals, e.g., the goal of "succeeding in life" will usually govern behavior in the form of subgoals such as, finish school, pass the exam, etc.

## 5. CYTOLOGY AND CONNECTIVITY WITHIN THE STRIATUM

**Editors:** In the chapter by Rebec, there is the implication that the motor-related neuronal population (about 80% of thesample) in the striatum corresponds to medium spiny cells, but these are believed to comprise 90–95% of striatal neurones. This raises the following questions: (a) Is there a sampling bias towards tonically active neurones?; and (b): Is there a population of medium spiny neurones that are not included in the sample because they are essentially silent?

**Rebec:** Yes, on both counts. We have recently completed a series of experiments in which we have used continuous glutamate iontophoresis at low currents while searching for striatal units (Rebec and Kiyatkin, 1997; Kiyatkin and Rebec, in press). We have found many such glutamate-activated units. They become "silent" (< 0.5 spikes/s) upon termination of the glutamate current. Although we do not yet have quantitative data, these "silent" units seem much more numerous than tonically active cells (we estimate a silent:active ratio of at least 10:1). Many of these "silent" units, moreover, become active during spontaneous movement, indicating that these units are "silent" only under certain conditions. This raises another important point: Striatal activity is likely to depend on behavioural state. Thus, the responsiveness of individual units to dopamine and glutamate, which in awake animals is sensitive to basal activity levels, also depends on the state of the animal. Models attempting to describe dopamine-glutamate interactions in the striatum, therefore, must include behavioural state as a critical variable (see section 2, above).

**Editors:** We note, amongst the anatomical detail reviewed by Bennett and Wilson, that it is mentioned that a minority of medium spiny neurones have axonal collaterals which ramify quite a distance from the soma. In addition some of the interneurones have quite wide ranging axonal trees. In another chapter Joel and Weiner describe a variety of ways in which different basal ganglionic-thalamocortical circuits have the anatomical possibility of interacting with one another. However, all of the links they describe are outside the striatum. Therefore two contributors were asked whether the longer-range collaterals within the striatum are common enough and widespread enough to help in integrating parts of the striatum which are concerned with different functional representations.

**Bennett/Wilson:** The axonal arborisations of two of the three major populations of neostriatal interneurones, specifically cholinergic cells and somatostatin/neuropeptide Y/NADPH-diaphorase co-containing cells, must, by virtue of the volume which they fill, innervate several microzones/matrisomes involved in the processing of distinct information e.g. hand and arm microzones in the somatosensory-recipient sector of the neostriatum. In this sense, the interneurones are integrative. Furthermore, the dendrites of interneurones do not observe compartmental boundaries, indicating that synaptic inputs directed to both patch and matrix compartments can influence spiking. This suggests that the interneurones themselves are integrative, because their output is influenced by the integration of inputs directed to both compartments. Spiny cells with widely dispersed intrastriatal axon collaterals are also likely to provide an integrative input to functionally distinct regions of the neostriatum, in a manner analagous to that described above for interneurone axon collaterals. However, in contrast to interneurones, spiny cell dendrites and axons do not cross patch/matrix boundaries.

All of these integrative connections within the neostriatum are short range and non-directional. They are not organised along functional lines, as are, for example, the associational connections of the cerebral cortex. It should also be noted that there is no evidence that any of the intrastriatal connections arising from the cholinergic or somatostatin-containing interneurones or the spiny cells are mediated by fast synaptic transmission, suggesting that they do not influence spiking in a moment-to-moment fashion. Rather it seems likely that these widely dispersed axonal arborizations are designed to influence the electrical behaviour of functionally diverse groups of spiny neurones through a neuromodulatory action. Comparison with the neocortex and hippocampus suggests that the neostriatum is unique, by virtue of the absence of synaptic interconnections between functionally diverse regions mediated by fast glutamatergic transmission, which is a common theme in the organization of cortical structures. Conversely, the GABAergic, fast-spiking, parvalbumin-containing cells within the neostriatum appear to be similar in many respects to their neurochemical counterparts in cortical structures and mediate fast GABAergic feed-forward inhibition (Johnson *et al.*, 1997).

**Joel/Weiner:** Intrastriatal connections have been suggested to play a role in integrating information processing between basal ganglia-thalamocortical circuits (Groenewegen and Berendse, 1993; Groenewegen *et al.*, 1990). However, as detailed by Bennett and Wilson in Chapter 6 although long-range collaterals within the striatum may play a role in integrating information between matriosomes processing distinct information within a functional subdivision (motor, associative, or limbic) of the striatum, they are unlikely to contribute to integration of information between the different functional subdivisions (motor, associative, and limbic). In this context, it is of interest to point out that information processing at several levels of the basal ganglia-thalamocortical circuitry appears to be characterized by a common principle, namely, a high degree of integration within a functional subdivision (e.g., convergence of corticostriatal and striatopallidal projections within a striatal and a pallidal subdivision, respectively) but relatively limited connections between functional (motor, associative, and limbic) subdivisions (e.g., segregation of corticostriatal projections arising from motor and sensory cortical regions versus association cortices). This principle may also hold for intrastriatal connections, which appear to exhibit a relatively high degree of integration within striatal functional subdivisions, but a relatively low degree of integration between functional subdivisions. Returning to the question of between-circuit interaction, it should be noted that the above principle had provided the basis for the idea that the basal ganglia-thalamocortical circuits are functionally segregated. However, although this principle holds for part of the connections comprising the basal ganglia thalamocortical circuitry, there are additional principles of organization which do enable between-circuit integration, and these are incorporated in our open-interconnected model.

**Editors:** We noted the fact mentioned by Bennett and Wilson, that only about 4% of spines on medium spiny neurones are contacted directly by dopaminergic terminals. We asked what this might mean for the function of dopamine as a transmitter. Is it compatible with the view that dopamine acts, like classical transmitters, at precise points of synaptic contact? Or does it fit better the view that dopamine has widespread and diffuse actions, modifying the chemical environment of the extracellular space, and perhaps modulating synaptic processes widely, even at a distance from the release site?

**Bennett/Wilson:** It is unclear how to interpret the observation that relatively few (4-7.5% see Ingham *et al.*, 1998, for a recent estimate) dendritic spines receive a dopaminergic input in terms of what this means for dopamine as a neurotransmitter. However, what is clear is that it certainly does not mean that dopaminergic inputs act to veto cortical signals as originally proposed by Freund *et al.*, (1984). Conversely, from a purely quantitative standpoint, there cannot be a spine-by-spine dopaminergic regulation of the cortical input to spiny neurones. Rather we propose that dopaminergic synaptic input is more likely to be involved in changing the electrical response of a segment of dendritic membrane, so that, for instance, plastic alterations in synaptic responses might be generated.

**Editors:** Rebec was asked for comments about the relationship between gene expression and physiological activity of striatal neurones.

**Rebec:** We have some electrophysiological evidence consistent with immediate-early-gene expression in response to D2 agonists. In behaving rats, quinpirole, a D2 agonist, both inhibits striatal neurones and excites neurones in globus pallidus (Hooper *et al.*, 1997). Parallel changes in immediate-early-gene expression have been obtained in these same regions with D2 agonists. Such consistency, however, is rare, and belies what appears to be a much more complex picture of the relationship between D1 and D2 receptors and striatal output pathways. The inhibitory effects of quinpirole, for example, occur throughout the striatum. Unless our data are obtained exclusively from striatopallidal projection neurones, an unlikely assumption, then quinpirole also inhibits striatonigral neurones, which immediate-early-gene data link to D1 receptors. Moreover, striatal dopamine depletion, which has been used as a basis for assuming a segregation of D1 and D2 receptors on the direct and indirect pathways, respectively, complicates the electrophysiological data by altering the neuronal response to quinpirole in both striatum and globus pallidus (Hooper *et al.*, in press). Collectively, our results support a co-localization of D1 and D2 receptors on the majority of striatal neurones, making it difficult to reconcile electrophysiological data with data based on immediate-early-gene expression. We have no easy answer for the discrepancy.

## 6. THE NATURE OF THE SIGNAL IN THE MEDIUM SPINY NEURONES

**Editors:** We noted (in the chapters of Rebec, and Bennett and Wilson) the detail that large numbers of excitatory synaptic activations are required before a medium spiny neurone moves from the "down" state to the "up" state, as well as the biophysical detail provided about the "up" state. This is relevant to another question, about the nature of the signal in medium spiny striatal neurones which contains information for cybernetic operations and transmission to other sites. In the cerebral cortex, recent research suggests that pyramidal cells may be coincidence-detectors, with the timing of every impulse having its cybernetic significance. Can one conclude from the available evidence that medium spiny neurones do *not* encode precise temporal patterns of afferent impulses in the millisecond range? If so, there may be a fundamental difference in cybernetic function between these neurones and cortical pyramidal cells. Comments were invited from three contributors .

**Rebec:** Observations of striatal units during episodes of spontaneous behaviour indicate that the motor-related change in firing rate is usually not linked to a specific movement, but occurs in response to a wide range of movements (e.g., grooming, head bobbing, locomotion, and/or rearing). Many of these units also respond to various sensory stimuli such as auditory clicks, changes in house illumination, and touching of the dorsum or vibrissae. Although it is difficult to designate any of these events as exclusively motor or sensory, it appears that, rather than encoding precise patterns of afferent activity, many striatal neurones respond with generalized changes in firing rate, perhaps related to the maintained plateau depolarization of these cells, thought to reflect the integrated activity of many cortical afferents (Wilson, 1993). Consistent with this view, we find many striatal units that increase activity to both the stimulus and response (White and Rebec, 1993; White *et al.*, 1994). Such units, moreover, are found throughout the striatum, further suggesting an integrative function of striatal units. Similar results have been obtained in the striatum in response to a relatively high dose of D-amphetamine (5.0 mg/kg). The behavioural response to such treatment includes an initial period of generalized behavioural activation (locomotion and rearing) followed by a period of highly focused movement (intense head-bobbing and oral behaviour). Many striatal units respond to this amphetamine dose with further increases in firing rate, despite the shift in behavioural pattern (Rebec *et al.*, 1997b). Thus, the striatum appears to encode relatively broad aspects of behaviour, related perhaps to intensity or temporal sequence, which may require convergent input from many cortical regions.

**Bennett/Wilson:** Action potentials are generated when the membrane voltage at the spike initiation zone reaches a critical threshold value. Coincidence detection is a function of the temporal window during which synaptic inputs can sum to produce a depolarization which causes the membrane potential to cross threshold. Therefore all neurones, except cells in which unitary EPSPs trigger spikes, are coincidence detectors, and the only variables are the length of time over which inputs are integrated and strength of the individual inputs, in terms of the magnitude of the voltage fluctuation which they produce. The membrane time constant (which is not a constant but rather a dynamic variable continuously altered by synaptic inputs interacting with non-linear membrane properties) determines the duration of the window.

Both neostriatal spiny neurones and neocortical pyramidal cells exhibit spontaneous membrane potential fluctuations between two preferred voltages, a relatively hyperpolarised Down state and a more depolarised Up state from which action potentials are triggered. In both types of neurones, state transitions require the coincident input from a large number of excitatory inputs. Perhaps pyramidal cells require somewhat fewer inputs as a consequence of activation of subthreshold inward currents, but it seems likely that the same order of magnitude of inputs is required in both cell types to produce a Down-to-Up state transition. Furthermore, the fact that pairs of spiny neurones can exhibit state transitions which are highly correlated, exhibiting synchrony on the millisecond time-scale, argues that spiny cells are excellent coincidence detectors, and that the threshold for state transitions is finely tuned (Stern, Jaeger and Wilson, 1998). Once in the Up state, action potentials are triggered by depolarisations which cross threshold. Such depolarisations require the coincident activation of many fewer excitatory inputs than are required for state transition. Hence, the level of coincidence detection in both spiny cells and pyramidal neurones is a function of whether the inputs are causing state transitions, or spike generation within the Up state. It therefore seems likely that spiny cells and pyramidal neurones are essentially similar in terms of their coincidence detection, and that both are able to encode precise temporal patterns of action potentials from afferent impulses on the millisecond time-scale.

**Editors:** Can medium spiny neurones encode precise temporal patterns of afferent impulses on the millisecond time scale? Are medium spiny neurones different from cortical pyramidal neurones, which seem to function as coincidence detectors, with timing of every impulse having its cybernetic significance?

**Surmeier:** This is a very difficult question to answer given our current level of understanding of how medium spiny neurones integrate synaptic information. However, there are several considerations from my perspective that might give us a clue. Based upon their electrophysiological properties and their anatomical relationship to afferent fibres (as shown by Wilson's group and others), it would seem that medium spiny neurones require strong temporal and spatial convergence of inputs to move into the depolarized "up-state" where spiking is possible. This clearly argues against a model in which the temporal structure in any one, or a small number of afferent fibres is subjected to some simple transformation and passed on (except for a trivial null transform). However, in the up-state, temporally-coincident synaptic inputs will activate voltage-dependent conductances capable of boosting dendritic signals and evoking temporally-correlated spikes. The exploration of these dendritic processes in medium spiny neurons is lagging behind that in other more experimentally favourable cells, like pyramidal neurones ,but it would be very surprising if the principles were fundamentally different in the striatum. In pyramidal neurones, a variety of conductances are making their appearance on the dendritic stage, including $Na^+$, $Ca^{2+}$ and $K^+$ conductances. There are some studies already showing that $Ca^{2+}$ conductances are important regulators of e.p.s.p. amplitude in a state mimicking the up-state in medium spiny neurons.

**Editors:** The general view present in the chapter by Salinas, Romo *et al.* of interactions between cortex and basal ganglia can be likened to a spiral or a "corkscrew", which engages both cortical and subcortical structures several times in sequence, as one passes from preparation of a program for behaviour, to sequential execution of its various stages. We have formed much the same view of the functional interactions between cortex and basal ganglia. However, at a lower level of organization, there is a question of what is the exact nature of the signal carrying information within the striatum. There have been several hints in other chapters that the "up" state of striatal spiny neurons represents the "envelope" of activity in such a large number of afferent fibres that it cannot represent the input and timing of specific afferent fibres, but rather an overall context. In this respect the striatum may be rather different from the cortex, where pyramidal cells appear to be "coincidence detectors", with the timing of every impulse having cybernetic significance. This issue also impinges on the larger scale function of cortico-basal ganglia circuits. If it is true that the striatum cannot represent exact timing, the sequential activation of striatal output neurones in preparing and executing a behavioural sequence should be a matter strictly of sequence, rather of precise metrical timing. If this idea could be supported, it would be a major advance in understanding the striatum.

**Romo:** This comment makes the implicit assumption that all or most pyramidal neurones work as coincidence detectors, or have the capacity to transmit information through spikes that are very accurately timed. This point is hotly debated in the field (see for example Shadlen and Newsome, 1998). In our opinion, it seems that in most cases it is the cortical neurones in primary sensory areas which truly encode the metrics and timing of a stimulus; we take somatosensory neurones as an example (Mountcastle *et al.*, 1990; Romo *et al.*, 1998). Primary somatosensory responses are quantitatively related to the characteristics of tactile stimuli, such as motion speed (Romo *et al.*, 1996), vibration frequency (Mountcastle *et al.*, 1990) and texture (DiCarlo, Johnson and Hsiao, 1998). Typically, these representations are not affected by the context of the task, that is, they are the same whether the animal makes use of the evoked activity to perform a task or not (Mountcastle *et al.*, 1990; Romo *et al.*, 1996; but see also Hsiao *et al.*, 1993). Some types of stimuli, such as vibrotactile stimuli, have a temporal structure that is exquisitely encoded—that is, with a temporal jitter of a few milliseconds—by the primary mechanoreceptive afferents (Johnson, 1974) and, slightly less accurately, by primary somatosensory neurones (Mountcastle *et al.*, 1990; Romo *et al.*, unpublished results). However, this initial representation is only the starting point leading to perception (Romo *et al.*, 1998; Zainos *et al.*, 1997).

What we see in subsequent processing stages are neuronal responses that are not as tightly related to the physical properties of the stimulus (Salinas, Hernández and Romo, in preparation). For example, the responses in secondary somatosensory cortex to vibrotactile stimuli—the same stimuli that produce neatly timed spike trains in the primary area—show no temporal pattern. Information about the stimulus seems to be encoded by the mean firing rate of these neurones, computed over a hundred millisecond timescale. It seems to us that if high temporal precision were the main business of pyramidal neurones this loss of temporal structure would not be seen. This activity is usually context-dependent; it changes depending on whether the animal is making use of this information or not; it is affected by attention (Hsiao *et al.*, 1993); and frequently it represents only some aspect of the stimulus that is important in the context of the task, rather than a full description of the stimulus. We have observed these higher-order, context-dependent responses in several cortical areas. In tasks where the speeds of tactile motion had to be categorized, these responses seemed to represent the output of the decision process (Romo *et al.*, 1997; Salinas and Romo, 1997). They were neither accurately timed – for example, their latencies were highly variable – nor directly associated with the physical properties of the stimuli. They were indistinguishable from the responses recorded in the striatum (Merchant *et al.*, 1997). It is possible that large differences arise when other tasks or other types of stimuli are used, but considering the current evidence it seems that spike timing is not an adequate criterion to distinguish between cortical and striatal neurones.

On the other hand, the switching between "up" and "down" states of medium spiny neurones from the striatum may reflect certain types of cortical events, such as synchronization of several inputs from a cortical module, or coactivation of different converging modules. But whatever its functional significance, there are neurones in prefrontal and premotor areas of the frontal lobe that behave very similarly; indeed, Wilson (1995) has a made a strong case about this issue. In terms of 'the exact nature of the signal carrying information within the striatum', the question may just as well be applied to many non-primary cortical neurones (see Shadlen and Newsome, 1998).

We agree that sequencing could be an important functional principle for the cortico-basal ganglia circuit; we mentioned results from several laboratories that support this idea, both in cortex (Roland *et al.*, 1980; Tanji and Shima, 1994) and basal ganglia (Kermadi and Joseph, 1995; Mushiake and Strick, 1995). But, if neurones from cortex and striatum have similar processing properties, what could be the differences between them in terms of sequencing? We speculate on the following scenario: The striatum may be in a position to select a subset of the cortical signals, or potential actions; this selection may depend on the context, and on internal and external demands. A sequence of selections can be made, where the selection moves by the pausing effects of the pallidum and the thalamo-cortical circuit, where program release occurs. This is along the lines of the computational model proposed by Berns and Sejnowski (1998), which relies on the anatomical arrangement of the basal ganglia, rather than on intrinsic neuronal properties, to learn sequences. It also agrees with the results of Cromwell and Berridge (1996), who suggested that the striatum is crucial in the implementation of stereotyped motor sequences actually produced by central pattern generators in the midbrain. In this view, the role of the striatum is to control or trigger such sequences within the normal behavioral context of an animal.

## 7. STRIATAL OUTFLOW PATHWAYS AND DOPAMINE

**Editors:** With regard to the functional role of D1-like and D2-like receptors we are aware of two relevant dichotomies: (a) In terms of projections there is evidence that neurones of the direct output pathway from the striatum via *pars reticulata* or entopeduncular nucleus tend to label for D1 receptors or its D1 receptor mRNA; while neurones of the indirect pathway, via globus pallidus tend to label for D2 receptors or their mRNA. This point is made by both Gerfen and Surmeier though in a less categorical way by Surmeier than by Gerfen. (b) In terms of behavioural pharmacology, there is a dichotomy, referred to by

Beninger and Olmstead that the motor performanace effects are mainly D2 effects, while the reward/incentive effects are mediated by D1 receptors. This case can be made not only from behavioural pharmacology, but also from studies of synaptic plasicity in the striatum, which is mediated by D1 receptors or cAMP rather than D2 receptors (chapter by Wickens). The issue on which we asked for comments is about the relation between these two dichotomies. For instance, could the reward/performance dichotomy somehow map against the direct *vs.* indirect pathways? Or is this implausible because of lack of supporting evidence from behavioural pharmacology? Did contributors have any other comments that can help clarify the role of dopamine receptor types for function at the macroscopic level?

**Beninger/Olmstead:** We know of no direct evidence for a mapping of reward/performance onto the direct/indirect pathways. However,some relationships can be specified. On the performance side, the evidence, especially from behavioural pharmacological studies with dopamine receptor antagonists, suggests that both D1- and D2-like receptors are involved in the control of motor activity; however, agonist evidence clearly shows greater motor effects with D2-like receptor agents. On the reward side, support for a critical role for D1-like receptors is strong and growing. If D1-like receptors are localized exclusively to the direct pathway, it would be reasonable to speculate that neurones of the direct pathway are the locus of the putative dopamine-mediated changes in glutamatergic synaptic effectiveness that underlie incentive learning. Similarly, if D2-like receptors are localized exclusively to the indirect pathway, it might be reasonable to think of it as primarily motor. However, D1-like receptors, although found in greater density on the neurones of the direct pathway, also are found on the neurones of the indirect pathway; similarly, D2-like receptors predominate on the indirect pathway but also are found on the direct pathway. Therefore, it would be premature to attribute a specific behavioural function, viz., reward or performance, to one of the direct or indirect pathway on the basis of these data.

Joel and Weiner (Chapter 12) make comments relevant to the possible differential function of the two pathways. They cite a number of authors supporting their statement that facilitation of appropriate responses (those producing biologically important outcomes, e.g., food reward) is attributed to striatal neurones of the direct pathway, while striatal neurones of the indirect pathway suppress unwanted responses. This differential function is in agreement with the idea that the D1 receptor-rich direct pathway mediates reward.

**Editors:** Bennett and Wilson also mention the direct and the indirect outflow pathways from the striatum. One of the ideas that has been suggested to make sense of this distinction (mentioned again "in the chapter of Joel and Weiner's") is that the direct pathway facilitates the output of behaviour, while the indirect outflow, with one more inhibitory synapse in the sequence of relays to the outside world, acts to suppress unwanted behaviour. However, another way in which suppression of unwanted behaviour might be achieved is via mutual inhibition between neighbouring neurones actually within the striatum (assuming that the local GABAergic collaterals can be effective in providing mutual inhibition—an assumption which has not yet been established satisfactorily). In the article of Bennett and Wilson, it is mentioned that there are interconnections via local collaterals between neurones with direct and indirect outflow projections. We therefore asked what the significance of such connections might be? For example, if such connections between direct and indirect projection neurones were functionally inhibitory, the end result at the output stage would be facilitatory in both cases (because the indirect pathway would them have *two* additional sign reversals). This would argue against the idea that the indirect pathways at the same time suppresses unwanted movements. One might rather expect the neurones of origin of the two outflow pathways to function synergistically rather than in mutual antagonism. Following this line of reasoning, the finding cited by Bennett and Wilson might then be used as an argument against the view mentioned above about the functional segregation between direct and indirect pathways. We invited comments.

**Bennett/Wilson:** Although there is anatomical data for synaptic interactions between spiny neurones of the direct and indirect pathways, there is no physiological evidence for these interconnections being functionally inhibitory. These negative observations do not rule out the possibility of surround inhibition mediated by spiny neurones in the neostriatum, but are highly suggestive that it is not a powerful influence. The observation that spiny neurones do not possess substance P receptors provides additional puzzlement, because if direct pathway cells do not influence other spiny neurones through GABAergic transmission, or via neuromodulatory actions mediated by substance P, then why are they synaptically interconnected? Whatever the role of the synaptic interactions between spiny cells, it seems likely that these neurones function as parallel processing elements, on the fast time scale, and that there is no strong coupling—synergistic, antagonistic or otherwise—which would serve to increase the redundancy across neurones in the neostriatum.

**Editors:** In the chapter by Salinas, Romo *et al.*, the interesting question is raised of why there are two inhibitory links in sequence in the outflow from the striatum. Executive functions, such as motor acts, larger scale behaviour, speech, and probably the thought required in planning these things, need to be performed one program at a time, rather than several programs running in parallel (which could interfere with one another). In addition, it is possible that thalamic relay neurones are part of cell assemblies which encode the above programs (Miller, 1996). If so, it should be necessary to arrange that the output from the striatum to the thalamus allows only one program to be activated at once. This might make sense of the fact that the immediate input to the thalamus from the basal ganglia is inhibitory, and is tonically active. Selective release from this inhibition of an individual program for execution would then be achieved by inhibition coming from activity in specific neurones in the striatum. Thus one would ensure that many programs are not activated together, and one could account for the strange arrangement of two inhibitory synapses in sequences. We invited comments.

**Romo:** A similar hypothesis has been put forward by Mink (1996), who proposed that the role of the basal ganglia in motor functions could be to select among a variety of motor programs, inhibiting the ones not selected and which may interfere with the movement to be executed. This picture may very well be correct; however, it is somewhat vague—and thus difficult to test. For example, artificial neural networks capable of selecting a single input (or producing a single output) have been studied for a number of years, both in relatively abstract models (see for example Hertz, Krogh and Palmer, 1991) where they are known as 'winner-takes-all' networks, and in more biologically realistic models (Ambros-Ingerson, Granger and Lynch, 1990; Salinas and Abbott,

1996). Winner-takes-all mechanisms are not complicated. They certainly do not require two inhibitory synapses in sequence, although they could be constructed in this way. It is certainly likely that the basal ganglia implement some form of winner-takes-all architecture, but their complexity suggests that there are other kinds of neural processes interacting.

What may be necessary to make sense of the functional role of the basal ganglia, apart from a large body of experimental results, are realistic computer models in which these kinds of ideas can be made more specific and can be tested. Some steps have been taken in this direction by Berns and Sejnowski (1998). They constructed a computational model based on the known anatomy of the basal ganglia. The word 'computational' here means that each unit in the model is really a somewhat abstract representation of a pool of neurones, where the detailed cellular properties are not considered; the network performs a computation: it learns a sequence and, after learning, it is able to reproduce it. The model incorporates the modulatory effects of dopamine and indeed assumes that the globus pallidus operates in a winner-takes-all fashion. Interestingly, it proposes that the so called "indirect pathway" that involves the subthalamic nucleus acts as a short-term memory buffer. The model does learn arbitrary sequences and fits in with a good number of behavioural and lesion data. This model is, in general, consistent with the general picture described above, but in some ways it is much more specific, and it has some predictive power. Clearly, it ignores lots of details about the connectivity of the basal ganglia and its cellular properties, but it is an example of a concrete proposal that can be further improved.

Berns and Sejnowski also mention a couple of functions that the two inhibitory synapses in series may serve. (1) To gate the ascending information that reaches the thalamus and is sent to the cortex, allowing this information to flow only when required for action and otherwise inhibiting its transit tonically. (2) To obtain a faster thalamic response, useful in producing rapid movement sequences; the release from inhibition leads to a rapid post-inhibitory rebound, and this makes the thalamic response faster than with direct excitation. These are interesting possibilities that are not necessarily incompatible with other functional aspects of the circuit. For example, it is possible that a mechanism to trigger motor sequences based on selective release from inhibition (Chevalier and Deniau, 1990; Miller, 1996) is more convenient, as suggested above, than a mechanism based on direct excitation. This is the kind of hypothesis that may benefit from a realistic model that includes the dynamics of sequential movement generation.

**Editors:** How strong is the evidence that there are two groups of striatal neurones, one which responds primarily to D1-class agonists and another that responds to D2-class agonists?

**Surmeier:** Obviously, this is a very appealing hypothesis. It simplifies dramatically the striatal efferent system in our thinking. As I stated in our chapter, there is something to the idea, but we need to be careful. A very large segment of the evidence in support of this view is circumstantial. The work using indirect indices, such as immediate early gene expression, peptide expression and behavioural changes, can, I think, be explained readily with a model that assumes substantial co-localization, or one that assumes segregation. Why? Even if we assume that D1- and D2-class ligands have completely disjoint sets of actions, this would not imply that the receptors activated by these ligands were on separate sets of neurones. It simply means that the physiological consequences of receptor activation are different. Since we know that D1- and D2-class receptors couple to completely different G-proteins, second messenger systems, and ion channels, it is patently obvious that the impact of these receptors on neuronal activity is quite different. As a consequence, it is easy to imagine that D1- and D2-class agonists could have quite different cellular and behavioural effects, while acting on the same neuronal ensembles. It should be noted that there also is a substantial evidence for cooperativity between D1- and D2-class receptors, that is very difficult to explain without them being on the same cells (some of the most recent work using TTX infusions are good examples). It is also worth noting that differences in cortical input could create heterogeneity in these indices, without any differences in receptor expression. After all, we know full well that dopamine acting on its own is not going to do much—it is a neuromodulator, not a neurotransmitter in the classical sense. Furthermore, it is not productive to view striatal efferent pathways as simple switches—turning on or off particular behaviours; this idea is implicit in the arguments for segregation derived from this work. The concept is a dark remnant of labeled line theories of brain function that we should dispense with. Having set this data aside, there is more compelling evidence for the model in my estimation. The best of it is the *in situ* hybridization data, arguing that D1 and D2 receptors do not colocalize. These data, and the corroborating immunocytochemical data make a strong case that D1 and D2 receptors themselves are largely segregated. What seemingly stands diametrically opposed to this model is the almost universal observation of electrophysiologists that the vast majority of striatal neurones respond to both D1- and D2-class agonists. This is true with isolated neurones, as well as with neurones in more intact preparations. Furthermore, the physiological evidence for colocalization is derived not just from studies looking at short term effects, but also from studies looking at long-term changes in synaptic efficacy. It was this paradox that compelled us to develop and apply to this question the most sensitive method currently available for looking at coordinated expression of mRNAs in single cells . In my view, our results have largely resolved the issue. The subset of neurones that colocalized D1 and D2 receptor mRNAs in our hands was about 20% of the total—well within the experimental limits of the *in situ* hybridization studies, especially given the difficulty of these experiments. The key finding was the detection of less abundant mRNAs that were not seen by *in situ* hybridization, particularly those for the D3 and D5 receptors. The D3 receptors were detected in a substantial segment (*ca.* 40%) of the SP expressing subset of medium spiny neurons. Recently, Schwarz's group has shown that the abundance of D3 mRNA and protein is dramatically elevated in this population of neurones following dopamine-depleting lesions and the introduction of exogenous l-DOPA. Our results showing the presence of D5 receptors in enkephalin-expressing medium spiny neurons further elevates the percentage of neurones colocalizing D1- and D2-class receptors (I also suspect that we have underestimated the frequency of D5 expression in this population). These results are consistent with the *in situ* hybridization studies given their limited sensitivity. They also provide a means of explaining two aspects of the electrophysiological data: the response incidence (*ca.*, 70–80%) and response heterogeneity within a particular pharmacological group. Now, it is tempting to view the expression of these less abundant receptors as so much biological noise. Maybe, but I think this is a dangerously simplistic view, especially in light of data like that showing rapid upregulation of D3 receptors in a condition mimicking the early stages of Parkinson's disease. Insofar as direct and indirect pathways are concerned, I think Kawaguchi and Wilson's data on this point provide a strong motivation to rethink this model. This data argue that there isn't a direct pathway that leaves GP alone. Certainly those neurones that project just as far as the GP in the rat (GPe), express high levels of D2 receptor, and those that project to the GP and beyond express high levels of D1 receptor. Based upon retrograde labeling, a subset of this latter group also express high levels of D2 receptor. What are the functional implications of this pattern? I think that this is far from clear. The Alexander/Crutcher/DeLong (ACD) model makes the argument that the direct pathway should allow movements, while the indirect pathway should suppress them. I don't want to comment too much on this idea because there are other authors that have more interesting things to say than I do. What I will comment on is the role of dopamine in this model. First, we have virtually no idea what postsynaptic D2 receptors do to medium spiny neurones. None of the direct electrophysiological data from striatal recordings are compelling in the least; the recordings from GP neurons following systemic administration of dopaminergic ligands can be interpreted in several ways. As a consequence, the suggestion that D2 agonist inhibit striatopallidal neurones is unfounded at this point in time. Furthermore, it ignores the fact that dopamine is not a classical transmitter. It is not going to do anything on its own. The cortical/thalamic input to D2 receptor-expressing neurones is a crucial variable in this equation that cannot be ignored. Furthermore, data from MPTP lesioned monkeys suggests that alterations in GP spike patterning (rather than gross changes in spike rate as predicted by the ACD model) are the most important consequences of D2 receptor activation in striatopallidal neurons. What about the 'direct' pathway? Fortunately, we know much more about these neurons. Activation of D1 receptors should effectively make these neurones less responsive to inputs lacking temporal or spatial coherence. At the same time, the response to strong, coherent maintained inputs should be enhanced. This inference is based upon solid electrophysiological data derived from a wide variety of *in vitro* and *in vivo* preparations. In my view, this D1 receptor effect should serve to focus striatal activity in just those neuronal assemblies necessary to accomplish initiation of a movement, while suppressing it in those assemblies tied to the initiation of competing or unnecessary movements. In many versions of the ACD model, this is a function of the indirect pathway, but I think there is good reason to believe that it is a consequence (in part at least) of D1 receptor stimulation of the direct pathway.

## 8. MUTUAL INHIBITION BETWEEN NEIGHBOURING NEURONES AND THE "DOMAIN HYPOTHESIS".

**Editors:** An important issue referred to by several contributors is the "domain hypothesis", the implication of this hypothesis being that mutual inhibition occurs between neighbouring medium spiny neurones in the striatum, when they are within reach of each other's local axon collaterals (i.e. within a "domain"). Following from this, it is inferred that there is a dynamic of intense competition between neighbouring medium spiny neurones, which in turn can impose competitive dynamics on firing of neuronal populations in other regions of the forebrain connected with the striatum. This hypothesis is dealt with in the chapters by Wickens and Oorschot, Oorschot, and in that by Plenz. There is however some controversy about whether the supposed mutual inhibition really occurs. We asked Rebec the following questions: "Have you found any evidence for inhibitory effects in striatum that may match excitatory effects? Could there be a sampling bias in preventing you from seeing such reciprocal effects?"

**Rebec:** We find occasional amphetamine-induced inhibitions in the striatum, but the small number (<20% of total responses) argues against this response as a mirror-image reflection of a neighbouring cell excited by amphetamine. Moreover, these amphetamine-induced inhibitions do not usually occur among cells inhibited by movement; rather they tend to occur among cells that show no obvious relationship to spontaneous movement (so-called nonmotor-related cells). The number of units showing an actual movement-related inhibition is very small (<5%). Although we cannot rule out a sampling bias, our data indicate that the inhibitions tend to occur among fast-firing units, which should be relatively easier to detect than the slow-firing or "silent" units that seem so common in striatum.

We feel that if there is some type of mutual inhibition between neighbouring striatal cells, it may depend on complex interactions between ongoing behaviour and basal firing rate. We have found, for example, that striatal neurones are extremely sensitive to GABA iontophoresis (removal of the retaining current on the GABA barrel often causes an inhibition), but this effect is highly dependent on discharge rate, with the fastest firing cells showing the strongest sensitivity (Kiyatkin and Rebec, 1998). Interestingly, however, iontophoretic GABA seems less effective at inhibiting striatal neurones activated by movement than neurones simply discharging at a high rate during quiet rest. Further work with the behavioural preparation seems essential to clarifying the role of inhibitory circuits in striatal function.

**Editors:** Dieter Jaeger (who, with collaborators, has conducted one of the few studies looking directly for mutual inhibition between medium spiny neurones) was also invited to contribute to the discussion of the hypothesis of domains of mutual inhition. He provided the following commentary

**The domain hypothesis: Time for a farewell?**
**D. Jaeger, Dept. Biology, Emory University**

As Wickens and Oorschot, Chapter 7, state, "the domain hypothesis proposes that mutual inhibition among spiny projection neurones is a central principle of neostriatal organization". The functional significance of this principle is proposed to be the prevention of cocontraction among antagonist muscles.

One advantage of the domain hypothesis is that it makes a clear prediction that is experimentally testable, namely that there is strong mutual inhibition between medium spiny neurones. Unfortunately, when we tested this hypothesis *in vitro* by dual intracellular recordings between medium spiny neurones in close proximity, and *in vivo* by antidromic activation of medium spiny neurones, no such strong mutual inhibition was found (Jaeger *et al.,* 1994). The authors of the domain hypothesis spend much of their chapter in trying to find a way around this negative finding. I will spend the rest of my commentary to argue that it might be better to bid farewell to the domain hypothesis as stated.

One argument the authors bring forward is that the probability of inhibitory interactions between neurones in close proximity might be only of the order of 3% and could have been missed in our study. This argument is based on findings of low local connectivity in hippocampus and neocortex. Curiously, in a previous publication, the same first author argued that the specific anatomy of the striatum results in a very high probability of neighbouring neurones contacting each other (Wickens *et al.,* 1995). A computational model of the domain hypothesis was built resting on this assumption. While this model showed that with full local inhibitory connectivity the striatal network might obey the rules of the domain hypothesis, it seems unlikely to me that this would be the case with a connection probability of only 3%. The low rate of spiking typical for medium spiny cells would appear to work even more against strong inhibitory interactions. Nevertheless, dynamic systems are hard to predict by thought experiments, so in my view, if the authors were to seriously defend the argument of sparse connectivity, at least they owe us a computer model where a winner-take-all interaction takes place within a sparsely connected network of rarely spiking cells. A well presented anatomical and statistical study of mutual connections in striatum would also be invaluable.

A second argument brought forth by the authors is that neuromodulators might turn on and off local inhibitory interactions in striatum. For this to account for our negative findings of local inhibitory interactions between medium spiny neurons, the modulatory state both *in vitro* and in anesthetized animals would have to be such that the interactions were turned off. In particular, the authors propose that a low level of dopamine and a high level of acetylcholine might decrease the amount of inhibitory interactions. Both substances are thought not to act by a direct effect on $GABA_A$ transmission, but by changing other conductances. It appears questionable that a strong mutual inhibition could be masked completely by such mechanisms. A weak mutual inhibition, on the other hand, would not appear to be sufficient to support the domain hypothesis.

This is not to say that there are no inhibitory interactions in striatum. In fact, strong IPSPs are frequently seen in medium spiny neurons, and are likely to be due to the activity of local interneurones (Kita, 1993; Johnson *et al.*, 1997) and possibly also due to inputs from GP (Sato *et al.,* 1997). While these input pathways explain the IPSPs observed in medium spiny neurones, they do not support the domain hypothesis as stated. Local interneurones would be activated by other sources than medium spiny neurones, and could not lead to mutual inhibition between medium spiny neurones. GP neurones would be suppressed by medium spiny activity, and in turn would inhibit other medium spiny neurones even less, contrary to the domain hypothesis. The presence of strong IPSPs from these sources also demonstrates that the modulatory state of the medium spiny neurones *in vitro* can not be such that $GABA_A$ receptors are ineffective. This argues against the suppression of mutual inhibition between medium spiny neurones by neuromodulators.

Under functional aspects, the probability of the domain hypothesis being correct should be evaluated by the criterion of whether it explains aspects of basal ganglia function that are otherwise hard to account for. This does not appear to be the case. First, the function of preventing co-contraction between antagonist muscles is not clearly associated with basal ganglia activity and might well be carried out by completely different circuits. Second, if the basal ganglia were responsible for suppressing co-contraction, this function would have to express itself for sure at the output stage of basal ganglia processing, i.e. the entopeduncular nucleus (EP or GPi in primates) and substantia nigra *pars reticulata* (SNr). In the rat there are 2.8 million neurones in one side of striatum and only 30 thousand neurones in EP/ SNr (Oorschot, 1996). To me this suggests that microdomains of striatal function are not likely to be faithfully represented in EP/SNr, but that many striatal microdaimns converge onto a single EP/SNr domain. Maybe to tease the authors of the striatal domain hypothesis, strong mutual inhibitory interactions may well occur in EP/SNr, and, in addition, activity in these structures is closer to movement control and thus the prevention of co-contraction than striatal activity.

Should the domain hypothesis, then, not better be applied to the output stage of basal ganglia processing? While this is tempting, much experimental evidence would be needed to make this *ad hoc* transfer of hypotheses credible. There are too many functional tricks a neural network may perform in order to jump on a hypothesis that does not have solid experimental evidence for it, that does not make correct predictions, and for which the function that it explains may not be present.

**Miller:** The issue of whether the striatum can or cannot be conceived as consisting of domains within which mutal inhibition and interneuronal competition occurs is an important one. If domains with such properties exist, it would help explain not only the shifting relationship between coactivation and mutual antagonism in the muscular apparatus: There are many parallels between sensory-motor processing and cognitive processing. Thus the same dynamic relationship may explain some aspects of selective attention, in which there is also a competition between sources of information. This has been discussed by many authors. Examples where this concept has been formulated, or used as an assumption are Miller and Wickens (1991), Joel and Weiner (Chapter 12). As mentioned by Joel and Weiner (1999; this volume) a process of selection by competition between representations has also been proposed, entirely on psychological grounds, to explain orderly generation of one program for behavioural output at once (see Norman and Shallice, 1986). Is there any evidence that such competition is generated specifically in the striatum, thus imposing competition between different corticostriatal circuits? A variety of data suggests that the striatum (or at least the basal ganglia in a wider sense) are involved in selective aspects of attention. Bradyphrenia and impairment in other aspects of attention are thought to occur in Parkinson's disease (e.g. Lees, 1994). EEG studies in Parkinson's disease suggests that normal competition between rival representations is lost in Parkinson's disease (Pulvermueller *et al.*, 1996). Electrophysiological studies at the single unit level give evidence that striatal neurones respond in a highly selective manner: When the experimental paradigm permits a variety of stimulus-response relationships, each neurone is related to only one such relationship (Kermadi and Boussaoud, 1995; Boussaoud and Kermadi, 1997). Thus, in understanding brain function on a scale larger than the striatum itself, there is an apparent need for a strong process of competition between coactive cell assemblies. It is difficult to see how this could be provided in the cortex itself, because the cortex seems to be an large associative network, where excitatory interactions between neighbouring neurones dominate over mutual inhibition (Wickens and Miller, 1997). Thus, it is likely that elements of competition or "dissociation", demonstrable in the cortex, are imposed from connected structures outside the cortex. The striatum is an obvious site where such mutual inhibition might occur, imposing its competitive dynamic on cortical cell groups, via the well known corticostriatal loops. The fact that attempts to provide direct evidence for a competitive dynamic in the striatum have so far failed should thus be regarded as a major paradox, requiring further attention. It may be that there are technical reasons why the experiments of Jaeger *et al.* (1994) failed to provide evidence of mutual inhibition between medium spiny neurones. Suggestions of such reasons are made by Wickens and Oorschot (Chapter 7). However, Jaeger would seem to be correct in saying that functionally useful mutual inhibition within a domain cannot occur if the connectivity between medium spiny neurones is as sparse as suggested by Wickens and Oorschot. (They cannot have it both ways!) Alternatively it may be that the striatal domain idea needs to be revised in such a way as to be compatible with the evidence of Jaeger *et al.* (1994), allowing the generation of competitive dynamics needed to explain some larger-scale aspects of brain function, but in a manner different from that envisaged in the domain hypothesis. The third alternative, which, if true, would represent a significant change of direction in the theorizing of many people, is that the striatum/basal ganglia are *not* implicated in the selective aspects of attention, so that the evidence referred to above requires a radical revision of proposed explanatory concepts.

**Wickens.** It is apparent from the comments of Jaeger and Miller that several key points of the domain hypothesis have not been explained very well, and that further clarification is required. The first of these concerns the probability of inhibitory interactions among spiny projection neurones within the same domain. As originally formulated "... the calculation for the number of neurones in a domain gives only an absolute upper limit, because it assumes that within the volume of a domain, every neurone is connected to every other. More realistically, one should think of several relatively disjoint sets forming superimposed domains" (Wickens, 1993; p. 42). Then, as now, there were no quantitative data on the probability of a synaptic contact between one spiny projection neurone and another (the missing value is the number of synapses made by the local collaterals of each spiny projection neurone). However, this probability cannot be determined from the data of Jaeger *et al.* (1994) because we have no data on the rate of "false negatives" in that study. Based on a comparison with the initial studies conducted in the cerebral cortex, this rate may be very high (i.e. 7:1). Thus, even when the actual probability of connections is of the order of 20% (as in the cortex), excitatory interactions were detected in only 3% of 1163 pairs studied using dual intracellular recording by Deuchars *et al.* (1994). Thus, while the dual intracellular studies are technically impressive, failure to detect inhibitory interactions in 35 pairs of neurons (Jaeger *et al.*, 1994) is hardly compelling evidence that inhibitory interactions are "weak or non-existent".

It is also important to define formally what is actually meant by the terms "sparse" and "strongly connected". Is a probability of 20 % sparse? In comparison with another area of the brain in which local interactions are assumed to be functionally important, such a probability is high. In graph theory, "strongly connected" has the following formal definition: If for any two vertices v and w of the graph D there is a path from v to w, then D is called strongly connected. A set of, say, 2000 neurones connected with a probability of 20 % is virtually certain to be strongly connected in this sense, even though not every neurone is connected to every other. Inhibitory interactions within such a set are bound to produce dynamical effects such as mutual inhibition or disinhibition as characterised previously (Wickens *et al.*, 1991; Alexander and Wickens, 1993; Plenz *et al.*, 1995). Being overly simplistic about connectivity is unlikely to lead to a truer picture of the dynamics of interaction among spiny projection neurons.

Finally, a key point of the domain hypothesis is that competition among spiny neurones may be turned on or off by neuromodulators such as dopamine or acetylcholine. We originally proposed postsynaptic mechanisms for this, but the data of Jaeger *et al.* suggest another intriguing possibility: that inhibitory synapses might exist that are functionally silent under the conditions that prevail in slices or in anaesthetised animals. In this connection it is interesting to note the following comment: "It is tempting to speculate that the release of GABA from the collateral axons of spiny neurons is tightly regulated by some neuroactive substances in the striatum. We were able to evoke IPSPs in the spiny projection neurons by antidromically activating their descending axons when 4-aminopyridine (i.e. a K-channel blocker which increases neurotransmitter release from terminals by increasing the duration of the action potential) was infused into the recording site (Kita, unpublished data)" Kita (1993, p. 69).

## 9. THE SPECIAL ROLE OF THE STRIATUM IN CORTICO-STRIATAL INTERPLAY

**Editors:** With respect to the chapter by Salinas, Romo and colleagues, it is important to integrate their approach based on systems neurophysiology with other approaches to the striatum, from psychopharmacologists (for instance the chapter by Beninger and Olmstead), or from researchers of synaptic plasticity based on slice preparations (chapter by Wickens). The results described in the chapter by Salinas, Romo *et al.* show that, with few exceptions, the response properties of neurones in various parts of the cortico-basal ganglia circuits are very much the same. This could be taken to indicate that there is little functional differentiation between the components of this circuit. However, most of the experiments they describe are based on study of animals with patterns of learned behaviour which are already established. It is rather difficult to study the actual process of learning in such experiments. From the point of view of psychopharmacology, dopamine appears to be a major component of the reinforcement system, mediating various types of learning. The experiments of Wickens and co-workers, using slice preparations, show that dopamine may mediate a type of synaptic strengthening of excitatory inputs to the striatum. This may be the synaptic basis of dopamine-mediated reinforcement at the macroscopic level. These actions of dopamine are likely to be exerted mainly in the striatum. We therefore asked whether these actions of dopamine in the striatum may be necessary in the learning process, for establishing critical links between cortex and striatum, permitting thereafter the circulation of activity in completed cortico-basal ganglia circuits. This would provide a distinctive role for the striatum, although it would be one not easily demonstrated in the sort of experiment described by Salinas, Romo *et al.* We asked whether there are any objections to this view?

**Romo:** Memory is typically subdivided in two parts: information storage and information recall. A similar distinction can be made for learning in general, and it is useful to think of the neurophysiology of the basal ganglia in these terms. Dopamine is necessary for information processing in the striatum, and the evidence implicates it in learning. The relevant studies in this respect are those in which basal ganglia neurones are studied precisely while animals learn a task, but most studies in behaving animals have been made in well-established learned sensorimotor tasks in which the recall processes strongly dominate. Aosaki *et al.* (1994a) have investigated a full learning process. Their experiments showed that the putative interneurones of the striatum undergo stereotyped changes in their firing, during Pavlovian conditioning. These neurones directly reflect the association between an arbitrary stimulus and a reward. Temporal coincidence between cortical and dopamine inputs into the striatum may be important for establishing such associations (Schultz *et al.*, 1995). This would suggest that dopamine somehow enables the synaptic changes that underlie this type of associative learning. We should also mention another important result in this direction. Stimulation of the corticostriatal fibres produces long-term depression at the synaptic terminals in the striatum, and this effect depends on the presence of dopamine at those terminals (Calabresi *et al.*, 1992). This is an example of dopamine-dependent synaptic modification, although its functional meaning is difficult to determine.

On the other hand, dopamine may also be crucial for the execution of life-long established functions, such as reaching toward a goal: It is well known that dopaminergic neurones show complex context-dependent responses (Romo and Schultz, 1990; Schultz and Romo, 1990). This idea is supported by the simple observation that monkeys with impaired dopamine transmission react poorly to known sensory cues that would normally guide their behaviour. Aosaki *et al.*, (1994b) studied the responses of the putative interneurones from the monkey striatum after conditioning. They showed that the acquired response modulations were strongly diminished when the dopaminergic input was eliminated, but this effect could be reversed by applying a dopamine agonist. This suggests that the ability to recall an established association is conditional on the presence of dopamine. An impending question relating to both storage and recall is the role of the most numerous medium-spiny neurones of the striatum during learning.

A final consideration is the role of dopamine in information processing in the cortex, as opposed to information processing in the striatum. A large portion of the frontal cortex is also innervated by midbrain dopaminergic cells, so the question is whether dopamine plays similar functional roles in the two structures. A study by Williams and Goldman-Rakic (1995) in behaving monkeys has shown that dopamine gates memory-related responses in prefrontal cortex. So there are some indications that dopamine may act as a gating signal in both striatum and cortex, enhancing or suppressing information flow in these circuits.

In summary, there is some experimental evidence indicating that: (1) Some striatal neurones undergo synaptic changes underlying the learning of associations. (2) These synaptic changes in the striatum probably depend on precisely-timed dopaminergic inputs. (3) In both cortex and striatum dopamine may control the transit of information. These results are certainly consistent with the proposed idea. Whether dopamine also underlies some forms of synaptic plasticity in cortex is still unknown. If it does not, this would certainly provide a distinctive role for the striatum, but the relevant experiments have not been done yet. Perhaps the actions of dopamine are similar in the two structures, both for storage and recall, but the inputs to these circuits, which shape any given function, are different.

**Editors:** The above comments indicate, from an electrophysiological perspective, that dopamine has a dual role, in both learning and recall performance. This dual role is supported by the large psychopharmacological literature. This evidence, is very complex and some of the complications are unresolved. Nevertheless, it points increasingly to the view that the learning functions of dopamine are mediated by D1-like receptors, while the performance functions are more closely related to D2 receptors (e.g. Miller *et al.* 1990). Clarification of this might be possible in electrophysiological experiments such as those described by Salinas *et al.* It would be predicted that the neuronal activity related to recall performance would be attenuated by blockers of D2 receptors, rather than by those of D1 receptors.

## REFERENCES


Aosaki, T., Tsubokawa, H., Ishida, A., Watanabe, K., Graybiel, A.M. and Kimura, M. (1994a) Responses of tonically active neurons in the primate's striatum undergo systematic changes during behavioral sensorimotor conditioning. *Journal of Neuroscience*, **14**, 3969–3984.

Aosaki, T., Graybiel, A.M. and Kimura, M. (1994b) Effect of nigrostriatal dopamine system of acquired neural responses in the striatum of behaving monkeys. *Science,* New York **265**,412–415.

Alexander, M.E. and Wickens, J.R. (1993) Analysis of striatal dynamics: the existence of two modes of behaviour. *Journal of Theoretical Biology*, **163** : 413–38.

Ambros-Ingerson, J., Granger, S. and Lynch, G. (1990) Simulation of paleocortex performs hierarchical clustering. *Science,* New York, **247**, 1344–1348.

Apicella, P., Ljungberg, T., Scarnati, E. and Schultz, W. (1991) Responses to reward in monkey dorsal and ventral striatum, *Experimental Brain Research,* **85,** 491–500.

Barto, A.G. (1995) Adaptive critic and the basal ganglia. In *Models of information processing in the basal ganglia,* edited by J.C. Houk, J.L. Davis, and D.G. Beiser, Cambridge: MIT Press, pp. 215–232.

Berns, G.S. and Sejnowski, T.J. (1998) A computational model of how the basal ganglia produce sequences. *Journal of Cognitive Neuroscience*, **10**, 108–121.

Booussaoud, D. and Kermadi, I. (1997) The primate striatum: neuronal activity in relation to spatial attention versus motor preparation. *European Journal of Neuroscience* **9**, 2152–2168.

Burton, H. and Sinclair, R.J. (1990) 2nd Somatosensory Cortical Area in Macaque Monkeys .1. Neuronal Responses to Controlled, Punctate Indentations of Glabrous Skin on the Hand, *Brain Research,* **520,** 262–271.

Calabresi, P., Maj, R., Mercuri, N. and Bernardi, G. (1992) Coactivation of D1 and D2 dopamine receptors is required for long-term synaptic depression in the striatum. *Neuroscience Letters*, **142**, 95–99.

Chevalier, G. and Deniau, J.M. (1990) Disinhibition as a basic process in the expression of striatal functions. *Trends in Neurosciences*, **13**, 277–280.

Cromwell, H.C. and Berridge, K.C. (1996) Implementation of action sequences by a neostriatal site: a lesion mapping study of grooming syntax. *Journal of Neuroscience*, **16**, 3444–3458.

Davidson, T.L. (1993). The nature and function of interoceptive signals to feed: Toward integration of physiological and learning perspectives. *Psychological Review*, **100**, 640–657.

Davidson, T.L. (1998). Hunger cues as modulatory stimuli. In *Occasion Setting: Association Learning and Cognition in Animals,* edited by N.A. Schmajuk and P.C. Holland, Washington, D.C.: American Psychological Association, pp. 223–248.

DiCarlo, J.J., Johnson, K.O., and Hsiao, S. (1998) Structure of receptive fields in area 3b of primary somatosensory cortex in the alert monkey. *Journal of Neuroscience*, **18**, 2626–2645.

Dickinson, A. and Balleine, B. (1990) Motivational control of instrumental performance following a shift from hunger to thirst. *Quarterly Journal of Experimental Psychology*, **42B**, 413–431.

Dickinson, A. and Balleine, B. (1994) Motivational control of goal-directed action. *Animal Learning and Behavior*, **22**, 1–18.

Dickinson, A., Balleine, B., Watt, A., Gonzalez, F. and Boakes, R.A. (1995) Motivational control after extended training. *Animal Learning and Behavior,* **23**, 197–205.

Dickinson, A. and Dawson, G.R. (1987) Pavlovian processes in the motivational control of instrumental performance. *Quarterly Journal of Experimental Psychology*, **39B**, 201–213.

Deuchars, J., West, D.C. and Thomson, A.M. (1994) Relationship between morphology and physiology of pyramid-pyramid single axon connections in rat neocortex *in vitro. Journal of Physiology,* London, **478** : 423–435.

Freund, T.F., Powell, J. and Smith, A.D. (1984) Tyrosine hydroxylase-immunoreactive boutons in synaptic contact with identified striatonigral neurons, with particular reference to dendritic spines. *Neuroscience* **13**, 1189–1215.

Groenewegen, H.J. and Berendse, H.W. (1993) Anatomical relationships between the prefrontal cortex and the basal ganglia in the rat. In *Motor and Cognitive Functions of the Prefrontal Cortex*, edited by A.-M. Thierry, J. Glewinski, P.S. Goldman-Rakie and H. Christen, Springer Verlag, Heidelberg, pp. 51–77.

Groenewegen H.J., Berendse H.W., Wolters J.G. and Lohman A.H.M. (1990) The anatomical relationship of the prefrontal cortex with the striatopallidal system, the thalamus and the amygdala: evidence for a parallel organization. In: The prefrontal cortex: its structure, function and pathology (*Progress in Brain Research,* 85) edited by H.B.M. Vylings, C.G. Van Eden, J.P.C. de Brain, M.A. Corner and M.G.P. Feenstra, Amsterdam, Elsevier, **85**, 95–118.

Haracz, J.L., Tschanz, J.T., Wang, Z., Griffith, K.E. and Rebec, G.V. (1998) Amphetamine effects on striatal neurons: implications for models of dopamine function. *Neuroscience and Biobehavioral Reviews,* **22,** 613–622.

Hertz, J., Krogh, A. and Palmer, R.G. (1991) *Introduction to the Theory of Neural Computation.* New York: Addison-Wesley.

Hooper, K.C., Banks, D.A., Stordahl, L.J., White, I.M. and Rebec, G.V. (1997) Quinpirole inhibits striatal and excites pallidal neurons in freely moving rats. *Neuroscience Letters,* **237**, 69–72.

Hooper, K.C., Harbeson, C.J. and Rebec, G.V. (1988) Dopamine depletion alters the electrophysiological effects of D1 and D2 agonists in the basal ganglia of behaving rats. *Society for Neuroscience Abstracts.*

Houk, J.C. (1995) Information processing in modular circuits linking basal ganglia and cerebral cortex. In *Models of information processing in the basal ganglia,* edited by J.C. Houk, J.L. Davis, and D.G. Beiser, Cambridge: MIT Press, pp. 3–9.

Hsiao, S.S. O'Shaughnessy, D.M., and Johnson, K.O. (1993) Effects of selective attention on spatial form processing in monkey primary and secondary somatosensory cortex. *Journal of Neurophysiology,* **70**, 444–447.

Ingham, C.A., Hood, S.H., Taggart, P. and Arbuthnott G.W. (1998) Plasticity of synapses in the rat neostriatum after unilateral lesion of the nigrostriatal dopaminergic pathway. *Journal of Neuroscience* **18**, 4732–4743.

Inoue, M., Oomura, Y., Aou, S., Nishino, H. and Sikdar, S.K. (1985) Reward related neuronal activity in monkey dorsolateral prefrontal cortex during feeding behavior, *Brain Research,* **326,** 307–312.

Jaeger, D., Kita, H., and Wilson, C.J. (1994). Surround inhibition among projection neurons is weak or nonexistent in the rat neostriatum. *Journal of Neurophysiology* **72,** 2555–2558.

Jiang, W., Chapman, C.E. and Lamarre, Y. (1991) Modulation of the cutaneous responsiveness of neurones in the primary somatosensory cortex during conditioned arm movements in the monkey, *Experimental Brain Research,* **84,** 342–354.

Johnson, K.O. (1974) Reconstruction of population response to a vibratory stimulus in quickly adapting mechanoreceptive afferent fiber population innervating glabrous skin of the monkey. *Journal of Neurophysiology,* **37**, 48–72.

Johnson, L., Koos, T., Zaborsky, L., Moore, K. and Tepper, J.M. (1997) GABAa receptor-mediated inhibition of fast spiking interneurons in rat neostriatum. *Society for Neuroscience Abstracts* **23**, 504.10.

Kermadi, I. and Joseph, J.P. (1995) Activity of caudate nucleus during spatial sequencing. *Journal of Neurophysiology,* **74**, 911–933.

Kermadi, I. and Boussaoud, D. (1995) Role of the primate striatum in attention and sensorimotor processes: comparison with premotor cortex. *NeuroReport* **6**, 1177–1181

Kita, H. (1993) GABAergic circuits of the striatum. In: *Chemical Signalling in the Basal Ganglia, (Progress in Brain Research, vol. 99),* edited by G.W. Arbuthnott and P.C. Emson (eds.), North Holland, Elsevier, pp. 51–72.

Kiyatkin, E.A. and Rebec, G.V. (1996) Dopaminergic modulation of glutamate-induced excitations of neurons in the neostriatum and nucleus accumbens of awake, unrestrained rats. *Journal of Neurophysiology,* **75,** 142–153.

Kiyatkin, E.A. and Rebec, G.V. (1997) Iontophoresis of amphetamine in the neostriatum and nucleus accumbens of awake, unrestrained rats. *Brain Research,* **771**, 14–24.

Kiyatkin, E.A. and Rebec, G.V. (1998) Heterogeneity of ventral tegmental area neurons: single-unit recording and iontophoresis in awake, unrestrained rats. *Neuroscience,* **85**, 1285–1309.

Kiyatkin, E.A. and Rebec, G.V. (in press) Differential modulation of striatal neuronal activity by glutamate and GABA: iontophoresis in awake, unrestrained rats. *Society for Neuroscience Abstracts.*

Lees, A.J. (1994) The concept of bradyphrenia *Revue de Neurologie* **150**,823–826

Merchant, H., Zainos, A., Hernández, A., Salinas, E. and Romo, R. (1997) Functional properties of primate putamen neurons during the categorization tactile stimuli. *Journal of Neurophysiology,* **77**, 1132–1154.

Miller, R. (1996) Cortico-thalamic interplay and the security of operation of neural assemblies and temporal chains in the cerebral cortex. *Biological Cybernetics.,* **75**, 263–275.

Miller, R., Wickens, J.R. and Beninger R.J. (1990) Dopamine D-1 and D-2 receptors in relation to reward and performance: a case for the D-1 receptor as a primary site of therapeutic action of neuroleptic drugs. *Progress in Neurobiology* **34**, 143–183.

Mink, J.W. (1996) The basal ganglia: focusing selection and inhibition of competing motor programs. *Progress in Neurobiology,* **50**, 381–425.

Montague, P.R., Dayan, P. and Sejnowski, T.J. (1996) A framework for mesencephalic dopamine systems based on predictive Hebbian learning, *Journal of Neuroscience,* **16,** 1936–1947.

Mountcastle, V.B., Steinmetz, M.A. and Romo, R. (1990) Frequency discrimination in the sense of flutter: psychophysical measurements correlated with postcentral events in behaving monkeys. *Journal of Neuroscience,* **10**, 3032–3044.

Mushiake, H. and Strick, P.L. (1995) Pallidal neuron activity during sequential arm movements. *Journal of Neurophysiology,* **74**, 2754–2758.

Nakano, Y., Lenard, L., Oomura, Y., Nishino, H., Aou, S. and Yamamoto, T. (1987) Functional involvement of catecholamines in reward-related neuronal activity of the monkey amygdala, *Journal of Neurophysiology,* **57,** 72–91.

Oorschot, D.E. (1996). Total number of neurons in the neostriatal, pallidal, subthalamic, and substantia nigral nuclei of the rat basal ganglia: a stereological study using the cavalieri and optical disector methods. *Journal of Comparative Neurology* **366,** 580–599.

Plenz, D., Wickens, J.R. and Kitai, S.T. (1995) Corticostriatal interactions. *Proceedings of the 4th Annual Computation and Neural Systems Meeting,* CNS* 95, July 11–15, Monterey, California, U.S.A.

Pulvermuller, F., Lutzenberger, W., Muller,V, Mohr, B., Dichgans, J. and Birbaumer, N. (1996) P3 and contingent negative variation in Parkinson's disease. *Electroencephalography and Clinical Neurophysiology,* **98,** 456–467.

Rebec, G.V. and Kiyatkin, E.A. (1998) Striatal neuronal activity and responsivity to dopamine and glutamate after pharmacological blockade of D1 and D2 receptors. *Society for Neuroscience Abstracts,* 565.9

Rebec, G.V., Christensen, J.R.C., Guerra, C. and Bardo, M.T. (1997a) Regional and temporal differences in dopamine efflux in the nucleus accumbens during free-choice novelty. *Brain Research,* **776,** 61–67.

Rebec, G.V., White, I.M. and Puotz, J.K. (1997b) Responses of neurons in dorsal striatum during amphetamine-induced focused stereotypy. *Psychopharmacology,* **130,** 343–351.

Rebec, G.V. and Kiyatkin, E.A. (in press) Striatal neuronal activity and responsivity to dopamine and glutamate after pharmacological blockade of D1 and D2 receptors. *Society for Neuroscience Abstracts.*

Roland, P.E., Larsen, B., Lassen, N.A., Skinhøj, E. (1980) Supplementary motor area and other cortical areas in organization of voluntary movements in man. *Journal of Neurophysiology,* **43,** 118–136.

Romo, R. and Schultz, W. (1990) Dopamine neurons of the monkey midbrain: contingencies of responses to active touch during self-initiated arm movements. *Journal of Neurophysiology,* **63,** 592–606.

Romo, R., Hernández, A., Zainos, A. and Salinas, E. (1998) Somatosensory discrimination based on cortical microstimulation. *Nature,* London, **392,** 387–390.

Romo, R., Merchant, H., Zainos, A. and Hernández, A (1997). Categorical perception of somesthetic stimuli: Psychophysical measurements correlated with neuronal events in primate medial premotor cortex. *Cerebral Cortex,* **7,** 317–326.

Romo, R., Merchant, H., Zainos, A. and Hernández, A. (1996) Categorization of somaesthetic stimuli: sensorimotor performance and neuronal activity in primary somatic sensory cortex of awake monkeys. *NeuroReport,* **7,** 1273–1277.

Salinas, E. and Abbott, L.F. (1996) A model of multiplicative neural responses in parietal cortex. *Proceedings of the National Academy of Sciences, USA,* **93,** 11956–11961.

Salinas, E. and Romo, R. (1998) Conversion of sensory signals into motor commands in primary motor cortex. *Journal of Neuroscience,* **18,** 499–511.

Sato, F., Levesque, M., Nakamura, Y. and Parent, A. (1997). Axonal projections of single cells from the external pallidum in monkeys. *Society for Neuroscience Abstracts* **23, part 2,** 196.

Schultz, W. and Romo, R. (1990) Dopamine neurons of the monkey midbrain: contingencies of responses to stimuli eliciting immediate behavioral reactions. *Journal of Neurophysiology,* **63,** 607–624.

Schultz, W., Apicella, P., Scarnati, E. and Ljungberg, T. (1992) Neuronal activity in monkey ventral striatum related to the expectation of reward, *Journal of Neuroscience,* **12,** 4595–4610.

Schultz, W., Apicella, P., Ljungberg, T., Romo, R. and Scarnati, E. (1993) Reward-related activity in the monkey striatum and substantia nigra. In: *Chemical Signalling in the Basal Ganglia, (Progress in Brain Reseach, Vol. 99),* edited by G.W. Arbuthnott and P.C. Emson, Amsterdam, Elsevier, pp. 227–235.

Schultz, W., Romo, R., Ljungberg, T., Mirenowicz, J., Hellerman, J.R. and Dickinson, A. (1995) Reward-related signals carried by dopaminergic neurons. In *Models of information processing in the basal ganglia,* edited by J.C. Houk, J.L. Davis, and D.G. Beiser, Cambridge: MIT Press, pp. 233–248.

Schultz, W., Dayan, P. and Montague, P.R. (1997) A neural substrate of prediction and reward, *Science, New York,* **275,** 1593–1599.

Shadlen, M.N. and Newsome, W.T. (1998) The variable discharge of cortical neurons: implications for connectivity, computation and information coding. *Journal of Neuroscience,* **18,** 3870–3896.

Stern, E.A., Jaeger, D. and Wilson, C.J. (1998) Membrane potential synchrony of simultaneously recorded striatal spiny neurons *in vivo. Nature,* London, **394,** 475–478

Strecker, R.E., Steinfels, G.F. and Jacobs, B.L. (1983) Dopaminergic unit activity in freely moving cats: lack of relationship to feeding, satiety, and glucose injections, *Brain Research,* **260,** 317–321.

Tanji, J. and Shima, K. (1994) Role for supplementary motor area cells in planning several movements ahead. *Nature,* London, **371,** 413–416.

Trulson, M.E., Crisp, T. and Trulson, V.M. (1983) Dopamine-containing substantia nigra units are unresponsive to changes in plasma glucose levels induced by dietary factors, glucose infusions or insulin administration in freely moving cats, *Life Sciences,* **32,** 2555–2564.

Tschanz, J.T., Griffith, K.E., Haracz, J.L. and Rebec, G.V. (1994) Cortical lesions attenuate the opposing effects of amphetamine and haloperidol on neostriatal neurons in freely moving rats. *European Journal of Pharmacology*, **257**, 161–167.

White, I.M. and Rebec, G.V. (1993) Responses of rat striatal neurons during performance of a lever-release version of the conditioned avoidance response task. *Brain Research*, **636**, 71–82.

White, I.M., Miller, D.W., White, W., Dike, G.L., Rebec, G.V., and Steinmetz, J.E. (1994) Neuronal activity in rabbit neostriatum during classical eyelid conditioning. *Experimental Brain Research*, **99**, 179–190.

Wickens, J.R., *A theory of the striatum*. 1993, Oxford: Pergamon Press.

Wickens, J.R., Alexander, M.E. and Miller, R. (1991) Two dynamic modes of striatal function under dopaminergic-cholinergic control: simulation and analysis of a model. *Synapse*, **8**, 1–12.

Wickens, J.R., Begg, A.J. and Arbuthnott, G.W. (1996) Dopamine reverses the depression of rat corticostriatal synapses which normally follows high-frequency stimulation of cortex *in vitro*. *Neuroscience*, **70**, 1–5.

Wickens, J. and Kötter, R. (1995) Cellular models of reinforcement. In: *Models of information processing in the basal ganglia*, edited by J.C. Houk, J.L. Davis, and D.G. Beiser, pp. 187–214. Cambridge: MIT Press

Wickens, J.R., Kötter, R., and Alexander, M.E. (1995). Effects of local connectivity on striatal function: stimulation and analysis of a model. *Synapse* **20,** 281–98.

Williams, G. V. and Goldman-Rakic, P.S. (1995) Modulation of memory fields by dopamine D1 receptors in prefrontal cortex. *Nature,London*, **376**, 572–575

Wilson, C.J. (1993) The generation of natural firing patterns in neostriatal neurons. In *Chemical Signalling In The Basal Ganglia, (Progress In Brain Research, vol 99)*, edited by G.W. Arbuthnott and P.C. Emson (eds), Oxford, Elsevier, pp. 277–297.

Wilson, C.J. (1995) The contribution of cortical neurons to the firing pattern of striatal spiny neurons. In: *Models of Information Processing in the Basal Ganglia,* edited by J.C. Houk, J.C., J.L. Davis and D.G. Beiser, Cambridge, MA: MIT Press, pp. 29–50.

Zainos, A., Merchant, H., Hernández, A., Salinas, E. and Romo, R. (1997) Role of primary somatic sensory cortex in the categorization of tactile stimuli: effects of lesions. *Experimental Brain Research*, **115**, 357–360.

# POSTLUDE
# Striatal Circuitry: Categorically Selective, or Selectively Categorical?


## Charles J. Wilson

*Department of Anatomy and Neurobiology, University of Tennessee, Memphis*
*855 Monroe Avenue, Memphis, TN 38163 USA*
*Tel: (001) 901-448-5965; FAX: (001) 901-448-7193, e-mail: charlie@savage.utmem.edu*


## 1. ANATOMY BEGETS THEORY IN THE BASAL GANGLIA

Most proposals for the function of the basal ganglia are presented as statements about the connections between the various basal ganglia nuclei. Despite the depth of current knowledge of basal ganglia neuroanatomy, dramatic new anatomical discoveries still occur, and they are the main sources of new functional ideas in the basal ganglia. For example, the Albin, Young and Penney model (1989), around which so much recent work has been organized, was based on the discovery of two independent and apparently antagonistic outputs of the striatum, the direct and indirect pathways. Similarly, ideas about the internal functional relationships between neurones within basal ganglia regions have been based largely upon anatomical rather than physiological observations. The observation by Ramon-y-Cajal (1995) that the most numerous neurones in the striatum, the spiny neurones, form a network of local axon collaterals is probably the most influential of these. The rationale for believing the spiny neurones to be an inhibitory has undergone dramatic changes, but since the time of C. and O. Vogt (1920) the striatum has been thought to function primarily as a mutually inhibitory network formed by the spiny cells. Many changes in our thinking were forced by the discovery that the spiny neurones are the projection cells, but the interpretation of local collateral connections of spiny neurones did not change. The inhibitory effect of the spiny neurones on their targets, the discovery that they are GABA-containing and release that inhibitory transmitter at their synapses in the globus pallidus and substantia nigra, and the direct demonstration that they make synaptic contacts amongst each other in the striatum has verified the anatomical essentials of the view advanced by the Vogts. A second fundamental anatomical observation was the presence of the corticostriatal pathway, also described by Ramon-y-Cajal. When in recent times this projection was shown to be topographically organized and excitatory, its superimposition on the apparently homogeneous mutual inhibitory network of the neostriatum suggested a role for inhibition in the striatum in *spatially* sharpening and sculpting the dynamic representation of cortical activity. The third anatomical feature of the striatum that has been heavily featured in our collective thinking about this structure is the convergence between the dopaminergic and the corticostriatal pathway. The dopaminergic projection to the striatum from the midbrain was discovered relatively recently, and its convergence upon the same spiny neurones that receive cortical input was discovered more recently still. The notion that dopaminergic input may be required for meaningful synaptic transmission in the corticostriatal projection is suggested by the motor impairments that accompany loss of dopaminergic neurones in experimental animals and in Parkinson's disease in humans.

Thus three anatomical observations: (1) the continuous topographical representation of the cortex in the neostriatum, (2) the inhibitory interconnections between the spiny neurones, and (3) the essential role of dopamine in processing of cortical inputs by the striatal spiny neurone, provide the basis of most functional neostriatal models of the past and present. While the interaction between dopaminergic and cortical inputs on the spiny neurone is alive and healthy, both of the first two ideas have suffered major setbacks in recent years, with the discovery that the corticostriatal projection is fundamentally discontinuous and the mysterious absence of surround inhibition among spiny neurones.

## 2. WHAT CONSTITUTES A MODULE IN THE STRIATUM?

In a continuously connected network, as the striatum was traditionally believed to be, the definition of a module is open to interpretation. A diagram showing a simple homogeneous model of the striatal intrinsic organization is shown in Figure 1. The cortical input is continuous, meaning that nearby regions in the cortex project to nearby regions in the striatum. "Nearby in the cortex" could be defined in functional rather than spatial terms, but "nearby in the striatum" must be interpreted spatially in this scheme. The important thing is that a small increase in the distance between selected regions of the cortex (however that distance is defined) will produce a small increase in the distance between their projection zones in the striatum. The volume of the striatum innervated by a cortical axon is a measure of the resolution of the cortical representation in the striatum. In this view, a striatal neurone ought to receive inputs from more than one cortical neurone, because the cortical cell's axonal arborizations ought to overlap (this producing the continuity). The cortical axons that converge upon one striatal neurone will all be near neighbours in the cortex (however this is defined). Nearby striatal neurones will receive nearly, but not completely identical input from the cortex.

To define a module in the continuous view of the striatum requires a decision. The volume occupied by the arborization of a single corticostriatal axon could be viewed as a striatal module. In this definition, however, each striatal neurone would be a member of as many modules as there were cortical axons arborizing in its dendritic field. It would be the centre of the module defined by the axon that makes the most synapses on it. Alternatively, a module might just as well be defined by the volume occupied by the axonal arborization of a single striatal spiny neurone. Spiny neurones within that volume may be inhibited by that one axon. In this scheme as well, each spiny neurone is a member of many modules, because each spiny cell axonal volume is slightly offset from its neighbour, and each spiny neurone could receive inputs from many other cells. Other inputs such as the thalamic input to the striatum, could be used to define a modular unit, as could the local axonal arborizations of other striatal neurones, such as interneurones. One interneurone with a large axonal field is shown in the diagram in Figure 1. In a continuous network it is possible to define arbitrarily a module, but the network actually

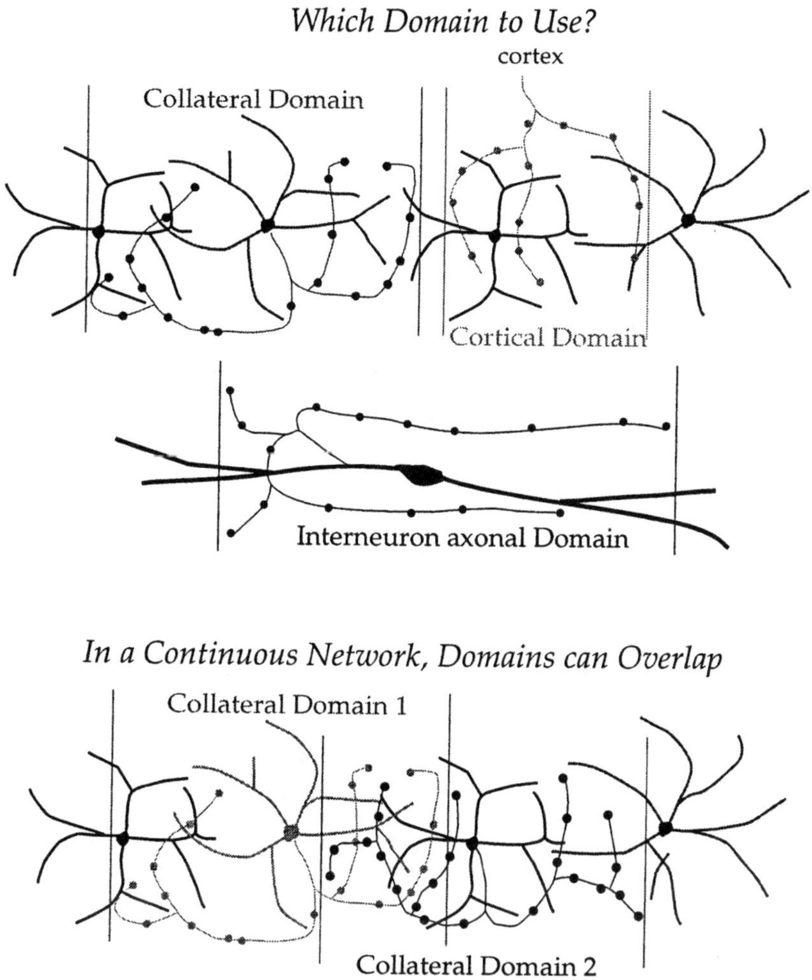


**Figure 1.** In the traditional view of the striatum as a neuronal continuum, there are no modules. In that view, each afferent system, the recurrent collaterals of spiny neurones, and each interneurone system has a functional domain of action. In this continuous view, there are as many domains as there are neurones forming the domains. For domains defined by the recurrent collaterals of spiny neurones, for example, each spiny cell is at the center of the domain defined by its axon, and also a member of the domains centered on all the spiny neurones whose dendritic and axonal fields overlap its own (about 2800 in rats). This is not so for a discrete modular arrangement, like that of the striosomes, for which afferent axons, local collaterals, and spiny cell dendrites all observe a confining boundary and share a single common space.

functions as a single field, rather than as a set of modules. This kind of organization is therefore best described as a set of *domains*, rather than modules. A smoothly varying spatial distribution of cortical input to the spiny neurones is guaranteed by the overlap of axonal arborizations of nearby cortical neurones and the interneuronal interconnections to produce a smoothly varying output. The shapes of the axonal and dendritic arborizations and the nature of the interconnections within the striatum will impose a continuous spatial and temporal filtering function on the striatal representation of the input. If the inhibition were very strong then the most strongly excited neurones would inhibit the rest, and only a small subset of cells would be active (assuming no means is provided for the other cells to escape, such as synaptic depression at the inhibitory synapses). Temporal factors would also interact with the spatial filter. For example, in the case of strong intrastriatal inhibition, the first neurones to respond to a sudden change in the excitatory input would inhibit the others and form an active assembly even if they were not the most strongly excited overall. The size and shape of the successfully excited subset of cells would depend primarily on the organization of inhibitory contacts. Small differences in the strength and latency of excitatory inputs could be accentuated in this case to produce subsets of active neurones. If individual cortical axons connected to the striatal population very sparsely, then differences among the cortical neurones might be emphasized, with striatal cells sharing particular afferent axons or particular small subsets of afferent axons having the advantage because their inputs are stronger, more correlated or have a shorter latency. If the local inhibition were weaker, then the local inhibition would act only to reduce the gain of the response to synaptic inputs. The winner-takes-all mode of synaptic inhibition is the basis for the domain model as originally described by Wickens (1993). It relies on strong local inhibition and spatial patterning between connected neurones, although the coupling need not be all-to-all, and the spatial patterning of the inhibition need not be the simple type described here. The domain model as described in this volume applies primarily to a continuous network view of the striatum, although it can also be applied within discrete modules if they are large in comparison to the domains of individual neurones, or across discrete modules if there are afferents or neurones whose axons disregard modular boundaries.

An alternative to continuous topography of the neostriatum has gradually been forced on basal ganglia investigators by an accumulation of anatomical data. Studies of the spatial organization of the corticostriatal projection, starting with Goldman and Nauta, (1977), Jones *et al.* (1977) and Künzle (1975) have suggested that inputs to the striatum do not lay out a continuous representation of the cortex either along spatial or functional lines. Taken together these experiments, and a variety of subsequent ones, show that single corticostriatal neurones make multiple discrete arborizations in the striatum, and that the arborizations from functionally similar regions overlap but may be immediately adjacent to those from functionally unrelated and spatially distant regions. The clearest case is that of the striosomes, in which the boundaries between related and unrelated projections can be visualized using a variety of cytochemical markers (Graybiel, 1983). Cortical projections to immediately adjacent cells in the striatum but across the striosome/matrix boundary often arise from distant and unrelated regions of the cortex (Graybiel *et al.*, 1994). The dendrites of striatal spiny neurones generally do not cross the boundaries between these compartments, and so the discrete arrangement of the cortical inputs is preserved in the synaptic input to individual cells of each compartment (Kawaguchi, Wilson and Emson, 1989). Likewise, striatal spiny neurones do not send axon collaterals to make synaptic connections across this boundary, so local interactions among these cells are restricted along the same boundaries as the input. Thus the striosomes, which occupy only about 15% of the striatum (Johnson *et al.*, 1990), constitute a set of discrete modules with clearly defined boundaries observed both by the inputs and by local interactions among spiny cells. Further, the axonal arborizations of individual cortical axons are about the same size as the striosomes, indicating that there may be no continuous gradient of the cortical innervation within the striosome (Kincaid, Zheng and Wilson, 1998). Likewise, spiny cells' dendritic fields and axon collateral arborizations are about the same size as the striosomes, so these cells may sample input throughout the module and have a uniform opportunity to interconnect with neurones everywhere within the module. Spatial gradients like those that arise in the classical interpretation of lateral inhibition (Ratliff, 1965) are not operational in a module such as this.

It is possible that the striatum is composed of two compartments, one representing the cortex in a discrete fashion (the striosomes) and one in a continuous fashion (the matrix), but there are a number of observations that suggest that the matrix may also be composed of discrete processing modules. Graybiel *et al.* (1994) have shown that projections from small functionally defined regions of the cortex can form discrete clusters in the matrix on the same scale as those seen to project to the striosomes. Likewise, individual cortical cells projecting to the matrix often form several small discrete arborizations of approximately the same size as those in the striosomes (Kincaid, Zheng and Wilson, 1998). The idea of a discrete modular organization of the matrix, called matrisomes, has been proposed by Graybiel *et al.* (1994; also Graybiel and Kimura, 1995) on the basis of the overlap of cortical inputs from functionally related structures. The boundaries of matrisomes are not so clear as those of striosomes. They cannot be demonstrated using any known cytochemical markers, and it has not been possible so far to determine whether the dendrites or axons of spiny neurones cross them. It is not even completely clear that cortical axons forming focal arborizations in the matrix are obeying pre-existing boundaries or are simply making small and partly overlapping arborizations (Flaherty and Graybiel, 1993). In either case, the degree of overlap between projections labelled by tracer injections in functionally defined cortical areas are strongly suggestive of a discrete discontinuous input. Also, the dendrites and axons of spiny neurones in the matrix have been reported to show distortions like those of cells at the edges of striosomes, suggesting that the cells do obey some otherwise invisible boundaries in the matrix neuropil (Kawaguchi *et al.*, 1990; Walker and Graybiel, 1993). Interneurones of all types apparently can interconnect the striosomes with the surrounding matrix, and are likely to ignore matrisomal boundaries as well (Kawaguchi, Wilson and Emson, 1995).

These observations suggest a discontinuous modular arrangement of the striatum, in which small changes in the origin of inputs in the cortex produce no change in the location of striatal arborizations until they cross a functional boundary, at which point the striatal projection jumps discontinuously to a non-overlapping module. Within a striatal module of this scheme, the inputs and the spiny cell interconnections of all the cells would share a common volume. The impact of this arrangement for the striosomes of the striatum (which are about the same size as a spiny cell's dendritic and axonal field) are illustrated in Figure 2. Interconnections could be all-to-all, or random, or they could connect according to a non-spatial rule based on the history of coactivation, or they could follow rules based on predetermined affinities among subclasses of neurones (Kincaid, Zheng and Wilson, 1998). However, they do not form a continuous spatial gradient. For theories of striatal function, this discrete modular scheme creates two different spatial scopes for intrastriatal interactions. Within modules, spiny cell interconnections could mediate winner-takes-all or other forms of interaction strictly within the module. If interconnectivity were random or all-to-all, strong inhibition would create a competition for control of the entire module. If the spiny neurones within the module formed small independent groups of highly interconnected neurones, then competition among neurones would be for control of these subgroups. Interneurones might mediate interactions between such subgroups, as well as among modules. It may be instructive to consider the numerical possibilities for such interactions within a striatal module. Focal corticostriatal axonal arborizations and spiny cell dendritic and axonal fields are on the order of 0.4 mm in diameter. Within such a volume, in the rat, there are approximately 2840 spiny neurones (Oorschot, 1996), and 32,000,000 asymmetric synapses (Ingham *et al.*, 1998 ), most of which are on dendritic spines. Accurate estimates of the number of spiny neurone collateral synapses are not yet available, but Ingham *et al.* (1998) counted the proportion of synapses that were asymmetric *versus* the symmetric ones, and estimated that approximately 20% of synapses were of the symmetric type. This indicates that there are 8,000,000 symmetric synapses, or about 2817 per spiny neurone in the volume. Among these 2800 synapses must be included all of the dopaminergic innervation of the module, all of the cholinergic innervation, the innervation by parvalbumin containing interneurones and by the SOM/NOS containing neurones (Kawaguchi *et al.,* 1995), and symmetric synapses made on interneurones, as well as the recurrent collaterals of the striatal neurones contacting other striatal neurones. The relative numbers of these are not known, but allowing the collateral synapses to be 1 sixth of the total is a good approximation, and yields about 500 synapses per striatal spiny neurone. This is consistent with preliminary work on the number of synaptic boutons formed by individually stained spiny neurones (Wilson, unpublished observations).

In the discontinuous network, mutual inhibition among spiny neurones would operate differently from the lateral inhibitory pattern seen in the continuous cell arrangement. At most, each spiny neurone could contact about one sixth (500/2840) of the neurones in its module (this assumes each presynaptic cell made no more than one contact on each postsynaptic cell). If excitation were uniformly applied to all the cells, and so strong that one cell firing would be able to silence all of the cells with which it was connected, at most about 6 spiny neurones would continue firing after inhibition was established. On the other hand, if inhibition were weak, all cells would experience about the same degree of inhibition as excitation was increased. The number of active inhibitory contacts would be distributed among cells along a binomial distribution whose mean and variance would depend upon the number of spiny neurones firing. For example, if 10% (284 of 2840) spiny neurones were firing, the binomial distribution based on 500 collateral connections per cell predicts that 14 neurones would receive 33 or fewer active inhibitory inputs, while 14 neurones would have 67 or more active inhibitory inputs, and the remaining 99% will have between 34 and active 66 inhibitory inputs. This uniformity in the strength of inhibition arises in the event that inhibitory inputs are not strong enough to reduce the active population to a small fraction of the total. If inhibitory synapses were scattered at random across the population, it would prevent the formation of independent assemblies of local winners, as would happen in the case of stronger inhibition and sparser activity as described above. Of course if inhibitory connections were not uniform, but occurred specifically among special subgroups of neurones (for example because of pre-existing affinity groups, or if connections have been sculpted by the action of a growth rule), the precise pattern of that connectivity would determine the organization of the network. Assuming such mutually inhibitory subgroups are not too small, the connectivity of collateral inputs should be relatively high and experimentally identifiable. If each spiny neurone had an equal probability of receiving an input from any of its neighbours, as many as 1/3 of spiny cell pairs should be connected in one or both directions, and if synaptic strength were relatively strong, winner-takes-all inhibition could result.

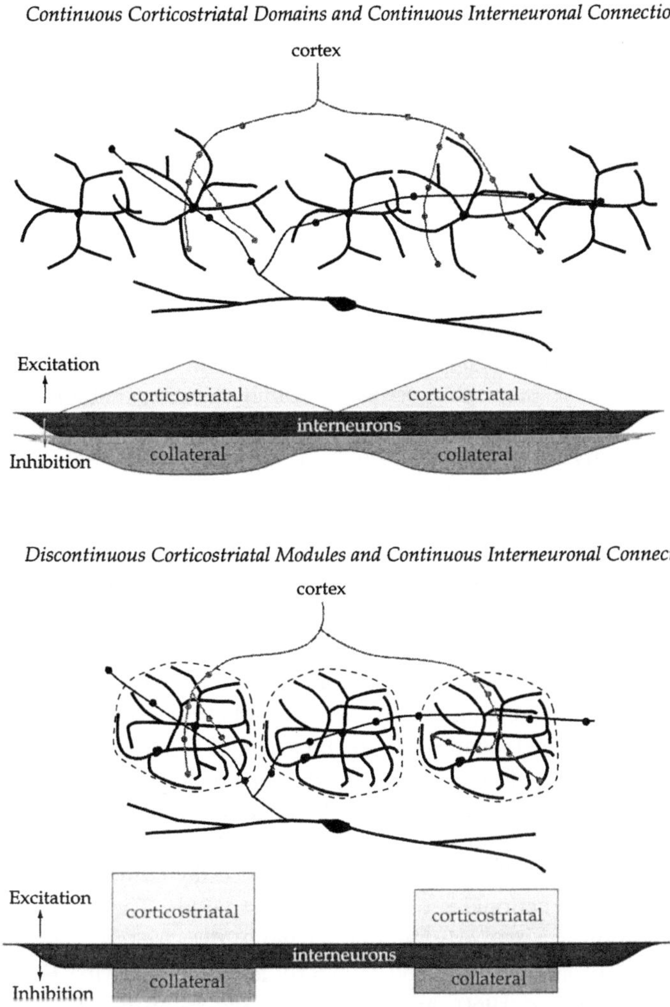


**Figure 2.** In the continuous view of striatal circuitry, recurrent inhibition can perform a spatial function, restricting the spread of excitation. Excitation spreads spatially according to the convolution of the axonal distribution and the density distribution of dendrites within the dendritic tree. Inhibition, spreading as the convolution of excitation with the axonal distribution of the spiny neurone, is more widespread. Thus collateral inhibition can produce an inhibitory surround around foci of strong excitation. In the modular organization suggested by the the internal organization of striosomes and perhaps matrisomes, inhibition and excitation are spatially coextensive. Only interneurones can perform cross-modular synaptic interactions.

## 3. A BRIEF HISTORY OF INTRASTRIATAL INHIBITION

Powerful local inhibition was an expectation of investigators performing early neurophysiological studies of the striatum, and it is not surprising that they interpreted much of what they saw in that context. The primary finding was that excitatory synaptic responses to stimulation of the cortex, thalamus or the region of the substantia nigra were followed by large and long lasting hyperpolarizations. This ubiquitous pattern of response in striatal neurones was called an "EPSP-IPSP sequence", and was considered analogous to similar sequences seen in spinal motoneurones, cortical and hippocampal pyramidal cells, and Purkinje cells (Purpura and Malliani, 1967; Buchwald, *et al.*, 1973; Kitai *et al.*, 1976). The hyperpolarizing secondary component of evoked synaptic potentials was assumed to arise from the inhibitory connections among spiny neurones, and this received support from studies in which it was blocked by systemic administration of a GABA antagonist (Bernardi *et al.*, 1975). There were inconsistencies in this interpretation however, including the inability of local GABA antagonists to block the hyperpolarization (Bernardi *et al.*, 1976) and the failure of the late response to reverse during hyperpolarization (Wilson, Chang and Kitai, 1983; Wilson, 1986). Subsequently, the hyperpolarizing phase of the response to excitatory afferent stimulation in the striatum was shown to be due to the transient removal of excitation, rather than to the action of an inhibitory transmitter (Wilson, Chang and Kitai, 1983; Wilson, 1986). This finding did not end the rule of spiny cell collateral inhibition in the minds of students of the basal ganglia, partly because studies of inhibition in striatal slices made available at the same time (Misgeld, Wagner and Ohno, 1982; Lighthall and Kitai, 1983; Kita, Kita and Kitai, 1985) showed a clear $GABA_A$-mediated short duration IPSP following the excitatory response to local stimulation. These results seemed to show the true mutual inhibition of spiny neurones, which had been present but masked by the complex excitatory events evoked *in vivo*. At about the same time, electron microscopic results demonstrated direct synaptic connections between spiny neurones and showed that spiny cells are the principal target of spiny cell axon collateral fields (Wilson and Groves, 1980; Somogyi, Bolam and Smith, 1981). It is understandable that the presence of GABAergic interconnections between spiny neurones and of GABAergic IPSPs was considered adequate evidence of powerful and pervasive mutual inhibition in the striatum. However, that conclusion ignored the presence of GABAergic interneurones, which represent an independent potential explanation for the IPSPs, and it also overstated the evidence for an important role of inhibition in the functional properties of the striatum. The original reasons for believing that inhibition was powerful and pervasive in the striatum was the low firing rate and long periods of silence exhibited by nearly all the cells (Albe-Fessard, Rocha-Miranda and Oswaldo-Cruz, 1960) and specifically the spiny neurones (Wilson and Groves, 1981). Intracellular recording studies in slices and *in vivo* subsequently showed that the silent periods in these cells were due to reduction in excitatory input, combined with the presence of powerful potassium channels active at the resting membrane potential (Wilson, Chang and Kitai, 1983; Calabresi, Misgeld and Dodt, 1987; Wilson and Kawaguchi, 1996). Furthermore, *in vivo* studies using intrastriatal injection of GABA antagonists had shown that these drugs do not abolish the silent periods in striatal neurones, and only moderately increase the firing rate of the neurones during their periods of activity (Nisenbaum and Berger, 1992).

The discovery that spiny neurones are the projection cells of the neostriatum raised the opportunity of employing the classical test for recurrent inhibition, using antidromic activation at threshold (Eccles, Fatt and Koketsu, 1954). At threshold for antidromic response for a particular neurone, on average about half of the responding neurones should be antidromically activated on each trial. On those trials in which the studied neurone does not respond antidromically, a recurrent IPSP should be present at about half maximal strength. An early extracellular recording study was reported by Katayama, Miyazaki and Tsubokawa, (1981). Antidromically activated neurones did not fire spontaneously, and so could not be tested for recurrent synaptic effects. For an unidentified set of caudate neurones that did show spontaneous firing (presumably not spiny neurones projecting to the entopeduncular nucleus), peristimulus histograms aligned to stimulation of the entopeduncular nucleus in cats showed a period of reduced spontaneous firing following the stimulus. This was sometimes interrupted by a brief excitatory response. This result was interpreted as support for recurrent inhibition of striatal output neurones, but the uncertainty of cell identification limited confidence in the results. In a more recent study employing intracellular recording (which does not require the cells to be spontaneously active), antidromic activation of striatal efferents was not accompanied by recurrent IPSPs, either *in vivo* or in slices (Jaeger, Kita and Wilson, 1994). In addition, paired recordings from spiny neurones were made in slices to allow direct measurement of any synaptic connections between them. In these cases, neurones were injected with biocytin and stained to determine that both cells in the pair were spiny cells, and that they were each in range of the axonal field of the other. In a subsequent study, simultaneous recording of pairs of neurones were obtained *in vivo* and tested for synaptic interconnections (Stern, Jaeger and Wilson, 1998). These studies also yielded no indication of synaptic inhibition among striatal neurones. These studies suggest that recurrent collaterals of spiny neurones may be functionally silent under most experimental conditions, but if this idea seems too abhorrent to the reader, there are some alternatives. The most simple of these is to assume that the connectivity among spiny neurones may be highly selective. Given the preceding calculations, if the spiny cells were uniformly interconnected, cells located very near to each other should have about a 1 in 6 chance of being connected in either direction. A randomly selected pair should have a connection in one or both directions about 30% $(1-(5/6)^2)$ of the time. If each spiny neurone made all of its 500 or so contacts on one other cell in the population this probability would go down to about 1% $(1-(2039/2040)^2)$. In that case, the paired recordings might not have revealed a connected pair. Because the estimate of 500 contacts per spiny cell is only an order-of-magnitude estimate and not a measurement, and because we do not know the rules that govern interconnections of the spiny neurones, caution should be applied to the interpretation of the paired recordings. However, making assumptions about the connectivity of spiny cells does not provide an escape from the failure of the antidromic activation test. Of course, a conclusive demonstration that these synapses are or are not functional would require a more direct test, using cells with anatomically verified synaptic connections. But regardless of the biophysical issue surrounding silent synapses, these results refute the traditional notion of a general and pervasive mutual inhibition among spiny neurones within a module. Furthermore, powerful mutual inhibition among spiny neurones in the neostriatum is not required to explain any of the remaining unexplained neurophysiological properties of the circuit. The anatomical fact that spiny cells make synaptic contact with each other is the only remaining empirical observation suggesting such a relationship. Of course, as pointed out by Wickens (Chapter 4), there are theoretical advantages in having powerful recurrent inhibition. It may be useful to examine the functions for which inhibition is considered important, to determine if some other neurophysiological mechanism could relieve the embattled inhibitory synapse in providing the necessary function.

## 4. IF NOT MUTUAL INHIBITION, THEN WHAT?

The benefits of mutual inhibition in the domain hypothesis, and in other similar models of the neostriatum are two. One is in generating permanent antagonism among activity patterns in the striatum. This would be beneficial if there were cortical input patterns that could occur simultaneously but should be behaviourally antagonistic. The most coherent cortical pattern could suppress expression of the others in the striatum, and would be reinforced via the cortico-basal ganglionic loop. The other theoretical use of mutual inhibition is in the categorization/selection function, as discussed in this volume. As described in the paper by Plenz and Kitai (Chapter 9) , this function does not necessarily imply mutual inhibition. It could also be achieved by a suitable sorting of the inputs to spiny neurones. If inputs were arranged so that spiny neurones within a module received little of their input in common, and required a relatively large proportion of their total input to be active in order to fire, then any particular pattern of active inputs to the striatum would activate only a small proportion of spiny neurones. If the sorting of inputs were sufficiently categorical, there would be no need for striatal neurones to compete via inhibition for the privilege of representing the active pattern of cortical inputs, because so few cells would be activated by it in the first place. Perhaps it is hard to believe that the corticostriatal projection could be sorted so thoroughly that cells within a module have very little synaptic input in common. Their dendritic fields occupy a common volume, and the cortical axons arborizing around them have axonal arborizations as large as, or larger than the module volume (Kincaid, Zheng and Wilson, 1998). Although at first look this degree of input sorting seems intuitively unlikely, a quantitative analysis of the corticostriatal innervation at the single axon level (Kincaid, Zheng and Wilson, 1998) seems to indicate that it does exist. In that study, individual corticostriatal axons were found to make few synaptic contacts within the volume occupied by a single striatal neurone, requiring that that volume (approximately the size of a striosomal or matrisomal module) must be innervated by a very large number of cortical axons (~ 380,000). As a result, each axon innervates a very small fraction of the neurones in the module, and each striatal neurone's cortical innervation is unique. This occurs regardless of whether single cortical axons make single or multiple synapses on single striatal neurones (which is not known). Indeed, only a small proportion of the possible cortical input combinations available within a module can be expressed, due to the small number of striatal projection neurones within the volume (only about 2840). The most important determinant of corticostriatal integration, i.e. which combinations of cortical inputs are represented by the striatum, is made at the anatomical level by the convergence of inputs from single axons. In the paper by Plenz and Kitai (Chapter 9) the authors show that if such an arrangement of inputs arose by pruning connections from a fully connected network by a competitive mechanism, redundant representations would be prevented within a module, and the categorization of inputs by striatal neurones could be relatively independent of inhibitory mechanisms. It is important to note, when thinking about such neural network abstractions, that innervation of the striatum by the cortex is sparse in the anatomical sense. That is, it is not a fully connected input with the weights of a large proportion of inputs adjusted to a low value. If there is a dynamic competitive process responsible for the selection of which combinations of cortical inputs are represented in the striatal output, it must make new synapses and destroy others, not merely alter the strengths of existing ones. By substituting probability of a connection for synaptic weights (so that a synaptic weight near zero is equivalent to the absence of any connection at all) results similar to those of the neural network models could be obtained.

Could this sorting of synaptic inputs serve the categorization and selection function which makes competitive inhibition so desirable for so many authors? Of course this mechanism could not provide competition between mutually exclusive but simultaneously active cortical patterns. Perhaps that competition may function in a cross-modular fashion via feed-forward interneurones, or at other levels in the cortico-basal ganglionic loop. But can input sorting provide a categorical decoding of the corticostriatal projection in the striatum? According to Kincaid, Zheng and Wilson (1998), each striatal neurone receives about 5000 synapses of cortical origin, arising from nearly that many different cortical axons, and shares on average about 75 of those with each of the other spiny neurones in its immediate vicinity. The selectivity of the spiny neurone for a particular pattern of cortical activity will depend upon the number of cortical inputs required to activate the striatal neurone. The set of cells in the cortex that provides input to a particular neurone could be considered the cortical assembly encoded by that striatal neurone. If a large proportion of the inputs were required to activate the spiny cell, then that striatal neurone would fire only when nearly all of its cortical assembly was active. If, on the other hand, a single cortical input were sufficient to fire the striatal neurone, then that cell would always be coactivated with up to 40 other cells in its vicinity (because each cortical axon contacts a maximum of about 40 spiny neurones in the region of one cell). Putting aside for a moment the fact that connectivity in the corticostriatal pathway may not be uniform (i.e. nearby spiny neurones may not have an equal probability of receiving a connection from any particular cortical axon), it is useful to consider the statistics for uniform connectivity. If 75 active inputs were sufficient to fire a spiny neurone, then complete activation of one cell's cortical assembly would activate the entire module (because each cell shares about 75 inputs with each of its neighbours), and partial activation of the assembly would engage a fraction of the assembly. It is not known how many active inputs are required to bring a spiny neurone to threshold, and the role of the thalamic input and of interneuronal inhibition in that process has not been studied. Taking a simple computational approach of simulating striatal neurones, values in the hundreds of simultaneous inputs have been obtained (Wilson, 1992, 1995). This is likely to be an underestimate (because it does not include inhibition) and is already well above the 75 inputs that are shared by nearby cells if synapses are formed at random. Because the value 75 is an average of the distribution of sharing expected in the population, it might be that some pairs of cells have many more inputs in common and some have many less. It is useful to consider the variance of the distribution of shared inputs, because it leads directly to the distribution of activation across neurones in the module when the cortical assembly for a single neurone is activated. That is, if all 5000 cortical axons projecting to a single striatal neurone were activated, how many striatal neurones would have 1, 2, 3...k active inputs. This distribution is shown in Figure 3A. It is very narrow. This means that for the sparse connectivity characteristic of the corticostriatal inputs, full activation of the inputs to one cell will result in a nearly uniform background activation of the others. Nearly all other cells in the module would experience between 50 and 100 background active afferents if a single striatal neurone's afferents were maximally activated. Whether this level of background will actually fire any of the other cells depends upon the threshold afferent input required to fire a spiny neurone. This value is not known. The distribution in Figure 3B shows the number of striatal neurones activated in a module when one neurone's cortical assembly is fully activated (and no other cortical inputs are active), for a number of assumed values for the number of active cortical inputs required to a fire a striatal cell. This shows that if hundreds of cortical inputs (even 100) are required to fire a spiny neurone, then simply the sparseness of the corticostriatal connectivity will ensure a completely categorical representation of the cortical pattern in the striatum without the intervention of lateral inhibition. The noise immunity of this arrangement is impressive. To activate a spiny neurone accidentally, a very large proportion of the cortical input to the module is required, as shown in Figure 3C. When cortical background activity is non-specifically increased, spiny neurones are not activated until a large proportion of cortical cells are activated. At that point, nearly the entire striatal module will respond. The level of cortical activity required to achieve the complete activation of the module depends upon the threshold of activation of striatal cells, and is shown for a relatively low threshold of 300 inputs (3% of asymmetric synapses).

It is more likely that the corticostriatal input is not uniform, but was established on the basis of some growth rule or some set of cell affinities active during development (and perhaps continuing to be active throughout life), like the mechanism described by Plenz and Kitai (Chapter 9). The degree to which striatal neurones share input after such a systematic arrangements of synaptic inputs is entirely dependent upon the mechanism determining the connectivity. Speculation about this mechanism is still constrained by the observation that synaptic input from single cortical neurones is sparse. Because of this, any developmental mechanism that increases input sharing (and redundancy in corticostriatal connections) does so at the expense of the number of cortical combinations that are encoded. Competitive mechanisms active during development would act to minimize redundancy, and to establish striatal encoding of cortical assemblies that are active together. It may be significant that synaptic inhibition is much easier to demonstrate in the striatum during development than it is in adults (Tepper and Trent, 1993). The main difference between the connectivity created by a competitive growth rule and random connectivity is appreciated by considering the total number of combinations of cortical inputs possible within a single striatal module. Because there are about 380,000 axons, and 2840 neurones with about 5000 cortical inputs each, the number of combinations is 380000!/(500! (380000–5000)!), which is about $10^{7000}$. Of these, only 2840 different cortical combinations are represented in the module's output. If the connectivity were uniform, all of these possible combinations would be equally likely to be represented (or not represented) in the striatal encoding of the cortical outflow. This makes a strong case against a random connectivity of the corticostriatal projection. Random connectivity with dynamic adjustment of synaptic strength according to the Hebb rule has been shown to be useful for self-organizing neuronal encoding in many situations. In the case of the anatomically sparse corticostriatal projection, however, it would mean that the chances of any particular cortical assembly of functional interest being encoded in the striatum would be approximately $2840/10^{7000}$ (nearly zero), and that functionally meaningless and perhaps never coactivated combinations of cortical input would be equally likely to occupy the corticostriatal category space. In the case of a competitive growth rule, the choice of which cortical patterns are encoded would be made on the basis of occurrence

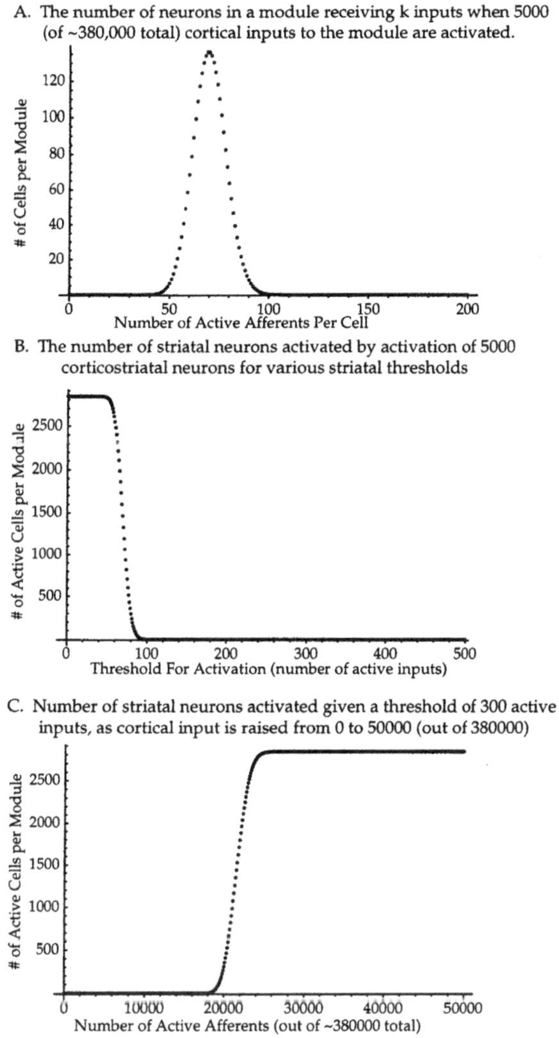


**Figure 3.** The binomial distribution governs sharing of cortical synaptic inputs among striatal neurones, on the condition that cortical axons make primarily single contacts on striatal neurones. It is not required that synaptic contacts be randomly distributed on striatal neurones, in order for the binomial distribution to be used in this case. **A.** The distribution of shared inputs predicts the distribution of excitation in a striatal module when all the cortical inputs to a single striatal neurone are active, and only those afferents are active. While one striatal cell (the one whose inputs are all active) receives maximal input (5000 active inputs), nearly all other neurones receive between 50 and 100 active inputs. **B.** The number of striatal neurones responding to this input depends upon the threshold for activation of striatal cells. If 50 or fewer inputs are required to activate a striatal neurone, then complete activation of the cortical assembly projecting to one neurone will activate all cells. If the threshold is over 100, then only a single striatal cell will be activated when its cortical assembly is fully active. In the absence of inhibition, the striatal module with completely overlapping dendritic and axonal trees can identify the cortical cell assembly and signal it with a single striatal neuronal activation if the threshold for activation of striatal neurones is moderate or high. **C.** The same arrangement of synapses has a high immunity to noise. As cortical activity is gradually increased as a background on the activation of a single cortical assembly, the noise is ignored until the background activity involves a relatively large proportion of the total cortical input. At that point, the entire striatal module is activated. The noise immunity increases with increasing threshold for the striatal cell (the curve shown is for a threshold of 300 active inputs).

of specific activation of those patterns during development. Patterns which never occurred in nature would have no striatal representation. It would not alter the specificity or noise immunity considerations described above.

## 5. SPINY NEURONE FIRING IS A TWO-STEP PROCESS

Estimating the number of active inputs required to make striatal cells fire for purposes of showing that is an important quantity, as done above, oversimplifies the transition to firing in these cells. Spiny neurones have resting membrane potentials 30–40 mV more negative than the firing threshold, and synaptic excitation can produce very large depolarizations that do not elicit action potentials. *In vivo*, these cells exhibit very large synaptically driven depolarizations, that can maintain the membrane potential in a just- subthreshold depolarized level for periods of a few milliseconds to several seconds (Wilson and Groves, 1981). Action potentials arise from additional synaptic depolarizations riding on the prolonged depolarizations. While the transition to the depolarized state is required for firing, it is not sufficient. In both anaesthetized and unanaesthetized animals, some cells showing the depolarizing episodes remain silent (Wilson and Groves, 1981; Wickens and Wilson, 1998). The difference between firing and silent neurones is primarily in the level of depolarization achieved during the depolarizations, rather than the size of the noisy depolarizations which directly trigger action potentials (Wickens and Wilson, 1998). Causing a striatal spiny neurone to fire is in some sense a 2 step process, in which convergent synaptic excitation must bring the cell into the depolarized but subthreshold state adequate for action potential generation, and then the noisy depolarizations superimposed upon the depolarized state must trigger individual action potentials. Perhaps the same inputs create both the overall episode of depolarization and the noisy fluctuations that trigger action potentials. This is suggested by the fact that both processes have the same spectral composition (Stern, Kincaid and Wilson, 1997). Even so, voltage dependent currents active during the depolarizations resist further depolarization, so that it takes substantially more synaptic excitation to trigger firing in the spiny neurone than it does to achieve the depolarized membrane potential state. Inhibitory neurones apparently are activated approximately in phase with the spiny cells (Plenz and Kitai, 1997), and GABA antagonists increase the firing rate of spiny neurones during firing episodes without increasing the rate or duration of the episodes themselves (Nisenbaum and Berger, 1992). For these reasons, it is most likely that inhibition in the striatum acts to modulate firing during the depolarizations, rather than to prevent or delay the depolarized episodes themselves.

## 6. ACROSS-MODULAR INTERACTIONS

If categorization and selection can occur in a module without recurrent feedback inhibition, then what role can interneurones play in the striatum? There are a number of different interneurones, as pointed out by Bennett and Wilson (Chapter 6), and very little can currently be said about the effects of any of them on the striatal output neurones. One feature apparently held in common by all of the interneurones is their failure to obey the striosomal boundaries. Presumably, the dendrites and axons of these cells also cross the boundaries separating matrisomes at will. Perhaps the interneurones form a spatially continuous network in the striatum, unlike the discrete and discontinuous nature of the corticostriatal projections to the spiny neurones. If so, this is curious, because the anatomical experiments on the corticostriatal projection do not suggest that functionally similar modules are located near each other. Interneurones may create an inhibitory surround around a highly activated striatal module, but it is difficult to guess what will be the functional consequences of that. Modules representing the motor and sensory projections of a digit, for example, are widely separated in the striatum, and will be adjacent to projections to other body parts. Given the projection patterns of interneurones, it may be very important how such discrete cortical representations are arranged in the striatum, which can be adjacent, and which never are.

Interneurones may subserve interactions between the focal regions innervated by single cortical regions. Although these are not adjacent, they are generally in nearby regions of the striatum, and it is possible that they could be specifically interconnected by interneurones. Single corticostriatal neurones can innervate several such foci, and so these may to some degree be duplicate representations of a cortical area. If they had inhibitory interconnections it could function to prevent them from producing redundant representations of cortical assemblies. Graybiel and her coworkers (1994) have proposed that within these separate foci, different aspects of the cortical inputs are extracted, and have shown that their outputs at least sometimes reconverge on a single region of the globus pallidus. They have suggested that the separate foci may act as local experts in a parallel distributed processing of cortical inputs. At least four different cellular interpretations of this proposal are consistent with the existing anatomical facts. (1) If each striatal focal arborization of a corticostriatal neurone is innervated by exactly the same assembly of cortical cells, then these are effectively a single module receiving ~380,000 cortical inputs but broken up spatially into several locations. In that case, the module effectively contains 2840 times the number of foci spiny cells. As pointed out by Plenz and Kitai (Chapter 9), the absence of inhibition among these foci is likely to produce redundancy there (because the same growth rule acting with the same input is likely to produce the same set of connections). Duplication of connections in the various foci would make a set of identical local experts, which is not a desirable outcome for the scheme proposed by Graybiel *et al.* (1994). (2) If interneurones caused interactions between neighbouring modules, and modules were arranged in a (so far undiscovered) functionally meaningful pattern, then the otherwise identical modules would have different neighbours and so be under different local interneuronal control. (3) If the various modules receiving identical input could inhibit each other via interneurones, the resulting competition might rule out redundancy across foci, solving the problem of redundancy . (4) The modules need not all receive the same input. Functionally related cortical regions have been shown to converge on some regions of the striatum, and not others. If each cortical neurone from a functionally defined region projected to a subset of all the striatal representations of that cortical area, each module would have a unique input, and each corticostriatal neurone would converge with a systematically different set of other cells in each innervated focus. This possibility raises a question about the nature of convergence in the globus pallidus, as illustrated in Figure 4. While

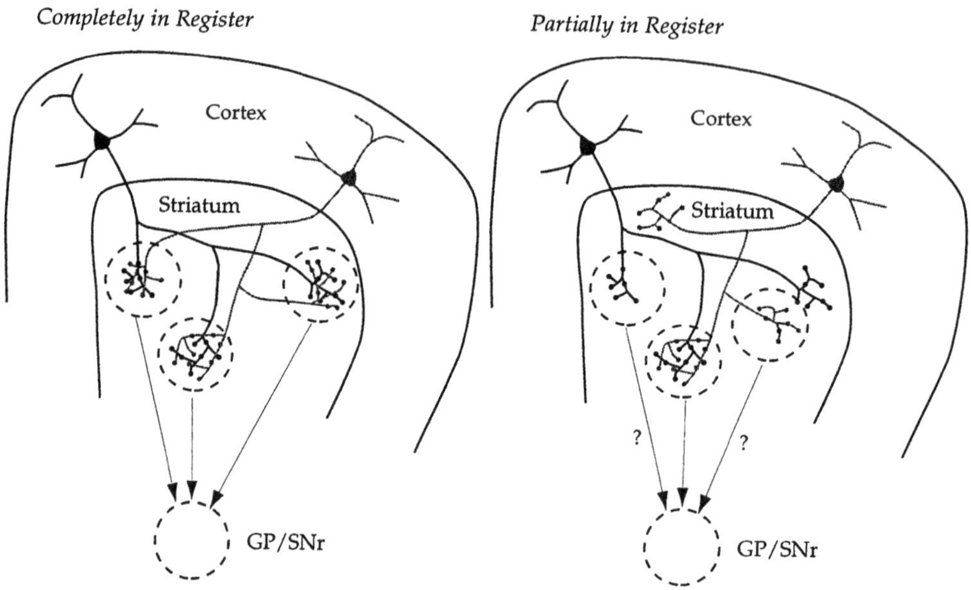


**Figure 4.** The degree of register between multiple modules innervated by single corticostriatal neurones is not known. For functionally related cortical areas, a substantial degree of register must occur, to account for the axonal tracing data (Graybiel *et al.*, 1994). Striatal modules receiving in register cortical input also have been shown to converge in the globus pallidus (Graybiel *et al.*, 1994). If all axons project in register, however, then the individual modules are identical, and will produce redundant output. The convergence of these redundant modules in the globus pallidus cannot reap the benefits of the parallel distributed processing. If axons are partly in register, each module innervated by a cortical region will have a unique combination of cortical inputs, allowing for local differences in information processing in each module. Recombination of these modules in the globus pallidus can perform a useful computational function, but it raises the question of how similar the input to two modules must be if they are to be part of the same functional unit in the globus pallidus.

each module receiving input from a cortical region is able to extract a somewhat different aspect of the cortical input, it is not clear which striatal region should converge in the globus pallidus to recombine the information. This situation, if it occurs, has some of the features of the split loops described in the chapter by Joel and Weiner (Chapter 12), but is transposed into the globus pallidus, rather than to the thalamocortical projections. It raises the possibility that information in the cortico-basal ganglionic circuit may have several such opportunities for splitting into parallel subloops with offset projections.

Attributing complex information processing functions to striatal neurones poses a number of difficult questions. As pointed out in the chapter by Bennett and Wilson (Chapter 6) , only one of the interneurone types, the parvalbumin/GABA interneurone, is known to have straightforward fast inhibitory effects on the projection cells, and this cell usually has a short range of action. Understanding the interactions between modules in the striatum will probably require a knowledge of the physiological mechanisms of interneuronal action on spiny cells. These cells apparently work primarily via neuromodulatory mechanisms that are only beginning to be understood.

## REFERENCES


Albe-Fessard, D., Rocha-Miranda, C. and Oswaldo-Cruz, E. (1960) Activités évoquées dans le noyau caudé du chat en résponse à des types divers d'afferénces. *Electroencephalography and Clinical Neurophysiology*, **12**, 649–661.

Albin, R.L., Young, A.B and Penney, J.B. (1989) The functional anatomy of basal ganglia disorders. *Trends in Neuroscience*, **12**, 366–375.

Amari, S. 1977 Dynamics of pattern formation in lateral-inhibition type neural fields. *Biological Cybernetics* **27**, 77–87.

Bernardi, G., Marciani, M.G., Morocutti, C. and Giacomini, P. (1975) The action of GABA on rat caudate neurones recorded intracellularly. *Brain Research*, **92**, 511–515.

Bernardi, G., Marciani, M.G., Morocutti, C. and Giacomini, P. (1976) The action of picrotoxin and bicuculline on rat caudate neurons inhibited by GABA. *Brain Research* **102**, 379–384.

Buchwald, N.A., Price, D.D., Vernon, L. and Hull, C.D. (1973) Caudate intracellular response to thalamic and cortical inputs. *Experimental Neurology* **38**, 311–323.

Calabresi, P., Misgeld, U. and Dodt, H.U. (1987) Intrinsic membrane properties of neostriatal neurons can account for their low level of spontaneous activity. *Neuroscience*, **20**, 293–303.

Eccles, J.C., Fatt, P. and Koketsu, K. (1954) Cholinergic and inhibitiory synapses in a pathway from motor-axon collaterals to motoneurones. *Journal of Physiology, London* **126**, 524–562.

Flaherty, A.W. and Graybiel, A.M. (1993) Two input systems for body representations in the primate striatal matrix: experimental evidence in the squirrel monkey. *Journal of Neuroscience* **13**, 1120–1137.

Goldman, P.S. and Nauta, W.J.H. (1977) An intricately patterned prefrontocaudate projection in the rhesus monkey. *Journal of Comparative Neurology* **171**, 369–386.

Graybiel, A.M. (1983) Compartmental organization of the mammalian striatum. In: *Progress in Brain Research vol 58, Molecular and Cellular Interactions Underlying Higher Brain Function,* edited by J.-P. Changeau, Amsterdam, Elsevier, pp. 247–256.

Graybiel, A.M., Aosaki, T., Flaherty, A.W., and Kimura, M. (1994) The basal ganglia and adaptive motor control. *Science, New York,* **265**, 1826–1831.

Graybiel, A.M. and Kimura, M. (1995) Adaptive neural networks in the basal ganglia. In: *Models of Information Processing in the Basal Ganglia,* edited by J.C. Houk, J.L. Davis, and D.G. Beiser, Cambridge, MA, MIT, pp. 103–116.

Ingham, C.A., Hood, S.H., Taggart, P. and Arbuthnott, G.A. (1998) Plasticity of synapses in the rat neo-striatum after unilateral lesion of the nigrostriatal dopaminergic pathway. *Journal of Neuroscience* **18**, 4732–4743.

Jaeger, D., Kita, H. and Wilson, C.J. (1994) Surround inhibition among projection neurons is weak or non-existent in the rat neostriatum. *Journal of Neurophysiology* **72**, 2555–2558.

Johnson, J.G., Gerfen, C.R., Haber, S.N. and van der Kooy, D. (1990) Mechanisms of striatal pattern formation: Conservation of mammalian compartmentalization. *Developmental Brain Research*, **57**, 93–102.

Jones, E.G., Coulter, J.D., Burton, H. and Porter, R. (1977) Cells of origin and terminal distribution of cortico-striaal fibers arising in the sensori-motor cortex of monkeys. *Journal of Comparative Neurology* **173**, 53–80.

Katayama, Y., Miyazaki, S. and Tsubokawa, T. (1981) Electrophysiological evidence favoring intracaudate axon collaterals of GABAergic caudate output neurons in the cat. *Brain Research* **216**, 180–186.

Kawaguchi, Y., Wilson, C.J. and Emson, P.C. (1989) Intracellular recording of identified neostriatal patch and matrix spiny cells in a slice preparation preserving cortical inputs. *Journal of Neurophysiology* **62**, 1052–1068.

Kawaguchi, Y., Wilson, C.J. and Emson, P.C. (1990) Projection subthpes of rat neostriatal matrix cells revealed by intracellular injection of biocytin. *Journal of Neuroscience* **10**, 3421–3438.

Kawaguchi, Y., Wilson, C.J., Augood, S.J. and Emson, P.C. (1995) Striatal interneurones: chemical, physiological and morphological characterization. *Trends in Neuroscience* **18**, 527–535.

Kincaid, A.E., Zheng, T., and Wilson, C.J. (1998) Connectivity and convergence of single corticostriatal axons. *Journal of Neuroscience* **18**, 4722–4731.

Kita, T., Kita, H. and Kitai, S.T. (1985) Local stimulation induced GABAergic response in rat striatal slice preparations: intracellular recording on QX–314 injected neurons. *Brain Research* **360**, 304–310.

Kitai, S.T., Kocsis, J.D., Preston, R.J. and Sugimori, M. (1976) Monosynaptic inputs to caudate neurons identified by intracellular injection of horeseradish peroxidase. *Brain Research* **109**, 601–606.

Künzle, H. (1975) Bilateral projections from the precentral motor cortex to the putamen and other parts of the basal ganglia. An autoradiographic study in Macaca fascicularis. *Brain Research* **88**, 195–209.

Lighthall, J.W. and Kitai, S.T. (1983) A short duration GABAergic inhibition in identified neostriatal medium spiny neurons. *In vitro* slices study. *Brain Research Bulletin*, **11**, 103–110.

Misgeld, U., Wagner, A. and Ohno, T. (1982) Depolarizing IPSPs and depolarization by GABA of rat neostriatum cells *in vitro. Experimental Brain Research*, **45**, 108–124.

Nisenbaum, E.S. and Berger, T.W. (1992) Functionally distinct subpopulations of striatal neurons are differentially regulated by GABAergic and dopaminergic inputs. 1. *In vivo* analysis. *Neuroscience* **48**, 561–578.

Oorschot, D.E. (1996) Total number of neurons in the neostriatal, pallidal, subthalamic and substantia nigra nuclei of the rat basal ganglia: a stereological study using the Cavalieri and optical dissector methods. *Journal of Comparative Neurology* **366**, 580–599.

Plenz, D. and Kitai, S.T. (1997) Up and down states in striatal medium spiny neurons simultaneously recorded with spontaneous activity in fast-spiking interneurons studied in cortex-striatum-substantia nigra organotypic cultures. *Journal of Neuroscience* **18**, 266–283.

Purpura, D. and Malliani, A. (1967) Intracellular studies of the corpus striatum. I. Synaptic potentials and discarge characteristics of caudate neurons activated by thalamic stimulation. *Brain Research* **6**, 325–340.

Ratliff, F. (1965) *Mach Bands: Quantitative studies on neural networks in the retina* Holden-Day Inc, San Francisco, pp. 77–142.

Ramón-y-Cajal, S. (1995) *Histology of the Nervous System of Man and Vertebrates Vol II.* (Translated by N. Swanson and L.W. Swanson), Oxford Uniiversity Press New York, pp. 416–428.

Somogyi, P., Bolam, J.P. and Smith, A.D. (1981) Monosynaptic cortical input and local axon collaterals of identified striatonigral neurons. A light and electron microscopic study usint the Golgi-peroxidase transport-degeneration procedure. *Journal of Comparative Neurology* **195**, 567–584.

Stern, E.A., Jaeger, D. and Wilson, C.J. (1998) Membrane potential synchrony of simultaneously recorded striatal spiny neurons *in vivo. Nature, London,* **394**, 475–478.

Stern, E.A., Kincaid, A.E. and Wilson, C.J. (1997) Spontaneous subthreshold membrane potential fluctuations and action potential variability of rat corticostriatal and striatal neurons *in vivo. Journal of Neurophysiology* **77**, 1697–1715.

Tepper, J.M. and Trent, F. (1993) In vivo studies of the postnatal development of rat neostriatal neurons. In: *Progress in Brain Research, Vol 99, Chemical Signaling in the Basal Ganglia*, edited by G.W. Arbuthnott and P.C. Emson, Amsterdam, Elsevier, pp. 35–50.

Vogt, C. and Vogt, O. (1920) Zu lehre der erkrankungen des striären systems. *Journal for Psychology and Neurology*, **25**, 628–846.

Walker, R.H. and Graybiel, A.M. (1993) Dendritic arbors of spiny neurons in the primate striatum are directionally polarized. *Journal of Comparative Neurology* **337**, 629–639.

Wickens, J.R. (1993) *A Theory of the Striatum.* Oxford, Pergamon Press, pp. 31–43.

Wickens, J.R. and Wilson, C.J. (1998) Regulation of action potential firing in spiny neurons of the rat neostriatum, *in vivo Journal of Neurophysiology* **79**, 2358–2364.

Wilson, C.J. (1986) Postsynaptic potentials evoked in spiny neostriatal projection neurons by stimulation of ipsilateral and contralateral neocortex. *Brain Research* **367**, 201–213.

Wilson, C.J. (1992) Dendritic morphology, inward rectification and the functional properties of neostriatal neurons. In: *Single Neuron Computation*, edited by T. McKenna, J. Davis and S.F. Zornetzer, Boston, Academic Press, pp. 141–171.

Wilson, C.J. (1995) Dynamic modification of dendritic cable properties and synaptic transmission by voltage-gated potassium channels. *Journal of Computational Neuroscience* **2**, 91–115.

Wilson, C.J., Chang, H.T. and Kitai, S.T. (1983) Disfacilitation and long-lasting inhibition of neostriatal neurons in the rat. *Experimental Brain Research*, **51**, 227–235.

Wilson, C.J. and Groves, P.M. (1980) Fine structure and synaptic connections of the common spiny neuron of the rat neostriatum: a study employing intracellular injection fo horseradish peroxidase. *Journal of Comparative Neurology*, **194**, 599–615.

Wilson, C.J. and Groves, P.M. (1981) Spontaneous firing patterns of identified spiny neurons in the rat neostriatum. *Brain Research* **220**, 67–80.

Wilson, C.J. and Kawaguchi, Y. (1996) The origins of two-state spontaneous membrane potential fluctuations of neostriatal spiny neurons. *Journal of Neuroscience* **16**, 2397–2410.

# Index

*For Product Safety Concerns and Information please contact our EU representative GPSR@taylorandfrancis.com Taylor & Francis Verlag GmbH, Kaufingerstraße 24, 80331 München, Germany*

* 9 7 8 9 0 5 7 0 2 4 7 8 8 *

T - #0009 - 280426 - C0 - 246/174/18 [20] - CB - 9789057024788 - Gloss Lamination